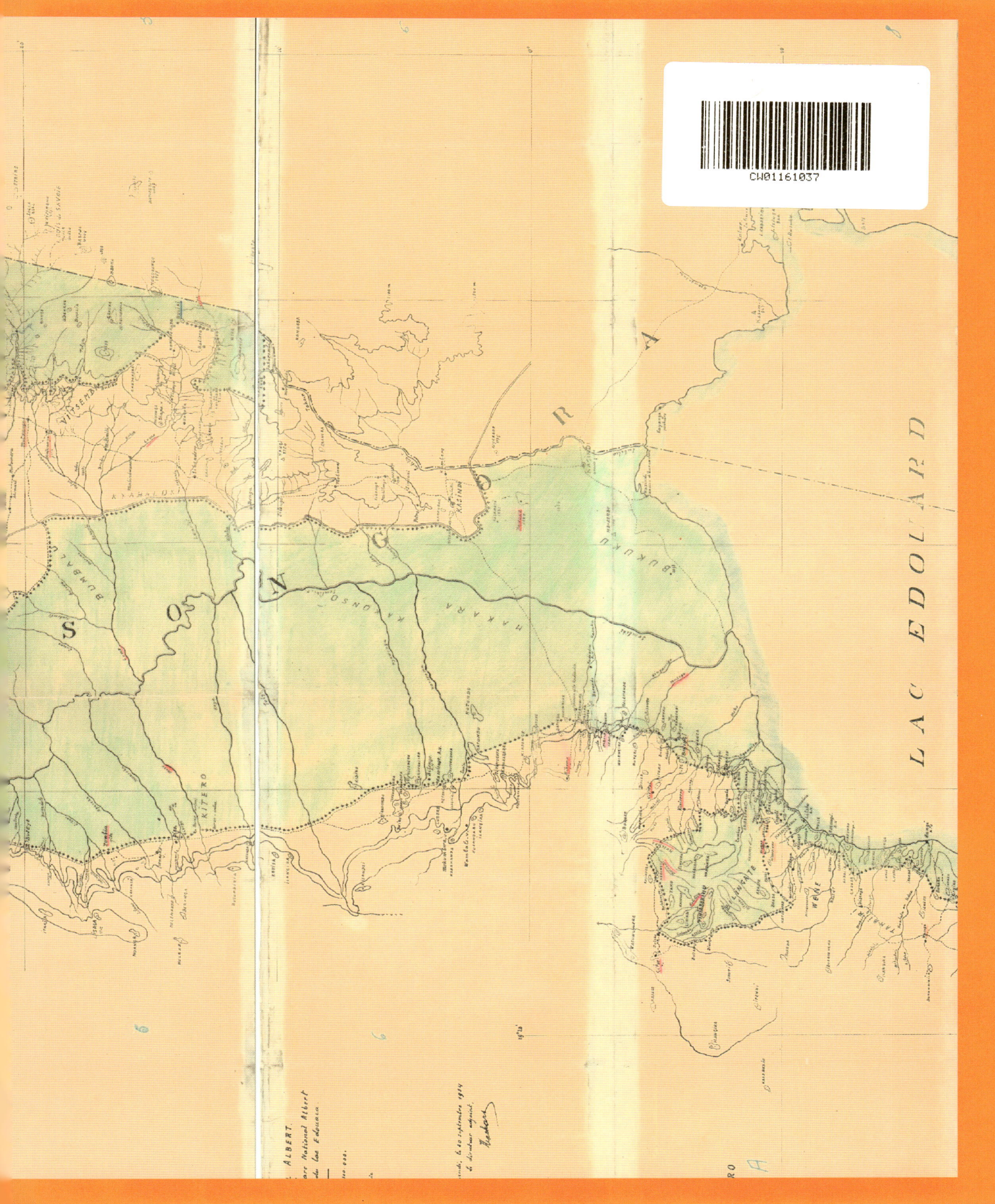

Virunga

The Survival of Virunga
Africa's First National Park

edited by
Marc Languy, Emmanuel de Merode

lannoo

WARNING

The publication of this book was made possible by financial support from WWF, the European Union and UNESCO.

All statements and ideas expressed in this book are those of the authors and do not reflect the positions or policies of the organisations or people who have supported its publication.

Likewise, the content of each of the chapters solely reflects the opinions of the authors directly concerned, and not those of other contributors to this publication, or its editors.

International boundaries shown on maps printed as part of this publication may vary in accuracy, and do not reflect any specific opinion on the part of the authors or editors.

Taxonomic protocols used are based either on commonest practice, or the most recent accepted standards. Certain recent changes may have been omitted, in particular with respect to the botanical classifications. For example, we have maintained the traditional use of Acacia, and have disregarded its recent reclassification.

This book, originally published in French, may have developed minor differences between the two editions. These differences are solely provoked by translation constraints. The French version is the original.

(p. 1)
The Mountain Gorilla is one of the iconic species of the park and is an important source of revenue for the region.

(pp. 2–3)
The Rwenzori massif reaches over 5100 metres and is a major tourist attraction.

(pp. 4–5)
Elephants have an important ecological role in maintaining the park's ecosystems; their ecological requirements make it essential that effective transboundary management systems are put in place as part of the Greater Virunga Landscape.

(p. 10)
ICCN rangers during one of the many regular gorilla monitoring patrols

The paper in this product comes from well managed forests, independently certified in accordance with the rules of the Forest Stewardship Council.

In referencing this book, please use:
Languy M. and E. de Merode (eds.) 2009.
Virunga: the Survival of Africa's First National Park.
Lannoo, Tielt, Belgium. 352 pp.

All additional information relating to this publication can be obtained at the following address:
WWF-Belgium
90 Boulevard E. Jacqmain,
B-1000 Brussels

Graphics and formatting: Antoon De Vylder
Type setting: Set in DTL Prokyon and DTL Antares by De Diamant Pers, Herentals
Printing and binding: Printer Trento, Trento

All reproductions or other extractions from the contents of this book are subject to international copyright laws

© Lannoo Publishers, Tielt, 2009
ISBN: 978-90-209-6562-9
D/2009/45/156

PREFACE by Louis Michel
European Commissioner

25 million Congolese citizens exercised their right to vote in the country's first democratic elections in almost half a century. Over 80 percent of these voters also ratified the country's new constitution, thereby establishing the foundations for the rule of law. Thus, while the Democratic Republic of Congo's war caused immeasurable damage, there remain many assets to build upon: the country's human and natural resources are massive. This book bears witness to that fact.

The book presents the remarkable struggle of a team of Congolese professionals in service to their country, under the most difficult conditions imaginable. The most extraordinary aspect of the history of Virunga National Park is without a doubt the effort and sacrifice that was invested by the wardens and rangers of the *Institut Congolais pour la Conservation de la Nature* (ICCN) during the conflict years, between 1996 and 2003. I feel compelled to express my complete admiration for their achievement. What is encouraging is that these are not isolated cases of dedication and devotion to a cause. In many sectors, such as health, education, civil administration and justice, Congolese public servants continued to do their duty. They continued their work despite the extreme hardship of their working conditions, often without remuneration for years at a time.

In Virunga, the commitment of these professionals has been exceptional, and has had tangible results: Virunga can still regain its former splendour, which is why this book is also a celebration of what is unquestionably one of the most precious of the world's ecosystems.

The international dimension to the work in Virunga is an important component of the story. While the natural resources of the park are integrally part of the Congolese national heritage, the efforts to protect and enrich those resources are increasingly, and rightly, being supported by the international community. The European Union was the first major donor to offer significant support to the conservation and development efforts in Virunga. The European Union has been working with ICCN for many years, a collaboration that began with the *Programme Kivu,* in the late 1980s. This support was resumed in 2002, and Virunga is once again the subject of a major European Union funded programme amounting to over 5 million euros. Thanks to that, field activities are better resourced.

This support also allows the wardens and rangers of Virunga to pursue more ambitious goals. Today, economic development is a central component of any viable conservation strategy, and nowhere more so than in Virunga National Park, situated at the heart of North Kivu. The human population density is very high and people's needs are considerable. The need to develop new sectors of economic growth, such as tourism, is also considerable in a province where agricultural activities alone will not adequately provide for its growing human populations. But while embarking on new sectors of economic activity, it is imperative that peace is consolidated in the North and South Kivu provinces. As outlined in several chapters of this book, it is perhaps on this issue that ICCN and its partners have the most to offer. A government institution that has succeeded in maintaining its integrity during the war, and that has successfully demonstrated its neutrality, while effectively maintaining the rule of law in a particularly challenging and isolated environment must have a positive role to play in a post-conflict democracy. It is certainly my heartfelt wish.

Preface by Pasteur Cosma Wilungula
Administrateur Délégué Général
of the *Institut Congolais pour la Conservation de la Nature*

As Virunga recently reached its eightieth anniversary, we can marvel at the wonders of this extraordinary national park, but it is perhaps better to reflect on all that has contributed to the survival of this park. Equally, it is an opportunity to explore those areas where we could have done more to limit the considerable damage that has been inflicted on this exceptional national and world treasure.

As we examine the survival of this park we must, first and foremost, give homage to the 97 rangers who gave their lives to protect our national park during one of the most difficult periods in our country's history. Their motivation was not driven by material rewards, because at the time of their deaths, there were none. Yet they remained committed to the protection of the country's resources in the face of extreme danger. In praising the sacrifice of those that died, one must not forget that Virunga's rangers who continue to protect this park continue to face these dangers: rangers continue to be killed on active service in the context of the ongoing armed conflicts in the national park. Yet the work continues, and there are a growing numbers of success stories: the recovery of invaded land at Kibumba in 2004, followed by Mavivi and Karuruma, together with the training programme at Ishango are two of a growing number of positive steps that have been achieved in Virunga. It is with enormous satisfaction that we witness the day-to-day strengthening of ICCN's ability to manage and protect our national park. This, of course, relates to the renewed support that ICCN has received from its international partners. The European Union has renewed its historical support for Virunga, support is now coming from the US and other bilateral organizations, and the conservation groups, many of whom remained by the sides of their ICCN colleagues throughout the war, continue to provide invaluable support to the park.

But, as is rightly highlighted in the book's conclusion, while there have been some positive advances in the protection of Virunga and we can allow ourselves some optimism for the future of this park, the real challenges are ever present. Virunga remains at the epicentre of the armed conflicts in the east of the country. Insecurity is compounded by the 'boom and bust' cycle that fails to provide it with the long-term financial stability that it needs. Donor funding may dry up, as it has in the past, and there are still no sustainable funding systems to keep the park alive. ICCN as an institution needs to reassess its structure and performance so that it can continue its role into the twenty-first century. With a growing human population in Kivu province, the threats that surround the park are ever present, and many of these can only be addressed when the park begins to generate tangible benefits to the local and national economy. With those perspectives in mind, it is important to remember that Virunga's greatest strength remains in the community of people who believe in its importance and are compelled to ensure that its valuable resources are protected for future generations. This book is a reflection of that community of people and provides a basis for the future of this exceptional national park.

This book was written by a group of specialists in their field, including 12 Congolese professionals and various international partners. It is symbolic of a reality that ICCN holds very dearly: the establishment of positive and constructive partnerships for the strengthening of ICCN's own capacity to manage its network of protected areas. This book does far more than bear witness to the past and present efforts to protect a world heritage. It is also an invaluable record on the current state of the park, which provides a basis for measuring our progress in rebuilding the park. Indeed some of this progress can already be seen on the ground. The book also gives us some important perspectives on future strategy, which I invite you to consider as best you see fit.

Preface by Dr James Leape
Director General of WWF

Virunga National Park is, without doubt, the centre of all superlatives: the oldest of Africa's national parks, the first World Heritage Site on the African continent, the protected area that is richest in vertebrates, but it is also one of Africa's most threatened ecosystems.

As such, it comes as no surprise that WWF began providing support to the Congolese authorities as early as 1964, in their efforts to protect this invaluable site. This support was increased substantially in 1987 with the launching of the *Programme Environnemental autour des Virunga,* and in 1991 with the International Gorilla Conservation Programme, in partnership with the African Wildlife Foundation and Fauna and Flora International.

In 2000, a number of partners joined forces with ICCN and WWF to combine their efforts and protect this UNESCO World Heritage Site, listed under the category of sites in danger in 1994. On the ground, partnerships formed between WWF, the London Zoological Society, the Wildlife Conservation Society and the Frankfurt Zoological Society in support of ICCN. These partners, receiving funds from the European Union, USAID and other donors, made it possible to re-establish conservation activities within the national park, thereby reducing some of the threats. The achievements in this regard are considerable, given the chronic insecurity of the region.

Nevertheless, as is shown in this book, the major challenges lie ahead: thousands of tons of charcoal are produced every year in the park with impunity, and elephant, hippo and other many other species have reached an all time low in their numbers, even if the decline appears to have stabilized. The Kilolirwe region is still occupied by settlers, as are the western shores of Lake Edward. These threats to the natural resources have implications that go beyond the park's boundaries, and threaten the livelihoods of thousands of people who are dependant on fishing and forest resources for their survival. The provision of ecological and economic services by the forest ecosystems of Virunga is critical to the local economy.

Virunga is an issue that concerns us all, whether we work in conservation, development, tourism or another private sector; whether we are local or national public servants or simply global citizens. With its mountain gorillas, its chimpanzees, and its remarkable assemblage of other species and habitats, Virunga National Park protects a vital part of humanity's environment that no one should stand by and watch disappear.

Thus, the process of establishing an inventory of the status of the park, of assessing the challenges and of exploring the options for the future is long overdue. This is the objective that the 36 authors of this book set themselves. Initially conceived as a compilation of lessons learned within the context of the WWF programme in and around the park, the manuscript gradually evolved to include the experiences of a wide range of other players, placing the work in its wider historical context.

As a celebration of the park's unparalleled diversity, the book is also a further example of the tight bond that unites the various conservation groups with ICCN, and their expression of need to maintain the conservation effort in the face of numerous future challenges.

This book is an appeal without ambivalence to all those concerned for the safeguard of Virunga. ICCN alone cannot be expected to overcome these challenges, no more than any of its partners. In demonstrating the unique value of the park and in recounting the achievements of the past and the threats that remain, this publication beckons us to provide what support we can, in whichever way possible, to this remarkable national park.

Table of Content

7	Preface by the European Commissioner	Louis Michel
8	Preface by the Director of ICCN	Pasteur Cosma Wilungula
9	Preface by the Director General of WWF	Dr. J. P. Leape
13	Introduction	M. Languy & E. de Merode
21	A brief overview of Virunga National Park	M. Languy & E. de Merode

The Past

65	1 Albert National Park: the birth of Africa's first national park (1925–1960)	P. Van Schuylenbergh
75	2 Life in Albert National Park (1925–1960)	J. Verschuren
85	3 Rebirth of the National Park (1960–1991)	J. Verschuren & S. Mankoto Ma Mbaelele
95	4 The crisis years (1992–2006)	J. Kalpers & N. Mushenzi
105	5 The history of scientific research in Virunga National Park	J. Verschuren
113	6 The history of volcanic activity in Virunga National Park	J. Durieux
123	7 The vegetation – 80 years of evolution	J. Verschuren, J. van Gysel & M. Languy
141	8 Dynamics of the large mammal populations	M. Languy
153	9 Changes in land use on the periphery of Virunga National Park	M. Languy, C. de Wasseige, B. Desclée, G. Duveiller Bogdan & S. Laime
165	10 The history of the COPEVI: the use and management of Lake Edward	F. Kasonia & N. Mushenzi
171	11 Virunga National Park: a jewel for tourism in the DRC and in the Great Lakes region	A. Lanjouw, M. Languy & J. Verschuren

The Present

185	12 The status of Virunga's large mammals	E. de Merode, A. Plumptre, M. Gray, A. McNeilage, K. Fawcett & M. Languy
197	13 The pressure of legal and illegal fisheries on Virunga National Park	M. Languy & D. Kujirakwinja
205	14 The supply of wood in the areas adjacent to Virunga National Park	M. Languy, S. Boendi & W. Dziedzic
213	15 The capacity of the *Institut Congolais pour la Conservation de la Nature* to manage Virunga National Park	H.P. Eloma & J.P. d'Huart
221	16 Political challenges to conservation in North Kivu	E. Balole, N. Mushenzi & E. de Merode
227	17 Involving local communities in the protection of Virunga National Park	E. Balole & S. Boendi
237	18 Managing a National Park in crisis	R. Muir & E. de Merode
245	19 Recent evolution of the threats to Virunga National Park and a synthesis of lessons to be learned after 80 years	J.P. d'Huart, A. Lanjouw & N. Mushenzi

The Future

253	20 Achieving better co-ordinated international support for the protection of a World Heritage Site	& V. Tshimbalanga
259	21 The need for an institutional reorganisation of the ICCN at the end of the conflict period	J.P. d'Huart & J. Kalpers
267	22 Transboundary Natural Resource Management in the Virunga Region of the Albertine Rift	A. Lanjouw, A. Kayitare & A. Plumptre
279	23 Towards a comprehensive research and monitoring program for Virunga National Park	A. Plumptre, M. Gray, M. Languy, L. Mubalama, D. Kujirakwinja & D. Mbula
289	24 The contribution of geomatics to the management of Virunga National Park: Social and ethical issues and perspectives for the future	S. Leyens, C. de Wasseige & M. Languy
299	25 Towards the resolution of boundary conflicts and a modern definition of the limits of Virunga National Park	M. Languy, P. Banza & Z. Maritim
311	Conclusions: 80 years of effort and experience – a vision for the future	E. de Merode & M. Languy

321	Appendix 1: Legal statutes on Virunga National Park's boundaries
325	Appendix 2: List of mammals of Virunga National Park
330	Bibliography
338	Index
342	Biographies
346	Acronyms
348	Acknowledgements
349	Photo credits

Introduction

Marc Languy & Emmanuel de Merode

As the Democratic Republic of Congo experiences its first democratic process since independence, and as world confidence is restored in this remarkable country after one of the bloodiest civil wars in history, the people of Congo will take stock of the many valuable resources, human and natural, they can draw on to rebuild their country. High on that list will be its National Parks, the pillars of a future tourism industry and a critical legacy for the generations to come. All eyes will turn to Virunga National Park, Congo's most valuable protected area, and a flagship, not just for Congo, but also for the whole of Africa. It is a park for which superlatives abound: it is the oldest National Park in Africa, and the most threatened; it has the most species of mammals, birds and reptiles, and the greatest diversity of landscapes and habitats ... and so the list continues. While the hardships of recent decades have caused enormous damage to the park, much of its biological and aesthetic value remains intact, and it is evident that the park can be rebuilt. As such, Virunga has survived. This has been no accident, however, or stroke of good fortune. Virunga was, after all, at the very centre of a ten-year conflict, serving as a battleground for warring factions on many occasions. Were it not for the perseverance and courage of the rangers and wardens of the *Institut Congolais pour la Conservation de la Nature,* who maintained a constant vigil over the park during the darkest period in its history, future generations would have been deprived of a truly great natural treasure.

This book was born out of the need to document the collective effort through which Virunga National Park has survived, and to demonstrate the enormous value it represents, both to Congo and the world at large. This effort spans almost a century and has involved a great many people, many of whom cannot provide direct testimony, either because they have passed away, or because they have left inadequate written records. The writers of the chapters in this book are all people who lived and worked in Virunga, who have contributed to conservation in the park, and who are well placed to record the efforts of others.

Virunga begins with a presentation of the area and a summary of all existing knowledge about the park. It is based on a review of the very considerable scientific work that has been undertaken in Virunga since the early days of its management. Since few parks have attracted such a high level of sustained scientific interest, it has been difficult to cover all the published material. For this reason, we have also included a bibliography listing the many published sources of information on Virunga.

The body of the book is divided into three main sections. The first is a review of the past, from Virunga's creation to the end of the civil war in 2003, which

Muhavura, on the border between Rwanda and Uganda, is the easternmost of six extinct volcanoes in the Virunga Massif. Its slopes, along with those of nearby Gahinga, do not extend into the Democratic Republic of Congo. Of the other four extinct volcanoes, three – Sabinyo, Visoke and Karisimbi – have their north-western flanks in Virunga National Park, while a fourth, Mikeno, lies wholly within the park.

includes the main ecological processes witnessed over this period. We then examine the present, summarising the status of the park by way of conveying consolidated baseline data upon which the park's future can be built. Finally, we review the management approaches now being deployed as part of the strategy for restoring and sustaining the park into the future.

THE PAST

The first four chapters are historical, documenting the creation of Africa's first national park and the subsequent efforts to sustain it through Congo's turbulent history. Patricia Van Schuylenbergh, drawing on several years' personal research into the early history of Congo's national parks, reviews the historical records for the earliest period covering the park's creation and the delimitation of its boundaries. This period is now the subject of some controversy, yet her analysis balances all the evidence, presenting an objective account of the legal and social processes surrounding the park's initial establishment and subsequent expansion. The second chapter adopts a very different approach to documenting the events of the past. Few people alive today can lay claim to a more detailed understanding of the events of the time than Jacques Verschuren, whose life has for more than 50 years been intimately tied to the development of Virunga as a national park. As a young researcher, a warden, chief park warden and ultimately as the director of the *Institut Congolais pour la Conservation de la Nature,* Verschuren recounts the personal experiences and memories of those who formed the park in its earliest days. Verschuren has also written the third chapter, dealing with the post-colonial history of the park during its golden age between 1960 and 1989. His co-author, Sami Mankoto wa Maembele, followed a remarkably similar career path, succeeding Verschuren as director of the *Institut*. As such, he lends added insights to a chapter outlining an extremely important and positive period in the park's history. The last of the historical chapters, contributed by José Kalpers and Norbert Mushenzi, documents what was unquestionably the most challenging period in the park's history. Both authors were intimately involved in the survival of the park during this period – one as the coordinator of a leading conservation programme focusing on the protection of Virunga's Mountain Gorillas, the other as the park's chief warden.

Various themes are then examined from a historical perspective. Verschuren recounts the history of scientific research in Virunga and reminds us that such research was, and remains, one of the three *raison d'être* of the national park. Few parks have been as intensively studied as Virunga, although the focus has changed over time from fundamental to increasingly applied research. Chapter 6 explores the park's geology, with a focus on the geological processes of the past 80 years. *Ibirunga* means 'volcanoes' in Kinyarwanda and Virunga is indeed dominated by an unusual volcanic geology that largely explains the region's exceptional biodiversity. Jacques Durieux, who has worked in North Kivu as part of the Goma Volcano Observatory, has since 1971 been studying volcanology of Nyiragongo and Nyamulagira, two of the world's largest active volcanoes. He can thus provide authoritative insights into their likely future dynamics. Jan van Gysel and Jacques Verschuren then describe the dynamics of the vegetation in the park over the past 80 years. Their accounts are drawn from a combination of field techniques. Van Gysel's work is more intense and local, being confined to the Rwindi and Lulimbi areas south of Lake Edward. Verschuren's work based on comparisons of site photographs taken and re-taken at intervals, but showing the same representative tracts of the park's various habitats, is complementary in presenting the wider picture. Mammal population changes and trends are presented by Languy. While regular mammal surveys have been undertaken over the past 40 years, the techniques used have often differed. This is hardly surprising, given the uncommonly wide range of different habitat types that are presented. Chapter 8 provides the most appropriate comparative analysis of population change over time, documenting the principal causes and exploring the wider implications for the whole mammal assemblage.

The last three chapters in this section outline the history of human-related impacts on the park. Languy *et al.* examine the changes in land use over the park's history. This work relies heavily on remote sensing techniques, so a quantified assessment of land use change can begin only with the first systematic aerial photographic surveys of the 1950s. Differences between these and more recent satellite images provide an accurate record of the growing social and economic pressures around the park. Economic pressures are particularly intense around Lake Edward, which although legally a part of the park, has today controlled fishing concessions. This follows the hand-over to local communities, under the supervision of the ICCN, of the CoPeVi *(Coopérative des Pêcheries des Virunga),* the body established to manage the controlled fishing. Fittingly then, we include a chapter written jointly by an ICCN director, Mushenzi, and the CoPeVi director, Fasonia. Lanjouw *et al.* then examine the history of tourism in the Virunga, documenting the change of focus from the Rwindi savannas to the Mikeno gorillas, along with the changing fortunes of Congo's tourism industry.

THE PRESENT

The second section of this book takes the form of a comprehensive analysis of the current state of the park, highlighting all the major issues that need to be addressed. To this end, we make use of the latest data on all the themes that best define the park. First, in their chapter on the park's mammal populations, de Merode *et al.* draw on a complex database compiled using aerial count techniques, rigorous ground samples and opportunistic observations. The savanna mammals and the Mountain Gorillas are the only taxa for which there have been accurate numerical estimates over the recent years. We can, however, present good assessments as to the current status and approximate abundance of a number of other species. Chapter 12 also provides a benchmark against which the performance of many of the park's new projects will be measured. Fish stocks and fishing and its consequences are then examined, amid detailed coverage of the settlements, many of them illegal, now strung out along the shores of Lake Edward. Languy *et al.* present a similar analysis of the illegal wood extraction within the park. Much of this industry is driven by the demand for charcoal in Goma, for which the park is now the only ready source. The authors share a decade of experience in tackling

The Okapi occurs only in the Democratic Republic of Congo. It frequents the lower Semliki Valley, in the Northern Sector of Virunga National Park.

The park was once home to the world's largest hippopotamus population. The rebuilding of this population, all but obliterated during the war years, is now a priority, given the critical role the species plays in the productivity of Lake Edward's fisheries.

Marc Languy & Emmanuel de Merode

this problem, and here put forward possible strategies for solving it.

With reference to the threats facing the park, Eloma and d'Huart assess the ICCN's current managerial capacity in Chapter 15, and present recommendations for strengthening the institution. The chapter that follows analyses the political context within which the park will have to survive. A park located at the heart of the Great Lakes Region, on the borders of three countries formerly at war, cannot expect to survive without addressing a number of political realities. As such, the conservation community of managers and scientists, though trained largely in the biological and earth sciences, have had to reinvent themselves as political activists, protecting the interests of a well-defined set of values in a deeply complex and volatile political environment. The two subsequent chapters – 17 and 18 – describe the main tools used in applying this new paradigm for conservation management, whose focus is on mobilising local interests through community intervention and on capacity building to manage unforeseen crises, a perennial problem during periods of conflict. Both of these chapters describe a unique response, developed in Virunga, to the conservation challenges posed by armed conflict. In this section's concluding chapter, d'Huart *et al.* review the lessons learned from challenges encountered in 80 years of conservation management.

THE FUTURE

The final section of the book assesses the future of the park, exploring the approaches and the tools that will be needed to ensure its survival. An important asset for the park has been the international attention it attracts and the resources it can draw on as a result. This is unlikely to diminish in the years to come, but will require careful management and coordination. Dubonet *et al.* explain the protection effort in the international context and recount the experience of mobilising and coordinating support from a wide range of individuals, organisations and public institutions. This analysis is developed in Chapter 21, which assesses the need for an institutional restructuring of the management authority. The analysis is presented in the context of a national review of the Institute concluded in 2006, and whose recommendations are being implemented to restructure the organisation. An invigorated Institute is a fundamental pre-requisite if the country's national parks are to be rehabilitated and are to survive into a post-conflict era replete with new challenges, over and above the potentially numerous unresolved political issues. Future challenges arising in the wider political context are revisited in Chapter 22, which explores transboundary approaches to protective management, strategies in whose formulation Virunga National Park has long played a pioneering role.

The park's monitoring and research programme remains important, not only in terms of its contributions to science, but also in terms of the guidance it offers to managers and decision makers. In Chapter 23, Plumptre *et al.* describe the current research and monitoring programme, and make recommendations for its future development. The focus, in Chapter 24, is on a particular discipline and set of tools that has taken on immense significance in conservation management in general, and whose efficacy will continue to prove invaluable to the restoration and future management of Virunga. The spatial sciences, including telemetry, remote sensing and geographic information systems, now occupy an integral part of the standard toolkit used in both day-to-day management and the longer-term planning of park activities. Leyens *et al.* outline the methods now in use and assess their value, while providing a few caveats concerning possible misuse. This discussion leads on to a chapter, presented by Languy *et al.*, describing the re-establishment of the National Park's boundaries, in a process that draws heavily on the advanced new imaging technology. Extensive recourse, meanwhile, to local knowledge through consultation has brought far greater legitimacy and acceptance to the status of the protected area.

All the lessons that emerge in the course of the book underpin our final chapter, which addresses the immediate goal of restoring the basic ecological and economic functions of the park, and the longer term goal of establishing a sustainable management structure. The end result, we hope, is a constructive prognosis drawn from the collective experience of all those who have been intimately associated with the park's management over the decades.

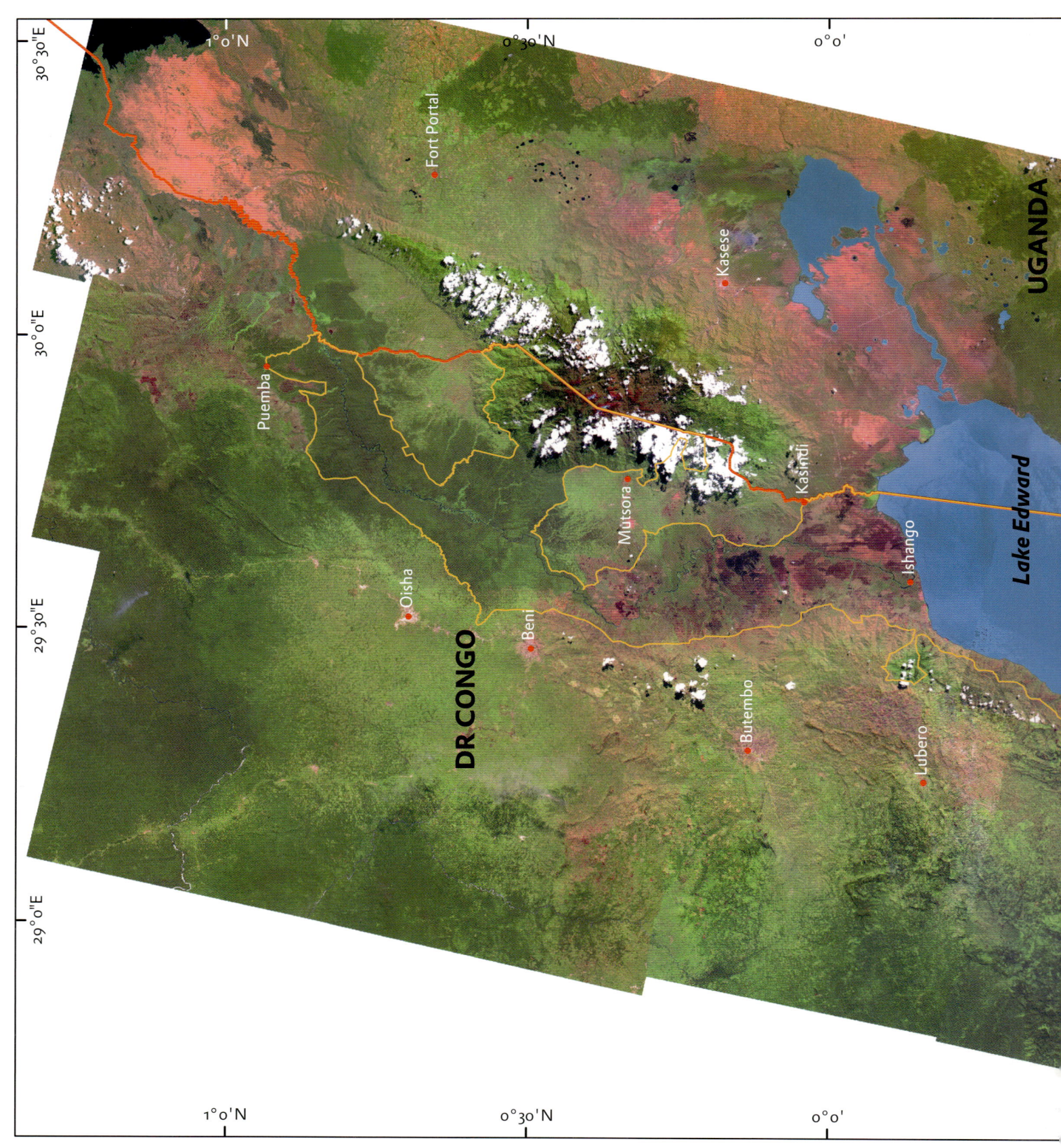

0°30'S
1°0'S
1°30'S
29°0"E
29°30"E
30°0"E
30°30"E
Lulimbi
Bushenyi
Vitshumbi
Ishasha
Rwindi
Rutshuru
Kabale
Rumangabo
Masisi
Musanze
Sake
RWANDA
Goma
Lake Kivu
0
20
40
80
Km
Town/Patrol post
International border
Virunga National Park boundary

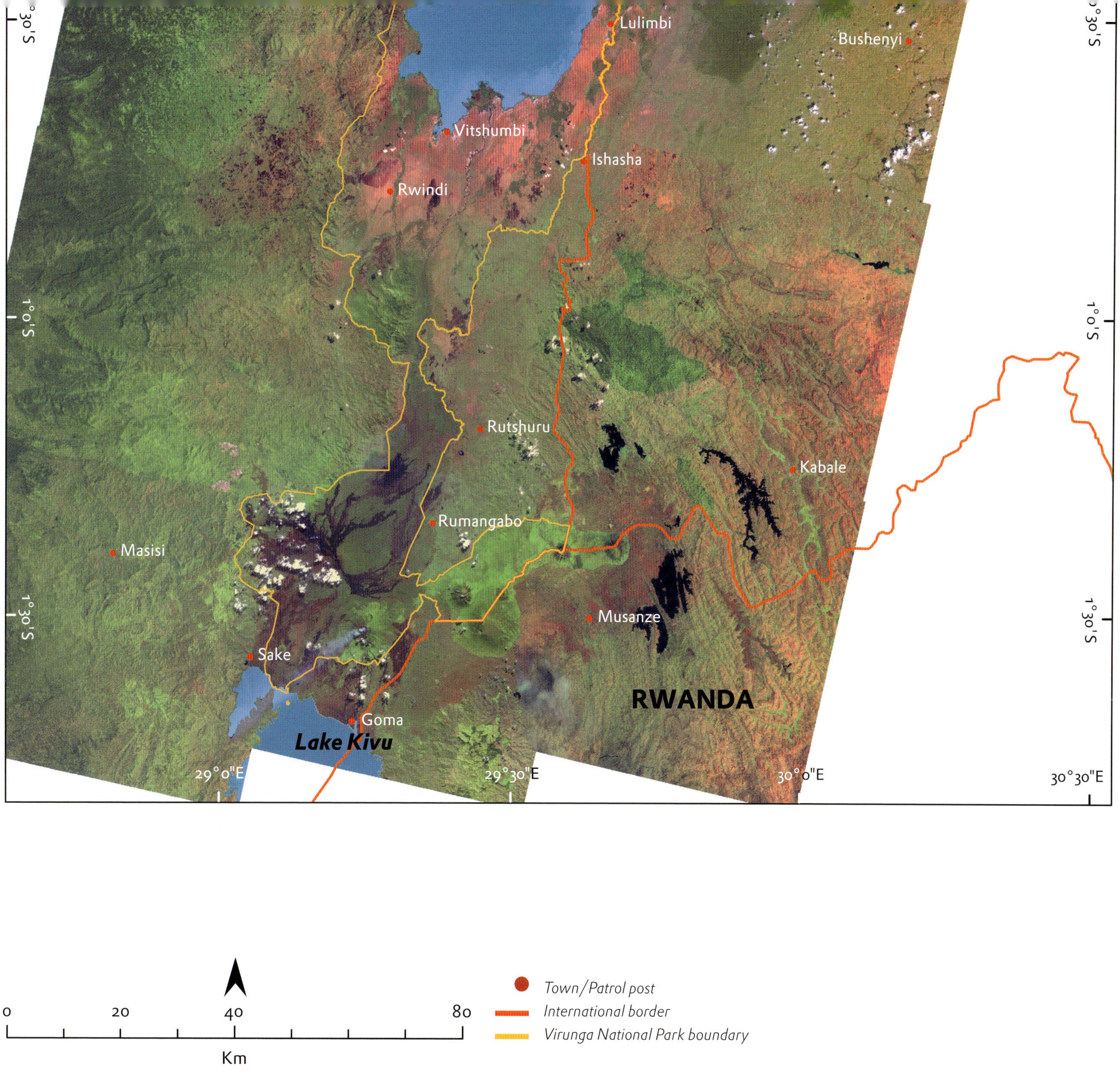

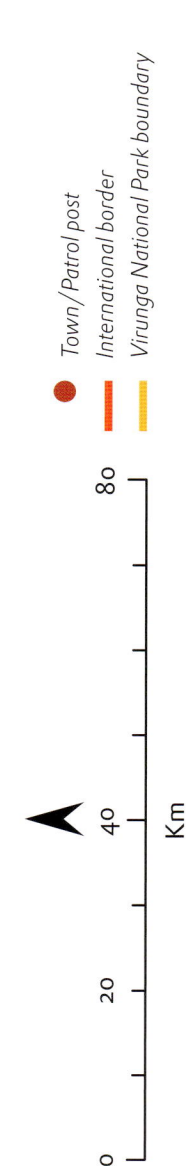

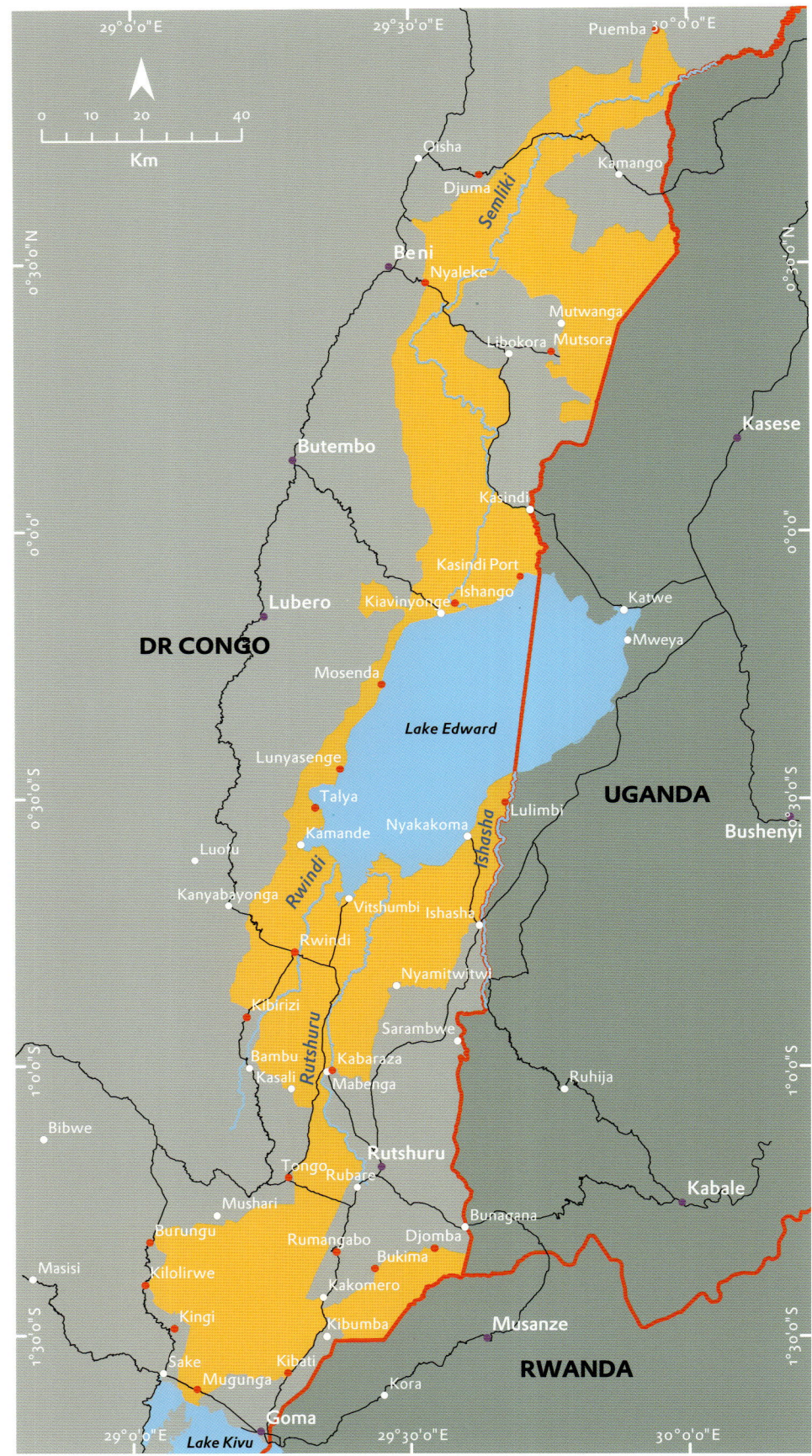

0.1
General map of Virunga National Park, showing the principal locations.

0 A brief overview of Virunga National Park

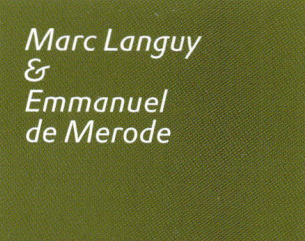

Marc Languy
&
Emmanuel
de Merode

0.1 LOCATION

Virunga National Park lies on the equator, in the east of the Democratic Republic of Congo (DRC), along the borders of Rwanda and Uganda. The park covers an area of 784,368 hectares, disposed north-south in a long, narrow, unwieldy shape. In the north (00°56'N), the park is bordered by the River Puemba and in the south (01°39'S) by Tshegera Island in Lake Kivu. Most of the park lies on the base of the Albertine Rift, the western arm of Africa's Great Rift Valley. The park extends for about 300 km north to south, but averages just 23 km east to west (with a minimum of only 2.3 km to the south of Mabenga). In view of its complex shape, the park has an exceptionally long boundary – of 1150 km.

The highest point in the park is Margherita Peak at 5119 metres above sea level (which is also the highest point in the country and the third highest on the African continent), while the lowest point is at the confluence of the Puemba and Semliki rivers at 680 metres above sea level.

0.2 CLIMATE (J. Verschuren)

Describing the climate of Virunga National Park is difficult, given the marked variations in altitude, even over relatively short distances. Indeed, the park might best be described as a mosaic of microclimates. The site with the most hours of sunshine, Ishango, is less than 60 km from the wettest and cloudiest site in Congo, the Rwenzori Mountains.

For many years, meteorological base stations have been managed from towns outside the park (such as Beni, Goma and Rutshuru), but their records bear little relation to the weather inside the park. Some meteorological stations were established within the park – at Mutsora, Ishango, Rwindi, Rumangabo and Lulimbi – and most of these have remained in service, despite the political upheavals.

Several authors, dating back to the 1920s (Bultot, Crabbé, Vandenplas), have studied Kivu's climate in depth, and various zoological and botanical publications give a general overview of the climate in their introductions (Bourlière, Cornet d'Elzius, Delvingt, d'Huart, Lebrun and Verschuren). None of these analyses extends beyond 1990, however. Data collected since 1990 should be treated with caution, owing to limitations in the condition and use of the measuring equipment. Some animals took to urinating in the rain gauges, which were sometimes blocked for long periods with hippopotamus excrement! Hyenas carried off some of the rain gauges; others were destroyed after elephants had rubbed against them.

0.2
ViNP is located at the heart of the Albertine Rift, the western branch of Africa's Great Rift Valley.

- ● Town
- ▬ International border
- ▬ Virunga National Park

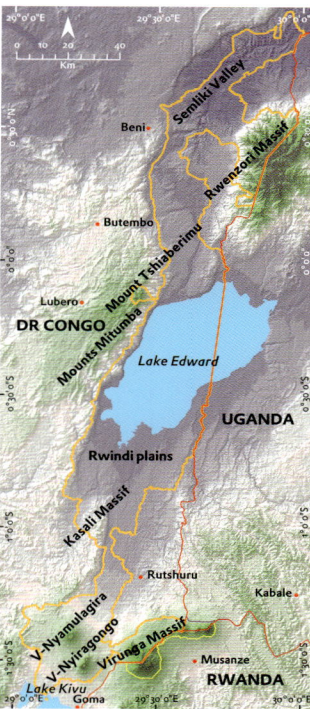

0.3
Main geographical characteristics of Virunga National Park.

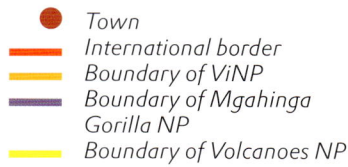

- ● Town
- ▬ International border
- ▬ Boundary of ViNP
- ▬ Boundary of Mgahinga Gorilla NP
- ▬ Boundary of Volcanoes NP

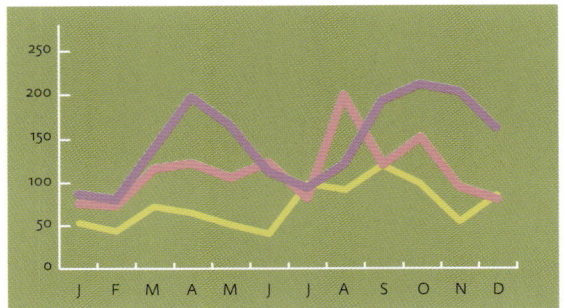
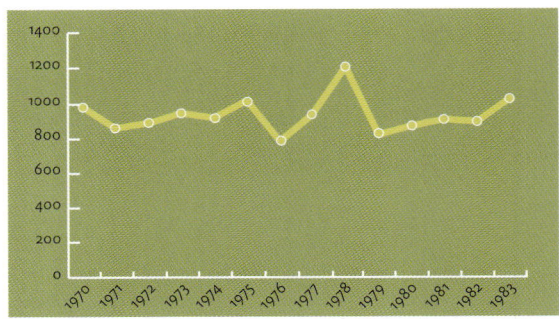

0.4
Average monthly precipitation (mm) at each of Virunga's three main stations (Mutsora, Rwindi, Rumangabo) over the six-year period, 1955–1960.

0.5
Monthly precipitation trends at Rwindi from 1970 to 1983, in mm.

0.2.1 Rainfall

Being located on the equator, the park is typically subject to two wet seasons and two dry seasons, with marked local variations in rainfall, but similar seasonal variations. An analysis of monthly precipitation for the park's three main stations (Mutsora, Rwindi, Rumangabo) between 1955 and 1960 shows a peak in April–May and again in September–October, with minimal precipitation in the months of February and July (Figure 0.4).

An analysis of the number of rainfall days at Ishango (900 metres asl), to the north of Lake Edward, and at Mutsora (1200 metres asl) on the foothills of the Rwenzori, confirms these seasonal patterns, but also illustrates the dramatic local variations in rainfall: Mutsora experiences more than four times as many rainfall days, and receives three times the amount of rainfall.

Ishango and Nyamushengero (lower Rutshuru) appear to be the driest locations in the park (and in the whole of the DRC, where they are rivalled by Banana, in the far west of the country).

The dry seasons are generally not very distinct, whether on the plains or in the mountains. Rainless months are very rare in Virunga, as are cloud-free skies. Yet, periods of continual rainfall lasting several days (as in parts of Western Europe) are not common either. Late afternoon showers tend to be the norm.

At Rwindi, 14 years of rainfall data are available. These allow for inter-annual precipitation analyses. Rwindi appears to be relatively stable (Figure 0.5), with an average annual precipitation, calculated over the 14 years (1970–1983), of 930 mm. Precipitation for the driest year was 783 mm, compared with 1203 mm (50 % more) for the wettest.

A comparison with data from before 1960, which reflect a yearly average of 885 mm, reveals that there has been no significant change in precipitation. Certainly, there is no evidence for a drying climate at Rwindi. Latterly, there have been several years in which annual precipitation has exceeded 1000 mm. If anything, the evidence points to a slight increase in overall precipitation.

We also have fragmentary data on Lulimbi, to the southeast of Lake Edward, where – on average – there are 99 rainfall days per year and where the average annual rainfall is 701 mm. Comparable data collected by d'Huart (1978) also shows a pattern of consistently high inter-annual rainfall variation.

1972	988 mm
1973	585 mm
1974	755 mm
Average	776 mm

Xavier Misonne has shown that the figure, close to 4000 mm a year, often cited for rainfall on the flanks of the Rwenzori Mountains is not correct. On the flanks of both the Rwenzori and the Virunga Volcanoes, precipitation and cloud cover are generally at their maximum at about 2700 metres asl (in the tree heather zone). Rainfall on the summits has been shown to be much lower. For instance, the gross data collected between 1930 and 1957 on the summit of Karisimbi at 4500 metres show an average annual rainfall of only 940 mm, but with considerable inter-annual variation: an annual minimum of 562 mm, set against a maximum that, at 1329mm, is more than double that quantity. A study by E. Roche on the Rwandan slopes of the Virunga Volcanoes indicates that rainfall increases with altitude up to about 2500 metres, where it peaks, before decreasing significantly beyond that point.

Periods of drought are extremely rare in Virunga National Park, although the Rwindi and Semliki savannas do sometimes appear to be very arid. No major river has ever been recorded as dry.

Disregarding recent years (the 1990s onwards) for which there are no reliable data, there do not seem to have been major variations in precipitation for almost 75 years. Sahelian drought cycles do not appear to have been a factor.

It is important to rehabilitate the park's network of meteorological stations, and to establish cumulative monthly rain gauges at patrol posts. The climate of Virunga National Park has not benefited from the kind of comprehensive analysis seen at Garamba, for example, where Noirfalise has carried out in-depth studies.

0.2.2 Temperature

Like the rainfall, temperature is extremely variable at Virunga. Figure 0.11 shows the average daily temperatures for the park's three main stations from 1955 to 1960.

On the mountains, the temperature gradient is about -0.7 °C per 100 metres at the equator.

At Rwindi, over 14 observation years, the annual minimum temperature averages about 15 °C under normal conditions, and 12–13 °C on exposed soil. A

0.6 Precipitation in the Northern Sector: number of rainfall days

Station	Year	J	F	M	A	M	J	J	A	S	O	N	D	Total
Ishango	1984	2	5	10	8	5	2	6	7	6	5	8	5	69
	1987	3	1	5	6	3	3	5	4	6	6	8	3	53
	1988	0	0	0	8	9	3	4	5	11	9	4	0	53
	1989	2	2	3	3	9	0	2	4	8	8	5	4	50
Mutsora	1984	9	16	12	12	6	20	21	19	15	22	24	12	188
	1987	25	13	18	20	20	19	12	15	16	23	24	12	217
	1988	16	8	19	22	18	13	22	27	25	22	19	20	231
	1989	2	3	19	19	21	13	17	23	20	21	18	18	194

0.7 Precipitation in the Northern Sector: mm

Station	Year	J	F	M	A	M	J	J	A	S	O	N	D	Total
Ishango	1984	14	10	53	20	12	3	32	30	7	33	42	27	283
	1987	15	4	57	35	26	37	29	47	29	14	55	7	355
	1988	0	0	0	86	128	5	34	57	110	72	19	67	578
	1989	5	15	66	15	28	0	2	19	36	41	13	56	296
Mutsora	1984	77	34	32	72	95	12	125	103	100	50	77	82	859
	1987	190	78	182	125	143	106	33	202	83	137	292	83	1,654
	1988	76	36	125	183	152	87	139	213	119	218	105	59	1,512
	1989	27	56	42	111	207	65	238	208	169	137	76,6	58	1,394

0.8
Mountainous regions of the park experience extremely high rainfall throughout the year. Coupled with economic stagnation and lack of maintenance, this has created major logistical challenges in the region.

A brief overview of Virunga National Park

0.9 Total number of sunshine hours at Rwindi, 1970 to 1979.

Years	Hours
1970	2582
1971	2437
1972	2050
1973	2088
1974	2322
1975	2356
1976	2190
1977	2370
1978	2042
1979	2397

0.10 Daily minimum and maximum temperatures on the flanks of the Rwenzori

Location	Habitat	Temperature (°C)
Semliki (lowland forest)	(730 m) Watalinga	19 to 32
Kalonge (montane forest)	(2,200 m)	14,8 to 23.5
Kalindere	(2,700 m) (bamboo forest)	11 to 19
Mahangu	(3,390 m) (heather)	5 to 14
Camp des bouteilles	(3,820 m) (heather)	0 to 8
Kiondo (alpine)	(4,200 m) (groundsels)	-0.3 to +8.1

Mutsora
Rumangabo
Rwindi

0.11
Average monthly temperatures at each of the three main stations of Virunga National Park (1955 to 1960).

record low of 9.4 °C was registered one February on a lawn.

The annual average temperature—23 °C at Rwindi—can also be established from measurements of the permanent soil temperature at a depth of 1.5 metres. Bush fires (rarely violent in Virunga) have only a limited effect on soil temperature.

Every night at Kiondo the ground freezes: X. Misonne and J. Verschuren have recorded a temperature of -8 °C on exposed ground at Kiondo.

On Margherita Peak, fairly constant temperatures of -3.5 °C have been recorded on the glaciers. Snowfall is frequent on the summit. Likewise, the summit of Karisimbi is often covered with snow, but this rarely holds for more than 24 hours. Analyses of the various temperature parameters (average, maximum and minimum), both at low altitudes and in the mountains, show that these have been fairly stable over several decades.

0.2.3 Insolation

Data on hours of sunshine at Rwindi over ten years are listed in Table 0.9. The table shows that in only half of all the years on record did annual hours of sunshine exceed 50 % of the available sunshine hours (roughly 4400 hours). Subjective impressions notwithstanding, insolation is generally low in Virunga National Park, even on the plains.

At Rumangabo, insolation occurs in less than 30 % of the total available hours: a rate comparable with that of Brussels in Europe.

Wind speeds, humidity and evaporation on the Rwindi plains and on the lava fields are examined in some detail by Lebrun.

To summarize the climate characteristics of Virunga National Park, the main pattern is one of extreme spatial variations that nevertheless have remained relatively stable over several decades until 1990, at least.

0.3 Virunga National Park, the Albertine Rift and Climate Change

More than 30-million years ago, Africa's Great Rift Valley formed from Mozambique in the south to the Red Sea (see Chapter 6). From its southern end, known as the Lake Malawi Rift, the valley separates into two branches. The eastern branch, sometimes referred to as the Gregorian Rift, extends through Tanzania and Kenya and includes a string of lakes: Natron, Magadi, Naivasha, Nakuru, Bogoria, Baringo and Turkana. The western branch is generally referred to as the Albertine Rift and stretches northward from the southern end of Lake Tanganyika to Lake Albert, to include Lakes Kivu and Edward as well. The two branches then rejoin to form the Great Ethiopian Rift, which extends northward to the Red Sea (Figure 0.14).

Virunga National Park is situated at the heart of the Albertine Rift, which is 1380 km long, but rarely more than 150 km wide. The Masisi Massif, the Mitumba Range and Mount Tshiaberimu make up the western flank, while the extinct volcanoes of the Mikeno Sector form its eastern edge. The northern shores of Lake Kivu are in one of the areas where the Albertine Rift is at its most narrow (less than 45 km), presenting one of the most dramatic landscapes in central Africa.

The Rwenzori Massif is not volcanic and is not part of the Rift's escarpments. It is a large massif that rose from the base of the Rift Valley some three million years ago.

Virunga National Park's geology is closely tied to that of the Albertine Rift. The Rift is still shifting, in response to tectonic movements that are splitting eastern Africa from the rest of the continent. Actively part of that evolution, Mounts Nyiragongo and Nyamulagira are among the most active volcanoes on the planet.

All these geological processes have had a major impact on the habitats and vegetation assemblages of the park. Vegetation successions on the lava fields in particular, have had a significant bearing on the area's plant diversity.

Climatic cycles are another major influence on plant diversity. There is now ample evidence of climate change accruing from both orbital cycles and human activity. A dramatic example in Virunga is the change in the surface area of the Rwenzori glaciers, which have retreated over more than 1500 metres in altitude, as demonstrated by the glacially eroded valleys at lower altitude. These cycles are independent of the climate change trends detected in recent years. The climatic cycles that provoked the great glaciations in Europe are the product of a combination of processes: major

0.12
Beans and bananas are the main crops cultivated at the foot of mount Nyiragongo, north of Goma.

0.13
The Rwenzori Mountains, with the third, fourth and fifth highest summits in Africa, are of non-volcanic origin, having risen from the base of the Albertine Rift two to three million years ago.

0.14
Virunga National Park lies in the western branch of Africa's Great Rift Valley, generally referred to as the Albertine Rift.

A brief overview of Virunga National Park

Marc Languy & Emmanuel de Merode

0.15
Virunga corresponds to the word 'volcanoes' in Kinyarwanda. Volcanic processes, past and present, are a major feature of the region, providing evidence of the great tectonic forces at work in creating the Albertine Rift, separating eastern Africa from the rest of the continent. This volcanism also explains the soaring relief in the Southern Sector of the park, and the presence there of montane forests that have nurtured exceptional numbers of unique species. The Nyiragongo and Nyamulagira volcanoes are among the most active in the world, and have become a major tourist attraction. This photograph was taken during the Kimanura eruption of 1989.

A brief overview of Virunga National Park

cycles, driven by astronomical factors lasting 413,000 and 100,000 years, and minor cycles lasting 23,000 and 19,000 years. Both are tied to the orbital equinox, and are responsible, among other things, for creation of the polar ice caps. Another cycle, lasting 41,000 years, is tied to the oblique shape of the planet and its rotational angle: factors that causes day length and seasonal intensities to vary.

These global temperature variations have had different effects on different locations on the planet. In the region covered by this book, a temperate humid climate led to an expansion of lowland moist forest and a lowering of the altitudinal limit of the montane forest. Over the past two million years, intermittent arid and humid periods have brought continual changes to the landscape of what, today, is Virunga National Park.

At the end of the Tertiary Period, two million years ago, a combination of the humid temperate climate and the area's geographical relief resulted in formation of a massive lake, Kaisian, which covered all of what are now Lakes Albert, Edward and George. The upper limit of the sub-montane vegetation belt was then 1,700 metres, as compared with 2,600 metres today. A period of significant climate change then occurred about one million years later, during the mid-Pleistocene, when a prolonged period of aridity triggered a retreat of the moist lowland and montane forests, and allowed for the growth of grassland savannas. Over the same period, volcanic activity blocked drainage from the highlands into the Kaisian Lake, and the watershed between the Nile and the Congo Rivers shifted. With a drier climate, the moist lowland forest that had dominated the landscape was gradually superseded by xerophilous assemblages and by flooded savannas left by Kaisian Lake's retreating.

Towards the end of the Pleistocene, humidity increased and the moist lowland forest began to re-establish itself, although patches of savanna remained. There followed an exceptionally dry period broken only by a few intermittent humid periods. The last 200,000 years have been characterized by a succession of arid and humid periods: aridity was extreme at around 200,000 BP, humid at 100,000 BP, and arid at around 20,000 and 10,000 BP. Between 10,000 and 5,000 years BP, the forests of the central African region were in full expansion, with a warm, humid peak at around 6000 BP. There followed a reduction in rainfall, with a peak of aridity (resulting in accelerated growth of sclerophyllous plants) at around 4000 BP, and then a final short dry period at around 2500 BP. Human impacts aside, the forests are currently expanding at the expense of the savannas.

An understanding of both the geological history and the climate patterns is essential in being able to explain today's broad spectrum of vegetation mosaics and species assemblages in Virunga National Park.

The vegetation assemblages are a response to the climate fluctuations, which have seen intermittent expansion and reduction of forest areas, and a raising and lowering of the base of the afro-montane forest zone. The diversity of Virunga's ecology then, is derived not only from today's varied micro-climates and habitats, but also from the climatic variations and geological processes of the past. The persistence of sclerophyllous vegetation, and of small forest blocks east of the Congo Basin during even the driest periods, has provided for a relatively rapid recolonisation of savannas by forest species during the more humid periods, probably leading to speciation.

A final, yet all-important, factor is the massive altitudinal variation within the park, which has allowed forest to persist throughout its long-term history. Climate changes have brought about a corresponding movement in the forest base, which has 'climbed' or 'descended' the hillsides accordingly. It has been established that these montane forests were once connected to those of other mountains, some as far away as Cameroon, before being cut off by periods of aridity. Today, the Afro-montane forests of the Albertine Rift, dominated by Virunga National Park, are the richest of their kind on the African continent.

0.4 The park's greater habitats.

Virunga National Park is made up of three sectors, the Northern, Central and Southern, as defined by park management requirements.

The following pages describe the greater habitats of the park, from north to south. Of necessity, these are very general descriptions, since an exhaustive account covering *all* habitats would be far too lengthy.

The habitats of Virunga National Park are exceptionally diverse, extending from the glaciers of the Rwenzori range to the grassland savannas of the Rwindi plains, through the great lowland rain forests of the Semliki Valley, the montane forests and the dry forests of Tongo, as well as through an array of extremely diverse aquatic habitats (high altitude vertical bogs, lakes, swamps, rivers and hot springs) to the vegetation successions on the lava fields. This landscape diversity is one of several reasons for Virunga National Park's designation as a World Heritage Site.

0.4.1 The Northern Sector: from the great lowland rainforests to the glaciers of the Rwenzori Mountains.

Covering an area of 299,523 ha (of which 22,324 ha are part of Lake Edward), the Northern Sector is the most extensive area of dry land within the park. The sector extends from the Puemba River to Lake Edward, and is defined by the Semliki Valley, the Rwenzori Massif and Mount Tshiaberimu.

0.4.1.1 The middle Semliki, as far as its confluence with the Puemba.

The dominant feature in this part of the park is the great tropical lowland forest massif through which the Semliki River has cut its passage in bearing water from Lake Edward to Lake Albert (Figure 0.16). The development of the dense lowland forest is determined largely by climate, altitude and edaphic conditions (average temperatures are 23 °C, precipitation in excess of 1500 mm is distributed over at least ten months of the year). While the lower Semliki Forest is part of the Nile Basin, its biological features are those of the Congo Basin. As such, it forms an ecological continuum with the Ituri Forest massif. Its assemblage of birds, mammals and plants are largely typical of the Congo Basin and okapis can be found on the edges of the Abia and Abatupi Rivers on the west bank of the Semliki.

Nevertheless, the lower Semliki is on the eastern

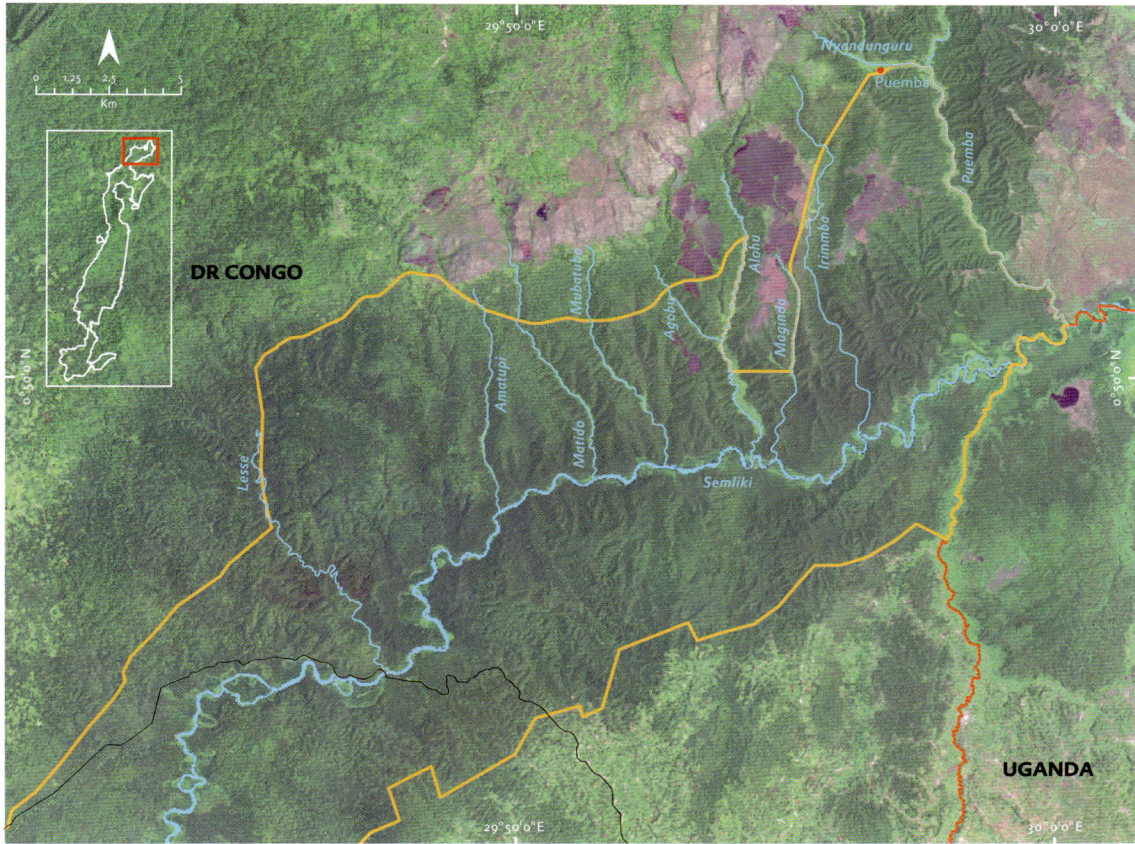

0.16
The northern reaches of Virunga National Park, with the Semliki River flowing south-west to north-east. The closed-canopy lowland forest is shown in dark green, the wetlands in lighter green. The pink and purple areas are grassland savannas, of which the largest extends seven kilometres from north to south.

- Patrol post
- International border
- Road
- ViNP Boundary
- Recently burned savanna
- Agriculture
- Dense forest
- Gallery forest
- Savanna

periphery of the great Congo Forest, and dense closed-canopy forest is found only in the valley bottoms. Secondary forests are generally found on the watersheds. Such areas are dominated by *Cynometra alexandri*, typically absent from closed-canopy forest. Unlike most of the Ituri, which reaches Virunga's boundaries, mono-specific stands of *Gilbertiodendron* are completely absent. Along the Semliki Valley are occasional 'pseudo' galleries of large acacias interspersed with *Cyperus papyrus* populations.

The lowland forest covers the Northern Sector of the park from the Puemba River to the bridge at Vieux-Beni. Some small patches of edaphic savanna grasslands are found in the northernmost reaches of the park (where they cover no more than just a few square kilometres).

0.4.1.2 The Rwenzori Massif

The Rwenzori massif straddles the border between Uganda and the DRC. It has protected area status in both countries: in Uganda's Rwenzori Mountains National Park to the east, and Virunga in the DRC to the west. The entire massif spans about 4800 square kilometres, of which 1493 km² lie above 7000 feet. The Rwenzori range is in fact a chain of mountains that includes the third, fourth and fifth highest summits in Africa. Margherita Peak, the highest at 5119 metres above sea level, is divided by the international border (Figure 0.22).

The chain is geologically quite young. The mountains rose from the base of the Albertine Rift a mere two to three million years ago, which partly explains the abruptness and height of their relief. As these formed, the mountains separated the Semliki River from Lake George, so that drainage on the Ugandan side of the range now flows east and then south, forming a semi-circle of rivers that eventually reach Lake Edward and continue down the Semliki, finally reaching Lake Albert and the Nile.

It is in this part of the park that the relief is most extreme: The middle Semliki, at just 800 metres above sea level, is only 30 km away from the summit of the Rwenzori (5119 m). Yet the most defining characteristic of this part of Virunga National Park is that the natural habitat is uninterrupted between these two extremes, so creating a unique corridor containing all possible habitat types. On the African continent, only Mount Cameroon can still boast such a well-preserved assemblage.

When climbing the Rwenzori Mountains from the park station at Mutsora, the succession of distinct vegetation types becomes readily apparent. After climbing beyond the mid-altitude secondary forest, at about 2000 metres, you pass through a dense montane forest. A bamboo forest marks the upper limit of this montane forest. On the Rwenzori Mountains the bamboo forest is fragmented, however, unlike that on the Virunga Volcanoes. This forest type is nevertheless very well represented in the northern part of the range, around the upper Ruanoli, where almost continuous stretches of bamboo forest can be seen between 2150 and 3130 metres – an exceptionally wide range for this vegetation type.

Hagenia abyssinica forest is much less homogeneous in the Rwenzori than on the volcanoes and is often

0.17
The plain referred to as Wasolda at the foot of Mount Nyiragongo was covered by a lava field during the 1977 eruption. This lava flow was extremely fluid, sweeping through the forest in waves that reduced the trunks of trees in its path to pillared stumps of solidified lava, which today bear stark witness to the successions of lava flows.

0.18
The park's numerous volcanoes and mountains have created unique assemblages of afromontane vegetation (including the giant groundsel illustrated here).

Marc Languy & Emmanuel de Merode

0.19
.The Rwenzori Mountains, also known as the Mountains of the Moon, dominate the Northern Sector of the park.

0.20
Frequent cataracts render the Semliki River, flowing through the middle of the Northern Sector, unnavigable.

0.21
The main vegetation types to the north of Lake Edward are acacia-dominated savannas interspersed with gallery forests.

A brief overview of Virunga National Park

0.22
The Rwenzori Massif is shared by the DRC and Uganda, representing the highest point in both countries. On the Congolese side of the border, there is still an uninterrupted continuum of habitats from moorland through to moist lowland forest: a rare phenomenon in Africa today.

0.23
On the northern side of Lake Edward, acacia savanna gradually gives way to tropical lowland forest by way of a mosaic of gallery forest interspersed with woodland savanna.

0.24
Intensive agriculture is the dominant land-use in North Kivu, especially in the Beni-Butembo region, along the western flanks of the Northern Sector.

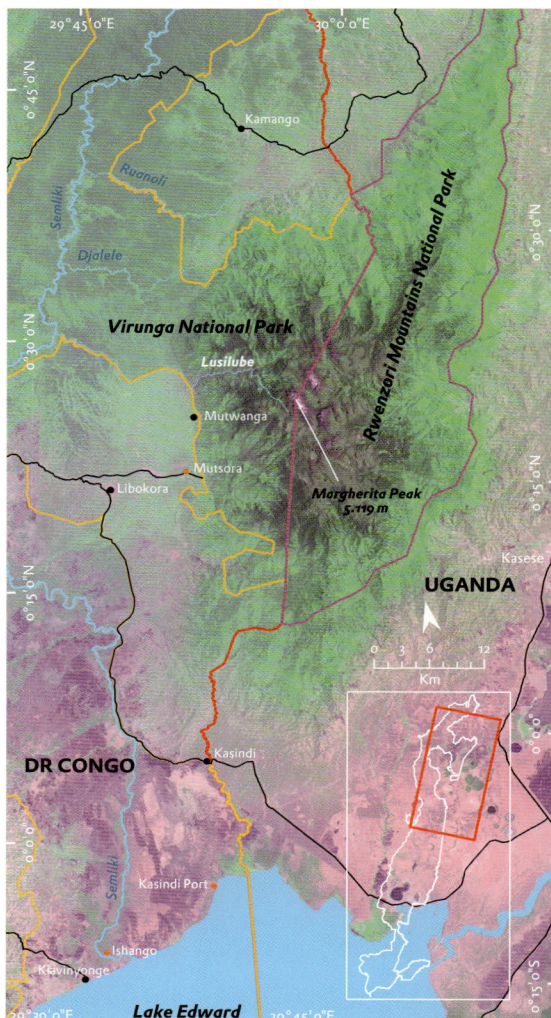

associated with a Myrcinacea, *Rapanea rhododendroides*. Less well established than on the volcanoes, this forest type is quickly replaced above 2600 metres by a tree heather that dominates above 3000 metres, but which vanishes altogether above 4000 metres. The biggest heather of the Rwenzori range is *Philippia trimera*. *Rapanea rhododendroides* and *Philippia trimera* dominate this vegetation zone on the Rwenzori, whereas on the extinct volcanoes the latter is quickly replaced by similar species of the same genus. These plants are easily identified by their tortuous trunks, while their branches are generally festooned with various lichens, mainly *Usnea spp.*, creating what appear to be large pale green beards. A moss, *Breutelia stuhlmannii*, usually covers the ground, producing a carpet of yellow.

As temperature and rainfall drop with altitude, the heathers gradually disappear, being replaced, at about 4000 metres, by the first giant groundsels and lobelias. The most common such species are respectively *Senecio johnstoni* and *Lobelia lanurensis*. Above 4500 metres, the afro-alpine flora is dominated by mosses and lichens.

The massive extent of ancient glaciers is reflected in the Rwenzori's many huge, U-shaped valleys, whose flat bases are today often covered by extensive marshes of *Carex monostachya* and *C. runssoroensis* sedges. These *Carex* complexes are interspersed with *Alchemilla* species, primarily *Alchemilla johnstoni*.

The Rwenzori's *Helichrysum*-dominated moorland extends over the rocky, well-drained slopes above the tree heather zone.

The rock and glacier landscapes of the Rwenzori are the highest habitats of Virunga National Park (Figure 0.25). The melting of the glaciers, first noted by Heinzelin de Braucourt in 1953, is unfortunately well advanced. The total surface area of the glaciers has decreased from 650 hectares in 1906 to 350 hectares in 1955, and to 108 hectares in 2005 (Figure 0.28).

0.4.1.3 The upper Semliki
Upstream from the bridge at Vieux-Beni, just above the falls of Lusilube, the valley is extremely steep, in places forming precipitous gorges, such as the Bomama Gorge. Spectacular waterfalls, such as those at the confluence of the Butahu and the Lusilube Rivers, are interspersed with long stretches of calmer water.

Moving south from Vieux-Beni (disposed at the same latitude as Beni town and the park station at Mutsora), the lowland forest gives way to some large gal-

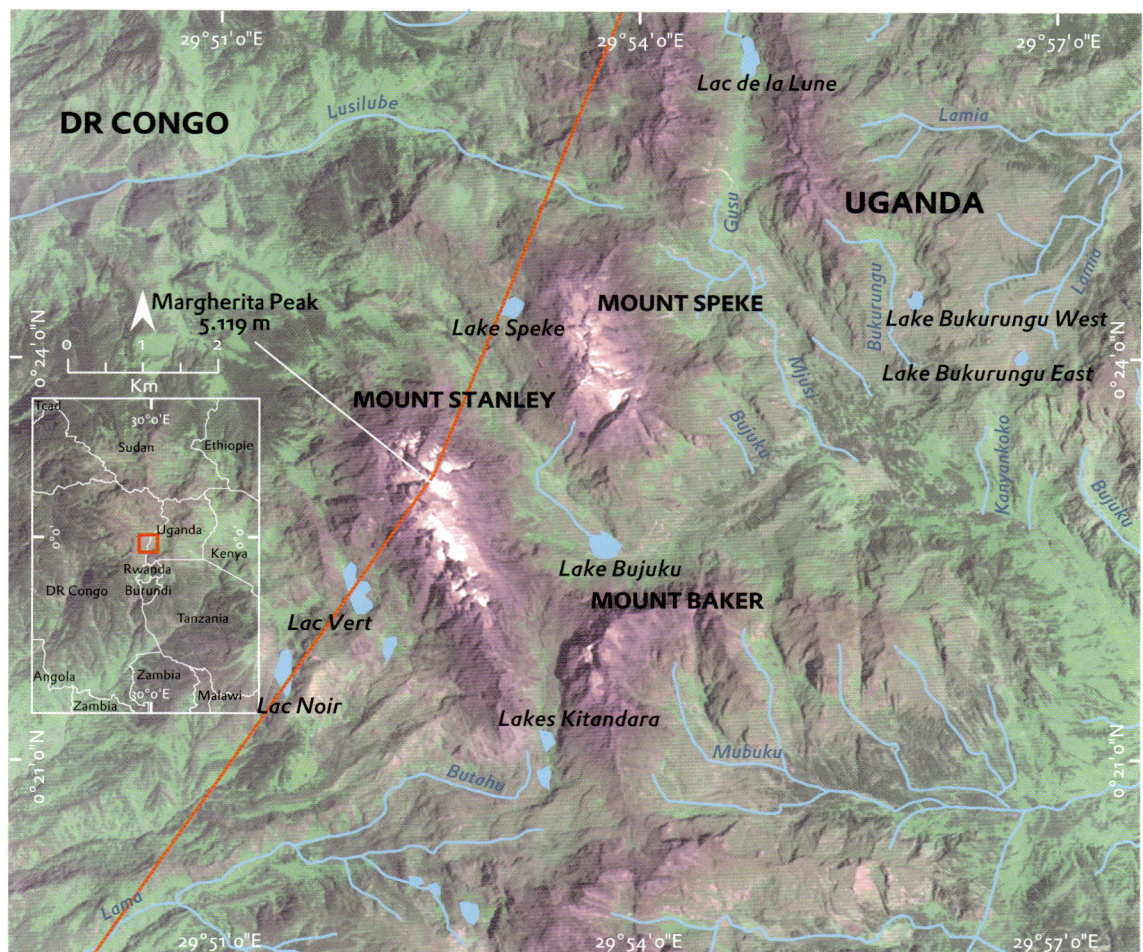

0.25
The glaciers and tarns of the Rwenzori Mountains. From both the Ugandan and the Congolese sides, there are several routes leading up to the bases of the main peaks.

— International border
— River

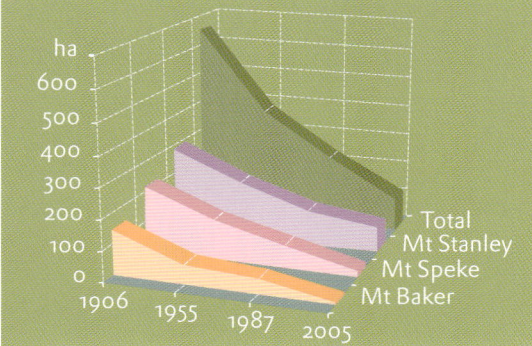

0.26
A giant groundsel in the Rwenzori's afro-alpine vegetation zone.

0.27
The upper reaches of the Rwenzori Mountains sustain many plant species that are adapted to the extreme temperatures and impoverished soils, such as these everlasting flowers, Helichrysum sp.

0.28
The melting of the Rwenzori glaciers, noted from the middle of the last century is continuing apace. Current projections are that the glaciers will have all but disappeared by the park's one-hundredth anniversary in 2025.

A brief overview of Virunga National Park 33

0.29
Papilio lormieri semlikana is a very narrowly distributed swallowtail sub-species, found only in the extreme eastern part of the Kivu region. This specimen was photographed in Virunga National Park.

0.30
Mount Tshiaberimu provides refuge for Virunga's only population of Eastern Lowland Gorillas. Situated to the north-west of Lake Edward, it is linked to the rest of the park by a narrow corridor.

- Village
- Patrol post
- International border
- Road
- ViNP boundary
- Recently burned savanna
- Agriculture
- Montane forest
- Clouds and shadows

lery forests. In this zone, stands of *Cassia siamea,* an invasive exotic of South American origin, are unfortunately displacing indigenous tree species.

An unusual type of gallery forest lines the northern part of the upper Semliki. Covering both riverbanks and extending for several hundred metres out to either side, are large acacias rooted in somewhat swampy ground. This habitat type is not easily penetrated by man and provides a refuge for elephants, buffaloes and Giant Forest Hogs. Large mammals are more abundant in this habitat than in the *Cynometra alexandri*-dominated primary forest.

Further south, towards Lake Edward on the plains of the upper Semliki, the vegetation is dominated by a *Cymbopogon* savanna, with a large number of *Borassus* palms. The southern boundary of the savanna, on Lake Edward's northern shore, is dominated by grassland. It is here, at Ishango, that the lake empties itself into the Semliki. Further east, the Lubilya Delta is a vast, semi-swampy lowland with representatives of the Mimosaceae very much to the fore.

0.4.1.4 Mount Tshiaberimu

Mount Tshiaberimu is best described as a small island of montane forest linked to the park via a narrow corridor known as *Mulango wa Nyama* in Kiswahili ('the gate for the wildlife') along the River Tumbwe (Figure 0.30). The altitudinal gradient is steep here in that the Tumbwe flows into the lake at 916 metres above sea level, while only 11 kilometres away the summit of Tshiaberimu rises to 3117 metres. This represents an overall altitudinal gradient of 22 %.

As with the Rwenzori, there are distinct montane vegetation zones between the foothills and the summit of Tshiaberimu. The afro-alpine zone is not well represented, however, and the tree heather is dominant only over the last few hundred metres. Bamboo forest, though, is widespread on Mount Tshiaberimu.

0.4.2 The Central Sector: Lake Edward and its Great Plains.

The Central Sector of Virunga National Park is made up of the western and southern banks of Lake Edward and the plains of Rwindi-Rutshuru through to Mabenga. This sector also includes the Lulimbi region, bordering the Ishasha River in the east, which forms the international boundary with Uganda. This zone is sometimes, erroneously, referred to as the Eastern Sector. The surface area of the Central Sector is 339,173 hectares, of which 144,548 ha are part of Lake Edward.

0.4.2.1 The western bank of Lake Edward

Running parallel to the western shores of Lake Edward, the boundary of the park corresponds roughly with the summit ridge of the Mitumba Mountains, thereby including within the park a band, three to five kilometres wide, from the ridge to the lake (Figure 0.32). Near

0.31
Elephants in the narrow band of protected land between the western shores of Lake Edward and the summit ridge of the Mitumba Mountains, linking the park's Northern and Central Sectors.

the summit, this zone is characterised by montane forest, while at lower altitudes acacia savanna is prevalent. Considerable human-induced habitat conversion over recent years has meant, however, that the summit ridge is now dominated largely by a high-altitude savanna.

Until recently, the riparian forest of Pilipili Bay, in the lower reaches of the Muyirimbo River, constituted the most intact of the forested areas bordering the lake. This forest can be defined as an evergreen forest, with very large trees of several species. Further south are vast assemblages of *Aeschynomene spp.*

Until very recently, wildlife could move freely between the Central and Northern Sectors along this belt on the western lakeshores. Yet, for some as yet unexplained reason, the topi antelope, *Damaliscus lunatus,* is completely absent from the Northern Sector.

0.4.2.2 Lake Edward

Like the Rwenzori range, Lake Edward is shared by the DRC and Uganda. However, an important distinction is that in the DRC the waters have protected area status, whereas in Uganda they fall outside Queen Elizabeth National Park.

The lake is 916 metres above sea level, and covers an area of 224,083 hectares, of which 166,872 ha (74 %) are in the DRC. The average depth is 33 metres. The lake is shallowest in the east, and deepest (at around 120 metres) at the foot of the Mitumba escarpment. Over most of its Ugandan waters, the lake is less than 30 metres deep (Figure 0.35).

Climate conditions are such that the lake's relatively shallow waters are subject to seasonal turbulence at the end of each main dry season (in August). The surface water cools and sinks to the bottom of the lake, displacing the nutrient-rich water below. Hippopotamuses daily enrich the lake with faecal matter con-

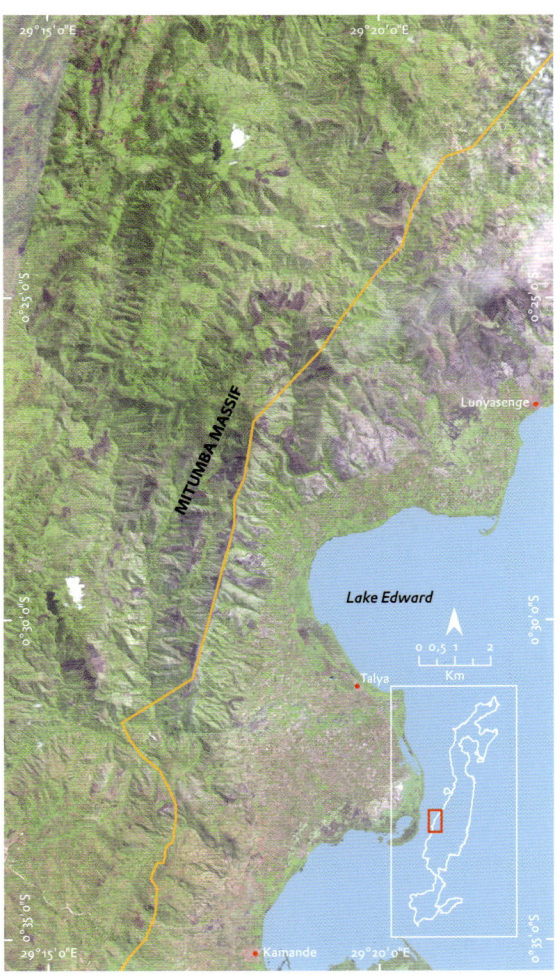

0.32
The Mitumba Mountains form a long north-south chain along the park's western boundary. Alongside Lake Edward, their summit ridge corresponds roughly to the boundary of the park, whereas the boundary further north is at the foot of the mountain chain.

● Fishing settlement
— ViNP boundary
▨ Agriculture
▨ Grassland savanna

A brief overview of Virunga National Park

0.33
Fishermen on Lake Edward.

0.34
Local people who depend on Lake Edward for the fish they eat also draw their water supplies from the lake.

taining nutrients from the tons of partially-digested lakeshore vegetation they consume during the night. These nutrients, coupled with the seasonal mixing of the waters, explain the lake's exceptional fertility and productivity in terms of fish yields.

With such quantities of fish available, it has been necessary to reconcile development and conservation priorities. This has led to the establishment of three fishing settlements on the Congolese shores of Lake Edward: Vitshumbi in the south, Kyavinyonge in the north, and Nyakakoma to the east of Vitshumbi. Nyakakoma was a later addition; the legal basis for this settlement remains unclear. The context within which these fishing settlements were created is explained in Chapter 1, their history is presented in Chapter 10, and the problems associated with them are discussed in Chapter 13.

0.4.2.3 THE PLAINS SOUTH OF LAKE EDWARD.

The area to the south of Lake Edward is characterised by grassland and wooded savannas, through which the Rwindi, Rutshuru and Ishasha Rivers flow (Figure 0.37). This assemblage of open habitats is a complex blending of several habitat types (which are described in detail in Chapter 7).

The grassland savannas are of different types. The lawns, or grazed grasslands, which typically include several species of *Sporobulus* or *Cynodon* are influenced by edaphic conditions in the presence of herbivores, which also favour the short grass savannas dominated by *Sporobulus pyramidalis*. *Themeda triandra*-dominated savanna represents another, very localised habitat type, while a final important type is the high-grass savanna dominated by *Imperata cylindrica*.

There are also several types of bush and wooded

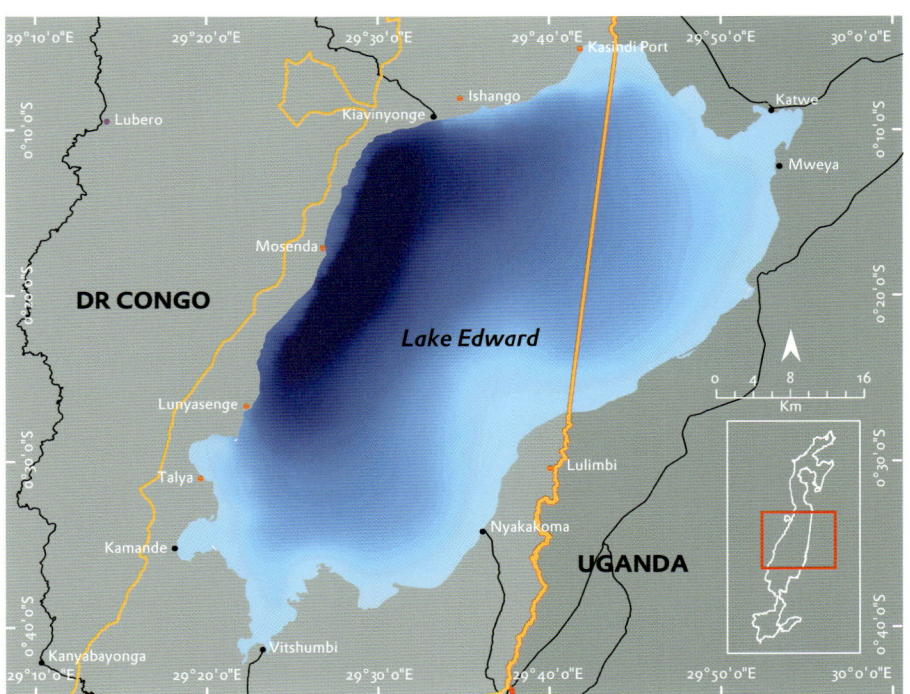

0.35
The Congolese part of Lake Edward is entirely contained within ViNP. The lake is not very deep, except along the base of the Mitumba range on its western shore, where it drops more than 100 metres.

A brief overview of Virunga National Park

0.36
The Rwindi Plains with the Kasali Mountain Range in the distance.

- ● Town
- ● Patrol Post
- ― River
- ― Road
- ― ViNP boundary
- Agriculture
- Grassland savanna
- Recently burned savanna

0.37
The plains south of Lake Edward, through which the Rwindi, Rutshuru and Ishasha Rivers flow, include savanna habitats of several types.

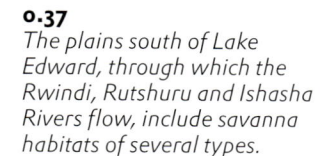

0.38
Euphorbia candelabrum is widely distributed across the savanna plains around Lake Edward.

savannas, dominated for the most part by a mixed assemblage of acacia species. On savannas in thin rocky soils, *Acacia gerrardii* is dominant, while in deeper soils there is a higher prevalence of *Acacia sieberiana,* with some trees reaching 15 metres in height.

Other habitat types can be found on the plains to the south of Lake Edward. These include, in areas where the dry seasons are most extreme, sclerophyllous forests dominated by *Euphorbia dawei,* which in places (near the Lake Kizi for example) are extensive and very dense. Another striking euphorbia, *E. candelabrum,* which can grow up to eight metres in height, is widely distributed across the plains.

Another type of arboreal vegetation, occurring at a very local level on the Rwindi-Rutshuru Plains, such as near the Lulimbi post, is that dominated by the aptly named *Rauvolfia vomitoria* (poison devil's-pepper).

An entirely different type of vegetation found in this area is gallery forest, which is particularly well represented along the Rutshuru River from Mabenga to the delta on Lake Edward. Dominating this narrow forest is the relatively small palm, *Phoenix reclinata* (the wild date palm) about five metres tall and found growing in association with representative of the Leguminosae, including *Sesbania sesban* and *Acacia kirkii.* Taller (10 to 15 metres) and more mature plant assemblages can be found in the upper-mid Rwindi and Rutshuru gallery forests. Here, the dominant species include two members of the Euphorbiaceae, *Croton macrostachys*

Marc Languy & Emmanuel de Merode

The soils of the Rwindi-Rutshuru-Ishasha Plains
M. Vanoverstraeten

Several studies have been carried out on the soils of the Rwindi Plains, making these soils the best understood in Virunga National Park.

Organic substances carrying biocenoses (interacting organisms), soils form an integral part of the landscape and its ecology. In Virunga National Park, soil type is greatly influenced by geological factors that have contributed to the region's geomorphology, as well as by drainage patterns and climate. The Rwindi Plains are part of a subsiding plate, the base of the Albertine Rift.

This geological depression formed as the result of a collapse of the underlying peneplain, together with an uplifting phase that created a mountain chain (average height 2,000 metres above sea level) that now forms the western wall of the Albertine Rift Valley. Around Rwindi, this mountain chain is referred to as the Mitumba Mountains. The altitude of the plains at the valley bottom averages 989 metres, but in the middle of these plains are the Kasali Mountains, which rise to 2220 metres.

The area's soil types are products of a combination of these tectonic processes and of ensuing climate conditions. At the base of the mountain chains are deposited accumulations of eroded rock displaced from higher on the mountains. These rocks represent the transition between mountain and plains habitats. The plains themselves are made up largely of alluvial deposits from both the lake and the rivers that have gradually levelled the valley bottom. The plains sink gradually eastward until they reach the eastern wall of the Albertine Rift, where their altitude steadily increases again in Uganda. On a much smaller scale, the soil units generated by the tectonic processes can be zoned as follows.

West to East:
- The Rwindi Plains, separated from the Rutshuru Plains by the Kasali Mountains.
- The Rutshuru Plains, whose eastern limit is defined by the Kizi Fault.
- The plains between the Rutshuru and Ishasha Rivers, bordered by the tectonic fracture that hangs over the valley of the Ishasha River.
- The Ishasha Plains, extending to the park boundary on the Ugandan border.

A detailed examination of the banks of the Rutshuru River reveal numerous strata, slanting eight degrees downward to the east and upward to the west, thus confirming the tectonic subsidence in the area.

0.39
A typical view of the Rwindi Plains.

Bedrock and soils:
biotopes of the Rwindi-Rutshuru-Ishasha Plains.

The bedrock underlying the soils of the Rwindi-Rutshuru-Ishasha Plains consists of fine sediments, wherein the ratio of argillous material to salts defines the soil's development. The soils here are built on fine argillous material. Their high proportion of exchangeable minerals is consistent with the area's sedimentary substrate.

Radiocarbon (Carbon-14) dating has shown that human communities occupied these plains long before the park was established. This will have had an impact on the vegetation – and consequently on the soil.

In marshlands, all formative processes are predominantly organic. Marshlands are by definition the result of frequent flooding, making them safe havens for wildlife, since it is difficult for humans to gain access to such areas.

0.40
The Kasali Mountains represent an island of sub-montane vegetation at the heart of the park's savanna-dominated Central Sector

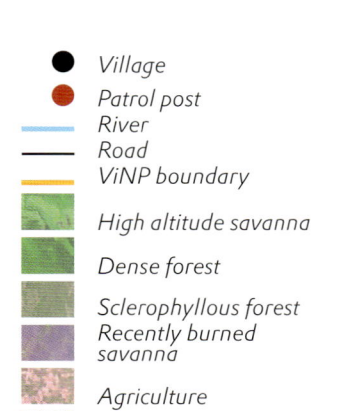

- ● Village
- ● Patrol post
- River
- Road
- ViNP boundary
- High altitude savanna
- Dense forest
- Sclerophyllous forest
- Recently burned savanna
- Agriculture
- Clouds and shadows

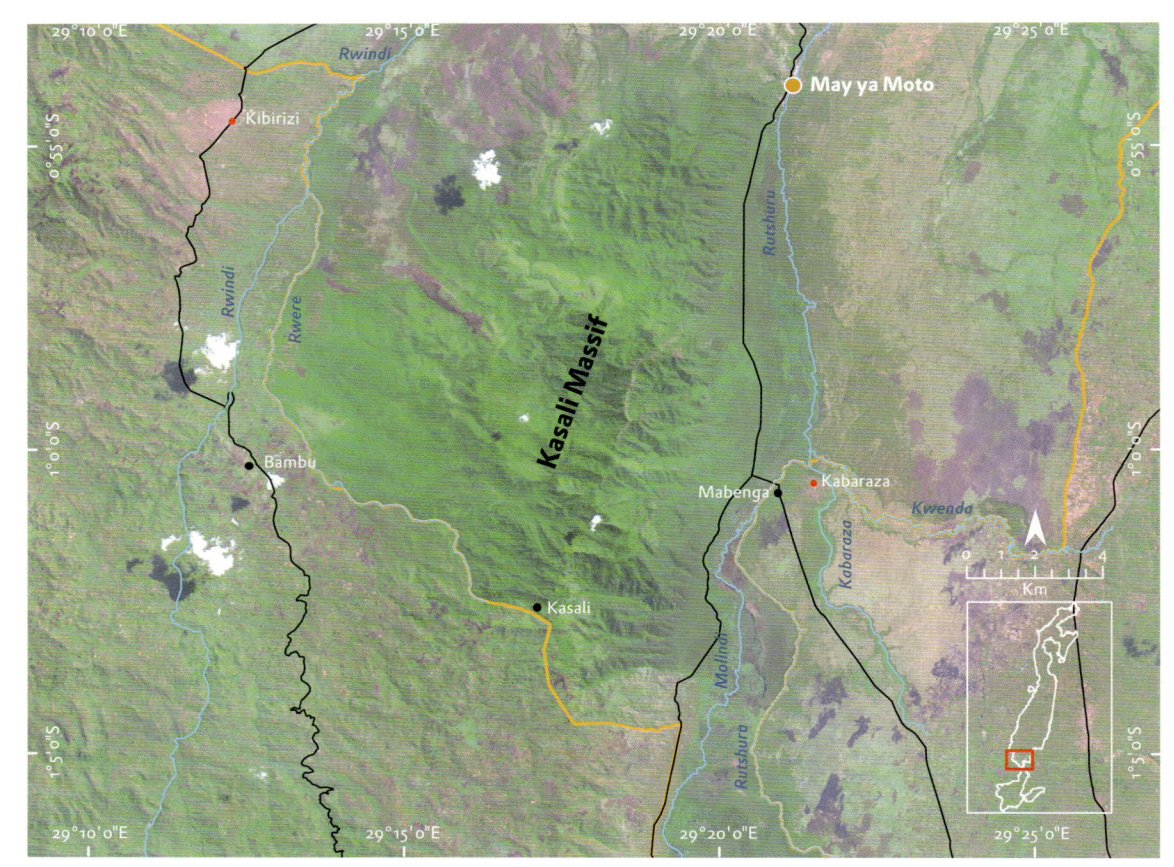

0.41
A gallery forest to the south of Lake Edward.

Marc Languy & Emmanuel de Merode

0.42
The Rwindi River, upstream from Rwindi Station, has carved a spectacular gorge in which gallery forest species are dominant.

and *Kigelia africana*. Locally, even taller species assemblages can be found along the banks of the Rwindi and the Ishasha Rivers and these are characterised by a Sterculiaceae, *Pterygota mildbraedii* (up to 25 metres tall).

Sclerophyllous vegetation dominated by the wild olive, *Olea europaea* (subspecies *cuspidatus*), occurs in the southernmost portion of the Central Sector, as well as along the middle reaches of the Rutshuru River, to the east of Mount Ilehe.

The famous hot water springs, known locally as *May-ya-moto*, support a highly specialised micro-habitat. The springs are a reminder of the region's underlying geological activity. The magma here is very close to the earth's surface. The hot water springs are highly alkaline and can reach temperatures as high as 95 °C. A microflora and microfauna have evolved in this marginal environment alongside algae and sulfobacteria.

The sub-montane forest covering much of the Kasali Massif, in the extreme south, constitutes one of the Central Sector's defining vegetation types (Figure 0.40). The massif is a branch of the Mitumba Range, disposed parallel to, and east of, the main escarpment, and extends for 15 km along its main north-south axis. On the slopes of the massif, an area of about 21,000 hectares, are extensive stands of sub-montane forest (as well as montane forest in certain limited areas), interspersed with large savanna grasslands along some of the ridges. At its greatest height, the Kasali Massif stands 2200 metres above sea level.

0.4.3 THE SOUTHERN SECTOR: THE ACTIVE AND EXTINCT VOLCANOES.

The Southern Sector of Virunga National Park covers an area of 145,672 hectares, stretching from the south of the Kasali Massif to the northern shores of Lake Kivu, thereby encompassing both active volcanoes, Nyamulagira and Nyiragongo, together with the extinct volcanoes of the Mikeno Sector.

The transition zone between the Central and Southern Sectors is of exceptional interest, owing to the extreme variations in habitat: Lake Ondo, the pools at Kibuga, the wetlands, the northern extremities of the lava flows, and the sources of the Molindi River. In an

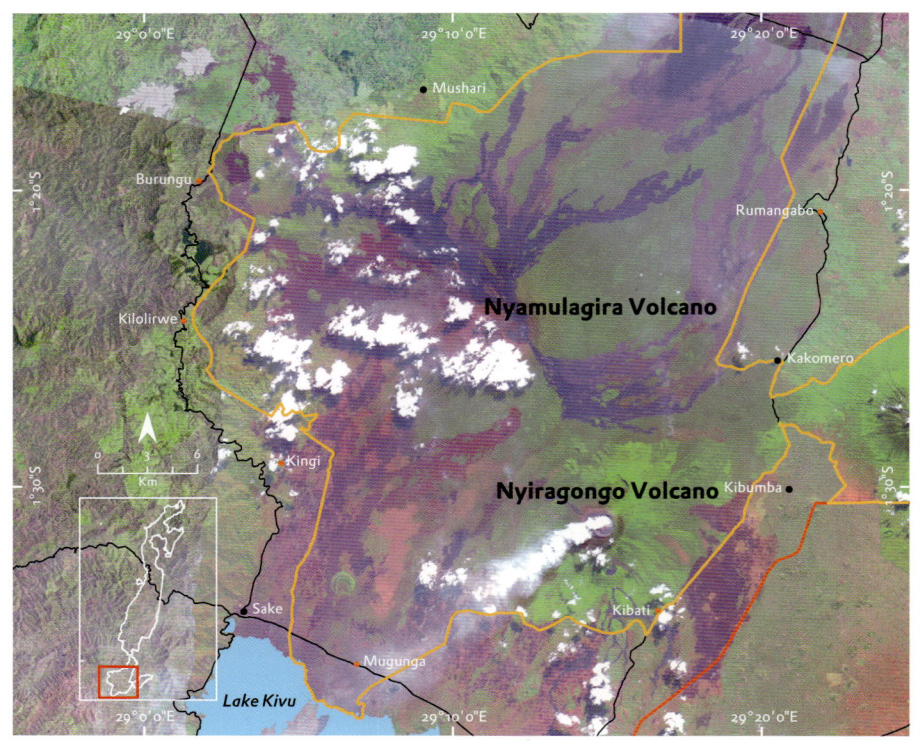

0.43
The Nyamulagira sub-sector of Virunga National Park is dominated by two of the world's most active volcanoes. On this SPOT satellite image taken in July 2004, the smoke plume of Mount Nyiragongo, blown over several kilometres by the prevailing north-easterly winds, is clearly visible.

- ● Village
- ● Park station
- ── Road
- ── International boundary
- ── ViNP boundary
- Lava fields
- Montane forest
- Agriculture
- Pasture
- Clouds and shadows

0.44
A lava flow on Mount Nyamulagira, two days after its passage.

area area where the topography is complex, to say the least, clusters of *Olea* are especially profuse around the many upsurges here of water seeping out from what are referred to as *pseudo-sources* of the Molindi River. Indeed, about 75 % of the Rutshuru River's water comes from the Molindi. Much of this water originates in the rainfall on the lava fields to the south, which is immediately absorbed. This area is the southern watershed of the Nile, where drainage takes place almost entirely underground.

0.4.3.1 The sclerophyllous forests of the northern slopes of Nyamulagira.

The sclerophyllous forests covering the relatively recent lava fields of Mount Nyamulagira are an important feature of the Southern Sector. These lava flows are widespread and extend as far north as the latitude of Rutshuru town. Given the porous ground and the very thin humus horizon on the lava fields, these forests have had to adapt to drier conditions. Repeated eruptions at different stages in recent geological history have produced a correspondingly rich array of forests at different stages of succession on the lava.

Within a year or two of an eruption, lichens and mosses form on the lava fields. Bracken follows, growing in the cracks between rocks, thereby creating some humus on which the first trees will begin to grow after 15–20 years. After 40 years, a sclerophyllous forest colonises the entire area with various olive and fig species, which in turn provide a niche for a number of mammals, such as chimpanzees. After 60 years, a paraclimax vegetation forms, dominated by *Bersama* and *Afrocrania*. These forests are typical of the Tongo Triangle to the north of Mount Nyamulagira, which for some unfathomable reason, was not included in the park when it was created.

0.4.3.2 The active volcanoes.

Mounts Nyamulagira (3056 metres) and Nyiragongo (3470 metres) are the most active volcanoes on the planet. As major features of the park, they are treated in considerable detail in Chapter 6.

The region forms what is known as the Nyamulagira sub-Sector of Virunga National Park (Figure 0.43). Since 1948, there have been eruptions almost every year from either, or both, of the two active volcanoes. On the flanks of these volcanoes, near the summits, the high altitudes support a sub-alpine flora. The east-north-easterly winds that predominate at this altitude mean that the south-western flanks of both volcanoes are affected by the acidity of the sulphur dioxide-rich rainfall. All vegetation within a 10 km radius of the summit is affected by the acid rain.

An important point is that most eruptions do not take place from the summit crater of these volcanoes, but through secondary craters on their lower flanks. As a result, there are mature forests at higher altitude, especially on Nyiragongo. Thus a whole spectrum of forest colonization can be found on the slopes of these two dramatic volcanoes.

Other unusual features are the small stands of bamboo forest that have formed in the extinct crater of Mount Shaheru, and the unique assemblage of sulphur-resistant flora that has become established within the vast crater of Mount Nyamulagira.

0.4.3.3 The extinct volcanoes

Six extinct volcanoes dominate the Virunga forest massif. Forming a crescent, these volcanoes are from west to east: Mounts Mikeno, Karisimbi, Visoke, Sabinyo, Gahinga and Muhavura. Only Mikeno is entirely contained within Virunga National Park. Karisimbi, Visoke and Sabinyo lie partly within Virunga National Park: the first two straddle the boundary between the DRC

0.45
Pioneer species, of hardy shrubs and ferns, colonise a fairly recent lava flow.

0.46
Vegetation near the summit of Mount Nyiragongo is dominated by groundsels.

0.47
The interior of Nyiragongo crater, seen at a time when the lava lake at the bottom had almost completely subsided.

0.48
One of the many small lava cones that lie scattered around the lower flanks of Mounts Nyamulagira and Nyiragongo.

0.49
Mount Nyamulagira and one of its many recent lava flows that has blazed through the forests on its lower flanks.

0.50
Mount Nyiragongo and the infamous lava flow of January 2002.

A brief overview of Virunga National Park

0.51
The view from Rwanda of (from left to right), Mounts Sabinyo, Gahinga and Muhavura.

0.52
Mount Mikeno is the only extinct volcano in the Virunga Massif that lies entirely within DRC.

Marc Languy & Emmanuel de Merode

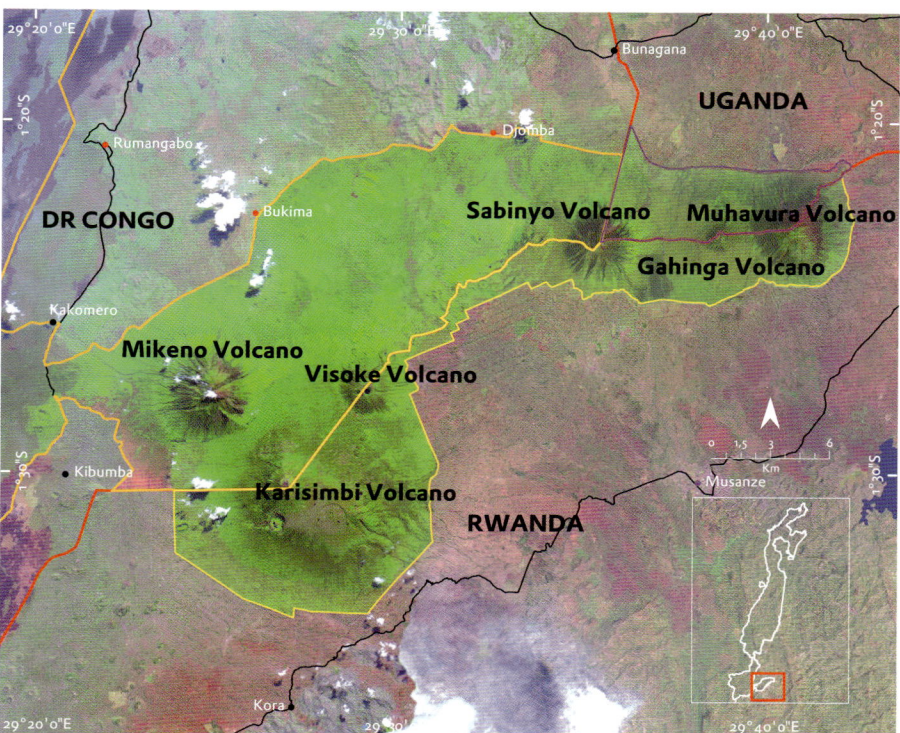

and Rwanda, while the summit of Sabinyo marks the intersection of the boundaries of all three countries. Gahinga and Muhavura are entirely outside Virunga National Park.

The park boundary on the Congolese side is much lower (about 2000 metres above sea level on average) than that of the *Parc des Volcans* on the Rwandan side, thereby giving it a far greater diversity of species and habitats.

The succession of montane vegetation types has been described extensively by Delvingt *et al.* (1990) in the visitors' guide to Virunga National Park, and as such we provide only a brief summary of this description.

On the African continent, montane forest types typically form at differing altitudes according to latitude and local climate conditions (the intensity and frequency of rainfall especially). In Virunga National Park, montane forest is found above 1800 metres.

Dense humid montane forest
This forest type is generally found between 1800 and 2800 metres and falls into three main categories:

Dense humid forest dominated by *Ficalhoa* and *Podocarpus*
is very diverse and species rich. Trees reach up to 25 metres in height and usually have smaller leaves than those growing at lower altitudes. *Podocarpus latifolius,* the park's only coniferous species is here found growing alongside a range of tree ferns. This forest is also particularly rich in epiphytes, harbouring a wide variety of orchids, mosses and lichens.

Secondary forest dominated by *Neoboutonia macrocalyx*
suggests that some forested areas now in the park were cleared before the latter's classification as a protected area. *Neoboutonia macrocalyx* is easily recognised from its large, rounded leaves, which are often riddled with small holes.

Bamboo forest is generally found on rich, humid soils between 2300 metres and 2600 metres. Bamboo can, as we have seen, reach altitudes above 3000 metres on the Rwenzori and Tshiaberimu. This is a mono-specific forest entirely dominated by *Arundinaria alpina.* Bamboos are giant grasses with extremely fast growth rates. Young bamboo shoots form a very important part of the Mountain Gorilla's diet. Except in clearings, this is not a very bio-diverse habitat type, and woody and grass species are rare. Exceptions are *Viola abyssinica* and *Clematis wightiana.*

The afro-subalpine vegetation
The afro-subalpine vegetation zone begins above 2600 metres, where precipitation begins to decline and the temperatures drop significantly. At around 3000

0.53
The Mikeno sub-sector of Virunga National Park incorporates the Congolese slopes of extinct volcanoes in the Virunga Massif that is shared with Rwanda (Volcanoes National Park) and Uganda (Mgahinga Gorilla National Park).

- Town
- Village
- Park station
- Road
- International border
- ViNP boundary
- Mgahinga Gorilla National Park
- Volcanoes National Park
- Extinct volcano
- Lava field
- Deforestation
- Montane forest
- Agriculture
- Clouds and shadows

0.54
Mount Visoke, seen from the slopes of Mount Karisimbi.

A brief overview of Virunga National Park 45

0.55
A typical Neoboutonia-*dominated secondary forest.*

0.56
Neoboutonia macrocalyx *is easily identified from its very large leaves. On most trees, the leaves are riddled with the perforation marks of insects.*

metres, the principal vegetation type is *Hagenia abyssinica* which forms an open canopy forest. Under these trees, which are typically about 10 metres tall, lies an understorey vegetation layer that is extremely diverse in grass species. *Peucedanum kerstenii,* a wild celery to which Mountain Gorillas are especially partial, also grows here. The most beautiful *Hagenia* forests in the park are to be found in the saddle between Mounts Karisimbi and Visoke. These forests are associated with a relatively open understorey dominated by the shrub *Hypericum revolutum.*

At higher altitude, a scrub forest takes over, made up primarily of *Philippia johnstonii,* which can reach up to 10 metres. On the drier slopes, *Erica arborea* dominates. A thick layer of moss typically covers the ground.

The afro-alpine zone begins at the upper limit of the *Hagenia*-dominated vegetation zone, and here large clearings are common at around 3700 metres. The plant species most prevalent at this altitude are the giant groundsels and lobelias, which can grow up to eight metres tall. On the edges of the depressions and crevasses is a bush habitat dominated by clumps of *Alchemilla,* while in other areas, such as in the extinct crater of Branca, there are enormous *Carex*-dominated swamps.

The cols between volcanoes ensure a continuity between vegetation types. There are small lakes cradled in the summits of Visoke and Muhavura, which intriguingly—unlike lakes found at similar altitudes on the Rwenzori—do not freeze.

The slopes of Mount Mikeno are extremely steep, and can be climbed only by trained mountaineers.

Marc Languy & Emmanuel de Merode

0.57
Young bamboo shoots unearthed and eaten by Mountain Gorillas.

0.58
Bamboo forest.

0.59
Stands of mono-specific bamboo forest can be difficult to penetrate.

A brief overview of Virunga National Park

0.60
Giant groundsels are common in the afro-alpine vegetation zone.

0.61
Above 3000 metres, the vegetation is dominated by tree heathers. These are often covered in lichens, such as Usnea *spp., which resemble large pale green beards.*

0.62
Lobelias are also common in the afro-alpine zone.

Mount Karisimbi, on the other hand, has very gentle slopes. On the summit of this extinct volcano, *Alchemilla* and grasses are comparatively rare, being found only in areas sheltered by rocks.

It is worth pointing out, in concluding this broad overview of the extinct volcanoes of the Virunga Massif, that there was an eruption as recently as 1957 (Mugogo). This eruption was followed by a massive displacement of toxic gases, on a scale far greater than that of the *mazuku* (Kiswahili word for 'evil winds'), which are dry gas vents emitting carbon dioxide in the active volcano sector. In reality then, the volcanoes in the Mikeno Sector are dormant rather than extinct.

0.5 THE BIOLOGICAL VALUE OF VIRUNGA NATIONAL PARK

It is often said of protected areas that they contain exceptional levels of biodiversity or that they possess many unique features. So overused are these superlatives, that today they are just meaningless clichés. In Virunga's case, however, the unique diversity evident in the species and landscapes simply cannot be overstated. And the same is true of the extraordinary biological and geological processes taking place in the park.

In terms of biological diversity, Virunga is by far the richest protected area on the African continent. The park is home to more than 700 species of birds

Marc Languy & Emmanuel de Merode

A brief overview of Virunga National Park

0.63 The total number of species, endemic species and threatened species known to occur in Virunga National Park, based on the main taxa

Taxon	Number of species	Number of species endemic to the Albertine Rift	Number of threatened species
Mammals	218	21	13
Birds	706	25	11
Reptiles	109	11	0
Amphibians	78	21	10
Fish	(100)	(71)	–
Butterflies	?	21	–
High order plants	2,077	230	10

0.64 Number of known mammal species in the park, based on order

Order	Confirmed species	Probable additional species
Tubulidentata	1	0
Hyracoidea	1	1
Proboscidea	2	0
Primates	22	0
Afrosoricida	2	0
Macroscelidea	1	0
Lagomorpha	1	0
Erinaceomorpha	1	0
Soriciomorpha	20	9
Chiroptera	45	8
Pholidota	3	0
Carnivora	24	1
Artiodactyla	22	0
Rodentia	73	7
Total	**218**	**26**

(twice as many as are found in the whole of Western Europe), and 220 species of mammals (also the highest for any African protected area). As such, no protected area in the world can lay claim to greater specific richness.

In addition to its unparalleled specific richness, the park includes a very large number of species that are endemic to the Albertine Rift. These species exist nowhere else in the world. The park also harbours an uncommonly large number of species listed as vulnerable or endangered on the IUCN Red List.

0.5.1 Mammals

Mammals of at least 218 species are known to occur in Virunga National Park. From studies of the park's habitats, however, and of the likely ranges of known species hitherto not recorded in the park, it is considered highly probable that a dozen or so insectivores, and an equal number of both bats and rodents, have yet to be observed (Table 0.64).

Dramatic habitat variations caused by the extreme altitudinal and climate gradients are largely responsible for the park's exceptional species richness. Thirteen of Virunga's mammal species have been classed as critically endangered. This puts Virunga at the top of the list of global priorities based on endangered mammal species.

While Virunga does have a number of iconic species, such as the lion *Panthera leo*, buffalo *Syncerus caffer*, and elephant *Loxodonta africana*, Virunga National Park is relatively poorly endowed with large savanna species. Yet, their abundance prior to the recent poaching onslaught was unequalled. As shown in Chapter 8, the total biomass of the main ungulate species in the Rwindi-Rutshuru-Ishasha plains (1250 km²) was estimated at 34,523 tons, or 27,619 kg per km², the highest in the world; much higher, for example, than that of the Serengeti. The main reason for such a high biomass was the presence of the biggest population of the hippopotamus, *Hippopotamus amphibius*, in the world. However, this population has in the recent years been decimated almost to extinction (see Chapter 8).

In addition to the extreme abundance and diversity of mammals, there is also an exceptional level of endemism, which carries much weight from a conservation and scientific standpoint. The list includes the Mountain Gorilla, *Gorilla beringei beringei*, Stuhlmann's Golden Mole, *Chrysochloris stuhlmani*, and the Western Rift Brush-furred Rat, *Lophuromys medicaudatus*, all of which are endemic to a very small area within the Albertine Rift. Other species such as Hopkins' Creek Rat *Pelomys hopkinsi*, are endemic to this area's papyrus swamps, while another local rodent, *Malacomys verschureni*, is known from only five specimens in the world.

Primates are without question the primary attraction that Virunga, with its 22 species, offers. They also add considerably to the conservation importance of the site. These species include a number that are typical of the Congo Basin, such as the Mona Monkey *Cercopethicus (mona) denti*, and some savanna species that are fairly widespread. Significantly though, they include a number of species that are endemic, or partially endemic, to the Albertine Rift, such as the Golden Monkey, *Cercopithecus kandti*. Virunga National Park is also the only protected area in the world with three taxa of great apes, namely the Mountain Gorilla *Gorilla beringei beringei*, the Eastern Lowland Gorilla *Gorilla beringei graueri*, and the Eastern Chimpanzee *Pan troglodytes schweinfurthi*. There are only an estimated 720 Mountain Gorillas left in the world, of which around 200 are in Virunga National Park. This species has been the subject of a plethora of scientific studies, and carries a particular weight among the international media. Mountain Gorillas are also the source of more than 90 % of the national park's tourism revenues.

Ungulates are clearly a key group of mammals in Virunga National Park. Not only do they make a vital contribution to a wide range of important ecological processes; they are also of enormous value in attracting tourism, while constituting a reserve of stock for neighbouring areas where hunting is permitted.

The ungulate that defines Virunga more than any other is the hippopotamus. Until recently, the park was home to the world's largest hippopotamus population (numbering roughly 29,000 animals), and to

0.66
The gentle monkey is relatively common in the lowland forests of Virunga NP.

0.65 The 22 primate species of Virunga National Park and their habitats

Family	Scientific name	English name	Habitat in ViNP
Loridae	*Perodictus potto*	Potto	Primary and secondary forest
Galagonidae	*Galago demidoff*	Prince Demidoff's Bushbaby	Dry forest, lowland and montane forest
Galagonidae	*Galago senegalensis*	Senegal Bushbaby	Secondary and montane forest
Galagonidae	*Galago thomasi*	Thomas's Bushbaby	Primary lowland forest
Galagonidae	*Otolemur crassicaudatus*	Brown Greater Galago	Montane forest
Cercopithecidae	*Cercopithecus ascanius*	Red-tailed Monkey	Gallery forest, lowland forest
Cercopithecidae	*Cercopithecus denti*	Dent's Mona Monkey	Lowland primary forest (Semliki)
Cercopithecidae	*Cercopithecus doggetti*	Silver Monkey	Lowland forest
Cercopithecidae	*Cercopithecus hamlyni*	Owl-faced Monkey	Montane and sub-montane forest
Cercopithecidae	*Cercopithecus kandti*	Golden Monkey	Montane forest
Cercopithecidae	*Cercopithecus lhoesti*	L'Hoest's Monkey	Montane forest
Cercopithecidae	*Cercopithecus mitis*	Blue Monkey	Gallery forest, lowland forest
Cercopithecidae	*Cercopithecus neglectus*	De Brazza's Monkey	Lowland forest
Cercopithecidae	*Chlorocebus tantalus*	Tantalus Monkey	Savanna and dry forests
Cercopithecidae	*Lophocebus albigena*	Grey-cheeked Mangabey	Humid primary and secondary forest
Cercopithecidae	*Papio anubis*	Olive Baboon	Savanna and dry forests
Cercopithecidae	*Colobus angolensis*	Angola Colobus	Montane forest
Cercopithecidae	*Colobus guereza*	Mantled Guereza	Gallery forests, lowland forest
Cercopithecidae	*Piliocolobus foai*	Central African Red Colobus	Lowland primary forest (Semliki)
Cercopithecidae	*Cercocebus agilis*	Agile Mangabey	Lowland primary forest (Semliki)
Hominidae	*Gorilla beringei beringei*	Mountain Gorilla	Montane forest (extinct volcanoes)
	Gorilla beringei graueri	Eastern Lowland Gorilla	Montane forest (Mount Tshiaberimu)
Hominidae	*Pan troglodytes*	Common Chimpanzee	Primary and secondary forest

0.67
The hippopotamus is a keystone species for the national park. In contributing to the high productivity of Lake Edward, it has a significant impact on both the savanna ecology and the local economy.

0.68
The Waterbuck is common on the savannas south of Lake Edward.

the world's highest density of the species. The sight of a large hippo pod plastered in the mud of the Bay of Mwiga remains an unforgettable image in the eyes of many. Unforgettable too are the sounds they make, contributing more than any other noise to the animation of nights on the Rwindi Plains.

Until recently, the park's second most abundant ungulate was the buffalo. With a population in excess of 20,000 individuals, buffaloes could be found in all sectors of the park, in both savannas and forests, and at altitudes both high and low. Uganda Kob, *Kobus thomasi,* is the most abundant of large mammal species on the savannas, and can be found in both the Central and Northern Sectors. Uganda Kob is one of the few species whose populations have remained fairly constant, despite the poaching onslaught of the recent past. Other savanna species around Lake Edward include the Waterbuck, *Kobus (ellipsiprymnus) defassa,* and the Bushbuck *Tragelaphus scriptus.*

Virunga is also renowned for its pigs. It is one of the few protected areas in which all three genera of African pigs are represented. The Common Warthog, *Phacochoerus africanus,* is relatively abundant, but only in the savanna areas around Lake Edward. They are easily observed with young in July and August. The Bushpig, *Potamochoerus larvatus,* is less easily observed, being a species that seldom emerges from the cover of the wooded savannas and forests. The most impressive of the pigs are the Giant Forest Hogs, *Hylochoerus*

Marc Languy & Emmanuel de Merode

meinertzhageni, which are relatively easy to find in the late afternoons or early mornings, when they emerge from the gallery forests. Along with Uganda's Queen Elizabeth National Park, Virunga is probably the best location in which to observe this species.

Another remarkable ungulate is the Okapi, *Okapia johnstoni,* which is closely related to the giraffe, but which lives exclusively in the forest and is endemic to the north-eastern DRC. The Okapi is also the emblem of the Congolese Protected Areas Authorities, the *Institut Congolais pour la Conservation de la Nature* (ICCN). A small, recently rediscovered population has survived on the west bank of the Semliki River, in the extreme northern reaches of Virunga National Park.

A small bovid, strictly endemic to the Rwenzori, is the Rwenzori duiker, *Cephalophus rubidus,* which is restricted to the *Hagenia* and bamboo forests and in the afro-alpine vegetation zones on the upper slopes of the mountains. There are several other duiker species in the park, of which some subspecies are endemic to the region.

Two other important groups, which cannot be ignored, are the **rodents** and the **shrews.** Indeed, these two groups account for the largest number of endemic species in the Albertine Rift. Some of these have been observed only in the Virunga or on the Rwenzori. It is probable that other species, new to science, will be found in the sub-montane and montane forests.

0.69
The Uganda Kob, with a population numbering some 10,000 animals, is the park's most common antelope species.

0.70
Servals are frequently encountered in the woodland savannas surrounding Lake Edward, tending to avoid denser forest altogether.

A brief overview of Virunga National Park

Marc Languy & Emmanuel de Merode

0.5.2 Birds

Virunga National Park boasts an exceptionally rich avifauna. Birds of no fewer than 706 species have been recorded so far, making this the richest protected area in Africa for bird species. As with the mammals, this species richness is a reflection of landscape and habitat diversity, although with birds the park's geographical location—at the intersection of several bio-geographical zones—is another major factor. Virunga National Park straddles the convergence of the central African avifaunal region (represented by the lower Semliki forests) and the eastern African avifaunal region, well represented on the Rwindi and Rutshuru Plains. There is, in addition, a cohort of afro-montane bird species, of which many are endemic to the Albertine Rift, while the rest are shared either with the Cameroonian mountains, or with the high mountains of East Africa (such as Mount Kenya and Mount Kilimanjaro). Then, finally, there are large numbers of Palaearctic migrants, including birds of prey, passerines and waders, which migrate from Europe and Asia to winter in Virunga National Park.

Twenty-five of the bird species found in the park are strictly limited to the Albertine Rift. Some of these species, such as Yellow-crested Helmet-Shrike, *Prionops alberti,* have seldom been observed outside the park. Eleven of the bird species that occur in Virunga National Park are threatened with extinction.

Numbering around 20 species, **herons** and **storks** are particularly well represented in the park, and are also easily spotted, especially in the Central Sector. From the Little Bittern, *Ixobrychus minutus,* to the Goliath Heron, *Ardea goliath,* most are confined to the water's edge. Among the rare species that occur here, the Shoebill or Whale-headed Stork, *Balaeniceps rex,* must be the most unusual and spectacular, although this species does not breed in the park.

Also well represented in the park are the **birds of prey,** which here include large numbers of hawks, eagles, buzzards, falcons and vultures. The African Fish-eagle, *Haliaeetus vocifer,* is a spectacular resident of the shorelines of Lake Edward and its surrounding rivers. Large in size, and with a stunning plumage of brown, black and white, this species is perhaps best known for its unmistakable call, which so characterises the lakeside savannas of east and central Africa. Five species of vultures are often seen together in Virunga, fighting over the remains of a carcass.

The **passerines** include a very large number of families. Species that are typically considered forest birds, such as bulbuls and robin-chats, are often concealed in the undergrowth, but are easily detected from their characteristic songs. Swifts and swallows, and other Palaearctic migrants, patrol the skies over all parts of the park.

One of the fascinating ornithological features of Virunga is that several of the bird families it supports are represented by an uncommonly large number of species. Good examples are the Bee-eaters, Meropidae, and the Kingfishers, Alcidinidae. Different species within either group occupy the same areas, exploiting the contrasting ecological niches on offer. It is not unusual, over the course of a single day, to see eight different bee-eaters—a feat that would be difficult to match in other protected areas.

Lake Edward is an extremely important site for

0.71
The Saddle-billed Stork is a very large denizen of marshes and of the shores of the park's lakes and rivers.

A brief overview of Virunga National Park

0.72
For the Lilac-breasted Roller, the tops of trees offer the perfect vantage point from which to prey on large insects.

0.73
The Little Egret is a fairly common species in the swamps and on the banks of rivers within the national park.

0.74
The Greater Painted-snipe is a seasonal visitor to the temporary marshes on the plains south of Lake Edward.

0.75
The Yellow-throated Longclaw is common throughout the savanna regions of the park.

Palaearctic migrants, such as the White-winged Tern, *Chlidonias leucoptera,* which gathers here in flocks sometimes numbering thousands of birds, along with waders of several species. Virunga occasionally serves as a refuge for flamingos, which at various times have tried (unsuccessfully) to nest near Lulimbi, and for other unusual species, including the African Skimmer, *Rynchops flavirostris.*

0.5.3 Reptiles

Virunga National Park shelters a large number of reptiles and amphibians, of which more than 30 species are endemic to the Albertine Rift.

There are relatively few **turtles,** although one species—the variable mud turtle, *Pelusios rhodesianus*—is commonly found on the Rwindi-Rutshuru-Ishasha Plains. William's mud turtle, *Pelusios williamsi,* an aquatic species endemic to eastern Africa, is common on Lake Edward and in adjoining rivers.

Lizards found in Virunga National Park include

0.77 The chameleon species of Virunga National Park

Species	Range
Chamaeleo adolfifriderici	Albertine Rift
Chamaeleo dilepsis	Widespread
Chamaeleo ellioti	Mountains of East Africa
Chamaeleo gracilis	Widespread
Chamaeleo ituriensis	Albertine Rift and Ituri
Chamaeleo johnstoni	Albertine Rift
Chamaeleo laevigatus	Widespread
Chamaeleo rudis	Mountains of East Africa
Chamaeleo xenorhinus	Rwenzori Mountains
Chamaeleo carpenteri	Rwenzori Mountains
Chamaeleo bitaeniatus	Widespread
Rhampholeon boulengeri	Mountains of East Africa

0.76
The Rwenzori Three-horned chameleon, Chameleo johnstoni, *is endemic to the Albertine Rift and can be found in montane forest throughout the park.*

0.78
The Nile Monitor is common along water courses and around swamps in the park, where it feeds on a wide range of prey species.

0.79
The Blue-headed Agama, Acanthocerus atricollis, *lives both in trees and on rocks. It is often seen on buildings, and is even found on lava flows at altitude of up to 2000 metres.*

geckos, agamas, chameleons, monitor lizards, skinks and 'real' lizards.

The most familiar of all the **geckos** is the ubiquitous Tropical House Gecko, *Hemidactylus mabouia,* found in and around buildings throughout the region. The clicking calls of these geckos can usually be heard in the evenings. The colour of this species, like that of many other geckos, varies widely from place to place, in keeping with the surrounding environment. In Virunga, there is another species endemic to the Albertine Rift. This is Sternfeld's Gecko, *Cnemaspis quattuorseriatus,* and is found in the montane forests of the Mikeno Sector.

Agamas are well represented in Virunga National Park. A particularly striking species is the Blue-headed Tree Agama, *Acanthocerus atricollis,* whose males, up to 30 cm long, are easily recognised from their large blue heads. The species is well distributed over a wide range of habitats, from tree stumps on the Rwindi Plains to the lava fields at 2400 metres above sea level.

The species diversity among Virunga's **chameleons** is exceptional. Of 12 recorded species of chameleons, two – *Chamaeleo carpenteri* and *C. xenorhinus* (both recently assigned to a new genus, *Kinyongia*) – are found only in the Rwenzori Mountains.

Monitor lizards are large in size. The Nile monitor, *Varanus niloticus,* common in Virunga, is most often encountered along riverbanks and along the shores of Lake Edward. This lizard has been known to exceed two metres in length, and is an excellent swimmer and climber, feeding mainly on invertebrates, but also on frogs, chicks and eggs. A female lays clutches of 20 to 50 eggs, usually in active termite mounds. A close relative from the forest, *Varanus ornatus,* is found in the Semliki Valley.

Virunga National Park is home, as well, to at least ten species of **skinks.** These are lizards whose limbs are typically withered. Of these species, *Leptosiaphos meleagris* is known only from the Rwenzori range and the Ituri Forest, while its relative, the Virunga Skink, *L. hackarsi,* is found only in the Bwindi forests and the Virunga Massif. This species therefore has the same distribution as that of the Mountain Gorillas.

Among the **snakes,** Virunga has some spectacular species, such as the python, *Python sebae,* which can be found in most of the park's habitats, but especially in the gallery forests and in marshy areas along the shores of Lake Edward. This species can reach lengths of seven metres and weigh as much as 40 kilos. Other impressive, but rarer, species include the Puff Adder, *Bitis arietans,* the Black-necked Spitting Cobra, *Naja nigricollis,* and Jameson's Mamba, *Dendroaspis jamesoni,* all venomous species that are widely feared, but which virtually never attack humans. The Montane Egg-eating Snake, *Dasypeltis atra,* endemic to the mountains of East Africa, is known to feed exclusively on bird's eggs.

The **Nile Crocodile,** *Crocodilus niloticus,* long observed in the Northern Sector of the park, had seemingly never extended its range beyond the cataracts of the Semliki. From 1986, however, the species was

A brief overview of Virunga National Park

0.80
The Nile Crocodile population underwent a sudden, massive expansion of its range towards the end of the 1980s, and can now be found throughout Lake Edward and its adjoining rivers.

observed for the first time at Ishango, on the northern shores of Lake Edward. Its range has since spread progressively southward and a crocodile was in April 2005 observed in the Rutshuru River, only a few kilometres from May-ya-moto.

0.5.4 Amphibians

At least 119 amphibian species are known to occur in the Albertine Rift. This represents about one-fifth of all amphibian species found on the African continent. A study of 27 sites in the Albertine Rift (Plumptre *et al.*, 2003), found Virunga National Park to be the richest location for amphibians, with 78 species represented in Virunga, of which more than one-quarter are endemic to the Rift. These species belong to 17 genera in 10 families and sub-families. Unlike the Eastern Arc mountains in Tanzania, the park does appear to harbour any caecilians, which are elongated, limbless amphibians that superficially resemble earthworms. The order of salamanders and their relatives (order Caudata) is also absent from the park, since that order is not found in Africa south of the Sahara. In Virunga, the class of amphibians is represented only by the order Anura (tailless amphibians), comprising frogs and toads belonging to two sub-orders: archaic frogs without tongues, and modern frogs, Neobatrachia, to which the toads belong.

Morphological variations are considerable across the species. The archaic frogs have neither tongues nor claws, and typically possess very flattened bodies, such as the Pipidae of the genera *Xenopus* and *Hymenochirus*. The 'classical' toads include the genus *Bufo*, and the massive *Bufo superciliaris*, which has a smooth skin (it is also listed in Annexe 1 of the Washington Convention). The park hosts a number of highly specialized toads, including the Dwarf Tree Toad, *Nectophryne afra*. Foam-Nest frogs, the Rhacophoridae, form a large, predominantly Asian family that in Virunga is represented by the Western Foam-Nest frog *Chiromantis rufescens*, a species with peculiar reproductive habits, making a nest out of a salival foam containing air bubbles.

More than 50 amphibian species have been found in the Northern Sector alone. These include the reed frog *Hyperolius xenorhinus*, which is endemic to tiny Mount Teye on the flanks of the Rwenzori, and *Rana ruwenzorica*, a river frog that is endemic to the Rwenzori massif and a few surrounding forest patches. With 47 species, of which 16 are endemic to the Albertine Rift, the Virunga Massif—hence the Mikeno Sector of Virunga National Park—is critically important for amphibians. Some species, such as *Phrynobatracus bequaerti*, are entirely restricted to the Virunga Massif.

0.81
Hyperolius castaneus *is one of many frog species found in the park.*

0.82
Virunga's insects and other invertebrates have hardly been studied at all since the 1960s. Their distribution and abundance, even their presence in many cases, has yet to be established. Some species have economic, as well as ecological value. The Goliath Beetle (shown here), for example, is exploited for international trade to the extent of being threatened over large parts of its range.

0.5.5 Fish

Fish are the least studied vertebrates of Virunga National Park. Estimates run to roughly 100 species, including several that are endemic to Lakes Edward and Kivu. The diversity of the ichthyofauna of these two lakes is low by comparison with that of Lakes Tanganyika, Albert or Victoria. With respect to Lake Edward, the low number of species (around 50, although their populations are extremely abundant), is probably a consequence of the lake's shallowness and its relatively limited habitat diversity. It is also possible that a long arid period between two rainy periods during the Pleistocene may have dried the lake, killing a large number of species. In this event, re-colonization of the lake would have been limited, owing to its isolation from Lake Kivu (by the volcanoes) and from the Nile (by the cataracts on the Semliki). With Lake Kivu, recent geological history accounts for the relative paucity in fish diversity (now limited to about 30 species).

One of the most important fish species in Lake Edward, in view the role it plays in the local economy, is the tilapia, *Oreochromis niloticus*. This species is heavily exploited on the lake. Other economically important fish include *Bagrus docmak*, *Barbus altianalis edouardianus*, *Protopterus aethiopicus* and *Clarias gariepinus*.

0.5.6 Plants

Virunga National Park boasts more than 2000 higher order plants. As with the animals, this high species richness is explained by the diversity of biotopes within the park and its geographical position. The park sits on the intersection of three great phyto-geographical regions: the Guinea-Congolese region, the afro-montane region and the central African great lakes region. The first of these regions is represented by lowland closed canopy forest in the lower Semliki. The second can be found in each of the three sectors: Rwenzori and Tshiaberimu in the Northern Sector, the Kasali chain of mountains in the Central Sector, and the active and extinct volcanoes in the Southern Sector. The third region is best represented by the savanna plains north and south of Lake Edward.

More than 200 of the park's plant species (about 10 %) are endemic to the Albertine Rift. Most of these are found in the montane forest. This is an exceptionally high level of endemism, and on the Rwenzori Mountains, which alone support 75 endemic species, endemism is as high as 13 %.

Species that typify each of the park's regions are summarised in the first part of this chapter. A simplified vegetation map has been produced recently by the World Wide Fund for Nature (WWF). Based on high resolution SPOT satellite imagery and the work of a wide range of contributors, this map (Figure 0.87) shows how the main vegetation types are distributed (according to the proportions shown in Table 0.84).

0.6 Management of the National Park

From an administrative perspective, the national park is located almost entirely in North Kivu province. Its extreme northern reaches lie in Ituri province, while Tshegera Island in the far south lies in South Kivu province.

As part of the network of protected areas in the

0.83
Lake Edward's fish populations represent an important source of protein and revenue for local communities.

0.85
Ferns are well represented in the montane forests of Virunga National Park.

0.86
The park's astonishing array of flowering plants cannot help but leave a lasting impression on all visitors.

0.87
Simplified vegetation map of Virunga National Park. There are spectacularly varied vegetation types, ranging from the 'tundras' of the afro-montane belt, through zones of dense humid forest, to the sclerophyllous forests and the dry savannas.

0.84 The surface areas of the major habitats of ViNP

Vegetation type	Area (%)
Afro-alpine zone	1.4
Afro-subalpine zone dominated by heathers	2.8
Montane forest dominated by bamboos	2.4
Montane forest dominated by *Podocarpus* and *Neoboutonia*	11.2
Drought-resistant sclerophyllous brush	12.3
Hagenia forest	0.4
Lowland humid forests	11.8
Gallery forests	1.4
Grassland and bush savannas	35.7
Recent lava flows	2.3
Lakes and rivers	18.3

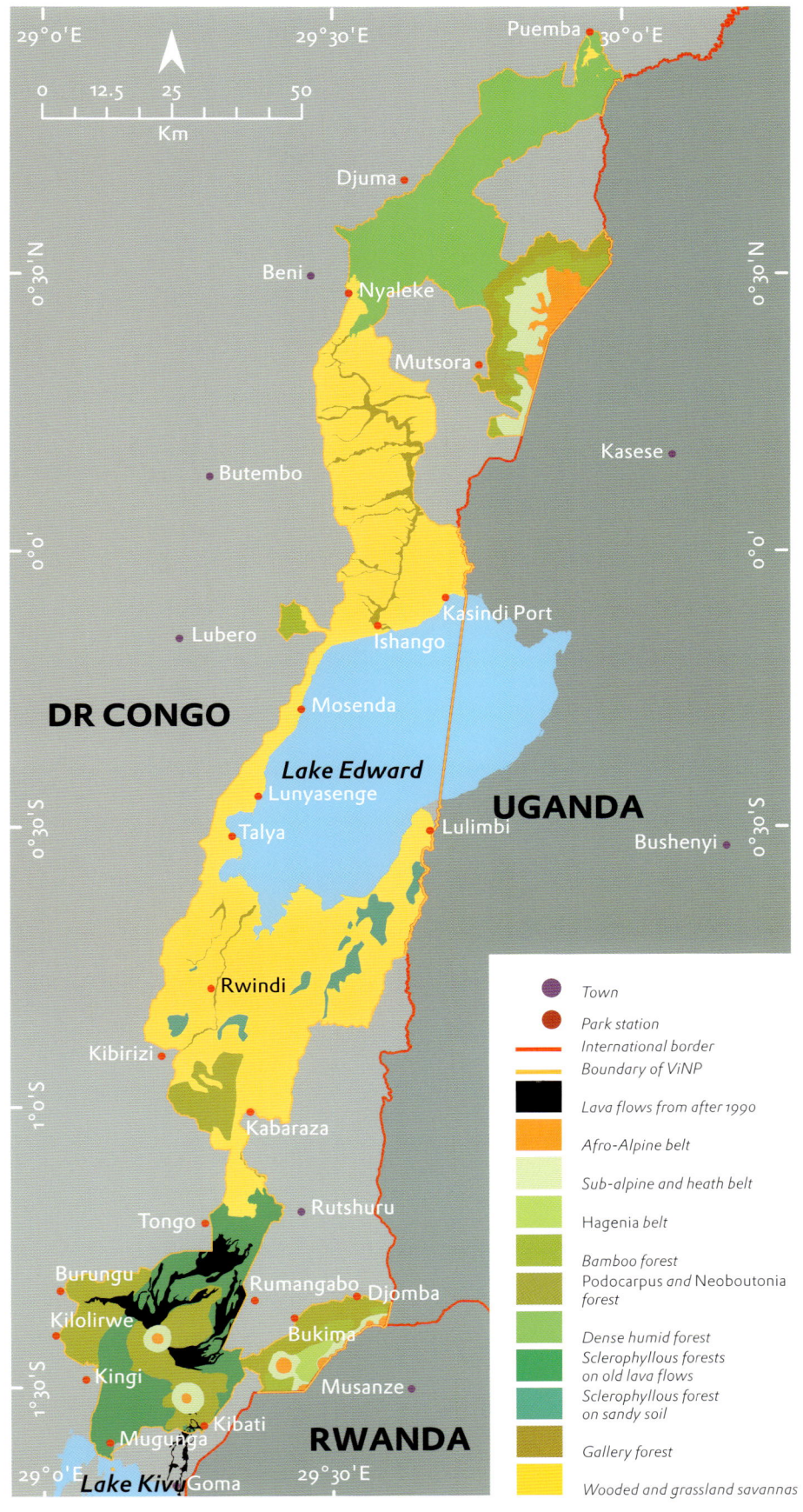

A brief overview of Virunga National Park

61

0.88
Virunga National Park has roughly 670 employees, of whom 500 are rangers.

Democratic Republic of Congo, the park is managed by the *Institut Congolais pour la Conservation de la Nature* (ICCN), under the Ministry for the Environment, Conservation of Nature, Water and Forests (MECNEF). The ICCN's organisational structure is described in detail in Chapter 15. Its headquarters are in Kinshasa and provide overall coordination for protected area management in the DRC, with a Provincial Direction for North Kivu based in Goma.

Virunga National Park has more staff than any other protected area in DRC. Of more than 670 employees in service on 1 July 2006, at least 150 were recruited during the war years without formal process. The park's management is organized into three sectors: the Northern, Central and Southern Sectors.

The Northern Sector is managed from Mutsora. Established in 1938, this station has 165 staff, including a director and two wardens, along with eight junior officers, 139 rangers and 14 auxiliary staff (with administration and maintenance portfolio). This sector includes the Semliki Valley, the Rwenzori, and two sub-sectors, Mount Tshiaberimu and Ishango. The Tshiaberimu sub-sector includes both the mountain and the narrow corridor linking it to the rest of the national park. It is the only site in the park where Eastern Lowland Gorillas are found. The Ishango sub-sector includes the archaeological site of Ishango, a popular tourist destination at the outlet of the Semliki River on the northern shore of Lake Edward. The Northern Sector has 22 patrol posts, of which only 18 are currently operational. These include Kilia, Kanyatsi, Nyaleke, Kambo, Gotongo, Lubilia I and II, and Karuruma. The geographical extent of the Northern Sector is from the extreme north of the park to the southern end of Mount Tshiaberimu.

The Central Sector, stretching from the Nguli patrol post through to Mabenga, is managed from the Rwindi Station. This station has been attacked and pillaged on numerous occasions over the last decade. As a consequence, the station's infrastructure is badly degraded. The station has 91 staff, including a senior warden, his assistant, a community warden, six officers, ten auxiliary staff and 72 rangers. Rwindi Station was particularly well known in the 1980s on account of then President Mobutu's frequent visits to the site, together with its high visitor rates. Its infrastructure, which includes Rwindi Hotel, is extremely valuable to the park. Rwindi Station oversees 13 patrol posts of which five are operational (Mabenga, BEVI/17, Nyamushengero, Vitshumbi and Kibirizi). The sector includes two sub-stations: Kabaraza and Lulimbi. Kabaraza sub-station was created as a training centre for rangers and has 36 permanent staff. Lulimbi was created as a research station. It has 84 staff including one warden, his assistant, two officers, ten auxiliary staff and 70 rangers. There are also seven operational patrol posts: Kihangiro, Nyakakoma, Birwa, Kinyonzo, Kasoso, Ishasha and Ruti.

The Rutshuru Hunting Reserve *(Domaine de Chasse de Rutshuru)*, created in 1952, is attached to the Central Sector. The reserve is staffed by 25 men, including a senior warden, his assistant, two officers, a clerical assistant and 21 rangers.

The Southern Sector is managed from Rumangabo Station, created in 1925. This Station used to be the headquarters of the *Institut des Parcs Nationaux du Congo Belge*. Much of its infrastructure, intact since the late 1920s, remains surprisingly in good condition. This includes administrative buildings, an infirmary, a garage, senior staff accommodation and rangers' barracks. The Southern Sector is staffed by 130 people, including a senior warden and his assistant, 11 officers, 112 rangers and 5 auxiliary staff. The sector includes two sub-sectors: the Mikeno sub-sector, where the extinct volcanoes are found, and which has four patrol posts (Jomba, Kimbumba, Kibati and Bukima), and the Nyamulagira sub-sector, which also includes Mount Nyiragongo.

Various aspects of the management of these sectors are examined elsewhere in this book, notably in Chapters 15 and 21.

Marc Languy & Emmanuel de Merode

0.89
In the 1960s there were an estimated 8,000 elephants in Virunga National Park. Today, their number is believed to be fewer than 400, although many may have migrated to Queen Elizabeth National Park in Uganda. The troubles of the late 1970s and early 1980s in Uganda saw elephants migrate in the opposite direction; so there is hope that, with a return to stability and a reduction in poaching, elephants might again return to Virunga in large numbers. Elephants are a flagship species in Virunga – not just for tourism, but also for the preservation of the park's natural habitats.

A brief overview of Virunga National Park

1.1
The first ever guide for tourists to the National Park was published in 1934. The gorilla on the cover is the work of artist James Thiriar.

64 Patricia Van Schuylenbergh

1 Albert National Park: the birth of Africa's first national park (1925–1960)

Patricia Van Schuylenbergh

In this chapter, the author explores the legal and institutional background to the park's creation and development. The historical context is important in understanding the social realities behind the formation of Africa' oldest national park (created a few months before South Africa's Kruger National Park), while clarifying misunderstandings that have persisted over the relationship between customary and national laws.

1.1 Virunga: paradise lost?

There is an extensive range of literature devoted to Virunga and its surroundings. Today, Virunga lies at the heart of an armed conflict that has wrought untold damage to the site, its people and its wildlife.

Virunga has a long and turbulent history. The first evidence of human occupation was on the northern shores of Lake Edward at Ishango between 18,000 and 13,000 BP (Chrétien, 2000). The populations that subsisted off these lands developed an agro-pastoral economy, but also lived from extensive trade networks linking the various lakes. Agricultural products, salt, copper and iron were traded, thanks to extremely favourable climatic, geological and pedological conditions.

In the more recent past, European colonial settlers celebrated the appeal of the region from a number of perspectives: political, being situated at the crossroads of three colonial empires—German, British and Belgian; economic, having very productive soils and a rich and varied geology; aesthetic, by dint of the exceptional landscapes and wildlife; scientific, being especially attractive to geologists, zoologists and botanists; and mythical, owing to its association with Ptolomy's source of the Nile on the Mountains of the Moon, and with the many legends of King Solomon's gold and gemstone mines.

Europeans first came to the region in the 1860s. Speke, searching for the source of the Nile, glimpsed the Virunga Volcanoes in 1861. Stuhlman provided the first detailed description of the Virunga, which he called the Mfumbiro Mountains, in 1892 (Stuhlman, 1894), and Count von Götzen also explored the Volcanoes in 1894 (von Götzen, 1899).

A series of topographic explorations in the Lake Kivu region followed. Their motives were political, in being essentially an effort to appropriate the territory. They also increased scientific understanding of the area, however, especially with respect to the flora and fauna. A European presence in the field had other consequences: notably the hunting of large mammals for sport, and for feeding large workforces engaged in agriculture, mining, or infrastructural development, together with a growing military corps.

1.2 Towards the creation of Albert National Park

It was in this context that Robert von Beringe, a captain in the German East African Imperial Army, shot two Mountain Gorillas during an expedition on the slopes of Mount Sabinyo (Anon., 2002). One of the two was sent to the Zoological Museum of Berlin by the zoologist Matschie, who described the species as *Gorilla gorilla,* sub-species *beringei*. This discovery, like so many others, provoked sharp rivalries among European nations. Each wanted to be first to announce a find and to display the specimens within its museums. During the first quarter of the 20th century, the Belgian Ministry for the Colonies was besieged with Mountain Gorilla hunting permit requests from museums and other scientific establishments around the world, especially from those in the United States. Largely scientific to begin with, the permit requests became increasingly commercial. A Scandinavian hunting expedition to the region led by Prince William of Sweden, received 14 permits to kill gorillas from the Belgian authorities, provoking a considerable international outcry. Prince William recorded a number of observations on the social behaviour of Mountain Gorillas (William, 1923), but probably the most significant outcome of his expedition was the pressure it brought to bear on King Albert of the Belgians to acknowledge the Virunga's biological value, and to have the area classified as a nature reserve to protect its wild flora and fauna.

That same year, an expedition organised by the American Museum of Natural History of New York travelled to Kivu to obtain specimens that would enable a family of Mountain Gorillas to be included in a display of African fauna. The expedition was led by the naturalist and taxidermist, Carl Akeley. An ardent conservationist, Akeley was a member of several prominent conservation lobby groups and organisations. He was also a staunch advocate of the ideas of the Museum's director, the palaeontologist Henry Fairfield Osborn, regarding the main vocation of the institution: that is, to provide a basis for collecting endangered species for the sake of public education, but also one of protecting those species in their natural environments (Akeley, 1923).

Akeley returned to the United States with the conviction that Mountain Gorillas in the area were rare, but as they were neither scared nor aggressive, their extinction was imminent. From this came the idea of establishing the Gorilla Sanctuary in the triangle formed by the three extinct Volcanoes: Mikeno, Karisimbi and Visoke. Believing there to be no time to lose, Akeley launched an aggressive lobbying campaign between 1922 and 1925. Through this campaign, he mobilised considerable support in scientific, financial and diplomatic circles in both the United States and Belgium.

1.2
Prince William of Sweden, pictured in 1921 during a hunting expedition in what was to become Albert National Park, where he shot several Mountain Gorillas. On returning to Europe, he submitted a formal request to the Belgian authorities, urging them to establish a nature reserve on the Rwindi Plains to protect the area's flora, fauna and soils.

From 1922, the Belgian Ministry of the Colonies began to implement wildlife protection measures in the area. Partly a result of Akeley's lobbying, this—perversely—was a spin-off of both Akeley's and Prince William of Sweden's published articles and conferences addressing the massive increase in requests for gorilla hunting permits from scientific institutions and private individuals.

Two hunting reserves were created in Kivu by decree of the Vice-Governor General: the Albert Reserve (on 24 February 1923), between the Rutshuru River and the southern shores of Lake Edward, as suggested by William of Sweden, and a second reserve (promulgated on 23 November 1923) to the northeast of Lake Kivu, between Mount Sabinyo and the Catholic Mission at Tongres Sainte Marie (now Rugari), set up to protect endangered animals, the Mountain Gorillas especially.

The colonial government's creation of these two reserves was by no means a simple matter. It was the outcome of a protracted debate on the protection of certain species, such as the African Elephant. The debate had its origins in the legislation passed in 1901 by the government of the Congo Free State, and in subsequent legislation passed in 1908 by the ministry of the colonies in response to the recommendations—agreed in London on 19 May 1900—of the International Conference for the Protection of the Animals of Africa.

The future of Congo's flora and fauna was also widely debated within the scientific establishments of Belgium. Within the Museum of the Belgian Congo, the *Bulletin du Cercle Zoologique Congolais* was established in 1924 by the zoologist Henri Schouteden, providing a forum for debate over the protection of wildlife in the Congo.[1] The journal documented the legislative commitments passed during that period at both a national and international level. Included, for example, were the commitments made during the first *Congrès International pour la Protection de la Nature* in Paris in 1923.

King Albert of Belgium became increasingly sensitive to many of these fairly innovative ideas, which he managed to have implemented, despite considerable reticence in certain circles, including a marked lack of interest on the part of Franck, then Minister of the Colonies. The King, together with his immediate entourage, was increasingly swayed by the positive approach of the Americans, and by the aggressive drive behind contemporary scientific initiatives, such as those he had seen during his visits to the United States and Brazil in 1919. He accordingly asserted a policy of supporting scientific development, and gave his full support to the conservation movement in eastern Congo. Following a positive meeting with Akeley and a number of exchanges with Prince William of Sweden, the King—having also read several American publications on the matter—not only backed Akeley's project, but also requested that the measures be extended beyond the gorillas, to include all the fauna and flora of an area extending on to the Rwindi Plains. To this end, he insisted that the word *National* be included within the Reserve's name.

Thus, on 21 April 1925, a protected area was created by royal decree and given the name Albert National Park. The new park covered an area of exceptional natural beauty incorporating the extinct volcanoes of Mikeno, Karisimbi and Visoke, and extending as far north as the Rwindi Plains, taking in a remarkable flora and fauna. Not only was gorilla hunting strictly prohibited, but so was that of all species, except in cases of self defence. The potential for scientific research was immense.

Another Decree—issued on 18 August 1927—then extended the national park to include the south-eastern slopes of Mounts Visoke and Karisimbi, situated in the Rwandan protectorate.

And so it was that Albert National Park, later to be renamed Virunga National Park, became Africa's first national park. Within less than one year, in January 1926, it would be joined by South Africa's Kruger National Park, when the latter's prior status—that of the Sabie Nature Reserve—was reclassified.

1.3   OBJECTIVES AND FIRST EFFORTS AT PARK MANAGEMENT

Delimitation of the park's boundary, as proposed by Akeley, was somewhat imprecise, given the lack, then, of information on the area's physical geography and human demographics. A provisional delimitation, gazetted by the Governor General of the colony (*Ordonnance loi* of 10 July 1925), included conditions to the effect that the needs of local human communities warranted close and careful considerations. Marcel Houtart, the then Minister for the Colonies, handed responsibility for undertaking the required field surveys to two American scientific expeditions. The two expeditions were led by the ornithologist James Chapin, and by Carl Akeley, then back in the field with the Belgian zoologist Jean-Marie Derscheid. The objective of both expeditions was to undertake a thorough study of Albert National Park in order to establish its exact boundaries, to enhance the understanding of the topography, to establish a basis for the park's management, and to collect and collate information on the distribution and abundance of its fauna and flora. The unexpected death of Carl Akeley in November 1926 at their camp at Kabara forced Derscheid to take over leadership of the expedition. Derscheid set out the essential principles of protected area management, defining the national park lands as areas of integral protection, where species conservation was the fundamental priority.[2] Such areas would exclude lands that were of limited biological value, even if unoccupied, as these would serve as timber reserves for local human populations living around the park, and as buffer zones for the park. The buffer areas would nevertheless be classed as nature reserves, subject to restrictions on hunting, fishing and timber felling. Where hunting, fishing and timber production could be shown to benefit local communities, these activities remained authorised.

The outcome of these two expeditions was the Decree of 9 July 1929 establishing an autonomous institution, Albert National Park, whit a legal status and with its headquarters in Brussels.[3] However, this remained under the financial control of the Ministry of the Colonies. It was governed by a 26-member board of trustees, made up largely of representatives from the Belgian and international scientific communities. The trustees were nominated by the King, who oversaw a board of Directors comprising seven, then nine, members. The Decree of 1929 also extended the park as far as Lake Edward, increasing its area from 25,000 to 190,000 hectares, divided into four sectors: Central (the extinct volcanoes including Sabinyo), Western (the active volcanoes, as well as Mounts Nyamulagira and Nyiragongo through to the northern tip of Lake Kivu), Eastern (areas including those portions of Mounts Gahinga and Muhavura to the south of the Ugandan border), and Northern (formerly the Kivu Hunting Reserve, extending as far as the southern shores of Lake Edward). Additional buffer zones added to the park included a strip of land around the Central and Eastern Sectors and a second strip of roughly the same area around the southern part of the Northern Sector, on land with a much greater human population density.

The first chairman of the board of trustees was Prince Eugène de Ligne. The vice-chairmanship was given to Derscheid, who campaigned both for the establishment of Albert National Park as the first in a network of African protected areas and for the forging of scientific cooperation at an international level – an important step in attracting funding, especially from the United States.

Conflicts soon developed among the various members of the board. The disputes were related partly to the nature of the first *in situ* developments within the national park. More contentious, though, was the freedom offered to foreign expeditions within the park. This issue led to the resignation, late in 1931, of both Prince de Ligne and Derscheid. They were replaced by Prince Léopold, Duke of Brabant, and Victor Van Straelen, the new director of the Belgian Museum of Natural Sciences, who was a close friend of the King's and a mentor to the Duke of Brabant. One of Van Straelen's first moves took form of a bid to curtail American influence on the affairs of the board and the park. After a joint visit to Albert National Park in 1932, Van Straelen and Prince Léopold established a set of broad principles for the reorganisation of the Park's administrative systems. They agreed to form an Institute for the National Parks of the Belgian Congo *(Institut des Parcs Nationaux du Congo Belge,* IPNCB*)*, as part of efforts to strengthen the surveillance and protection of the park. They further expressed the wish to extend the park northward to include the western slopes of the Rwenzori Mountains, along with a portion of tropical lowland forest known to contain a population of Okapis.

The status of Albert National Park as an 'integrally protected area' was reasserted during this mission, which clarified the objective of maintaining the absolute integrity of areas within the National Park by reducing human intervention to an absolute minimum, not just for aesthetic reasons, or for the promotion of tourism (as had been emphasised when creating both Yellowstone in the United States and Kruger Park in South Africa), but also for the sake of preserving a natural heritage as a basis for increasing scientific understanding. This objective was explicitly stated in the minutes of the International Conference of African Flora and Fauna on 8 November 1933, which asserted the notion of preserving natural environments as 'Humanity's common heritage' (IPNCB, 1935). The Duke of Brabant, in his speech to the Africa Society in London on 16 November 1933, also asserted a moral imperative to protect such a heritage 'for which we are the caretakers' (Brabant, 1933).

The national parks, then, were accorded a clear scientific function, being defined as open air laboratories in which researchers could observe the natural environment without human interference.

Scientific research then assumed two broad dimensions: first there was the establishment of systematic, multi-disciplinary inventories, and then there was the resolution of particular scientific problems well represented in, or even unique to, Albert National Park. From 1933, there were appeals – addressed to universities, especially – calling for scientific collaborations. The field inventories began with G. F. de Witte's herpetological expedition between 1933 and 1935, which focused on the taxonomy and ethology of those species. Studies in the second category followed, with expeditions such as Father Schumacher's among the Pygmies between 1933 and 1936. By the eve of independence, a total of nine major scientific expeditions had been completed in

1.3
Rangers at the Rumangabo Station in the 1930s. Note the absence of firearms. Silhouetted in the background is the cloud-topped Mount Nyamulagira.

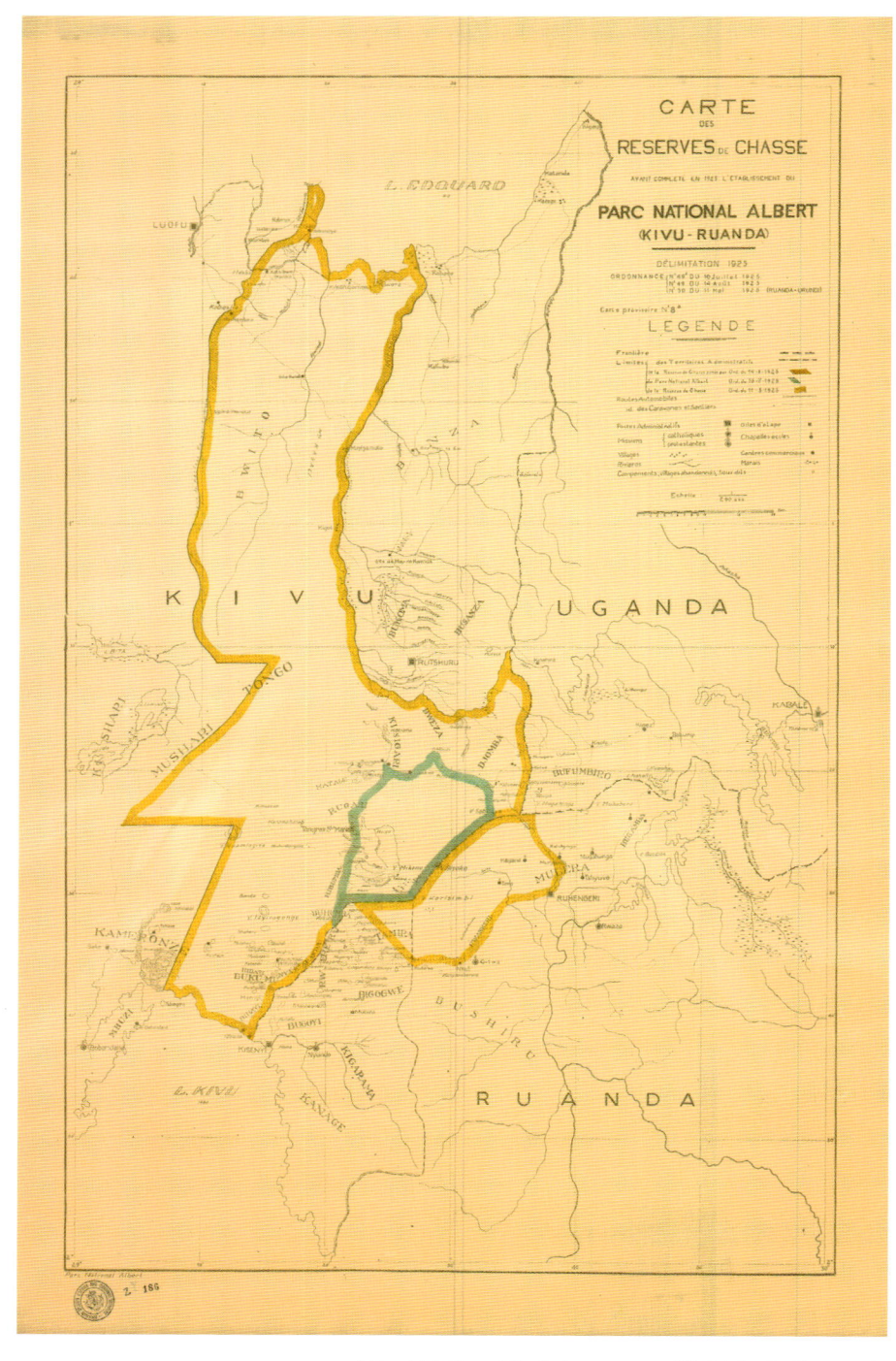

1.4
Two hunting reserves, already in place at the beginning of the 1920s, were integrated into Albert National Park only in 1929. The first Royal Decree (of 1925) established Albert National Park, assigning it an area of roughly 20,000 hectares, centered mainly around the extinct volcanoes in the Belgian Congo. These areas are bordered in green on the map.

Albert National Park, in a range of fields including zoology, hydrobiology, anthropology, geology and botany. Two further expeditions explored both the Albert and Kagera National Parks. Collectively, these expeditions produced numerous scientific papers, many of which were published by the IPNCB.

1.4 The management of Albert NP by the 'Institut des Parcs Nationaux du Congo Belge'

The Decree of 26 November 1934 provided for the creation of additional national parks through the establishment of the *Institut des Parcs Nationaux du Congo Belge*. The agenda behind this initiative was to create greater autonomy for the parks, particularly with respect to the recruitment of a special unit of wardens and rangers. This meant it was no longer necessary for all recruitments to be passed through local civil servants of the Ministry for the Colonies. The Institute had three broad objectives: the protection of wildlife, the promotion of science, and the development of tourism in certain parts of the park as a means of funding the first two objectives. Other than the subsidies provided by the Belgian Ministry for the Colonies for the management of Albert National Park, another source of funding was secured in 1934 through the establishment of the *Fondation pour favoriser l'Etude Scientifique des Parcs du Congo Belge*. Significant funding was secured from individuals such as Louis Empain, to cover the travel expenses and other costs of scientific expeditions to Virunga, together with the costs of publishing the findings of such expeditions.

The year 1934 was pivotal for Virunga: another Royal Decree was passed, re-adjusting the park boundary to minimise conflict with local people. This decree was a re-assessment of the Decree of 1929 regarding traditional land rights, while ensuring that areas to be gazetted as a national park could be thoroughly protected. Lands of less value for agriculture were retained within the park boundaries. The status of *Réserve Intégrale* (total protection) was assigned to the remaining lands, which included unproductive agricultural land in areas not previously included in the national park. This effectively increased the park's area to more than 390,000 hectares. A new, and significantly enlarged, national park now existed. This entire block, spanning the Southern Sector (Nyamulagira-Mikeno) and the Central Sector (Rwindi-Lake Edward), was to be managed under a regime of integral protection.

The following year saw the signing of the Decree of 12 November 1935, which added another large gazetted area to the National Park. Based on the extensive ground-work of Colonel Hackars, the decree extended the park by 470,000 hectares to the north. The new sector included all of Lake Edward, the plains and some of the forests of the Upper and Lower Semliki, the Congolese slopes of the Rwenzori Mountains, and Mount Tshiaberimu.[4] The area was divided into seven units defined by specific habitats and landforms. These included Mikeno and Nyamulagira in the Southern Sector, Rwindi-Rutshuru in the Central Sector, and Lake Edward, Upper Semliki, Lower Semliki and Rwenzori in the Northern Sector. Mikeno apart, the units formed a contiguous block from the northern shores of Lake Kivu to the northern reaches of the Rwenzori. Mikeno was separated from the rest of the park by the Goma-Rutshuru Road. The administration and conservation of the seven units were managed from two stations. Rumangabo, established to manage the Southern and Central Sectors, was run by a senior warden, a warden in charge of visits, and an under-officer. Rumangabo was also the African headquarters for the *Institut des Parcs Nationaux du Congo Belge*. A second station was established at Mutsora to manage the Northern Sector, under the authority of a senior warden.

From 1928 to 1934, park surveillance and protection operations were conducted under the leadership of senior warden René Hemeleers, who answered to the Belgian Ministry for the Colonies. Hemeleers' second-in-command was a Danish officer, Rasmus Hoier, who was a Colonel in the *Force Publique* (the police force in the Belgian colony). Colonel Hoier was answerable to the Board of Directors. After 1934, management of Albert National Park came under a single authority: that of senior warden Henri Hackars, who was no longer answerable to the Ministry for the Colonies, but directly to the IPNCB Board of Directors. It was Hackars who negotiated the new boundaries, purchased land titles for the extensions and managed the protection of the park from his headquarters at Rutshuru. He was then replaced by Colonel Hoier, who held the same position at Rutshuru between 1934 and 1946. Hoier was largely responsible for the management systems that were implemented in the Central Sector, and which became a standard for park management in Congo and elsewhere.

The Second World War saw a rupture in links between the wardens on the ground and the IPNCB headquarters in Brussels. Management of the park became essentially autonomous during the war years.[5] These were also difficult years for the relationship between the park authorities and local people. The level of tolerance accorded to local communities over the exploitation of protected land was greatly reduced – with far-reaching consequences in subsequent years. We examine this issue in the last section of this chapter.

After the war, the authorities acknowledged a need to restore relations with the communities of local people living around Albert National Park. To this end, Victor Van Straelen made several journeys to Congo. He put considerable pressure on Wigny, the Minister for the Colonies, to acknowledge the land rights of the local people. Included were the fishing rights to Lake Edward, a lake entirely contained within the national park. By 1947, Albert National Park had 132 rangers, most of whom were based at Mutsora, together with a large number of local workers hired for both skilled and manual labour. The number of rangers had increased to 234 by 1954 and remained at this level until 1960.

The 1935 boundaries of Albert National Park were again modified by Royal Decree on 15 May 1950. The new Decree, a comprehensive revision of the earlier Decrees of 4 May 1937 and of 17 May 1939, was based on the conclusions of the enquiry into the rights of the indigenous people. It modified the boundaries of the Southern Sector, while easing rights of movement through the Mikeno Sector for non-tourists.

The period following the Decree of 1950 also saw a marked increase in the number of European staff. This was largely a response to the sharp increase in responsibilities then facing the senior warden. Not only were VIP visits on the increase; there was also a lot of construction work in progress, and a dramatic escalation in the administrative and financial requirements imposed by the IPNCB.

1.5 THE NEGOTIATION OF LAND-RIGHTS WITH INDIGENOUS POPULATIONS

The colonial history of the national park was characterised by two categorically opposing visions: a wish to preserve park land in its natural state to the absolute exclusion of local economic interests and local pressure based on a rejection of the very existence of the park, which was seen as representing a major obstacle to local development. There was precious little middle ground, and little attempt made to integrate the protectionist and economic agendas. The situation became increasingly complex moreover, making it all the more difficult to find a compromise position, and an optimal form of management that could satisfy the need to preserve natural resources while strengthening the local economy. This analysis is too general, though, to explain the reality on the ground. There were other factors that prevented successive managers from finding a compromise between the needs of local communities and the protection of the reserve. Among these complicating factors was an increase in instances

1.5
The extent of Albert National Park in 1929, when it covered about 200,000 hectares, spread over four non-contiguous blocks. Only in 1935 were the Southern (volcanoes) and Central (Rwindi Plains) Sectors linked up to form a single block.

of human-wildlife conflicts, as the area's human population expanded amid a changing pattern of migration that also saw the arrival of a growing number of colonial settlers.

The conflicting interests of park managers and of the region's various economic players are too complex to be examined exhaustively in this chapter. So discussion here will focus only on the most important of these issues, given its lasting implications: namely, the rights of the indigenous populations and the settling of their land claims.

When Albert National Park was created, the delimitation of boundaries came under the Belgian Ministry for the Colonies. Some of the areas now included in the national park were indeed occupied by local populations prior to the park's establishment. Such populations had traditional rights to those lands, and those rights were important to their subsistence. Their claims were also recognised by the colonial administration. Yet, in the interest of being able to maintain an 'integrally protected' national park, these populations had to be moved, and an acceptable solution needed to be found to compensate them. It was decided that all human populations living within the boundaries of the national park, should be expelled and that the movement of people within the park should be forbidden. There was to be only one exception: the 300 Twa pygmies living in the area would be allowed to remain, as their presence in the park was considered non-threatening and potentially of interest to anthropologists.

The forced expropriation of land from local people would be followed by the purchase of their traditional land rights (in a process rigorously monitored by the colonial authorities and covered by colonial law). The legal basis for expropriation was first established under the Royal Decree of 26 November 1934. The Decree of 28 July 1936 then amended the resulting laws, under which local populations could be ordered to vacate their lands, but would have to be provided with an indemnity equivalent to the cost of both their relocation and the purchase of equivalent agricultural land outside the national park. In practice, these laws proved extremely difficult to implement, largely owing to dissenting positions within the colonial administration, some members of which questioned the moral acceptability of forced expulsions of local people for the sake of a protected area. For several years, a debate raged over the nature of the guarantees that should be offered to dispossessed farmers. Eventually, a settlement was reached allowing for outright purchase of the land, rather than subsequent allocation of new land. The landowner's written consent was a prerequisite for state appropriation of land.

Communities of local people living on park-adjacent lands were not subject to expropriation orders, but were forbidden from undertaking certain economic activities (hunting, fishing, fuelwood collection). Exceptions, based on traditional rights and subsistence needs, were made. Such waivers were often confusing and contradictory, however, for want of precise definitions for 'traditional rights', and 'subsistence needs'. These might vary enormously from place to place, depending on the local geography, the prevailing ecology and the ethnic background of the people.

In 1929, it was acknowledged that the boundaries of Albert National Park did not take into account the needs of local communities, both in terms of essential agricultural land and access to fuelwood and other resources. In response to this problem, the colonial administration set up a commission of enquiry chaired by Hackars to examine the land claims of local people. The recommendation of Hackars' commission resulted in the modification of the park boundaries in 1933, and in the restitution of tracts of densely settled land to the respective local communities. Most of the land returned to local people was in Rwanda and at Kibumba and Binza. At the same time, other lands – on the west coast of Lake Edward (in the Wanande-south), in the Semliki Valley, and on the Rwenzori massif (Wanande-north and Walendu) – were included in the park. Incorporation of the Wanande-north territory was controversial in that local government and the park authorities held opposing views on the issue of outright protection of the lowland forests of the Ituri and on extraction of fish resources from the Semliki River. The authorities were intent on accommodating the needs of the Nande people, who were considered to be the most important source of labour for the Province of Stanleyville.

The Commission's findings culminated in the signing of the Decree of 1934, which addressed many of the previous frictions with local people through providing them with fishing rights to a stretch of the Rutshuru River, above May ya Moto. Rights to exploit the vegetation – including lianas *(kerere)* and edible wild plants *(ubuzi)* – between the river Molindi and the Mabenga-Tongo footpath were also granted. Unrestrained movement on the Goma-Irumu road within the Central Sector, and on the Goma-Sake road in the Nyamulagira sector, was also allowed.

With respect to the contiguous roads outside the national park, which had been included as part of the protected area through the Decree of 12 November 1935, the commission ruled that people who relinquished their rights to hunt wildlife without demanding compensation could (in all but certain areas of the Semliki Valley) stay on. In exchange, they would be granted rights to fish in Lake Edward and along the Semliki in predetermined areas, to be established with the IPNCB, with local government, and with the provincial health services. Come 1937, new agreements between the colonial authorities and the traditional authorities had been passed. These agreements were intended to meet conditions set out in the Decree of 1934 for gazetting certain areas of Albert National Park that had hitherto not been fulfilled by the colonial administration. The frequent changes in senior administrative personnel had made it difficult to track the implementation of agreements between local people and the commission. Nevertheless the ratification of these agreements, key to making the National Park legitimate, was extremely important to the IPNCB, which could now assume a free reign over park management.

In the meantime, the resident populations had established new settlements in the surrounding reserves with the tacit understanding of the local administration. In early 1935, the fishing villages were also resettled, despite having been closed in 1925 because of suspected trypanosomiasis.[6] As such, the Decree of 1934 was impossible to implement on the ground, and the context remained very similar to that prior to 1934. A new commission was set up under the chairmanship of Hackars to oversee the settlement of land disputes. Under this commission, the local populations relinquished certain land rights they had claimed in 1934

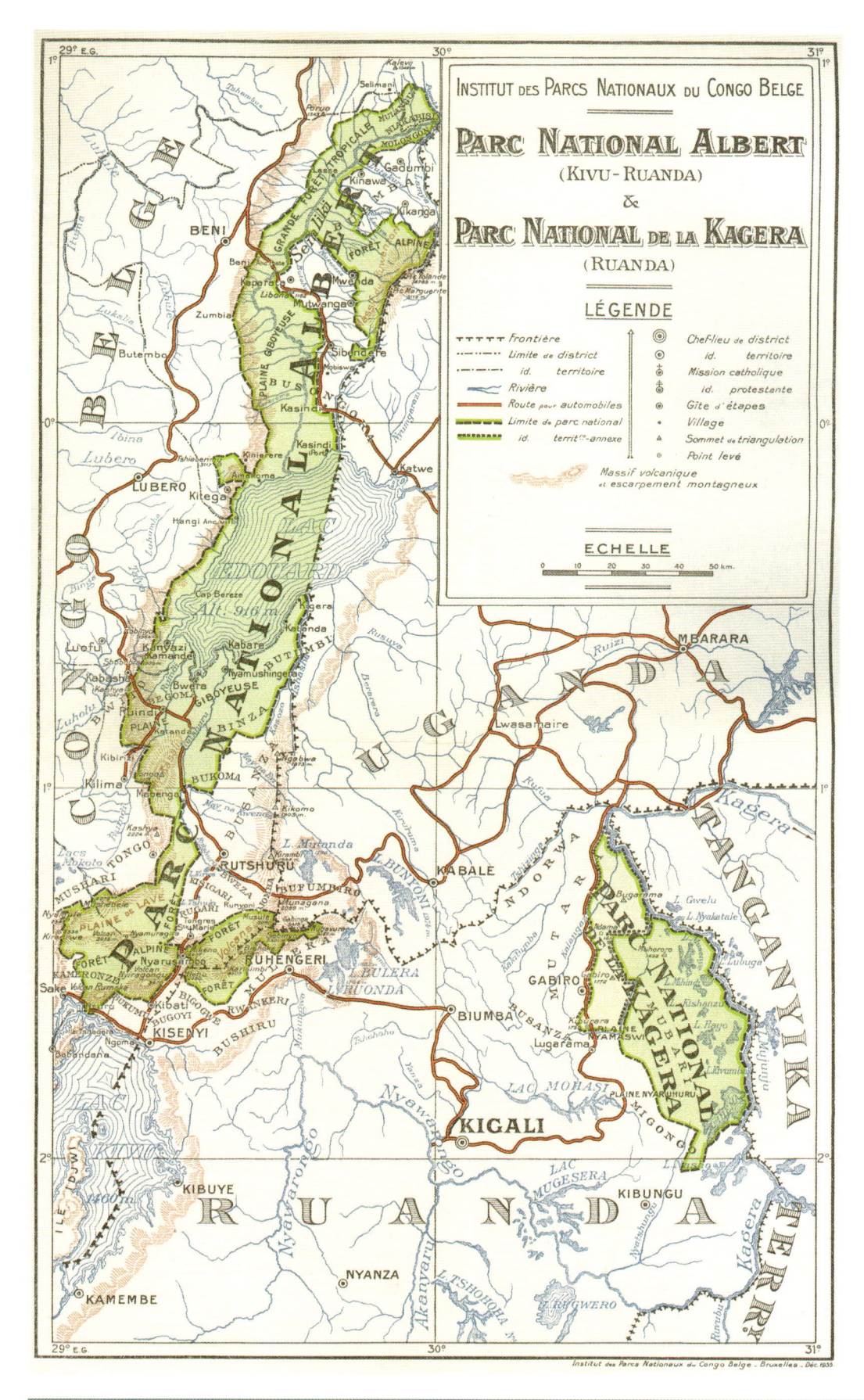

1.6
Since 1935, the shape of Albert National Park has remained broadly the same. What changes there have been are minor, and were introduced mainly in the wake of the Decree of 15 May 1950.

Albert National Park: the birth of Africa's first national park (1925 to 1960)

1.7
Colonel Henri Hackars in 1937. As Warden of Albert National Park between 1933 and 1934, Hackars compiled most of the data on which the Decree of 12 November 1935 was based, giving much of the National Park its present shape.

(including the right to reclaim lands vacated because of trypanosomiasis, hunting rights, and rights to salt collection), in exchange for new fishing rights on Lake Edward and along the Semliki. The health authorities never ratified those rights. Indeed, in 1937, the health authorities prescribed the evacuation of all the villages on the Semliki plains and along the shores of Lake Edward, and suspended a fishing agency of the *Société Minière des Grands Lacs Africains* (Great Lakes Mining Company), which had been extracting large quantities of fish to feed its workforce.

Compensation paid to local populations in the Central Sector of the park in 1937 was formalised in the colonial budget of 1938. After this date, the agreements with the local people were considered binding.

However, these agreements did not apply to land covered by the Decree of 1929 that had not been subject to any land adjudication, or to any formal submission of land rights. Likewise, the agreements did not ratify the 1934 extensions to the park. So, it became necessary to institute an enquiry into the submission of land rights, with the result that a new commission was formed.

The new commission faced a marked difference in the attitudes of local people, and a strengthening of their resolve not to relinquish their rights to land and to natural resources. Their resistance was especially marked when it came to negotiating the rights to exploit Kivu's bamboo forest, all of which lay within, and was effectively protected by, the National Park. Bamboo was considered an essential commodity for agricultural production. The local populations also claimed rights to pasture for their livestock, which during the dry season was in desperately short supply in Rugari, Bukumu, Bweza and Jomba, as well as fishing rights to the lower Rutshuru and Lake Edward. They further laid claim to fishing rights on the Bwito and Ngezi Rivers in exchange for relinquishing their hunting rights. Finally, they claimed the rights to cultivate at Vitshumbi, on the shores of Lake Edward, and at Kashwe, on the Rutshuru.

It became clear that a negotiated settlement could not be achieved. An increasingly complex situation drove the IPNCB to consider dispossessing populations that refused to relinquish their land rights. No such decision was taken, however, due to the onset of the Second World War. And so, by May 1940, no substantive solution had been reached regarding the status of populations living in the Northern Sector of Albert National Park. The Governor General, Pierre Ryckmans, meanwhile, had come out in support of the local populations, and of their entitlement to fish on Lake Edward and the Semliki and Rutshuru Rivers, which he deemed essential, given the social and economic importance of this activity. The Governor General saw to it, moreover, that lands in the Kigeri and Rugari-Kishigari areas of Rwanda were returned to their local populations.

During the war, local populations invaded the park at several places: fishing villages were established within the park, banana plantations grew up, bamboo was extensively exploited. The war created new markets and increased the demand for certain commodities: rubber was exploited within the park and there was an increase in iron extraction. Law enforcement in the park was relaxed, bringing about an associated increase in land claims, both from the indigenous and European populations. The colonial settlers often encouraged local people to exercise their land claims, with a view—later—to being able to incite the peasant farmers to produce cash crops like pyrethrum and timber. There were high prices on offer for bushmeat, moreover, and rifles were readily available following the defeat of the Italian troops.

Economic conditions after the war were such that the concept of an integrally protected reserve was difficult to justify. In the years that followed, the problem of reconciling biological imperatives with pressing demographic and economic priorities became increasingly intractable. The area around Lake Edward was especially controversial, owing to the high value of the lake's resources, and the anomaly reflected in the fact that Ugandan fisherman were allowed to fish from Katwe, on the shores of the same lake, and sell their catches in Congo, where fishing was prohibited.

Its position increasingly isolated, the IPNCB began to lobby the Ministry for the Colonies to provide it with unequivocal legislation protecting the Institution from a successive dismantling of the park's integrity. Efforts to consolidate the park also focused on diffusing the tension surrounding access to critically important resources. This was achieved through a compromise granting local residents some access to certain resources. Victor Van Straelen attributed many of the park's problems to ill-conceived policies on the part of the colonial administration. Such policies, coupled with the dramatic demographic growth seen during this period were undermining the local economy. Soaring demand was fast depleting available natural resources. The IPNCB repeatedly condemned the colonial administration's reluctance to adopt the various Decrees (including those of 1929, 1934 and 1935) providing a legal basis for the National Park. By contrast, the Governor General insisted that the Institute comply with colonial legislation and refrain from seeking to overstretch its authority.

In 1947, the King's Counsel, Louis de Waersegger set up a new Commission of Enquiry into indigenous rights. This commission succeeded in bringing establishing a greater degree of clarity to the issue of the status of land within the park. A number of concessions were offered. For example, rights to fish in the Rutshuru and Semliki Rivers were withdrawn, but in exchange two independent fishing systems were established at Rutshuru and Kiavinyonge, under the supervision of COPILE *(Coopérative des Pêcheurs Indigènes du Lac Edouard)*, together with rights of access to walk across park land. The new fishing settlements were intended to ensure that adequate fish resources could be made available to the populations of Kivu. In addition, these settlements were expected to provide a commercial supply of fish to the growing towns of Goma and Costermansville (Bukavu), as previously requested by the colonial administration. In the 1950s, however, the activities of the COPILE were expanding well beyond the bounds prescribed by law: there was a continual movement of people and vehicles within the park, excessive cutting of trees for fuelwood and construction, digging of quarries in the park, the introduction of livestock, fishing in prohibited areas, and use of illegal nets. The COPILE's illegal growth peaked in 1957, when—under pressure from the local authorities—the IPNCB reluctantly authorised the opening of markets on the western shores of Lake Edward, thereby further boosting economic activity in the region.

Other infringements in the 1950s, evident in both the Northern and the Southern Sectors of the park, included: the illegal cutting of bamboo, poaching, incursions by Ugandan fishermen and by Hema pastoralists in search of pasture, and the regular lighting of fires on the savannas. A number of laws were passed to counter these developments. Based on the Decree of 1950, the new laws imposed restrictions on fishing on the Semliki, on the cutting of lianas and *ubuzi* plants, on salt mining on the Rumoka Volcano, on harvesting palm nuts on the west bank of the Semliki, and on fuelwood collection from Mount Bukuku. Alongside this legislation, the park authorities began to take measures aimed at reducing friction with local people. Such measures included distribution of bamboo around Rumangabo and construction of water reservoirs in Rutshuru. A campaign to make local residents aware of the importance of the park, and of the work of the IPNCB, was also organised during this period.

In the run-up to Independence in 1960, a number of other problems arose within Albert National Park. One of the most intractable was the demographic growth around Jomba and Rugari, in Rutshuru Territory. The area's human and livestock populations increased rapidly after 1950, and predictions at the time were that the situation would become untenable within 25 years. The park authorities saw an increase in tension with local residents, who since the early 1940s had been complaining of a lack of available arable land. The colonial government responded by introducing a programme known as *glissement* (sliding) of the populations. The programme sought to provide incentives for people to move to less populated areas, such as Mushari and Bwito, while at the same time promoted the use of irrigation on the Rutshuru plains. The colonial administration also demarcated an additional 350 agricultural extension plots in Rutshuru in 1953 – each plot covering 3.5 ha.

In 1956, the National Park's monitoring teams found several thousand head of livestock in protected forests on the flanks of the volcanoes, especially on Mount Mikeno. Numbers increased in the dry seasons, and the result, according to George Schaller, was a worrying decline in gorilla habitat. The presence of pastoralists in the Mikeno Sector was believed to be associated with smuggling operations between Congo and both Uganda and Rwanda. Tackling this illegal presence in the park required the intervention of Jean-Paul Harroy, then Governor of Ruanda-Urundi (who, incidentally, was formerly chief warden of Albert National Park in 1937–38, and of Garamba National Park in 1947–48).

The issue of illegal use of pastoral land within the park became intensely political, largely because the pastoralists were faced with but two choices: either to enter the park, or to lose their livestock to famine. Thus it became essential to find new pastures in the Volcanoes region. In 1958, a provisional solution (pending the establishment of a definitive decree) entitled pastoralists to graze their livestock on a 6,000-hectare block of pasture in the Ruhengeri-Gisenyi area.

1.8
The Duke of Brabant, the future King Leopold III of Belgium, pictured in 1933 on the crater of one of the active volcanoes in the Nyamulagira Sector (known at the time as the Western Sector)

1. One of the primary objectives is the protection of wildlife, especially endangered species (Okapis, gorillas, the Northern White Rhinoceros). It became apparent that too many hunting permits had been issued for gorillas in Kivu. (See *Bulletin du Cercle Zoologique Congolais*, 8 March 1924, pp. 5–6).
2. The Commission of Enquiry recommended that all settlement, pasture, agricultural land, footpaths, roads and watering holes be kept outside the park boundary so as to minimise the effect of disruption on the local people.
3. Autonomy and parastatal status put the Institute on a legal footing, under the governorship of a board made up of representatives from the Belgian and international scientific communities. This audacious move effectively wrenched control over an extensive area of land away from the grip of the local colonial government, giving the Institute complete independence, while making it directly answerable to the Ministry for the Colonies, which provided the funds.
4. The boundaries of the protected area were somewhat modified following the Royal Decrees of 4 May 1937 and 17 May 1939.
5. The decree stipulated that all the president of the commission's powers, along with those of the commission itself, rested with the Governor, whereas those of the board rested with the park's warden, Hoier (No. 42/Agri, 30/10/1940).
6. The displacement of people was a deliberate attempt, on the part of the medical authorities of *Province Orientale,* to eradicate sleeping sickness in the Semliki Valley and along the shores of Lake Edward. There were three such displacements – in 1929, 1932 and 1934. Only the third was of land within the park. The period January to March 1934 saw the evacuation of 17 villages in the Northern Sector and one village on the western shores of Lake Edward, following the 1927 epidemic during which 7,000 people in these areas were infected.

2.1
The first photograph of live Mountain Gorillas at Kabara, taken by Jacques Verschuren in August 1960. Mountain Gorillas came to the notice of science only in 1902, attracting the interest of many American and European scientists. The protection of Mountain Gorillas and their habitat was one of the primary considerations behind creation of Albert National Park in 1925.

Jacques Verschuren

2 Life in Albert National Park (1925–1960)

Jacques Verschuren

In this chapter, Jacques Verschuren describes the experience of administrative staff and of rangers in the field during the first 35 years of the park's existence, from its creation until 30 June 1960 – the historic date of Congo's Independence.

Based at Rutshuru, the park's then management station, Verschuren travelled extensively on foot, exploring and sleeping rough in even the most remote and difficult terrain. In all, he spent more 2000 nights out camping in the bush.

The inspiration for this journey, and for much of the historical account that follows, was the testimony of a close relative, Jean Fontaine, who lived in Congo from 1909 to 1926, and who was known as *Bwana Kitoko,* the 'friendly man' – also the name given to King Baudouin in 1955. As a District Commissioner in Kivu, Fontaine was responsible for some of the reserve's first demarcations, marked with cairns in the Kasali area. Verschuren obtained additional information from discussions he had with the chief, Daniel Ndeze, in Congo, and also from various sources in Belgium. The recollections of some very elderly Congolese individuals were also used.

2.1 Written sources

The first five-year report (1935–1939) issued by the IPNCB *(Institut des Parcs Nationaux du Congo Belge),* proved to be a remarkable source of information. No subsequent annual report was ever published, although a retrospective did appear in 1976. Otherwise, the study of what few archive records remain proved to be a monumental task. The hundreds of people who contributed to the first report all died before 2008. Unfortunately, while several papers were published on the status of the Belgian Congo's parks until 1960, almost nothing was published over the key period 1960–2008, other than the author's own texts. It would be important if, in addition to Patricia Van Schuylenbergh's forthcoming thesis on the early history of the national parks and other protected areas of the Belgian Congo, historians were to document the more recent, post-colonial fates of these areas.

2.2 The issue of claims: perspectives from the field

How were the various claims of local populations perceived in the field? First, let us not confuse Albert Park and its modest problems with Upemba Park, where certain claims had, by the end of the past century, still not been completely settled. Large parts of Albert Park's Southern and Central Sectors, some unoccupied by man or volcanic, were offered by Chief Ndeze, a large land-owner, to his 'brother and friend' King Albert I. Land problems with the *Wahunde* (Hunde) ethnic group in the southwest of the Southern Sector, meanwhile, were settled relatively quickly.

In the Northern Sector, prior to the creation of the park, the *Wanande* populations, of their own accord, evacuated large areas on account of sleeping sickness, then rife and responsible for decimating the inhabitants. Likewise, areas around the important Beni Post were spontaneously evacuated. Until recently, one could still make out the ruins of the old Beni clinic (once a quarantine station, or *lazaret*) on savanna flanking the northern Semliki River.

Early land-use surveys carried out in the Northern Sector of the park, detected no sign of contemporary human occupation. A few claimants, opposed to the park's creation, were diplomatically persuaded, without pressure from the authorities, to settle on better lands, and were given generous financial compensation. There are no surviving eyewitnesses, however, and living descendants now know nothing at all about the events of that period. Most of the official documents detailing such transfers of financial compensation were destroyed during the troubles of 1960. The regions that appear to have been most affected were the lower slopes of the Rwenzori Mountains and the north-eastern base of the Mitumba Range (around Kasaka, Museya and Karurume in particular). In this part of the park, the boundary extended only to the base of the Mitumba massif, and not up to the peaks.

Throughout the 1950s, problems relating to land claims were the last thing park rangers ever had to deal. The few inhabited northern zones, such as the Watalinga Enclave, lay outside the reserve. In all the years I have spent roving the park, no Congolese associate of mine has ever said: 'This is the land of my forefathers; it is where they are buried (*Awa, adjali kabouli na tata na tata na bisu,* in Lingala). Nor, in all my wanderings, have I heard any Congolese national declare "The Park robbed us of our ancestral land"—or voice any claim to that effect'.

With hindsight, the absence of a definitive solution in the extreme south of the park seems understandable; for here—where the park reaches Lake Kivu (incorporating Tshegera Island and islets of human occupation)—extremist groups have long been in the habit of presenting claims to the authorities.

There undoubtedly exist, within the park's boundaries, signs of ancient temporary habitations, including graves, pit traps, the remains of forest villages, caravan tracks, and citrus fruits. These Bantu groups were not sedentary, however; after a few years of exploiting an area, they would migrate elsewhere. That some of the ancient names for their villages are still known today cannot be seen as proof of ownership, as the villages were no more than temporary structures. Even today, the region's forests are, for the most part, *terra incognita* (unknown territory); even the area's most respected trackers invariably got lost. It took me a long time to find the vernacular name for that huge expanse of water known as *Bwirina* Lake—an old meander of the Semliki.

In 1958, I had prepared, at the request of the management committee in Brussels, a detailed report on zones of the reserve that could be retroceded on account of relatively poor ecological interest. Fortunately, this report remained in the archives, unused. Congolese nationals who, much later, were shown this report were aghast: 'The park,' they exclaimed, 'must retain all of its 800,000 hectares!'

Minor claims, presented in years gone by, were seen as having been 'digested' a long time ago. There was no reason, even if it had become *politically correct* in 2008 to criticise the actions of the pioneers, to revisit such problems. This chapter was closed and—in the wider public interest—should remain so.

2.3 HISTORIC CHARACTERS AND EVENTS

The first personality to mention is the legendary American, Carl E. Akeley, genuine creator of the reserve, who died in 1926 in Kabara, in the middle of the *Hagenia* forest, 'the most beautiful place on earth'. He was buried there and a gravestone erected on the site kept his memory alive. His grave was subsequently robbed during the rebellions and other horrors. The other true creator of Albert Park was the illustrious Victor Van Straelen, director of the Institute of Natural Sciences, who was an uncompromising advocate of *integral conservation*. The Belgian King, Albert I, personally attempted the steep and demanding ascent of Mount Mikeno (4,400 metres above sea level), but was reluctant at first to lend his name to the National Park. King Léopold III, by contrast, turned down a subsequent request to have Garamba Park named 'Léopold Park', refusing to allow prominence to be accorded, in the naming of protected areas in the Belgian Congo, to a second member of the dynasty.

On becoming King of the Belgians in 1934, Léopold III, relinquished his position as president of the *Institut des Parcs Nationaux du Congo Belge*—a capacity in which he had been assisted by Victor Van Straelen, then the Institute's vice-president. The extraordinary speech Léopold gave in London, on 16 November 1933 remains topical even today, three-quarters of a century later. Here are some extracts:

If there is one subject of great magnitude, beyond human horizons, it is that of the protection of nature's eternal riches, for which we are temporarily responsible. Do we have the right to change the natural state of things, worrying only about the consequences that our current knowledge of phenomena allows us to foresee?...

If we contemplate the scientific aspect of the question, we notice that natural reserves constitute the essential continued work of laboratories. Until today, progress made in the area of natural sciences was a result of laboratory studies and work; only a small proportion was the fruit of directly observing aspects of nature... We want to flee our hectic towns. We feel imprisoned. We want clean air, light, space, earth, water and greenery: our generation has reconciled with nature...

This heritage, unique in kind, that nature has bequeathed us, involves a duty that is threefold. First, we have to see to its integral conservation with the utmost vigilance. But while ensuring this fundamental goal, we want to pursue the methodical exploration of our remarkable home. Which leads to our second duty: to make our institution actively contribute to the progress of knowledge. To this end, specialist assignments are dedicated to the study of scientific problems. Finally, without harming the principle of conservation in any way, we open certain parts of our reserve to visitors, since we cannot deprive humanity of such a source of emotions, joys and splendours...

It can be seen from these pronouncements that Belgium showed an interest in conservation in Congo long before modest conservation efforts were launched in Belgium itself. Until he died, in the early 1980s, Léopold III retained a keen interest in the development of Albert National Park, visiting the park on numerous occasions.

Queen Elizabeth of Belgium, who visited Albert Park at the beginning of 1958, was especially impressed by the 'indescribable beauty' of the birdsong. The author recalls having tried, in Ishango in February 1958, while the Queen was on the summit of a cliff overlooking one of the most beautiful places on Earth, to surround her with guards to protect her from possible lion attacks. Tellingly, her response had been: 'But risk, young man, is the spice of life.'

Other characters who played an important role in shaping the park include the genial conservationist Henri-Martin Hackars, who died in 1940, having covered thousands of kilometres on foot in order to establish the park's boundaries, which are almost as long as those of the whole of Belgium—in a venture the author of this chapter would go on to repeat a quarter of a century later. Then there was the Dane, Rasmus Hoier, who—left without orders from Brussels during the war—nevertheless stuck to the task, despite being forced, against his will, to give in to pressures concern-

2.2
Rangers assembled outside the Rutshuru station, which for many years was the headquarters of Albert National Park.

2.3
The start of one of many field expeditions mounted in the 1950s. Such expeditions called for plentiful supplies and considerable logistical support.

2.4
The visit of Elizabeth, Queen of the Belgians, in 1958.

Life in Albert National Park, 1925–1960

2.5
Park ranger Kanzaguhera, pictured at Rutshuru, in 1956.

Jacques Verschuren

2.6
One of thousands of field notes, complete with photographs, that were used to document the park's fauna and habitats. Accompanying this note is a photograph of African Wild Dogs taken by E. Hubert in 1938. Extirpated in the late 1950s, the African Wild Dog is the only mammal species on record as having disappeared from Virunga National Park.

ing the Vitshumbi fishing ground. There was Hubert, a first-rate populariser who, like the great naturalist G. F. de Witte,[1] was also an exceptional photographer. And finally there was Harroy, the future Vice-Governor General, who proved to be an efficient administrator of Albert Park before the war, replacing the conservationist Lippens, who would go on to establish the Zwin Reserve in Belgium.

No historical account of Congo's parks would be complete without a mention of the contributions made by Verhulst (in Kagera National Park), Ory (in Kagera and Garamba National Parks), and Bouckaert, Heine, Kint and Rousseau, all of whom were dedicated conservationists. Others who made a lasting impression on conservation in the region include: Major Van Coolst, future General-in-Command of the Belgian Army and Donis, who lobbied hard for the creation of new parks. More recently, Micha—now in his nineties—contributed much, as both senior warden and renowned cartographer, while also supervising the creation of neighbouring Rwanda's Kagera National Park. Count Cornet d'Elzius, currently in charge of the FFRSA *(Fondation pour Favoriser les Recherches Scientifiques en Afrique)*, was another major formative influence.

2.4 THE MANAGEMENT AND RUNNING OF THE PARK

One reproach for which there can be no excuse was the absence, before 1960, of any Congolese ranger on the staff employed in the park. Reflected in other sectors as well, this was in line with a general policy during the colonial era. Today, some of the actions of the Brussels management committee seem difficult to understand: one Belgian park official who, in the run-up to Independence in 1960, invited future leaders of the region to his home in Rutshuru to brief them on a visit he had organised for them to the National Park was given a stern reprimand! For all the power it wielded, the committee in Brussels was extraordinarily ignorant regarding African issues, despite the prestigious titles of some of its members. From a distance, most of the real management done was the work of just two people—the president Van Straelen and the Chief Administrator, Van der Elst—and of three or four of the executives working under them. Notable among the latter was De Saeger, mainspring of the Institute, a brilliant man known for his formidable efficiency who died only recently at nearly one hundred years of age. Nuyten too, and Houben especially, should also be singled out. After a lifetime spent managing the paperwork for the parks, Houben was finally able to visit Albert National Park in 1974—the fulfilment of a lifelong dream. At most, then, five people were responsible for managing the activities of thousands of operatives in the field, including rangers, guards and other workers. Later, a team of 50 or 60 employees was charged with supervising the parks from Kinshasa. Results were mixed, but there were a few notable successes: not the least, those of one exceptional man, Biwela, from Kinshasa. Worth mentioning is the set up in Kivu of a financial committee comprising prominent people from the province, including settlers and planters.

How were the parks run 'in the field'? The author first visited the reserve early in 1948, as a student collaborating with the ethologist, Hediger. First impression? That everything worked impeccably. The trip from Brussels had been extremely well organised, down even to including an interview with the Governor-General in Léopoldville, who had remarked: 'Lucky man! Not only are you going to visit a biological paradise, but—even if it is a bit of a state within the State—you will also experience the most efficient organisation[2] in the whole of Belgian Congo.'

Albert Park was divided into three sectors: the south, which was home to active and extinct volcanoes, with its base in Rumangabo; the centre, which included the southern plain of Lake Edward, with its

2.7
The lodge at Rwindi in 1934.

2.8
Mutwanga Hotel in 1950.

base in Rwindi; and the north, which protected the upper Semliki, the equatorial forest and the Rwenzori, and which was managed from Mutsora. This structure has never changed. In order to include all the management 'stations' in the park, some appendages were added to the original boundaries. One strange shape that emerged was that of the Tshiaberimu. Rutshuru station, an old management base built almost a century ago by someone close to the author, still belongs to the Institute, even if it is viewed with envy today by certain people in overpopulated North Kivu. Official documents supporting the IPNCB's land rights to the area were destroyed in 1960.

A warden managed each sector with assistance from a deputy and often a head of post. Many of the latter were former officers. Others were agronomists, or reserve officers. These administrators spoke the local vernacular: in Albert Park this was *Kiswahili*, albeit of a dialect different from the one spoken in East Africa. Some decisions were enunciated in French, but in most cases *Kiswahili* or sometimes *Lingala* took priority.

Every warden doubled as a Police Officer within a park's confines, which caused problems at times with the regional administration. Relations with these civil servants were for the most part cordial, however.

Strict surveillance and law enforcement were the main activities of a warden, who typically would spend two or three weeks a month in the field (camping, but relying too much on 4x4 vehicles to get around, and not enough on foot). Wardens also had to carry out administrative tasks, which they did with the collaboration of *karani*, native employees, often of high standing.

A warden compiled reports, and organised the construction of shelters and bridges. Every month, he had to send a detailed report to Brussels, including accounting details and biological observations. To begin with, European wives were few and far between. Parks seemed to be preserves solely for single people. Over the years, however, the presence of wives was tolerated, and then accepted. Some wives, such as Marie Huguette Ory in Garamba National Park, actively assisted their husbands. Others, often recently married, were never comfortable in wild environment and wished they could run away, within two weeks of their arrival in the bush. There were never any female wardens or rangers.

In most cases, relations between the wardens and their staff were harmonious. An expatriate's term was an uninterrupted 36 months, during which time there was no leave to return to Belgium. Often, expatriates were obliged to work at the same station for the entire period. One advantage was that they had time in which to get to the root of local problems. A disadvantage was that some wardens remained completely ignorant of other reserves. It was not a question of a 35-, 40-, or 80-hour week for wardens or for rangers: they were all on duty 24 hours a day. Once or twice a year, they would have to submit to an inspection by the big boss, Van Straelen, who would then analyse their problems. There was no managerial divide in those sectors of Albert National Park sectors that crossed both Congolese and Rwandan territory.

One of the wardens' duties was to receive authorities both from within the country and from abroad. In Albert Park, the task of visitor liaison fell to Hubert and Baert. Another activity, of far greater importance, was that of establishing good relations with the local people living on the outskirts of the park, even though vast stretches of land adjacent to the park were almost entirely uninhabited. For the most part, relations with the various local chiefs and elders were cordial (although it has become fashionable in 2008 to claim otherwise). Some chiefs demonstrated their loyalty by turning in poachers. Back then however, the Utopian 21st-Century idea of fostering community participation in the running of the park was simply not an issue. In reality, the park brought many benefits to neighbouring communities: well-paid jobs, free transport, admission to certain schools and clinics, and sometimes the licence to cull animals that left the park.

The mainstay of Albert National Park was undoubtedly its complement of rangers. Their number grew considerably over the years, exceeding 250 people at one point and rendering the park self-sufficient.

Jacques Verschuren

Besides rangers, the park employed workers at all levels, reflecting (as the following list shows) a remarkable diversity of skills:
- Assigned to visits
- Assistant biologists
- Assistant drivers
- Assistant rangers
- Blacksmiths
- Boatmen
- Boys
- Bricklayers
- Carpenters
- Catechists
- Cobblers
- Cooks
- Drivers
- Electricians
- Ferrymen
- Guards
- Guides
- Gunsmiths
- Heads of post
- Meteorologists
- Mountain carriers
- Nurses
- Orderlies
- Ordinary carriers
- Ordinary workers
- Painters
- Pit sawyers
- Plumbers
- Radio operators
- Ranger cadets
- Rangers
- Road menders
- Seconded military
- Tailors
- Taxidermists
- Teachers
- Telephone operators
- Track cutters
- Trackers
- Wardens
- Warehousemen
- Washerwomen
- Water carriers
- Welders
- Wood carriers
- Woodcutters

All these workers were housed, fed and equipped by the park.

A team of rangers, 'barracked' at each of the large 'core' stations, undertook regular 'heavy' patrols, in order to intimidate anyone breaking the law. There were, in addition, patrol posts spread at intervals along the park boundaries, each accommodating four or five rangers and one sergeant in housing built for them by the national park authorities. At these posts, the rangers and their families were almost completely self-sufficient. Each team patrolled its sector almost every day. Initially, there were few problems, but later, when poaching gangs started threatening the rangers' families, the authorities were forced into a change of policy, ruling that one, or in some cases two, rangers should remain at their posts during the day, to stand watch over the families.

The role of the roving sergeant, accompanied by one or two porters, was key. He would visit all the camps in his sub-sector at least once a month, bringing supplementary or repaired equipment to isolated rangers, as well as food and wages. A rotation system meant that rangers were transferred from post to post on a regular basis. Since most of their camps were located on the park boundaries, they and their families were often able to buy supplies from nearby villages. In some cases, their wives even had small plots of land nearby, where they could raise domestic animals, such as chickens and goats. The rotation system proved to be an effective way of pre-empting abuse – always a risk when any one team of rangers is allowed to mingle with a particular local community for an extended period. By and large, respect for the rangers bred and attitude of tolerance among local people towards wild animals they encountered outside the park: the baboons in Mosenda, for example, or the warthogs in Lunyasenge; the buffaloes then at large in Rwindi and in Lulimbi, and even the elephants that were in the habit, then, of wandering through both Vitshumbi and Rwindi.

Heavy surveillance patrols were launched without warning. As proof that the rangers were patrolling their sectors, they were obliged to perforate service cards with a pair of pliers, and to attach these to tree trunks in different areas of activity zones under their control. Each ranger had three kits: one for parade and two for use in the bush. The rapid wear and tear of uniforms was construed as a positive sign, implying that they had indeed been patrolling the forest, which tore their clothes. Back then, patrolling rangers carried only machetes and spears; later, they would be issued with rifles.

In staffing the surveillance posts, efforts were made to mix ethnic groups so as to avoid 'arrangements' that might lead eventually to corruption. Cases of rangers' either being 'bought off' or of killing animals, whether for food or to sell on, were unusual, however.

Poachers and other offenders captured by patrolling rangers were transferred to the central station. Typically, the rangers spent three weeks out of every month in the field. Well-paid, well-equipped and well-nourished, they came to be regarded as the aristocrats of Kivu and lawbreakers were terrified of them.

All rangers were given tents, raingear, boots and rucksacks. Most were literate and carried notebooks for recording observations. They had a vested interest, moreover, in managing their respective sectors effectively. Veterans were singled out for special care: Sergeant Bavukahe, hired when the park was created, died when he was almost one hundred years old. Widows, particularly those of rangers killed on duty, received preferential treatment. Old retired rangers were often given houses in which to live out their remaining days.

Large teams of rangers and porters were indispensable on hundreds of foot safaris that enabled scientists to explore remote and difficult habitats. Without their valuable assistance, much of the scientific research that we take for granted today would not have been possible. Individuals who deserve special mention include: the remarkable tracker, Senkwekwe; the rangers Joachim and Lubutu, and two of the *capita* (headmen) Kissa and Kasiwa.

Rangers and zoologists alike found themselves in what was both a paradise, on account of the environment, and a hell, owing to the Spartan living conditions. The string of hugely successful early scientific expeditions to this region is a tribute to the expertise of

2.9
A mountain hut in the Nyamulagira Crater, 1932.

the park's extremely well-trained rangers, and to the strict military discipline with which they operated as a force, even if, in 2008, it is politically correct to criticize this rigorous behaviour.

Unfortunately, the diligence of these rangers, who sometimes risked their lives while apprehending poachers and other criminals, was often not reflected in the 'justice' that was subsequently dealt with. All too often, those who had committed crimes would be released soon after their arrest. Many acts of poaching went unpunished. Some members of the judiciary still openly opposed the park's existence, while others – such as Goma's brilliant deputy prosecutor, Fontaine, whose support of Albert Park was always invaluable – persisted in seeking the scrupulous application and enforcement of existing anti-poaching laws.

2.5 OPERATIONAL FINANCING AND LOGISTICS

Going to the most conservative of estimates, the Institut des Parcs Nationaux du Congo Belge financed, in addition to a small staff in Brussels, about 20 wardens and other senior officers in Congo, along with at least 700 rangers and a further 700 workers. The costs of operating and supplying such a large workforce were immense, to say nothing of the scale of investment in infrastructure for the various national parks. The obvious question, then, is: who was paying for all this? From what I remember, the money came from Brussels, in the form of subsidies from the Ministry of Colonies.

After 1960, financial support was channelled through the Congolese Ministry of Agriculture in Léopoldville, then through the Presidency, and later trough the State Commission for the Environment, Nature Conservation and Tourism. Additional financial support came in the form of revenue from park visitors. Before 1960, foreign cooperation was almost non-existent, other than for lingering partnerships with the FFRSA *(Fondation pour Favoriser les Recherches Scientifiques en Afrique)*, and with the IRSNB *(Institut Royal des Sciences Naturelles de Belgique)*. Later, Germany and then the United States, Switzerland and the Netherlands all intervened, establishing the basis for wider international cooperation, both official and private.

The Albert Park's original infrastructure included, in addition to the three base stations – Rumangabo, Rwindi and Mutsora, all 'permanent' structures built to last 100 years – a number of ranger posts, some of which were also 'permanent' structures. Other solid buildings were constructed at the access barriers (Kasindi, Mabenga and Kakomero). Shelters were erected on the mountains (three on Nyamulagira, four on Rwenzori, one on Kabara), while other buildings were put up near Lake Edward to control the fisheries. The lakeside rangers proved to be accomplished sailors, which was just as well, since Lake Edward's waters can be treacherous. The Nyiragongo, Rukumi and Lulimbi shelters were built by the author in the 1970s.

A distance of at least 100 metres separated all permanent junior staff housing from that constructed for the park authorities. Sheds, clinics *(lazarets)* and schools were also built, and some of the people living outside the reserve would go on to benefit from these services. Permanent roads, though, were deliberately never built, and in the park tourist tracks were deliberately kept to a minimum. The Institute was not responsible for those routes that already crossed the park. To begin with, the main such route was the Goma-Beni road and its various branches. Later, in contravention of the law, a number of other roads and motorable tracks came into being, however. These multiplied uncontrollably after 1960, usually for reasons entirely unrelated to tourism. On the mountains meanwhile, there were some well maintained footpaths, although in many areas there were no paths at all.

Transport in Albert Park was a problem (and remains so in 2008). In the early days, people got around on foot. Each station generally had two trucks and two or three 4x4 vehicles, as well as a prestige vehicle for important visitors. The local drivers proved to be very capable and some were resourceful mechanics, able to repair even the most severe damage to vehicles. The surveillance stations had small garages, which would function independently of the larger installations (Salvi, Schalbroek) in Rutshuru and of course in Goma. Some of the drivers took to carrying passengers illegally in return for payment, and had to be tightly controlled. The efficiency of two technicians – Boné and Mazowa, both excellent drivers – was exemplary.

Several boats were used on Lake Edward for controlling the fisheries and for intercepting poachers. The Vitshumbi-Ishango boat trip allowed travellers to avoid the cumbersome road journey between Kayna and Beni, via Butembo, which held little of interest for naturalists. Shipwrecks were not uncommon on these deceptively calm waters. From time to time, hippopotamuses would overturn boats, causing the loss of human life. Inflatable canoes were rarely used.

In 1958, the author received permission – with Xavier Misonne – to travel down the middle and lower courses of the Semliki River in a vessel made from two dugout canoes joined at the sides. This expedition proved to be a somewhat perilous 'first', opening up access to imposing and unknown places, such as the extraordinary Sinda Gorges. Lower down on the Semliki, the crocodiles could be dangerous. Back then, the huge reptiles had not yet dispersed upstream to Ishango.

The use of aircraft, already common in parks across East Africa by the middle of the 20th century, was forbidden over Albert Park, in the name of… ecological purity: a good example, this, of how conservation could be taken to ridiculous extremes!

Communication between surveillance stations was another major problem, since back then radiotelegraphy did not exist. Instead, messages were transmitted by orderlies and by 'tam-tam radio'. A surprisingly efficient postal service, run by excellent Congolese technicians, went some way towards compensating for this weakness, however. P. O. Box 18, Rutshuru, assumed a quasi-legendary status. At one time, a letter sent from the heart of Kivu would get to Europe within just three or four days: a quite extraordinary feat!

2.6 The last days of Belgian control of the park

How did events unfold in Albert Park over the years and months preceding the 30 June 1960, Independence Day? Apart from the elections themselves, which were very calm (the author was an assessor in the bush at the time), almost nothing happened. North Kivu was fortunate in experiencing almost unbroken political peace until the end of colonisation (although of course it would start paying dearly for this 30 or 40 years later). No rioters ever confronted the author in North Kivu, where – in the years following Congolese Independence – the Public Forces were impeccable. Stone-throwing mobs of the kind seen elsewhere in Congo were unheard of in Kivu, whose inhabitants had little prior exposure to political issues, and who made no political demands.

Of far greater concern to park rangers at the time was that they saw as the shameful retrocession of the Tamira Zone, a Rwandan sector of Albert National Park. New moves aimed at paving the way for creation of the distant Salonga National Park were another pressing concern. Some rangers were sent to the 1958 Exhibition in Brussels, where they made a particularly strong impression. With just one notable exception, the park experienced no real upsurge in poaching, hunting or fishing in the run-up to Independence. I shall never forget having to swim across the torrential upper Rutshuru River – with a team of rangers, some of whom had to be helped across, as they were not used to the water – in order to capture the members of a poaching gang that seemed determined to eliminate the park's Giant Forest Hogs. Scientific research continued without incident, culminating in the discovery – on 29 June 1960, the eve of Independence – of the gigantic thermal sources.

Come Independence, some of the wardens' wives left the country of their own free will. Claims to the effect that the colonial administrators fled the country 'in a panic', sending all their personal effects and even some scientific collections overseas, are grossly inaccurate. The true picture was mundane by comparison: life went on peacefully, surveillance operations continued and Albert Park remained intact.

For 35 years, between 1925 and 1960, Albert Park was under colonial administration. Since Independence in 1960, the Congolese authorities have managed all the reserve's activities. And the consensus, to date (some 48 years later), is that they have – through difficult times – managed this vital aspect of their heritage astonishingly well.

On a historical scale, the troubles of the past decade and the damage they caused, while undeniably dire for conservation in the short term, may yet prove to be reversible.

No historical account of Albert National Park[3] is complete without reference to the core principles upon which the park was founded. And to this end, it is important to remember, in closing, that the park's sole objective at the outset – that of safeguarding Kivu's long-term ecological future, at all costs – has never once been compromised. The motivation behind the creation of Albert Park was indeed the conservation of exceptional fauna and biotopes, scientific research and, to a lesser extent, tourism. So, whereas it is easy, in retrospect, to find fault with some of the early methods used, the park's very existence today must be seen as a triumph for the vision and determination of its founding fathers, in their commitment to preserving an area of unparalleled biodiversity.

---

1 Shortly after the end of the war, in 1946, de Witte was sent on a mission to Albert Park to reaffirm the Belgian Institute's involvement and interest, after five years of isolation.
2 At the time l'INEAC (subsequently INERA) became established as a model institution. Shortly after that, l'IRSAC was established.
3 The author would like to suggest one change in terminology. The name 'Park' is not suitable. It is reminiscent of animal parks, the Royal Park or Woluwé Park in Brussels or even industrial parks. A better term would be 'Nature Reserve'.

Jacques Verschuren & Samy Mankoto

3 Rebirth of the National Park (1960–1991)

Jacques Verschuren & Samy Mankoto Ma Mbaelele

The history of the national parks in Congo before 1960 has been the subject of many detailed chronicles and analyses. Yet very little has been written on the subsequent history of these parks, over the crucial period extending from Independence in 1960 to the present. In this chapter, the authors hope – in covering the period 1960–1991 – to go some way towards correcting this imbalance.

3.1   The road to Independence

In Congo, as with many other African countries, Independence would have a huge implication for the management of national institutions, including national parks. Albert National Park experienced a tumultuous period just after Independence, followed by a revival, and then a slow but ineluctable decline of State support, subsequently eased in part by growing interest and support from the international community.

At Independence on 30 June 1960, the National Park was 35 years old. Predictions regarding its future were far from optimistic. In the words of one French journalist: 'The first thing to go in the Congo will be the national parks'. Yet, this view, as it turned out, was completely mistaken…

Until the final hours of the colonial period, Albert Park had never known a major problem. There had been very little poaching and what hostility there had been to the park's existence was minimal. Scientific research had flourished and there was a growing tourism industry. The park was fortunate, then, in being located in a region that, politically, was fairly stable. After the colonial flags were lowered on 30 June 1960, the park rangers at their various posts continued to dispatch their duties as before.

Attacks from outside the region began on 10 July 1960, with the mutiny of soldiers in the State Forces. The soldiers had counted on support from the rangers in perpetrating their savage excesses. They were mistaken, however; for the rangers having enjoyed 30 years of exacting discipline, were unaccustomed to anarchy.

The resulting stand-off took a tragic turn when, in the months that followed, the rebels – in a frenzy induced by drugs and alcohol – committed a wave of atrocities, forcing the rangers to leave. The wider Kivu population remained calm, however. The author, who chose to remain in the area, suffered repeated harassment at the hands of the rebels, but received nothing but support from the local population.

One date in particular stands out as historic: on 27 July 1960, between 200 and 300 rangers from all parts of the park (including even isolated posts) gathered at Rwindi to share their concern over the collapse – now apparent to all – of the park authority. Some, a minority, had become resigned already to the park's imminent disintegration and demise. The intervention at the meeting of the new administrator for Rutshuru territory, Habarukira, was to prove decisive, however. At great personal risk, given that he was surrounded by armed men, Habarukira delivered a passionate speech in both *Lingala* and *Kiswahili*, punctuated by the resounding refrain: 'The park will never die. Long live the national parks of the independent Congo.' From this, the rangers in the crowd understood that Albert

3.1
The ferry crossing at Nyamushengero on the Rutshuru River provided a link between the Rwindi sector and Lulimbi.

3.2
The burial in November 1960 of park ranger Valère Sauswa, murdered by Ugandan poachers.

National Park was to be reborn.

The need for such a crisis meeting, and for Habarukira's enlightened intervention, might have been averted, had the colonial authorities taken steps, prior to Independence, to nurture a cadre of local Congolese park wardens. The fact that at Independence there were no trained Congolese wardens and no Congolese department in Léopoldville from which to coordinate park management left a vacuum at the top. In making sure that control did not fall into the wrong hands, the rangers' meeting of 21 July 1960 effectively saved Albert National Park. The Brussels Management Committee, 7000 km away and oblivious to events on the ground, had in the mean time been acting as though it were still in charge of the Congo Parks!

Had this immense protected area foundered in the wake of Independence, when the eyes of the world were trained upon the Congo, the precedent—for conservation across parts of Independent Africa—would have been disastrous. The wildlife reserves and the national parks of States would doubtless also have collapsed. Fortunately, this did not happen, and the Congolese rangers, of whom some are still alive in 2008, quickly went back to work.

To all intents and purposes, the parks had been saved. Indeed, the first official act of Goma's District Commissioner, Ruyange, was to appoint a Congolese head warden, Mburanumwe, an agronomist.[1] Then, three national wardens were appointed. Munyaga, a man of conviction long since deceased, and Bakinahe and Kajuga, who in 2008 after 45 years of service (a record), are both still working in Virunga National Park. The legal basis for these appointments rested in the laws enacted before Independence, which were still being applied, and which also formalised park boundaries and other parameters. Kivu's new Provincial Governor, Miruho, representing the President of independent Congo, officially sanctioned the appointments.

Soon afterwards, the Department of Waters and Forests in Léopoldville made the revival of Congo's national parks official Government policy. The Agriculture Director, Biwela, who for a decade thereafter showed enormous courage in overseeing the affairs of the parks, deserves special praise, along with his colleague Mokwa. Management input from Brussels diminished steadily and then stopped altogether. That a smooth transition was achieved at the highest levels of government was to the dynamism of Van der Elst. It is fitting, here, to recall the death, in 1964, of Victor Van Straelen, who—as Director of the Royal Institute of Natural Sciences in Belgium, and as President of the parks in Congo from 1934—had been the Albert National Park's principal architect, creating the 'state within the State' that for many years was the envy of scientists and conservationists the world over.

In 1960, 1961 and 1962, the park rangers in Congo remained active and highly motivated. Many 'died for the elephants' in the course of fighting poachers, who were mostly Ugandans. Sergeant Sauswa was the first ranger to die in the service of protecting nature. The death of his colleague, Saambili, followed hard upon.[2]

Amid renewed rebel activities, the Mutsora warden, Buni, was heard—before he died—defiantly telling his murderers: 'I am willing to die for this Park, but the Park will never die'. The attacks, though, were not aimed at the park itself, but at authority in general. The same was true of subsequent rebel attacks, between 1990 and 2005. Life for the staff at the parks was far from easy, and there were serious financial constraints as well. In the absence of funding subsidies from Brussels, local operating budgets had been slashed.

Channelling what little funding did become available to remote and isolated sectors of the park was an extremely hazardous undertaking. The delivery in 1961 to the park's Volcanoes Sector of money that Biwela, the Director, had allocated in Léopoldville, together with private Belgian donations, called for a major operation involving a convoy of official 'smugglers' and an escort of Indonesian bodyguards from the United Nations. At the time, amid a new wave of rebel attacks, warden Mburanumwe and others (including the author) were accused of 'selling the park to the UN' and might easily have been condemned to death. The beleaguered staff in the Volcanoes Sector, meanwhile, were convinced that they had been forgotten—until, at last, the convoy got through and they could be paid.

Three wardens lost their lives in combat, or in active service against poachers: Buni in Mutsora, Mburanumwe in Rumangabo and Bilali—a warden in the central sector—on 23 October 1975. It seems that an aggressive hippopotamus, wounded by the poachers, caused Bilali's death. There is a monument at May-ya-Moto, erected in his memory.

3.2 FOREIGN INFLUENCES

In the vacuum that followed the end of the Belgian administration and the exodus of colonial park officials, there were attempts—originating in neighbouring Uganda, then still a British colony—to rename the Albert Park. The new name, Kivu National park, was proposed. The emphasis at the time, however, seemed to be on securing hunting concessions in the area! The efforts of the settler community proved invaluable in resisting such pressure. Some of the settlers cared deeply about 'their' reserve. Rwanda, which remained under Belgian administration until July 1962, concentrated on protecting Kagera Park, whose new young warden, De Leyn, had been murdered by political bandits. The Rwandan sector of Albert National Park, modest in size, would go on to become the *Parc des Vol-*

cans (Volcanoes National Park). The Albert Park's official handover, from the Belgian park authorities to their Congolese successors, was to have taken place at the end of July 1960, but the envisaged ceremony – to have been held on the Rwanda-Congo border – was aborted after major differences arose. Of the 'old guard', only the courageous Rwindi hotelier, Ballegeer, continued to operate, along with the author (Verschuren), who had become a park advisor, and who continued his surveillance and research safaris in the bush.

A hugely influential presence at this time was that of the American biologist George Schaller, who – while studying and protecting the gorillas – was instrumental in saving an entire sector of the park. At Arusha International Conference at the end of 1961, the global scientific community congratulated the Congolese authorities on having preserved the park and on having stuck to a policy of integral conservation – even at the cost of many human lives.

The renowned German conservationist, Prof Bernhard Grzimek, was a key influence in shaping the park's future. And in 2008, German cooperation was still evident in nearby Kahuzi-Biega National Park. The American, Harold Coolidge, of the US National Academy of Sciences (and later the New York Zoological Society), was another who contributed much to the renewal of Congo's parks.

The acclaimed British biologist, Julian Huxley, an avid promoter of conservation projects around the world, visited the Albert Park in 1961, but then controversially proposed that its policy of integral conservation should be discontinued.

That the Congo parks were able to withstand the traumatic events of 1960–1961 without forfeiting any land or being overwhelmed by poaching was a triumph in itself. Recovery, though, over the period 1962–1968 was severely hampered, owing to a lack of financial resources at a time when administration of the parks was the responsibility of a department in the Waters and Forests Service of Congo's Ministry of Agriculture, based in Léopoldville-Kinshasa.

3.3 Towards a revival of the national park

By 1968, conditions on the ground had deteriorated: the rangers no longer possessed uniforms; transport was non-existent. And, incredibly, surveillance operations – albeit greatly scaled back – were still being carried out. Teams of wardens and rangers had remained at their posts and were bravely defending the reserve, some as if their lives depended on it.

While still a priority in Léopoldville-Kinshasa, local political pressure was weakening Albert National Park. The illegal installation, for example, of a large fishery at Nyakakoma, which still exists in 2008, annihilated the unique xerophile forest bordering Lake Edward. Attempts to get rid of this installation proved fruitless. The impact of uncontrolled fishing from dugout canoes, even in the fish breeding grounds, was severe. The illegal fishing was then largely the work of semi-militarised gangs from Uganda.

The findings of a 1968 survey of the area were very worrying. Herdsmen from Rwanda with thousands of cattle had invaded the park's Gorilla Sector (Mikeno-Karisimbi). Their presence, while not irreversibly damaging the forest, was symptomatic of the breakdown of official contact between the Congolese and Rwandan park authorities. In the Central Sector, meanwhile, there was evidence of heavy elephant poaching, although this had not yet reached 2000 levels. Yet, in July 1968 the author counted 5272 hippopotamuses on the Rutshuru River (Central Sector), an increase of 1300 on the number counted ten years earlier. The Northern Sector, poorly controlled, was the park's most severely impacted area. There, large numbers of squatters had exterminated thousands of hippopotamuses, although again nothing like the scale of the slaughter of 1990–2005. The Rwenzori and the equatorial forest zones were not monitored at all at the time.

In 1969, the situation deteriorated further, despite the best efforts of both the directors and the wardens. Reports sent to the Congolese Head of State elicited a firm promise to the effect that drastic measures would be taken to revive the country's 'natural jewel'. By Order No. 69/041 of 22 August 1969, the President created the *Institut National pour la Conservation de la Nature* (INCN) – the National Institute for the Conservation of Nature. This order was later revised and re-issued in the form of a new official document – No. 72/02 of 21 February 1972, clarifying certain elements.[3]

Another official Order, passed in August 1969, confirmed the author's appointment as General Director of the new Institute. Before taking up the position, the author spent three weeks flying over the country in the presidential plane. His report on this reconnaissance ended with the verdict: 'Situation serious, but not in the least desperate.'

The new administrator, who had almost dictatorial powers, answered directly to the Head of State with, as occasional intermediaries, Kayenga Onzi Ndal, and later Dr Muema. Mobutu's Head of Cabinet, Bizengimana, played an ambiguous and sometimes harmful role with regard to the principles of integral conservation. The Congolese State Presidency financed the new Institute almost entirely. It also supervised the tackling of environmental problems through a Ministry of Environment: a first for Africa. Mrs Lessedjina was appointed the head of the new Ministry, following a 1972 Order.

Punishments for poaching became heavier. Hunters' dogs, which often carried rabies, were shot. Another far-reaching measure saw to it that rangers were no longer required to prove legitimate self-defense for having used their weapons against poachers. They were authorised to fire after issuing three warnings. The results were immediate – and positive. Courageous rangers who had been languishing in filthy jails for having apprehended criminals were immediately released and congratulated.

The Institute (afterwards referred to as IZCN – *Institut Zaïrois pour la Conservation de la Nature* – and later still as ICCN – *Institut Congolais pour la Conservation de la Nature* –) had its headquarters in Kinshasa, initially on Rue Dreypondt and subsequently on the Avenue des Cliniques. A number of other, mostly superfluous, regional administrations were later established. Having to liaise with a head office in faraway Kinshasa was never easy, despite advances in radiotelegraphy communications. To have established a general administration in Rumangabo (Kivu), at the heart of the country's oldest protected area, instead of 800 miles away would have made far greater sense. From 1969, the General Director embarked on a struggle against bureaucratic

3.3
An encounter between an elephant and a Landcruiser at Rwindi.

3.4
Virunga National Park has always been the jewel in the crown of Congo's network of protected areas. This promotional leaflet shows the Mikeno Volcano.

parasites who routinely absorbed as much as one-third of the global budget for the parks, while making no contribution whatsoever to conservation. Priority was given to fieldwork, which was supposed to engage 90 percent of all staff. Field staff were given very strict instructions and were closely supervised. Here are some examples:

Your primary activity, that of inspecting the bush, has not been carried out. Any warden found to have spent fewer that eight days of each month on foot and out camping in the bush will be demoted to a lower level.

Some of you are parasite rangers: you never go out on patrols. Patrolling for less than eight days a month in the field is a serious offence: dismissal will be immediate.

Do not beat up the poachers whom you capture, since they must be punished according to the law. Wardens are entitled to reward rangers for exceptional performance, through raising their monthly bonuses.

Patrols will be intensified on, and on the eve of, all public holidays. On 24 and 31 December, all rangers will remain in the field until 10:00 p.m. On full moons, night monitoring will be stepped up.

In what state is the annex of Mahangu's hut? Has it been rebuilt? Have you discovered any illegal farming activity near Mount Tshiaberimu? Any gold diggers? How many fishermen have you evicted from Nguli?

Be firm with foolhardy tourists, but otherwise maintain generally tolerant attitude. Straying from the set trails isn't always wrong. Burn and bury confiscated weapons. Seek the permission of village chiefs, then gather villagers and explain to them the usefulness of the park. Be sure to remind them of the benefits of the reserve. Never compromise with people who have broken the law, but try always to exercise the power of persuasion; we have to live in harmony with the national park's neighbours.

A 'hands-on' approach to management, introduced at this time and providing for direct problem-solving actions (as opposed to the ponderous, bureaucratic machinations of previous years), greatly speeded up the rehabilitation of park infrastructure. The General Director, on one occasion, noticing that a bush clinic had become run down, was able – through advancing 3000 US dollars from his own pocket, enough to finance half the work – to ensure that reconstruction began immediately. In this way, infrastructure and equipment that in some cases had been damaged and out of service for years was repaired within a matter of weeks. Repairs to vehicles alone (boats and aircraft included) increased the park's available transport capacity ten-fold.

Over the first three years, 1969 to 1972, the General Director acted alone, but, from 1973 he was assisted by a management committee. Despite cordial relations among the committee's members, this commendable initiative impeded direct action, as interminable discussions, agreements with bureaucrats and appeals for contributions again became the order of the day. A distressing immobility replaced direct management.

At the end of 1969, the General Director and Mokwa, his assistant, attended the IUCN general assembly in New Delhi. There, as in Arusha eight years earlier, Congo was congratulated on its positive conservation actions. In relation to the Albert Park, the Congo Government received particular praise for upholding integral conservation, a fundamental principle since the park's inception.

A period of sustained economic growth in Congo during the early 1970s breathed new life into the country's national parks. Indeed, such was the upsurge in park activities that in 1972 an Assistant Director General, Gahuranyi Tanganyka was appointed, as part of a dynamically expanding Congolese Institute for the Conservation of Nature. Influential figures at the time included Feyerick, Geurden, Lecrenier, Rollais, and – above all – Powis de Tenbossche, a remarkable if sometimes unpredictable character. And then of course there were Congolese Secretaries – Gasigwa, Magunga and Bobelhabi – who were a good deal more effective than some of their expatriate colleagues in Kinshasa.

The incidents of July 1960 excluded, the Albert Park had – come 1976 – been the subject of no post-Independence claims from neighbouring communities. The stand taken by the park's keenest local sympathiser, the great traditional chief Daniel Ndeze, was still widely respected. Ndeze, who lived in Belgium for many years, next door to the Tervuren Museum in Brussels, had offered up his lands to King Albert.

The battle against construction of illegal roads in the park remained a priority throughout the 1970s. There were, however, two major roads the park authorities were obliged to tolerate: one, known as *de Monseigneur*, extending from the heights of the *Nande* region into the Semliki Valley, near the Kasaka and Museya Rivers; and another – of later construction – cutting right across the park, bisecting the lava plain just south of the Rutshuru latitude and leading to Tongo.

3.4 FIRST SUPPORT PROJECT WITH BELGIAN COOPERATION

In 1970, a special Cooperation Agreement signed with Belgium contributed a little towards the investment and operating costs of Albert Park and other Congolese reserves. The agreement provided for the financing of the General Director's position as Head of Project, and for the hiring of five leading foreign experts: Willy Delvingt,[4] future university agronomy professor; Jean-Pierre d'Huart, scientific researcher; Alain Jamar, pilot; André Letiexhe, indispensable technician (having known Congo since 1950); and Jean-Pierre von der Becke, administrator. On being integrated into the IZCN management structure, each of these specialists worked closely with one Congolese (now Zairian) counterparts.

This project's many achievements included the launch of the IZCN magazine, *Léopard,* of which five issues were published between 1971 and 1977; the building of infrastructure (tourist huts, the Lulimbi research station, landing strips, patrol posts, trails and bridges, and so on, in all the parks (Letiexhe); original studies on the ecology of hippopotamuses (Delvingt) and of the Giant Forest Hog (d'Huart); the introduction of integrated park management (Mankoto); the organisation of tourism (von der Becke); the ringing of migratory birds (Delvingt, Bagurubumwe and d'Huart); the completion of periodic censuses of large mammals; the setting up of a reference herbarium; and the training of Congolese colleagues. Samy Mankoto ma Mbaelele was then a researcher at Lulimbi, who later – after completing his thesis on *Ecological problems of the Virunga National Park* at Laval University (Quebec, Canada) – returned to Congo as IZCN Scientific and Technical Director, before becoming PDG *(Président Délégué Général)* of the institute in 1985.

The project was to have lasted until 1975. Yet there were problems, arising in part from the Zairianisation policy of 1973, which – in the wake of Mobutu's change of the country's name to Zaire in 1971 – sought to nationalise all senior Government appointments. These problems did not compromise the project's achievements, however. And the agreement, while limited in scope, proved to be a model for expanded future national-international cooperation.

One less fortunate initiative, for which the author now feels responsible, was the introduction of a small ferry service on the lower Rutshuru River. Initially with visitors in mind, the intention was to open up access to the east of the plain, but alas the ferry was soon being used by invaders as well (in providing passage into the Nyakakoma and Lulimbi areas). The harmful consequences of this project, subsidised by WWF-Belgium, would not become apparent for some time. Then in 1972–1973, the Lulimbi station and research laboratory was built on unsuitable ground, subject to frequent heavy flooding, while being open to human invasion as well. As the canker lying at the root of all illegal settlement in this part of the park, the Lulimbi laboratory – still in place in 2008 – should be dispensed with altogether.

A number of promising young Congolese park officials, including Norbert Mushenzi, were sent to Garoua (Cameroon) and then to Mweka (Tanzania), to complete specialised courses at the International Institute for Warden Training, where they performed with distinction. By the end of 1976, a total of 19 Congolese wardens had completed this supplementary training. Congolese delegates were present at the IUCN General Assembly in Banff, Canada, and at the second Congress on National Parks at Yellowstone in the USA. Congo invited these global bodies to celebrations held in Kinshasa and Rwindi in 1975 to commemorate the fiftieth anniversary of Albert National Park.

In August 1970, the President of Congo invited Baudouin, the King of Belgium,[5] and his wife Fabiola to Albert Park. The two heads of state then spent nearly a week in the park. Both men were amazed by its vast expanses of unspoiled wilderness. After several subsequent visits, the President ordered that improvements be made to the Rwindi Lodge.

Christiane Linet, then President of WWF-Belgium, was exceptionally active at the time (and over later decades) in drumming up international support and recognition for the Congo parks. Indeed, it was she the Congo parks would have to thank for receipt of their first WWF-International gold medal for conservation excellence.

The status of those Pygmies living in the park was re-examined between 1970 and 1975. Previously tolerated as forming part of the forest ecosystem, some Pygmies were found to have made the transition from subsistence hunter-gatherers to active poachers, engaged in supplying outlying communities. A motion calling for their relocation was approved, but this was never enforced. Management of the Rutshuru Hunting Reserve, meanwhile, was entrusted to Albert National Park by Departmental Order No. 024 on 14 February 1974.

3.5 THE ZAIRIANISATION PERIOD

In 1974, amid strident Zairianisation and a general lack of discipline (although not on the part of the rangers), the General Director was obliged to participate in several rounds of talks aimed at pacifying potential rebels in North Kivu, while at the same time pursuing his scientific research. There were exchanges in central Semliki, where several known poachers were killed. And, while it may be regrettable to have to spill the blood of poachers, it should be remembered that, without recourse to such drastic measures, it is usually the rangers who give their lives.

In 1974, there were in Congo's parks 714 rangers monitoring 22 areas or territories. In 1975 the Cooperation Agreement with Belgium was broken off, but this did not stop scientists – both national and foreign – from carrying on with their activities. No one, come the beginning of the 1980s, could foresee the extent of the deterioration that lay in store for Virunga National Park. Several General Directors came and went. One turned out to be so corrupt as to engage in brazen large-scale poaching operations. He, though, was eventually arrested and ended up in prison. The rangers, by contrast, remained dedicated to their monitoring activities, refusing to be drawn into corrupt

3.5
The station at Mutsora, here pictured in the early 1990s, is an example of the park's original infrastructure, dating back to the 1930s. Solidly constructed, these buildings have proved remarkably durable. They are currently being rehabilitated with funding from the European Union.

3.6
A certificate issued in 1985 to commemorate the park's sixtieth anniversary and signed by the then Zairian Minister for the Environment, Conservation and Tourism, was given to every tourist who visited the park that year.

practices. The result was that, in 1975, Virunga National Park commanded overwhelming public respect and recognition throughout North Kivu.

3.6 The second phase of the Belgian cooperation project

Delays over signing the agreement between the AGCD (then Belgium's International Development Agency) and the government of Zaire meant that the second four-year phase of the Belgian cooperation-financed project could begin only in 1978. Another team of technical experts arrived. Based at Lulimbi, they were to support the management of Virunga National Park while carrying out their research. One of the newcomers, Jan van Gysel, was interested in the vegetation dynamics of the plains south of Lake Edward. Another, Mireille Vanoverstraeten, was interested in the pedology of the same region; while a third — Hadelin Mertens — was to concentrate on documenting ungulate population sizes and dynamics.

All this support was aimed at obtaining enough scientific data upon which to base a comprehensive plan — the first — to manage and equip the park. Sadly, this objective was never met, and the third project's envisaged phase never went ahead, on account mainly of a cooling in relations between Zaire and Belgium, following disagreements over aspects of Zairianisation.

The two phases of the AGCD support project nevertheless succeeded in significantly advancing scientific knowledge of Virunga National Park. Their other major contribution was a marked acceleration in the development and honing of local skills, through the training of Zairian researchers and technicians. With financial support and an improved infrastructure, tourism was able to recover, as news of the park's enormous potential began to spread. Film and television crews, university exchange programmes and visits from world-famous personalities all contributed to raising the park's international profile, and in 1979 UNESCO declared Virunga National Park a World Heritage Site. For the first time,

it became apparent at the end of the 1970s that tourism could attract enormous revenues, not just for the National Park, but also for the country's entire network of protected areas. Zairianisation, though, by putting brakes on foreign investment and allowing the recruitment of management staff to be based on loose technical criteria, slowed the development of certain aspects of tourism.

3.7 The FZS/WWF/IUCN programme (1984–1991)

Management difficulties in Kinshasa, coupled with the withdrawal of Belgian support, left Virunga National Park in need of a boost. Help, fortunately, was soon forthcoming, in the shape of the FZS/WWF/IUCN project. This had been negotiated in 1983 for implementation from April 1984, and coincided with some welcome changes in the IZCN's top management. The FZS project was able, with support from the WWF, to make huge strides in developing gorilla- and, later, chimpanzee-viewing tourism. A dramatic increase in visitor numbers saw Virunga National Park regain its status as one of the world's most appealing tourist destination. The financial implications were considerable. With more than 6500 visitors a year (for the gorillas alone) and with fees now payable in US dollars (rather than in the local currency, the Zaire), the IZCN was rejuvenated. Such was the potential for further growth that in 1988 the European Union incorporated the Virunga Programme into its larger Kivu Programme.

In 1985, Virunga National Park celebrated its sixtieth anniversary. Ceremonies took place in Kinshasa and Rwindi to mark the occasion. The Head of State again emphasised the park's 'vital importance' to Zaire.

In 1987, the FZS/WWF/IUCN's Gorilla Programme was expanded to include a chimpanzee component. With ZSF funding, a group of chimpanzees was habituated near Tongo, both as a tourist attraction and for purposes of environmental education. The WWF's Virunga Education Programme (PEVi, which later — in 1996 — was renamed the Virunga Environmental Programme), came into being at this time and remains active today. The International Gorilla Conservation Programme that — with AWF, FFPS and WWF involvement — would go on to play a decisive role, was then still in its infancy. In 1988, it was through the Kivu Programme that gorilla tourism developed.

3.8 The Kivu Programme financed by the European Union (1988–1991)

The European Union-financed Kivu Programme addressed a wide range of development-related activities. Priorities included improvements to the road infrastructure and expansion in both the health and agriculture sectors. There was, however, also a component focusing specifically on Virunga National Park. The aim was to build further on what had been achieved under the FZS/WWF/IUCN Programme and with earlier Belgian development (AGCD) support.

The Virunga Programme had the backing of four international experts, with Conrad Aveling as principal technical adviser. It committed funding support worth seven million ecus over four years to the development of park infrastructure (Rwindi, Rwenzori mountain huts and trails among others) and upgrading of park equipment, while boosting local community service amenities and livelihoods in a partnership with the WWF-run PEVi Project.

This programme helped IZCN to bolster the international image of Virunga National Park. The tourism industry flourished accordingly, amid glowing reports from some of the park's more illustrious visitors. Until the end of the 1980s, President Mobutu continued to invite fellow heads of state and other dignitaries to join him in his favourite park on visits that, inevitably, attracted much additional publicity.

A number of important studies and works of reference were completed and published during the course of the Virunga Programme. These included: Mackie's censuses of hippopotamuses (1989); Aveling's elephant and buffalo censuses (1990), the excellent *Guide du Parc National des Virunga* by Delvingt et al. (1990), and Vakily's groundbreaking study of the productive potential of Lake Edward (1989). For other researchers (ourselves included), these works have since proved invaluable.

The late 1980s marked the end of an era–and the beginning of a slump from which, 20 years on, the country is still trying to emerge. The rate of inflation, early in the 1990s, reached 3000 percent a year. Looting by the army was rife, in a climate of increasing corruption. Socio-economic and political decline inevitably followed. September 1991 became a milestone in the history of both the country and Virunga National Park: the Kinshasa riots signalled the end of cooperation with most Western countries and multilateral donors. The result was a premature withdrawal of the Kivu Programme, its Virunga component included. Subsequent events have dashed efforts, in 2000 and 2005 respectively, to celebrate the seventy-fifth and eightieth anniversaries of Virunga National Park. Some, determined to ensure that the Centennial in 2025 cannot be overlooked, are reportedly already working on the commemorative arrangements!

3.9 IZCN INTERNAL FACTORS IMPACTING ON THE EVOLUTION OF ViNP

3.9.1 Context

At the beginning of the 1980s, the morale of IZCN field staff–that of rangers in particular–was at its lowest ebb. The IZCN President Director General (PDG) had then just been arrested, on a ministerial order and imprisoned over the illegal sale of two tons of ivory to the son of an influential politician. Poaching was then rife in protected areas; despite the best efforts of the under-equipped and underpaid rangers, who looked on aghast as judges released one poacher after another, making a mockery of the trial process.

Further north, in Garamba, there had been a catastrophic decline in Northern White Rhinoceros *(Ceratotherium simum cottoni)* numbers: from a high of around one thousand to 250 in 1971, then to 11 in 1984. Faced with this catastrophic situation, the IUCN General Assembly meeting, in Madrid, resolved to send a high-level delegation to Kinshasa. Led by Prof Jean-Paul Harroy, First Secretary-General of the IUCN when (as IUPN) it was created in 1948, this first delegation–comprising Prof Jeffrey Sayer, Charles Mackie and Dr Kay Curry-Lindahl–duly travelled to Kinshasa, where it met President Mobutu in the presence of Victor Njoli Balanga, then Zaire's Minister of the Environment and Nature Conservation. The President declined the offer of relocation of rhinos proposed in the Madrid resolution, promising instead to 'pull out all the stops' in order to ensure the survival of the rhinos *in situ*.

It is in this context that Samy Mankoto, aged 38, was appointed head of the IZCN in April 1985, becoming the youngest ever Institute's General Director and President of the Administration Council. Cynics at the time referred to Mankoto as *petit pdg* (little director general)!

3.9.2 Institutional review, reorganisation and revival of the Institute

From the outset, PDG Mankoto embarked on some important reforms, using–as his launchpad–an impressive celebration held in Goma in October 1985 to mark Virunga National Park's sixtieth anniversary. The ceremony, featuring parading rangers sporting new uniforms and carrying modern weapons, succeeded in mobilising all Goma's inhabitants. The day itself had been declared a public holiday. The festivities were paid for by North Kivu's Vice-Governor, Koya Gialo Basete Gerengo. The gathering was remarkable, above all, for the many guests of honour who were present. In attendance, besides Ministers from

3.7
The late Roger Minne, Norbert Mushenzi and the late Adrien Deschryver are decorated during the ceremony held to mark the park's sixtieth anniversary.

3.8
At the park's sixtieth anniversary ceremonial parade in Goma are the ICCN's board of directors, including–on the left–General Director Mankoto and behind him, from the left, Joseph Mokwa Vankang-Izmithsho, Dr Thibassu, the late Biwela M'Finda and Albert Ngezayo, Prof Joseph Lumande Kasali, the late Anicet Mburanumwe Chiri wa Rutezo. Missing is the late Colonel Powis de Tenbossche.

3.9
Another milestone in the park's history came with the global recognition it received in 1979 on being declared a UNESCO World Heritage Site.

the Central African Republic, from Congo-Brazzaville and from Zaire itself, were ambassadors from several countries and a contingent of prominent scientists and global representatives, including Prof Jean-Paul Harroy; Prof Alexandre Prigogine; Dr David Kabala Matuka of UNESCO; the Director-General of the *Office Rwandais du Toursime et des Parcs Nationaux* (ORTPN) and Dr Jacques Verschuren, in his capacity as former IZCN General Director, as well as all the members of the Institute's then serving Administration Council and General Management staff. There had been an important national conference in Kinshasa beforehand, at which experts from several countries were invited to participate. Convened under the patronage of the President of the Republic, the conference was opened by Zaire's First State Commissioner (Prime Minister) Kengo wa Dondo. Animated exchanges prompted a review of fundamental problems besetting conservation in Zaire and the wider region, culminating in a tabling of new policy recommendations and strategies. After the conference, 150 people were transported by plane to Goma, and then driven to Rwindi. This logistically complex operation was accomplished without a hitch, thanks to perfect coordination between the PDG General Services and Protocol team, under Simon Sivi dia Yamba N'Suka (otherwise known as 'the bulldozer') and the State Protocols.

The Virunga Park's sixtieth anniversary celebration in 1985 was a triumph for IZCN. Another celebration, held in 1989 to mark the fiftieth anniversary of Garamba National Park, further boosted the profile of Zaire's parks in the international media. For all the country's parks—four of them already UNESCO World Heritage Sites (Virunga, Garamba, Kahuzi-Biega and Salonga)—this was the beginning of a new era: that of rehabilitation.

The results were not long in coming: Garamba's rhinoceros population tripled to 33 individuals in 1995. 'Zaire has saved the rarest animal on the planet...' one newspaper headline proclaimed. On the strength of this recovery, Garamba was taken off the list of Endangered World Heritage Sites in 1994. In Virunga, notorious poachers—including the feared *Mzee Tembo*—were arrested and transferred to Makala Central Prison in Kinshasa, far away from their home environments. The arrests followed the successful outcome of the legal battle enjoined by the IZCN General Management, with support from both the Minister in charge of the Environment at the time and the Prime Minister, who was an acclaimed lawyer and university Professor. News of Mzee Tembo's 'relocation', in particular, created havoc among poachers in the region, some of whom—in a surprising twist—resorted to furtively dumping their weapons in churches, particularly in Mutsora, as proof of their having renounced poaching.

Keen to speed up the restructuring and reinforcement process, PDG Mankoto ordered an institutional audit of the IZCN. Through Dr Jean-Pierre d'Huart and Roseline Beudels, he made contact with the European Commission in Brussels, presenting his project to Enrico Pironio, whom he had met in Rwindi in 1987, and to Tincani, the desk officer. After extensive data gathering, an institutional review was carried out between 1990 and 1991. There were two major components: an institutional study (organisation, management, human resources) financed by the World Bank, and another study on the IZCN's accounting and financial organisation, focusing on revenue-generating activities (ecotourism, management of properties such as the Rwindi Hotel), which was financed by the European Commission.

The review identified the following priorities:

a at the General Management level:
- Enhancement of existing ecotourism collaborations (with WWF-International, WWF-Belgium, WWF-USA, IGCP, GTZ, WCS, the UNESCO-IUCN-Frankfurt Zoo consortium and the European Union) in respect of ventures capitalising on the appeal of the Great Apes (Gorillas in Virunga and Kahuzi-Biega and chimpanzees in Tongo). Steps would have to include the introduction of a policy to keep track of revenue and of a system for doing RME (Foreign Currency Resident) accounting in compliance with National Bank regulations; the introduction of policy governing the distribution of revenues (taking into account both bonus incentives for guides and rangers and re-investment in village micro-projects benefiting communities living on the edges of National Parks); and formulation of a permit system under which film-making in the parks could be taxed.
- Reinforcement of technical and scientific staff capabilities, through using funding secured from donors to send young researchers abroad for training, or—in the case of rangers, assistants, and even some officers—to the Garoua Wildlife Training School in Cameroon;
- Computerisation of all accounting functions, to be followed by the recruitment of young graduates to be drawn from the Higher Institute of Commerce especially, who would be proficient in both use and all aspects of bookkeeping, accounting and financial management. The IZCN had, as part of this fundamental upgrade, already received a donation from the FZS in the shape of two minibuses for transporting General Management staff in Kinshasa (a most considerate gesture, under the circumstances).

b at the regional and field levels:
- Implementation of a decentralisation policy provid-

ing (as of 1991) for administrations and for retrocession of revenue to the parks following an adjustment – of 60 percent (parks) versus 40 percent (general management) – agreed by Administration Council;
- Nomination of regional directors, who (from 1991) would be accorded certain powers by the IZCN Administration Council;
- Launch of a programme to expand the training opportunities available to rangers following the establishment in 1986 of the Lulimbi Scientific Training Camp and issue of the Institute's first rangers' training manual. Early training sessions (1986) had lasted for three months. The instructors had included park wardens and senior officials from Rumangabo, along with high-ranking officers from Kinshasa, seconded through the Head of State and Commander-in-Chief of the FAZ (*Forces Armées Zairoises*, the national army). Colonel (later General) Kalume Numbi, who would go on to become Planning Minister under the Third Republic, had been responsible for coordinating the para-military aspects of the training sessions. The Kabaraza Elite Mobile Troupe, commanded by Lieutenant Fofolo, came from Lulimbi. Training sessions would conclude with an official ceremony in Lulimbi, at which representatives of the FAZ top brass, having travelled from Kinshasa with the PDG, would award diplomas to the successful candidates.
- Introduction of a policy of dialogue with local communities, with an emphasis on developing understanding between young people and the authorities through the Virunga Education Programme (PEVi). Concrete measures would include the broadening of existing community reafforestation projects and the launch, with WWF-GTZ funding support, of the popular *Kacheche* magazine. The WWF's audio-visual vehicle in Goma, the first of its kind, had produced psychological impact and was a resounding success in North Kivu. Indeed, this vehicle was then the envy of MOPAP (the Mobilisation, Propaganda and Political Promotions wing of the MPR – the State Party).

c at the national level:
- With annual revenues to the order of USD 650,000 generated mainly by the Virunga and Kahuzi-Biega National Parks, the IZCN would be upgraded from category E of public enterprises to category C, giving the Institute the same status as commercial Gecamines for example, and making it a source of great pride to the government;
- Inauguration of an IZCN contract-programme – complete with performance criteria and indicators, set by the Portfolios Higher Council. This important document was signed in 1994.

d at the international level:
- the globalisation of the IZCN, making it a role model for conservation management in other parts of Africa, notably Madagascar;
- Signing of the first framework agreement between the DRC and the WWF. This took place on 1 March 1990 in the offices or the Foreign Ministry, and coincided with the appointment of Dr Hadelin Mertens as first resident Representative of the WWF in Zaire. The wide political and legal scope of the agreement led to a marked strengthening of international cooperation with the IZCN. Earlier, during the twenty-fifth anniversary proceedings of the WWF in Assisi (Italy) in September 1986, the PDG had succeeded in getting Zaire rated among the 'priority countries' for WWF activities in Africa;
- Expansion of existing networks and partnerships. The World Bank Forests and Environment Project, the European Union, UNESCO, IUCN, WWF, GTZ, ZSF, NYZS (later WCS), IGCP, GIC and the Milwaukee Zoological Society had all helped elevate the IZCN's international profile, while contributing important extra-budgetary funds for Virunga National Park and other protected areas;
- Consolidation of the climate of trust that lay at the heart of all collaborations between the Institute and its international backers. The IZCN's willingness to host major international meetings, not only in Kinshasa but in parks in the interior of the country as well, had won many high-placed friends in conservation circles around the world. A regional seminar-workshop on training and orientation of rangers in the National Parks and protected areas, held in Rwindi in August 1989 under the stewardship of the Secretary of State for the Environment, attracted more than 60 delegates, including representatives from Burundi, the Republic of Congo and Uganda, among other African countries.
- Acceptance of the IZCN as an IUCN member with Public Law Organisation status.

3.10 Conclusions

This chapter has outlined the changes in the fortunes of the park from independence to the early 1990s. In particular, it showed how the determination of individuals, principally in the park, but also in the capital enabled Virunga to survive some of the most challenging years in its history. Also important, was the sustained financial support from the state, and from international donors. The success of this effort provides important lessons in the wake of the recent conflicts, and should be seen as a source of encouragement for those currently struggling to protect the park. The recent restoration of park land (see Chapter 25) and the evidence for a slight increase in animal populations (Chapter 8) are the outcome of the efforts of the team in the park, in Kinshasa and by ICCN's international partners. Time will tell if the conservationists on the ground will receive the necessary support to fulfil their work.

1   It was Mburanumwe who first championed the culture of nature conservation reflected in the traditional customs of elders in the region
2   Before 1970, 23 rangers 'died for the Parks'. Hundreds of other such deaths were to follow, including that of ranger Paluku Dunia – murdered by rebels in July 2005, while the authors were working on the draft text for this chapter.
3   On 22 July 1975, Zairian law 75-023 made the Institute answerable to the Department of Environment, Nature Conservation and Tourism: a serious step backward after the promising early years of the Mobutu Presidency.
4   In 2006, Delvingt energetically took charge of the Bombo-Lumene Reserve, near Kinshasa, where the capital's dwellers can discover nature, while d'Huart, now an independent expert after 23 years of service with the WWF, was in 2008 still actively engaged in resolution of conservation problems in Africa.
5   King Baudouin's father, Léopold III (who was Belgium's king from 1943 until his abdication in 1951), visited Albert National Park on several occasions, having developed a very keen interest in the park his father, Albert I, had created.

José Kalpers & Norbert Mushenzi

4 The crisis years (1992–2006)

José Kalpers & Norbert Mushenzi

From the beginning of the 1990s, Virunga National Park faced a succession of crises. Several armed conflicts took their toll on the park's biodiversity and on the staff responsible for its protection. This chapter reviews the period 1992–2006, describing the effects of these conflicts and outlining the restorative and preventative measures the national authorities and their conservation partners adopted.

4.1 The management of protected areas in situations of armed conflicts.

Even in times of peace, the management of protected areas over much of the African continent is a perpetual challenge. Fragile institutions lacking the support of national authorities, mounting demographic pressure and rural poverty are all factors that impose major constraints on the sustainability of conservation activities in sub-Saharan Africa. Sudden crises, such as armed conflicts or natural disasters, place exceptional demands on the managers of protected areas and on their institutions and partners. Globally, the impacts of armed conflicts on protected areas can be classified as follows (Blom, 2000):

1 **Direct impacts on biodiversity:**
a The parties to a conflict directly and deliberately damage the environment: this destruction usually has a strategic goal (as in, for example, the clearing, deforestation or defoliation of territory in order to limit – or prevent – ambushes and access to infiltrators).
b Natural resources are used to finance military operations, as in the extraction of ivory or rhino horn, for example.
c Refugee populations, fleeing combat zones, seek refuge in camps in or near protected areas, from which they extract resources in order to survive.

2 **Indirect impacts:**
a Logistical constraints: situations of armed conflict hinder conservation activities, sometimes causing human losses among surveillance personnel.
b Financial constraints: donors and partners often discontinue their support in conflict situations. Tourism revenue is heavily affected.
c Changes in priorities: in times of conflict, national authorities and supporting agencies tend to prioritise humanitarian aspects, at the expense of environmental and biodiversity conservation issues. Attention and resources are re-focused accordingly.

Before 1991, Virunga National Park already had serious structural problems (Kalpers, 1996). One major problem was the weakness of the Zairian Institute for the Conservation of Nature (IZCN), the official agency responsible for the management and conservation of protected areas in Zaire. This institutional weakness, reflected in the IZCN's general management in Kinshasa as well as in the field stations, was largely the result of a lack of technical and logistical resources, and was aggravated by the socio-economic and political crisis that had plagued the country for many years.

4.1
The forests in the southern sector of Virunga National Park suffered massive deforestation for fuelwood during the refugee crisis of 1994 to 1996.

Declining socio-political stability led to the withdrawal in September 1991 of most biodiversity-related projects, imposing a raft of constraints.

The following sections describe in detail the impacts of the various armed conflicts and crises on the biodiversity of Virunga National Park between 1990 and 2006, and show how these impacts shaped the park's development.

4.2 Unrest in the Northern Sector of ViNP (1986–1994)

The Northern Sector was the first area to be affected by armed conflict. From 1986, unrest – at first involving poaching gangs – flared up in the Rwenzori foothills. From 1990, when President Mobutu declared a three-party government and then a multi-party government, several rebel groups, based in Watalinga and near Mount Hoyo, formed in the region. Patrol posts were attacked, several rangers and members of their entourages were killed, and looting was on the increase.

At the same time, there was an upsurge of poaching in order to feed the rebels and finance their activities: elephants and buffaloes were slaughtered *en masse* and, near Ishango, the hippopotamus population was almost wiped out. Elephant and hippopotamus meat and ivory was trafficked locally and exported to Uganda. In 1991, assault rifles became increasingly available and attacks began on personnel from the IZCN and the park. After a relative lull, the situation worsened in 1994, with several raids on patrol posts in Mwalika and Vimbao, as well as on the Ishango sub-station. At this time the IZCN's Technical Administrative Director, Dr Simon Lulengo, was killed when his vehicle was blown up by a land-mine.

The IZCN gradually lost its grip on the northern reaches of the park, over which different rebel groups, plundering the region, eventually assumed control. This marked the beginning of a long period of insecurity for the Northern Sector of Virunga.

Further south, the IZCN retained control of the park's territory. However, the socio-economic decline, and particularly the lack of good governance, resulted in cases of external poaching, as well as poaching from within the IZCN, and serious direct threats on the wildlife. Thus, contrary to popular belief, by far the most significant loss of hippopotamuses (18,000 animals killed) dates from the period preceding the armed conflict (Languy, 1994).

The situation led UNESCO to carry out an evaluation in October 1993. The conclusions of this evaluation saw Virunga National Park included on the list of Endangered World Heritage Sites. This step was based largely on a report written before the Rwandan genocide and the subsequent refugee crisis, even though the formal decision was not made until the end of 1994 (UNESCO, 1995).

4.3 The war in Rwanda (1990–1994) and its impact on ViNP

The guerrilla war in Rwanda not only caused enormous damage to the conservation of biodiversity and to protected habitats in the region of the Virunga massif, but it also seriously affected the border regions.

From January 1991, the date of the first offensive by the Rwandan Patriotic Front (*Front Patriotique Rwandais*, FPR) in the northwest of Rwanda, and the first Ruhengeri attack, use of the Virunga massif for military operations began. Military chiefs were quick to appreciate the immense strategic importance of the Virunga massif, as the only forested area spanning Rwanda, Uganda and Zaire. As such, this area offers armed groups dense cover and a good line of fire. Here, soldiers can move freely, without attracting the attention of their enemies. In 1991, military operations affected both Volcanoes National Park in Rwanda and Mgahinga Gorilla National Park in Uganda. The FPR quickly adopted a tactic of skirting around Sabinyo Volcano (a mountain bordering on all three countries); as a result, military operations affected the Mikeno sub-Sector of Virunga National Park on Zairian territory. Between 1991 and 1994 the Virunga massif was subjected to repeated infiltrations of FPR troupes and patrols of the Rwandan Armed Forces (*Forces Armées Rwandaises*, FAR). The military presence in this vast forest intensified, as the numbers and actions of the rebel forces increased, along with those of the regular army. The FAR established permanent positions in strategic areas, along the edges of the forest and in cavities between the volcanoes in eastern Virunga (Sabinyo, Gahinga and Muhavura). Several hundred mines were placed on and around the massif, most of them on paths leading into the forest, as well as along the length of the Rwandan-Zairian border.

Despite the enormous impact of the 1994 genocide on the Rwandan population, the environment in general, and the protected areas in particular, were not severely affected during the one-hundred-day-long massacre. When the FPR made their final offensive, which liberated Rwanda in July 1994, several thousand people took refuge in Zaire, using the paths and trails through Virunga, and bringing with them large numbers of livestock. This massive exodus marked the beginning of one of the most acute humanitarian crises of the century, and was to have a dramatic environmental impact on Virunga National Park.

4.4 The Rwandan refugee crisis (1994–1996)

In July 1994, over just a few days, nearly two million people left Rwanda and took refuge in neighbouring countries. Most came to Zaire. In just one day, on 15 July, about 500,000 people crossed the border and arrived in Goma, where – in the days following – they were joined by 300,000 others. The refugees hoped the town of Goma would meet their basic needs for water, firewood and food: essential commodities that were all available in and around the southern part of Virunga National Park.

In July 1994, three refugee camps were built: Kibumba, Mugunga and Katale. Rwandan refugees continued to pour in. Humanitarian organisations built two more camps at Lac Vert and Kahindo (late in 1994 and in early 1995 respectively). Come the end of 1994, the refugee population was estimated at around 720,000 (Delvingt, 1994). In the years that followed, management of Virunga National Park was dominated by the urgent human needs of three quarters of a million refugees.

The refugee camps of the Goma region remained on the edge of the national park for two years. It took a new armed conflict to bring about their closure. By then, the presence of the camps had left many lasting impacts on the park.

4.2
One of many refugee camps that were established in 1994 along the national park boundary.

4.4.1 THE MAIN IMPACTS OF THE REFUGEE CAMPS

4.4.1.1 DEFORESTATION

Deforestation was one of the most visible and most documented impacts of the refugee crisis. Humanitarian agencies provided food and shelter for the refugees, but the refugees had to do the cooking. The cutting and collection of firewood rapidly became a major threat to the environment. Trees were cut down for firewood, for construction or even for commercial purposes: for example, the large-scale production of wood charcoal became a flourishing business. At the beginning of the crisis, an average of 40,000 people were going into the park every day in search of firewood (Tombola & Sanders, 1994). Deforestation soon became even more pronounced, however. On some days, as many as 80,000 people entered the park, cutting around 1,000 tonnes of wood daily (Languy, 1995; Henquin & Blondel, 1996). Over the 27 months that refugee camps existed on the edge of Virunga National Park, deforestation continued to intensify apace, particularly in the Nyamulagira sub-Sector (encompassing the active volcanoes).

The impact of the deforestation can be gauged from the following figures (Henquin & Blondel, 1997). Two years after the arrival of the refugees, 105 km^2 of forest had been affected, of which the razed zone equivalent — reflecting total deforestation — amounted to 63 km^2. Table 4.3 shows details of the deforestation by camp.

From a qualitative point of view, at least two thirds of the deforestation took place in forests on the lava plains – zones of relatively low biodiversity when compared, for example, with the sub-alpine virgin forest in other northern areas of Virunga National Park. Moreover, at least 50 percent of the zones that were either razed or seriously affected belonged to young forests composed of pioneer species in the first stages of re-colonising lava flows.

The most severe long-term damage was observed in the Mikeno sub-Sector, where some important tracts of prime forest in the Kibumba Camp zone were destroyed. Stands of the conifer, *Podocarpus latifolius*, in the mountain forest were particularly badly affected.

In the zones around the Katale and Kahindo Camps (accommodating 290,000 refugees in total), levels of deforestation remained relatively low in the first year, probably because the humanitarian agencies, when they set up the camps, had introduced programmes to protect the park.

By contrast, forested zones adjacent to the Kibumba Camp were subjected to intensive deforestation during the first year. Measures taken during the second year of this camp's existence proved decisive in protecting the zone's vital ecosystem from further damage; so much so that by 1996, the deforestation had virtually ceased.

The zones around the Mugunga and Lac Vert Camps (together sheltering some 200,000 refugees) suffered particularly high deforestation rates. Not only were the refugees in the zones amassing large quantities of fire-

4.3 Deforested areas two years after the arrival of the refugees

Area	Katale-Kahindo (2 camps)	Kibumba	Mugunga-Lac Vert (2 camps)	Total 5 camps
Area affected	14 km^2	35 km^2	56 km^2	105 km^2
Razed area equivalent	6 km^2	15 km^2	42 km^2	63 km^2

wood for their own personal use; they were also feeding a vast commercial enterprise that had been set up to sell firewood and wood charcoal to Goma (Languy, 1995). In the absence of any protection or security presence, this commercial activity flourished, with result that deforestation levels continued to rise throughout the second year.

A study carried out over two years (Henquin & Blondel, 1997) was able, through extrapolation, to calculate the total extent of the deforestation in the park for which the refugees had been responsible over the two years and three months of the camps' existence.

Zones affected by deforestation
in the Virunga National Park   Approx. 113 km^2
Zones completely razed   Approx. 71 km^2
Zones razed equivalent   Approx. 75 km^2

4.4.1.2 The cutting of bamboo

The illegal cutting of bamboo *Arundinaria alpina* was mainly organised and carried out by the Kibumba Camp refugees (Bremer, 1996). South of Lake Edward, bamboo is found in the park only in the upper reaches of the Mikeno sub-Sector where it forms an essential Mountain Gorilla habitat, and in limited areas of Mount Nyiragongo. The refugees had various uses for bamboo. These included the building of shelters and the weaving of baskets and mats. There was even a project—the brainchild of one of the international Non Governmental Organisations (NGO) working in the camps—under which the refugees were encouraged to make handicrafts from bamboo. It had not occurred to the NGO in question that the bamboo was coming from the park! In all, 192 hectares of bamboo were harvested in the Mikeno sub-Sector, where 50 percent of the affected zone was razed, meaning that one in every two saplings was cut (Henquin & Blondel, 1996).

4.4.1.3 Poaching

The two-year refugee presence in the region saw poaching intensify in the two southern sub-sectors of the park (Biswas & Tortajada-Quiroz, 1996). The poachers' main targets were bushbuck and duikers, but they would also poach forest buffaloes and elephants (Wathaut, 1996).

In July and August 1995, poachers killed four Mountain Gorillas—three male silverbacks and one adult female (Cooper & Cooper, 1996). This was the first gorilla massacre in Virunga in ten years (Weber, 1989). While there may have been no direct link between the presence of the refugees and the poachers who committed these acts, it is likely that this type of poaching was a consequence of the general state of disorder and insecurity attending the refugee crisis (Werikhe et al., 1998).

4.4.1.4 Disturbance due to refugees in transit

Of the hundreds of thousands of refugees who fled Rwanda in July 1994, several thousand crossed the Virunga massif to reach Zaire. Some even lived in the forest for several weeks, before emerging out on the Zairian side. Many made the journey with livestock—cows, goats and sheep. All this traffic had repercussions on the forest ecosystem; it also increased the risk of transmitting diseases to the wildlife.

4.4.1.5 Dumping of medical waste

During the first year of the camps, and especially during the first months, many refugees required urgent medical treatment. This produced significant quantities of medical waste. Some of the organisations providing medical support disposed of this waste simply by dumping it in the park. The waste included used syringes, human waste and materials containing human blood (Biswas et al., 1994). During the second year, the dumping of medical waste was largely halted.

4.4.1.6 Security

On Zairian territory, security deteriorated markedly during the period of the refugee camps. The presence of tens of thousands of former soldiers from the Rwandan army, who had brought modern weapons with them, greatly compromised the efficiency of the IZCN forest ranger patrols. Groups of the heavily armed former soldiers would confront and threaten the IZCN field staff. The IZCN lost control of two zones in the southern part of the park, one near the Mugunga-Lac Vert Camps in the Nyamulagira sub-Sector, and the other at the heart of the Mikeno sub-Sector (a zone located between the Mikeno, Karisimbi and Visoke Volcanoes).

4.4.1.7 General state of disorder and insecurity

Even in normal times, some people commit crimes. A deterioration of security invariably gives rise to a surge in illegal activity. In the Virunga region, some local people were quick to take advantage of the new opportunities the state of general disorder presented. The intensification of both poaching and production of wood charcoal are but two examples (Werikhe et al., 1997).

4.4.1.8 Collapse of tourism revenue

Due to the presence of a large number of refugees and the turmoil shaking the region, tourism was seriously affected for the duration of the crisis. Tourist visits to the gorillas continued however. Some visitors continued to observe the gorillas in Djomba, an excellent access point into the Virunga National Park for tourists.

4.4.1.9 Shortage of available natural resources for local communities

Much of the deforestation seen during the refugee crisis occurred in the park (Werikhe et al., 1997). Plantations forests outside the park were also damaged, however. In a region that already lacked firewood, the presence of the refugee camps not only aggravated the situation, but also threatened long-term capacity to meet the energy needs of the local population after the crisis.

4.4.2 Relief measures and conservation strategies

This section describes positive relief and conservation actions taken by humanitarian agencies intent on curbing the environmental deterioration sparked during the refugee crisis.

4.4.2.1 Supply of wood (Henquin & Blondel, 1997)

Wood supply operations were financed mainly by the United Nations High Commission for Refugees (UNHCR), in collaboration with certain executive agencies (including the *Gesellschaft für Technische Zusammenarbeit*, the German technical cooperation body for sustainable development, GTZ). Quantities of wood supplied to refugees in the camps tripled between July 1994 and July 1996. On average, 50 percent of the energy requirements of the camps were met during the second year. Available data suggest that environmental programmes developed over this period by various partners – not the least the planting of three million trees under the umbrella of the WWF project – allowed 4,000 hectares of forest to be saved in two years. In other words, had there been no intervention on the park's behalf, the scale of the deforestation would have been 1.65 times greater than it was (Blondel, 1997).

4.4.2.2 Regular meetings and coordination among the interested parties

From the beginning of the refugee crisis, the GTZ collaborated with the UNHCR's technical arm to create, in Goma, a pilot environmental unit (the Environmental Information Office) capable of collecting and disseminating information on the environment (Delvingt, 1994). At the beginning of 1995, the UNHCR incorporated a new official structure through creating the post of Environmental Affairs Coordinator (Leusch, 1995).

4.4.2.3 Improved wood stoves and other energy-saving measures

Several agencies, in particular the GTZ and the International Federation of the Red Cross, encouraged the use of improved wood stoves and practices for saving energy in some of the refugee camps (Delvingt, 1996). Aware that the application of improved techniques would not automatically reduce energy consumption, the GTZ did not focus only on improved wood stoves, but was responsible too for popularising other methods of saving energy. GTZ staff showed local populations how to dry, store and prepare firewood, as well as more efficient ways of preparing and cooking food.

4.4.2.4 Education and awareness measures

Education campaigns aimed at refugees in camps around the southern parts of Virunga National Park were carried out with the assistance of several international and local NGOs.

4.4.2.5 Protection and surveillance measures

The IZCN, the organisation officially responsible for the management of Virunga National Park, had been reduced to a relatively weak institution prior to the refugee crisis. Initiatives supporting IZCN operations in the southern sector of the park helped to raise the Institute's profile, while strengthening its ability to apply the law and to carry out its general surveillance work.

Several hundred soldiers were placed at the disposal of the Zairian Special Presidential Division (DSP). All the costs of the surveillance operations – including employee bonuses, communications, transport and operational expenses – were covered by the UNHCR.

4.4.2.6 Rehabilitation phase in the DRC (after November 1996)

After the Rwandan refugee camps were finally closed in November 1996, a number of agencies launched rehabilitation programmes in the region. The UNHCR embarked on an ambitious plan in northern Kivu targeting different sectors, including the environment. Many local organisations participated in this programme, helping with reafforestation and setting up tree nurseries. The UNHCR also took a hand in conservation projects run by long-term partners of the IZCN, the WWF Virunga Environmental Programme and the International Gorilla Conservation Programme (IGCP). Its involvement in the planning and execution of a number of joint rehabilitation activities came to an abrupt end in October 1997, when the region's new political authorities forced the UNHCR to stop all its operations in the Goma area. What rehabilitation occurred was limited in scope. Projects aimed at repairing hospitals, roads and markets had to be abandoned, amid frequent outbreaks of violence. Yet, despite these difficulties, both the ICCN – the *Institut Congolais pour la Conservation de la Nature*, as the Institute was called from now on – and the IGCP (financed by the African Wildlife Foundation, Fauna and Flora International and WWF), found alternative sources of funding and were able to continue their activities.

Another programme, the Virunga Education Programme – a WWF project – received funds from the UNHCR to intensify its reafforestation activities around the southern part of Virunga National Park and to continue with its community education and awareness programmes. This support allowed the WWF project

4.4
The area around Kilolirwe, in the western part of Nyamulagira sub-Sector, suffered some of the worst damage seen in the park during recent crises. This photograph shows cleared fields well inside the park to the east of Kilolirwe; only in some of the valleys do a few isolated strands of high-altitude forest remain.

to produce and plant – through different partnerships – nearly one million trees in 1996 alone.

The International Gorilla Conservation Programme, meanwhile, concentrated its rehabilitation efforts on the park and, in particular, on the restoring the abilities of the management authority, the ICCN. This programme contained several components (Kalpers, 1998) and was essentially aimed at surveillance and the reinforcement of institutional abilities (Kalpers & Lanjouw, 1998).

4.5 CIVIL WARS IN CONGO-ZAIRE (1996–2003) AND UNCONTROLLED ARMED GROUPS (2004–2006)

The two civil wars that shook the DRC had a number of direct impacts on Virunga National Park.

In October 1996, directly following the Rwandan refugee crisis, Congo-Zaire entered a protracted period of civil wars. The first 'liberation war' culminated in the taking of Kinshasa in May 1997 by the Alliance of Congo Liberation Forces (AFDL). After a period of relative calm, a second war broke out in August 1998 and went on until November 2003, affecting both North and South Kivu, as well as Ituri, especially badly.

ICCN staff bore the brunt of these savage wars. Many employees were killed or wounded by armed groups from all sides (Mushenzi, 1996). The infrastructure in the southern sector of the park (Rumangabo station and patrol posts in the Mikeno sub-Sector) was seriously damaged (Werikhe et al., 1997), either by local populations in search of construction materials, or by armed groups moving freely through the region inflicting systematic destruction.

Several gorillas were killed by armed groups or poachers (Cooper & Cooper, 1996). However, taking into account the seriousness of the various crises then shaking the region, it is almost a miracle that there were not more gorilla losses during this period. The gorillas may have benefited from the fact that local people generally do not eat gorilla meat, and from the perception of gorillas as an asset in view of their economic value for tourism.

Illegal activities such as wood-cutting and poaching were routine among the various military forces then present in the region: rebels, government troops, and even the forces of foreign allies.

Military units, along with several thousand refugees, carried out agricultural activities in the Mikeno Sector, growing exotic plants such as potatoes, tobacco, wheat and hemp (Rutagarama, 1999).

There was also strategic deforestation. The Mwaro ecological corridor, which links the Mikeno and Nyamulagira sub-Sectors, suffered badly from strategic deforestation. Several animal populations, including elephants, used the corridor to migrate from the extinct volcanoes to other parts of Virunga National Park, in the DRC. By clearing the verges of the Goma-Rutshuru road in order to limit the risk of ambush, the military forces controlling the region changed the situation. Initially 20 metres wide, the cleared verges were quickly broadened (to 50–70 metres in places) by local people, sometimes with the complicity of military officers. This had disastrous consequences for the ecological thoroughfares that elephants and other dispersing animals had always used in their seasonal movements. A pervasive insecurity continued to hamper conservation efforts. Armed rebel groups had occupied (and, in some cases, were still occupying, even in 2008) certain forest areas, particularly around the Mikeno Volcano and the Gatovu, Kibumba, Bukima and Bikenge patrol posts.

The military seizures, coupled with a dramatic decline in tourism revenues further eroded the authority of the ICCN, which was powerless to intervene.

In the Northern Sector of the park, the arrival in February 1997 of the Alliance of Congo Liberation Forces (AFDL) complicated an already unstable situation. All the IZCN staff were disarmed, ambushes and looting raids were common place, and the rebel army set up a training centre in Nyaleke. In January 1999, the Kasindi-Lubilia site suffered a serious blow when a car park was constructed for vehicles crossing the international border between the DRC and Uganda, and when several areas of park land were cleared for farming. Since December 2000, the Northern Sector of Virunga National Park had been the subject to continual incursions and attacks. Its limits had been violated in several places (Kyavinyonge, Kanyatsi, Bulongo, Balombi, Lume, Mayangose, Mavivi, Lubilia, Kasindi, Tshiaberimu); Hima shepherds had installed themselves and several thousand head of livestock, along the banks of the Karuruma River, and some poaching gangs were still operating freely across the sector. Only recently, in April 2006, were the problems of invasion largely resolved.

4.5.1 Relief measures and conservation strategies

In April 1999, representatives from several leading conservation organisations gathered in Naivasha, Kenya, for a seminar on Endangered World Heritage Sites in the DRC. Participants included government bodies such as the Congolese Institute for Nature Conservation; non-governmental organisations such as the WWF, the Wildlife Conservation Society, the Gilman International Foundation and the International Rhino Foundation, and groups from the development sector, including UNESCO and the GTZ. While providing a platform on which to assess the status of Endangered World Heritage Site-listed protected areas in the DRC, this seminar allowed delegates to devise strategies for the future conservation of these sites and to develop appropriate joint action plans (ICCN, 1999). The seminar attracted international attention and led to the drafting of an emergency funding proposal, which was sent to the United Nations Foundation (UNF) by UNESCO (UNESCO, 1999). This proposal was accepted in November 1999, but the project did not officially begin until June 2001. The project, named UNF/UNESCO/RDC, provided essential logistical and institutional support over a four-year period to the five Endangered World Heritage Sites in the DRC, including Virunga National Park. The aim of this programme was to garner financial and diplomatic support and to focus the international community's attention on the five sites – both with a view to enhancing the protective effort and in order to show the world what lessons can be drawn from the consequences of armed conflicts on a threatened environment. The budget for the first phase of the programme amounted to more than USD 4-million, of which around USD 2.9-million was taken on by the United Nations Foundation (Bishikwabo, 2000). Now in its second phase, with a budget of USD 5.6-million over four years, the programme provides for the direct involvement of all the ICCN partners, which actively collaborate over the administration of funds, coordination of technical inputs and mobilisation of supplementary funding for specific tasks.

4.6 Conclusion: the toll of 15 years of armed conflict in Virunga National Park

This section provides a brief analysis of interventions and other factors that have contributed to the success, or failure, of measures taken to limit or repair the damaging effects of a succession of environmental emergencies.

4.6.1 Concerns regarding security

The most serious threat, still only partially resolved in 2008, is the high level of insecurity in the park and its surroundings. This insecurity has prevented the ICCN from exercising control over the entire park during times of crisis. It has also thwarted efforts to lessen the environmental impact of the refugee camps.

4.6.2 Intersectorial collaboration mechanisms

Under normal circumstances, it is relatively easy to forge collaborations between the development and conservation sectors. In the Great Lakes region, several projects have been able, very quickly, to draw on such collaborations. Theirs has been a stable operating environment, however, without armed conflict and social chaos.

Traditionally, relations between the emergency relief and development sectors, whether in war situations or when discussing intervention strategies, have always been good. This has not been true, however, of relations between the conservation and emergency aid sectors, where collaborations have at times proved delicate, to say the least. Humanitarian agencies such as the UNHCR, the International Committee of the Red Cross (ICRC), Oxfam and Médecins Sans Frontières are, it seems, ever more sensitive to the question of incorporating environmental components into their intervention programmes.

Shared objectives, though, do not necessarily produce effective collaborations in the field, and least of all in crisis situations. Thus, at the beginning of the refugee crisis in Zaire, there was an obvious lack of coordination among the different organisations working in the sphere of environmental protection (Languy, 1995). Little by little, collaboration was organised around a common partner, the Congolese Institute for Nature Conservation (ICCN). Only through a series of last-ditch meetings were all the players engaged in the environmental effort able to discuss strategies, coordinate their operations and work together. Both the UNHCR and the European Union later set up environmental coordination units, which succeeded – during the period of refugee camps – in limiting the extent of the damage to some areas.

During the rehabilitation phase in the east of the DRC, the collaboration was achieved through UNHCR

4.5
Damage to Virunga National Park during the crisis years included an unprecedented increase in poaching, reflected (top) in this haul of snares recovered in the course of a single patrol in 2004, and in acts of vandalism that left the ICCN's equipment and infrastructure all but destroyed. These dilapidated buildings at Rwindi (centre and bottom) bear the scar of years of occupations by armed groups.

4.6
Attacks by militias and other bandits on the Rutshuru-Kanyabayonga road have become frequent over recent years. This lorry was one of three attacked, pillaged and destroyed by militias on the road to Rwindi on 18 April 2004.

contractual agreements providing partners for participating organisations on their respective environmental programmes. Partners were chosen according to their technical expertise and credibility. On this basis, both the IGCP and the Virunga Education Programme (WWF-PEVi) set up rehabilitation programmes that were funded and coordinated by the UNHCR.

In a crisis situation, it is important not only to coordinate field activities, but also to exchange information and knowledge. Such exchanges, whether taking the form of seminars, symposia or informal get-togethers, allow the different sectors to adapt and share their respective methods, requirements and mandates.

4.6.3 Cross-border collaboration in emergency situations

Virunga National Park is part of an enormous cross-border block made up of roughly ten adjoining protected areas, a characteristic that has long raised, and which continues to present, huge challenges for efforts to protect the region.

Maps and satellite images clearly show the isolated nature of the protected areas in the Great Lakes region, and particularly in the area of the Virunga Volcanoes, where the forests, as the only natural habitat, are completely isolated within an essentially human agricultural context. The Virunga forests are also the only areas with vegetation offering sufficient cover for armed groups to move around freely without attracting enemy attention. That these forests form part of a cross-border habitat considerably increases their strategic value in the eyes of military forces. It is not surprising, then, that the Virunga massif has always been quick to attract military forces operating in the region.

With its cross-border location, the Virunga massif also serves as a buffer zone. During periods of intense fighting, animal populations—elephants, for example, and gorillas—are able to flee to safer zones. When the fighting moves on, conservation projects (such as the International Gorilla Conservation Programme) can again focus their efforts on areas vacated by the combatants.

Cross-border collaboration among the agencies responsible for managing protected areas in the three countries dates back to the 1980s. This collaboration was informal, amounting to an exchange of ideas and information on conservation aspects affecting the whole of the massif.

The possibility of establishing a single, cross-border protected area on the Virunga massif was examined in detail by Kalpers and Lanjouw (1997). A number of possible names were mooted, including: 'park for peace', 'international park' and 'cross-border park'. The creation of such a cross-border park could have many benefits. The cross-border proposal is explored in detail in Chapter 22.

4.6.4 Other lessons and conclusions

We can draw one major lesson from recent diplomatic and political events: throughout the crisis that has rocked the Great Lakes region over the past ten years, the international community has, by and large, passed up the chance to play a positive role in the resolving the core problems.

The environment- and conservation-related mandates of United Nations agencies tend to be rigidly prescribed, so their mobilisation in crisis situations is often slow, even non-existent. It seems that neither the United Nations Environment Programme (UNEP) nor UNESCO has acted upon the recommendations that were made to them during the crisis years, and particularly at the time of the refugee crisis in Zaire (Biswas *et al.*, 1994). Only much later, in June 2001, did UNESCO, in collaboration with all active partners in the eastern DRC, launched a project that made a real difference on the ground, though providing the high-level political and diplomatic support the ICCN had then been lacking for almost ten years.

The vulnerability of conservation projects that are funded by official donors is another lesson that emerges from the events of the past decade. As soon as the political and/or military situation is perceived to be unfavourable, contributions from donors—often across a broad spectrum—are immediately suspended, or stopped. This is what happened in Zaire in 1991–1992, when both the European Union and USAID cancelled important projects in and around Virunga National Park.

Contrast this with the role played by those non-governmental conservation organisations that are active in the region. Unlike official donors, the conservation NGOs have little or no political constraints. They can dedicate themselves entirely to their mandate, which is conservation, and are free to adapt or to redirect their actions as circumstances dictate.

Conservation NGOs, despite their sometimes limited financial means, are able to have a very positive effect, not only though providing a degree of material and financial support to the agency responsible for the protected areas, but also through providing technical and even moral support to field staff who are sometimes disorientated by the events unfolding around them. Conservation NGOS also play an important role as communicators. By acting as a transmission tree for their partners in the field, NGOS can convey messages to the international community, identify and appeal to sources of funding, coordinate activities that are sometimes disorganised and put pressure on Western governments and international institutions.

In conclusion, the weaker the institution that is responsible for the protected areas is, the harder it will be to achieve coordination (if at all) among partners from the different sectors (Languy, 1995). Institutions that are strong before a crisis always are better equipped to handle such a situation.

Collaboration with military personnel can also be very productive in emergency situations. In the case of armed conflict, some of the classic consequences are: the sudden seizure of protected areas by the military, the isolation of conservation workers (for example, by prohibiting access to the protected areas or by confiscating their weapons) and the wanton exploitation of resources.

In this context, the nurturing of close ties with influential military officers in a region can benefit the administrators of protected areas. The administrators have a responsibility to make the military understand the role—and political neutrality—of conservation agencies, as well as to develop codes of conduct for the military, and to convince military personnel that field staff in protected areas, thanks to their extensive local knowledge, are a very precious asset.

5.1
ViNP is the only protected area in the world that harbours three taxa of great apes, man's closest relatives. These apes — the Chimpanzees and Eastern Lowland Gorillas, as well as Mountain Gorillas — are the subject of important ongoing research.

Jacques Verschuren

5 The history of scientific research in Virunga National Park

Jacques Verschuren

This chapter chronicles the rich story of scientific investigation within Virunga National Park, a veritable 'living laboratory'. Rather than a nostalgic journey, this history is a critical appraisal of the research efforts, highlighting – for the benefit of rangers, investigators and visitors – some of the more important scientific findings, while examining some of the trails that have needed to be blazed in order to make these scientific breakthroughs possible.

5.1. Research from the foundation of the park to Independence

The fact that better understanding leads to better protection is certainly not a recent discovery. Scientific research has been one of the core objectives of Albert National Park, ever since the park was founded in 1925.

Take this excerpt from the speech given by King Albert I on 15 October 1929 when he inaugurated the Albert National Park Management Commission: 'A new tendency has emerged in the concept of reserves, which is now pivotal to their development: scientific research has become the *raison d'être* of such institutions'. In its first five-yearly report, the Belgian Congo National Parks Institute expressed a wish 'to house one or two naturalists in each reserve'. The hiring of field naturalists would allow the Institute to integrate the scientific effort across a range of disciplines. Astonishingly, however, it was not until 1957 (23 years on) that the Institute finally took the plunge and hired its first naturalist – the author of this Chapter.

5.1.1 Early investigations: mainly taxonomic

Although Albert Park was founded in 1925, the first publication on its natural history was released only in 1937 (Robyns, 1937). Many other papers would soon follow, however. The work of the distinguished herpetologist and photographer, G.F. de Witte, who collected numerous zoological samples,[1] is still unparalleled (de Witte 1937, 1938). His pioneering research, conducted in 1937-38, focused mainly on the park's amphibians and reptiles. Robyn's monumental flora was another seminal work dating from this early period, as was J.G. Baer's study of the parasitic worms, the helminths, and H. De Saeger's study of the Hymenoptera (sawflies, ants, bees and wasps), along with S. Frechkop's investigations on the mammals (Frechkop 1938, 1941 and 1943). There were, at the time of Congo's Independence, in 1960, a total of 96 publications dealing exclusively with scientific research within Albert National Park.

Taxonomic research on the large mammals and rodents has since also been carried out, together with a major study of the insectivores, which is to be published shortly. The fauna, mainly rodents, living on the slopes of the Rwenzori Mountains have been the focus of a groundbreaking study by Misonne (1963).

The remarkable findings of a major scientific expedition undertaken in 1932 were, unfortunately, published independently of the IPNCB *(Institut des Parcs Nationaux du Congo Belge)* by de Grunne et al. (1937). While it was mandatory for the results of investigations carried out within Albert National Park to be published as part of the IPNCB series, there were a few other exceptions, notably including research done by Prigogine (1953) and de Schouteden (1938) on birds; Poll's work on fishes, which was published by the Tervuren Museum; and some texts by the Institute of Natural Sciences in Brussels. By 1972, this institution[2] had, in total, published more than 300 books.

Looking back, it is lamentable how the focus of the early research fell almost exclusively on taxonomic study of the fauna, essentially the invertebrates, which were investigated decades later by researchers from different nationalities. These groups were studied in isolation both from one another and from their wider natural environment. Such analytical works left no room for ecology or for ethology, both perspectives whose scientific value was understood only by handful specialists in the entire world – sometimes, just by one! Another point of concern, in retrospect, is the wastefulness of some of the early collectors: what has happened to the huge collections that were once scattered across the globe? Most, with the exception of those meticulously preserved at the Institute of Natural Sciences in Brussels, or at Tervuren, have vanished without trace!

5.1.2 Eco-ethological research and other types of investigation

Some investigations, thankfully, were not confined to pure systematics. Before 1940, Schumacher conducted research on the Pygmies living in the reserve. Lebrun (1947, 1960) published the findings of his remarkable studies of phyto-sociological associations on the plains surrounding Lake Edward and the lava fields. After the war, aspects of vulcanology featured prominently in various studies, including those of Hoier (1950, 1952), Verhaege (1958), Meyer and Egoroff. From 1954, Heinzelin, a geologist, studied the regression of the Rwenzori glaciers, a process first noted in 1937. He also studied the soil layers of the upper Semliki River valley. Being very rich in fossils, these provided the anthropologist Twiesselman with the opportunity to study the 'Ishango man'. Heinzelin's discovery, here, of what came to be known as the 'Ishango Bone' offered proof of a basic knowledge of mathematics among the early inhabitants of this region.

5.2
Park's rangers undergo thorough training on the various techniques used in conducting mammal surveys.

strangely, the Brussels Committee had prohibited the use of aeroplanes for counting purposes). Investigators concentrated on determining the sex and age ratios of resident species, notably hippopotamuses, which they counted one by one on foot. There were found to be 20,000 hippopotamuses. Today, that number has been reduced to less than 1,000. The measurement of skulls taken from animals that died from natural causes provided valuable biometric data. A senior warden, C. Cornet d'Elzius, supervised counting by the park's rangers. Details appear in his most recent publication (Cornet d'Elzius, 1996), which is a follow-up to his earlier book on the savanna types of Albert National Park (Cornet d'Elzius, 1964).

Though very late by comparison with similar research in other African parks and reserves, the Bourlière-Verschuren team calculated the biomasses of savanna ungulates in Albert Park, arriving at an aggregate figure of 24,406 kg/km^2, compared with 5,283 kg/km^2 for the Serengeti in Tanzania. Other studies, comparing animal and plant productivity, soon followed, as scientists set out to explain a biomass of such unparalleled richness. It is also worth mentioning the works by R. Hoier and Hubert, whose publications aimed at scientific dissemination.

The hydro-biological mission undertaken by Damas gave way to the K. E. A (Kivu, Edward, Albert) exploration under A. Capart, a gifted limnologist whose successors, including in particular P.D. Plisnier, are today still working in the region – on Lake Tanganyika.

Come the end of 1940, the Institute's park authorities realised that the research effort, by focusing on collecting samples and identifying species in order to enrich dusty collections, had lost its way. It was then, on the initiative of V. Van Straelen, that the distinguished Swiss ethologist, H. Hediger, led the first ethological mission to the area. Even today, 55 years later, his work stands out as exceptional. The success, meanwhile, of an ecological mission led by H. De Saeger in Garamba National Park between 1949 and 1952 made the authorities aware of huge gaps with respect to ecological research at Albert National Park. In 1957, F. Bourlière, a French gerontologist, persuaded the authorities that, in order to arrive at a better understanding of human populations, it was important to study other mammals living in a pristine environment, such as Albert National Park.

For several years, the team led by Bourlière analysed the habitats and fauna of the entire reserve (Bourlière and Verschuren, 1960). The position of biologist[3] was established in July 1957 and held by the writer of this text; it was a logical replacement of the position of researcher held by G.F. de Witte.

Other noteworthy investigations included a study by Curry-Lindahl (1956) on the park's overall ecology and the research by Verheyen (1954) on the ethology of hippopotamuses. A popular book, on protected animals of the Belgian Congo, was written by Frechkop (1941) for more general dissemination. Included in the four editions of this invaluable publication are the texts of the decrees that created the national parks.

5.1.3 First counts

Thirty-two years after the founding of Albert National Park, the area's first game counts commenced (although

5.1.4 Geographic and geological explorations

Geographical knowledge of the park and its boundaries was in 1940 still fairly rudimentary, despite the pioneering earlier surveys of H. Hackars and J. P. Harroy. The advent of Robyns' booklet and map (1948), then, was especially helpful; indeed, both are still valid today, their few errors notwithstanding. Between 1957 and 1961, a Bourlière associate travelled the length and breadth of the park on foot in order to establish its precise geography.

The region has long been a magnet for zoologists interested in studying the impact on fauna of volcanic activity. One phenomenon in particular – the discovery, here, of vast 'gas chambers' (with CO_2 content exceeding 55 percent) that are the cause of many deaths among animals, which are either anaesthetised immediately, or die afterwards of anoxia – has fascinated scientists. This phenomenon may explain the disappearance of elephants from some southern areas of the park. Although several publications have been dedicated to this topic, there is still a need for further investigations (see Chapter 6).

5.1.5 Bush-fire studies

Both naturalists and the park authorities became embroiled in the highly controversial issue of the lighting of savanna bush fires, which was officially prohibited. From 1937, the ban had sparked fierce opposition from some quarters. The Governor-General, P. Ryckmans was an especially forthright critic, arguing (in these words, quoted by Mankoto ma Mbaelele), that, 'By prohibiting bush fires, we are deliberately destroying the fauna of Albert National Park'. The Committee nevertheless remained intransigent on the issue. In 1970-71, amid a change of policy, the Director-General authorised the lighting of bush fires, bowing to the argument that occasional such fires, while undeniably destructive, were part of the natural order of things. Bushfire plans were then devised, in order to prevent mass migration of ungulates towards Uganda, where the deliberate use of bushfire was standard practice.

For experimental purposes, several isolation quadrats (small sampling areas) were demarcated; some enclosing old pasture on which no fire had ever been recorded, others enclosing pasture burned at varying intervals. Elephants, though, quickly destroyed these quadrats, the outlines of which can still be seen today.

Conspicuous among the research triumphs of the 1960s were the first scientific studies of Mountain Gorillas and their behaviour. Several decades after the American, H.C. Coolidge, had completed the morphological analysis that in 1902 confirmed the subspecies' distinctive status, it was another American – G. Schaller – who developed methods of approaching these large anthropoids. Before Schaller, few foreigners had caught so much as a glimpse of a living Mountain Gorilla. Schaller's brilliant research paved the way for later studies, conducted by J. P. von der Becke and D. Fossey, on the Rwandan slopes of the Virunga Volcanoes. In August 1960, the writer of this chapter was able, for the first time, to photograph a group of these Great Apes in their natural habitat. The image was taken at Kabara, 3,100 metres above sea level.

Other investigations under way at the time included studies of volcanic caves and of those on Mount Hoyo (outside park limits) in particular. These studies produced important data on the region's bats and guano fauna, while opposing any exploitation of guano, on the grounds that it too has a part to play in this exceptional ecosystem.

5.1.6 Methods
and techniques used before 1960

Until 1960, scientists working in the field carried only the most basic instruments: binoculars, cameras and altimeters, as well as jars of preservatives (denatured alcohol, formalin). What other equipment they possessed was largely hand-crafted. Sequences featured in some of the films sponsored by King Leopold III, such as *Lords of the Forest* and *Secret African Forest*, were shot in Albert National Park. Other sequences were filmed in what is now the Kahuzi-Biega National Park and were one of the topics of the famous *Exploration du Monde* series. At the Royal Institute of Natural Sciences in Brussels there is a collection, of more than 30,000 early photographs of both habitats and fauna, all taken in Albert National Park.

Before 1960, the use of aeroplanes – less harmful to wildlife than helicopters, except when flying very low – was prohibited. At that time, of course, there were no computers. Ubiquitous today (although not always put to good effect), computers have in recent years proved invaluable in streamlining the research effort.

Numerous sound recordings of the fauna in the park have been made since 1960. Owing to a lack of resources, none of these recordings has been processed, however. The result is that the DRC is now one of the few countries lacking a substantive natural sound collection on tape or compact disk. By contrast, ethnomusicologists from Tervuren have, from the DRC, assembled one of the world's most remarkable human 'phonic' collections.

Without the valuable contributions of Congolese technicians, taxidermists and entomologists, much of the research outlined in this chapter would not have been possible. The dedication of men like Kabwa, Kambere, Kanzaguera, Karibumba, Mbirinde, Tito Paluku, Vukuyo, and above all Lengelima and Nendika (both of whom retired only recently) has been nothing short of inspirational.

5.2. Research from Independence to the 1980s

Congolese Independence in July 1960 had no direct disruptive influence on research and conservation activities in Albert National Park. It did lead to a change of priorities, however, as protection – rather than science – became the dominant consideration. For several years thereafter, scientific studies were limited to this writer's own regular submissions and to the analysis of collections gathered over previous years.

In 1969, when the Congolese Head of State embarked on a campaign to revive the country's national parks, and Albert Park especially, scientific research enjoyed a welcome resurgence.

With only very modest Belgian co-operation, the President's office subsidised new research activities. Two experts, W. Delvingt[4] and J.P. d'Huart, embarked – with the Congolese colleagues whom they had trained – on eco-ethological studies of the hippopotamus (Delvingt, 1978) and the Giant Forest Hog (d'Huart, 1978). These investigations produced two outstanding doctorate dissertations.

A new generation of Congolese investigators – such as Muembo, who became Scientific Director, and Mankoto, who went on to become a senior official with UNESCO – came to the fore. The writer's ecological studies at Lulimbi benefited from the emergence of a wealth of new data (Mankoto ma Mbaelele, 1978, 1989). Several Congolese naturalists compiled bodies of research, the results of which, unfortunately, have not been published.[5]

The choice of Lulimbi as the site for the laboratory installed in 1971-1972 came to be seen as a poor decision, given the levels of human traffic through the area (see Chapter 3). Rudimentary laboratories then already existed in both Rutshuru and Mutsora. Today, the Lulimbi facility looks more like a surveillance base than a scientific laboratory. The ICCN has also struggled to control the built up, around the station, of settlements created by its own staff.

Between 1969 and 1975, priority was given to ornithological investigations. A new bird species, the Yellow-crested Helmet Shrike, *Prionops alberti,* had been discovered in high-altitude habitats in 1933. Work initiated by L. Lippens, who had studied the behaviour of Palaearctic migrants to the region, and whose work had been continued by Prigogine and Verheyen, was revived, and bird observations – especially at the Lulimbi Ringing Station, located on a major migration route – recommenced in earnest. After five years of intense work, Bagurubumwe, Sengemoya, Delvingt and d'Huart ringed more than 30,000 birds, mainly aquatic, of hundreds of species (d'Huart, 1977). Their work established that many of the birds that overwinter in the park come from Siberia. In 1974, a team sponsored by the Presidency prepared an important book on the subject (Lippens & Wille, 1976).

5.3
One of the more than one hundred scientific papers that the IPNCB and ICCN, between them, published between the 1940s and the 1970s. Since 1974, the Fondation pour Favoriser les Recherches Scientifiques en Afrique – FFRSA – has continued to publish original research carried out in Virunga National Park.

5.4
The intensity of the volcanic activity in Virunga (shown here are eruptions at the foot of Nyamulagira in 1993) has spawned a succession of forest types. These have yet to be studied in detail, and would make for a fascinating research topic.

Jacques Verschuren

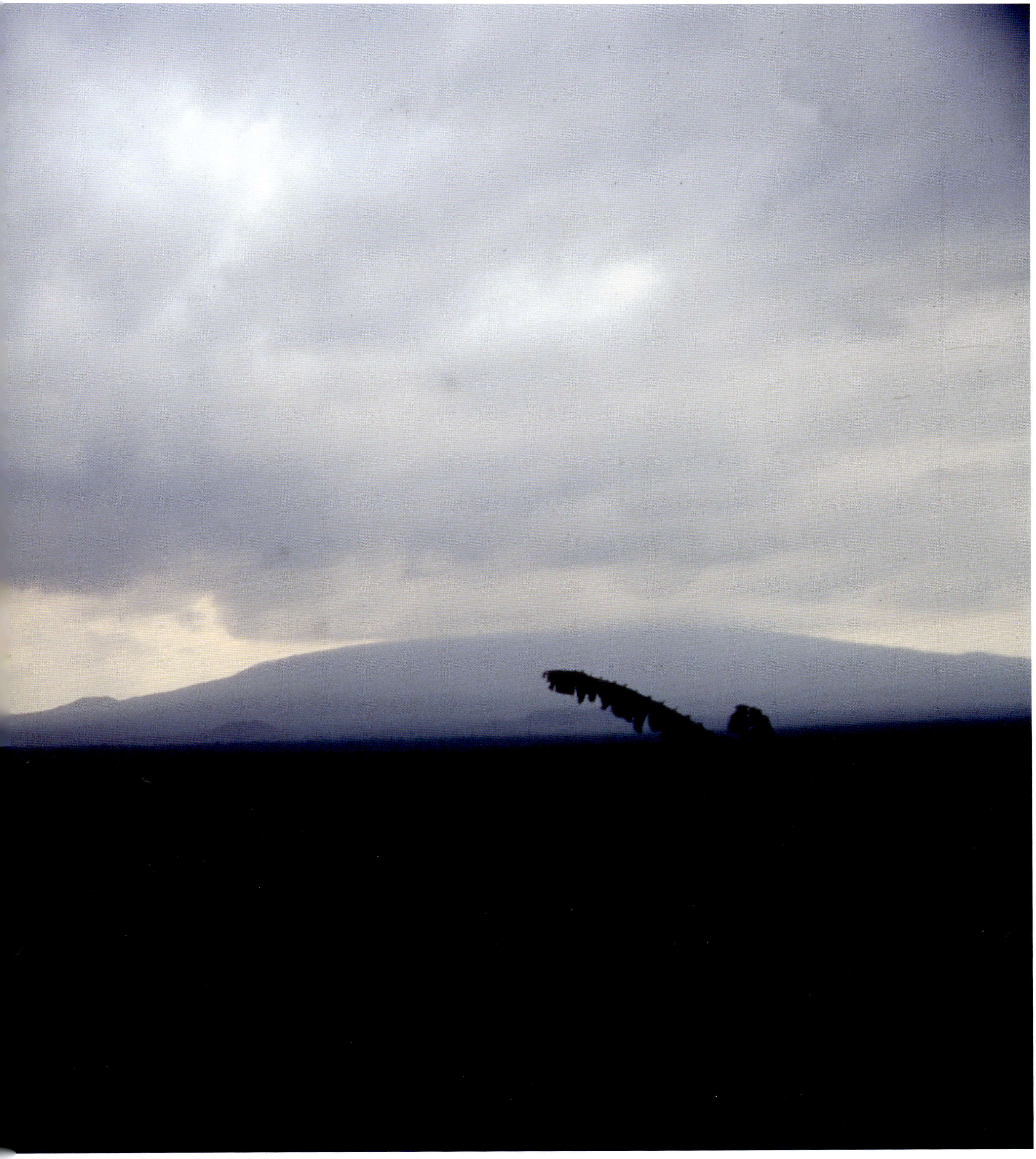

5.5
Virunga's location on the watershed of Africa's two greatest rivers, the Congo and the Nile, coupled with the park's vast expanse and hugely varied biogeography, make for an exceptionally rich research environment.

At this time, tragically, the park lost one of its most brilliant Congolese researchers, Lulengo K'kul Vihamba, who died in the line of duty. An outstanding field naturalist, Lulengo left an indelible impression and is sorely missed.

From 1972, the FFRSA (Foundation for Advancement of Scientific Research in Africa) published the findings of a number of studies conducted in the park, many of them ecology-focused. This foundation, now co-chaired by C. Cornet d'Elzius (President) and J.J. Symoens (Vice President), has over its long history funded a number of important field research projects in the region.

Important investigations undertaken in the 1970s, and from the Lulimbi station in particular, included – in addition to the landmark studies of J.P. d'Huart and W. Delvingt, on Giant Forest Hogs and hippopotamuses respectively – work on the pedology (natural soils) (Vanoverstraeten, 1989; Vanoverstraeten et al., 1984), on the vegetation of the Rwindi-Rutshuru Plain (Van Gysel and Vanoverstraeten, 1982), and on integrated conservation in Virunga National Park (Mankoto, 1978).

More recently, the Leopold III Fund for Nature Exploration and Conservation, presided over by Princess Esmeralda and managed by J. Van Goethem, has also played a part in furthering scientific research in Congo.

Between 1972 and 1976, the Institute published a magazine, *Léopard,* for general dissemination. Although described as a 'mine of valuable information', this magazine has since folded, owing to a lack of financial resources. A new version, launched in 2003, resembles the old magazine.

Over the period 1983-1989, the author's main focus was on monitoring of habitat changes, based on past periodic photographs (Verschuren, 1986, 1993), some of them dating back to the 1930s. The WWF project has since taken over this task, while providing additional photographs for more than 12 sites. Some of the results of this work, which has yielded valuable comparative data, are shown in Chapter 7.

5.3.   INVESTIGATION FROM 1990 TO 2008

With the 1990s came a long period of turmoil that saw research programmes either severely restricted or suspended altogether. There were of course other, more urgent, priorities during those tense years. No new investigations were attempted, as attention shifted instead to documenting the critical situation then facing the park.

For the WWF, M. Languy conducted in July 1994 an aerial count of hippopotamuses spread throughout the park, and of dugouts on Lake Edward (Languy, 1994). In 2003, de Merode carried out another count, which he expanded in 2005 to include large mammals as well, living on the savannas of Virunga National Park. Over roughly the same period, a consortium of partner Non Governmental Organisations completed an exhaustive count of the Mountain Gorillas in the Virunga massif, along with a quick inventory of other taxa (see Chapter 12). The most recent count of the park's main savanna mammals was completed in June 2006 (Kujirakwinja et al., 2006).

Over recent years, the park has benefited too from the enormous advances made in the field of cartography. The transition, over the course of a single lifetime, from the cumbersome ground surveys carried out on foot during the park's early years, through aerial photography (first applied in the park in 1958-1959), to satellite images of the kind the WWF is providing today, has been nothing short of astonishing! Field research remains crucial, however.

Since 2003, overall security in the park has improved, thanks to better monitoring through reinforced ranger patrols. It can only be hoped that, as peace returns to the country, so scientific research programmes will again be able to take centre stage; for there is much work to be done.

5.4.   CONCLUDING REMARKS

The store of knowledge gathered over many years of research in Virunga National Park is a tribute to the application of the scores of dedicated scientists who have worked here, often for long periods and under difficult circumstances. The following remarks are based on 70 years of research in Virunga National Park.

5.4.1   THEMES

Scientific research in the park has lagged behind in recent years through failing to take advantage of exacting modern techniques (molecular biology, DNA analysis). No mammals have ever been anaesthetised in order to extract blood, tissue or other samples.

For some years, there was a controversy, both practical and scientific, over there were too many hippopotamuses in Virunga Park. At one time, the park was home to over 25,000 of the animals, that is one-fifth of species' entire world population. The controversy has since been overtaken by events: brazen poaching has decimated this hippo population. Some ecologists had attributed the 'erosive' white spots seen on the Rutshuru plain to overgrazing by hippopotamuses. In fact, these spots seem to have an edaphic origin; their limits have not changed much over recent decades. The monitoring of such spots following the disappearance of the area's hippopotamuses would provide invaluable data to confirm or reject this explanation, as pictures taken by M. Languy in 2007 seem to indicate a reduction of such spots along the Rutshuru River.

In the past, no veterinary activity was undertaken within the park, even during epizootics (of anthrax and plague). Apart from some work on diseases that can be transmitted to or from gorillas, there have been very few investigations exploring the health implications of contact between wild fauna in pristine habitats and livestock, or between the few people living in the reserve and the wider human populations.

Within the context of climate change, continued monitoring of the retreat of the Rwenzori glaciers is important, as is the recording of altitudinal migrations (seasonal or long-term) of species moving from low altitude forests to mountain forests. The upper limits of species distribution need to be established and monitored on a long-term basis. There is scope, too, for a continuation, and broadening, of investigations into the lakeside ecology.

5.4.2 LIMITATIONS

The most serious obstacle to research in Virunga National Park is the fact that it is located in an area that has seen repeated major conflicts over the past 45 years. Understandably, some naturalists have chosen to amend their priorities and to focus on conservation activities. Science has inevitably suffered as a result. The fact that no global long-term research programme has been set up since the park's foundation is another regrettable aspect, although the ICCN and its current partners are now making commendable efforts to establish a long-term monitoring system. This is discussed in Chapter 23.

5.4.3 THE NATIONAL PARK AS THE LAST REFUGE TO INVESTIGATE NATURAL PHENOMENA

It is crucial, despite reckless, politically correct arguments to the contrary, that the protected area of Virunga should—over most of its present area, at least—be allowed to go on benefiting forever from its status as an integral reserve.[6] Integrally protected habitats provide the pristine reference specimen, that is essential to many different branches of scientific enquiry, enabling accurate deductions to be made concerning changes over time. In unprotected areas of Congo, especially in Kivu, all the original fauna has long since been massacred. Where such fauna remains, its preservation must continue to be a priority. Whimsical schemes cannot be allowed to jeopardise this heritage. Had Virunga National Park not been classified as a protected area 80 years ago, there would be no wild animals left in this part of Africa today.

Congo has always opposed pseudo-scientific cropping projects of the kind implemented in neighbouring Uganda, where half of that country's 14,000 hippopotamuses were 'legally' butchered in the Queen Elizabeth National Park, theoretically for research, but in reality for commercial purposes. One scandalous proposal in DRC—to the effect that hippopotamus meat should be 'harvested' for sales in butcheries—was instantly rejected. That proposal, drawn up under the influence of South Africans, and presented by Mr Bizengimana, Mobutu's Chief of Cabinet, met with fierce resistance among Congo's rangers.

5.4.4 THE FUTURE

The centennial, in 2025, of Virunga National Park is fast approaching. It is imperative, well in advance of this historic landmark, that a committee of national and foreign scientists, meeting on a regular basis, should agree on a comprehensive action plan, defining future research objectives and setting in place a sound monitoring programme. This, after all, is Africa's richest and most varied ecosystem, and while there may be signs in some areas of temporary degradation and neglect, the bulk of the protected area is still intact, and has the potential, after it has turned 100, to be more valuable than ever in global conservation terms.

It is crucial that limits established in the past be maintained, as they have been adequately positioned to include a large number of different biotopes. For the moment, squatters must be immediately driven out of the national park.

We hope, at the centennial in 2025, that the international community will recognise and proclaim the enormous achievements that, against all the odds, have made the Virunga National Park what it is today, and that it will redouble its support for continuing efforts to build further on this park's remarkable legacy—of research and understanding.

---

1. The early collections were often excessive. In the name of science, some investigators organised wholesale massacres. What possible justification was there for killing hundreds of colobus monkeys? Why was shooting a gorilla the first action undertaken by a mammologist sent to Albert National Park? To prove that such gorilla existed? Captures by investigators caused extraordinary damage.
2. A scientific section of the Belgian Congo National Parks was then based in the Natural Science Institute of Belgium, where several entomologists worked. After Congo's Independence, there remained a liaison office in Brussels. This was managed by A. Houben until the early 1980s.
3. The writer, a biologist, completed his training in ethology under H. Hediger, and in ecology under H. De Saeger, in Congo's parks and reserves.
4. Delvingt, Lejoly and Mankoto (1990) later published an excellent tourist guide on Virunga National Park, which included some important biological observations.
5. Of 300 publications that appeared before the 1990s, 299 were written by foreigners. Banahumere Baliene has compared the nature conservation in Congo with conservation in Wallonia. In 2008, there must be a conscious effort to prioritise the publication, in specialist magazines, of work by Congolese researchers. Congolese scientists had participated freely in colloquia, but few had ever published their papers.
6. Much has been said and written about how best to manage national parks. Often, in the case of large reserves such as Virunga National Park, the best form of management is simply to let nature manage itself.

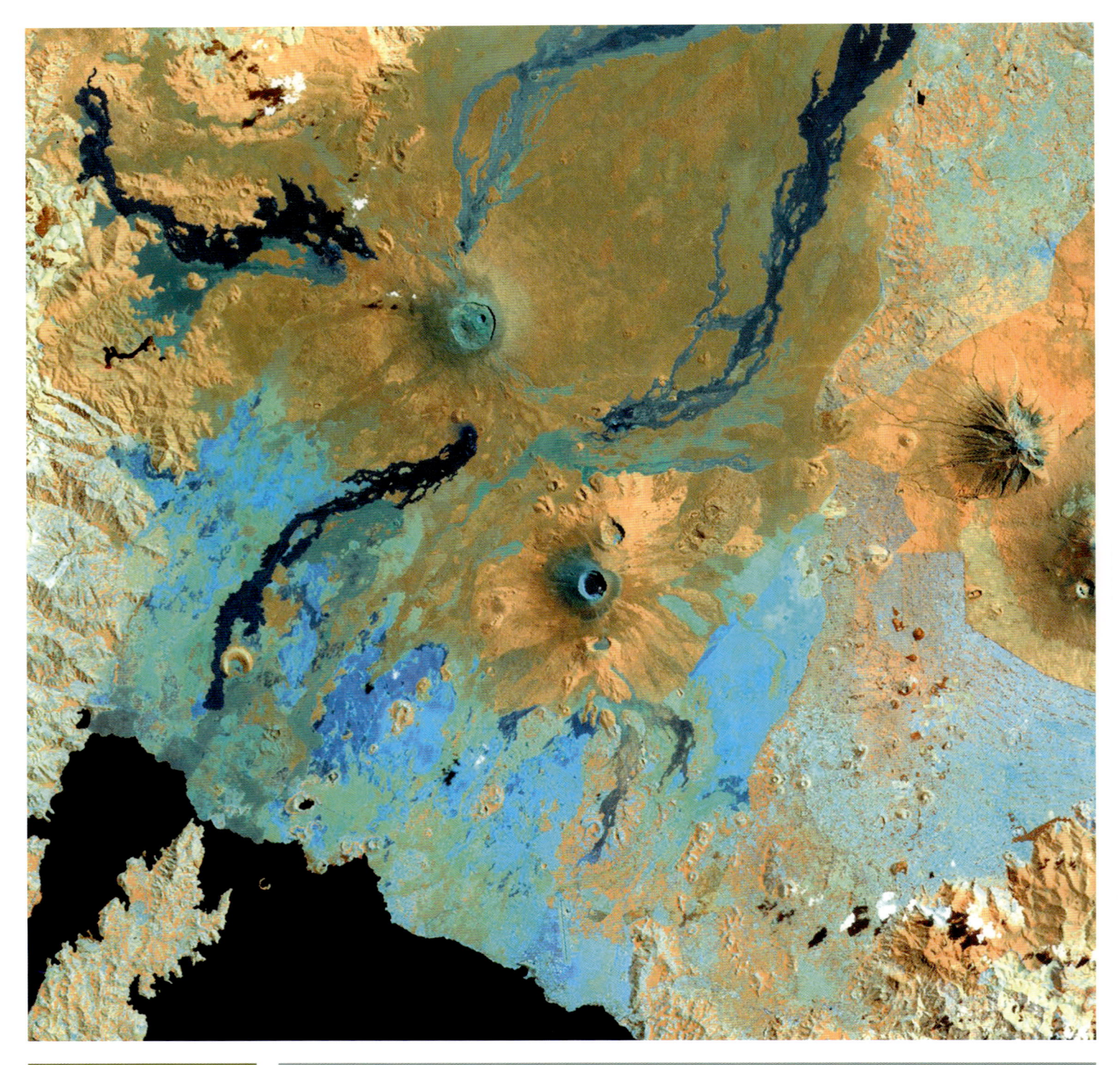

Jacques Durieux

6 The history of volcanic activity in Virunga National Park

Jacques Durieux

The volcanoes of Virunga National Park are, without a doubt, one of its best known features. However they present a double-edged sword for the park: they form an attraction for tourists, but the periodic eruptions of the two still active volcanoes are the cause of much danger and destruction. Bringing both riches and ruin, there is no getting away from the fact that the volcanoes are an essential part of the park, having shaped so much of it and provided the fertile backdrop for its exceptional flora and fauna. This chapter describes the ancient and recent history of volcanic activity within the national park, particularly in the Southern Sector where they predominate.

6.1 The East African Rift

The East African Rift is one of the largest geological features on the planet. Cosmonauts say it is visible from the Moon with the naked eye. It is almost 2,500 miles (4,000 kilometres) long and it stretches from the Red Sea to Mozambique. This giant is, according to geologists, relatively young, having occurred only 30 million years ago, and its formation is not yet complete.

Before the Rift developed, Africa formed a tectonic plate much larger than the one we know today, with the Arabian Peninsula an integral part of its eastern flank. Throughout 150 million years of tectonic tugging at its western flank, where the Atlantic Ocean opened up, the African plate held strong and together.

Then 30 million years ago a plume, an ascent of hot lightweight material moving up from the Earth's mantle deep below, struck the African plate. This burning plume pushed up on the Earth's crust, at the same time heating and shrinking it, distorting the mighty African plate. It bulged, inflated, shrunk and tore. The hot magma spewed towards the surface through a dense network of long fissures, eventually emerging in the form of giant lava flows.

The eruption lasted nearly 500,000 years. Lava flow piled onto lava flow up to a depth of nearly 3,000 metres, forming what we now know as the high plateau regions of Ethiopia, Somalia and Yemen. Pierced by spurts of magma, torn apart by fractures, shrunken and heated, the African plate was weakened and began to break up, splitting down three branches, one branch formed the Gulf of Aden, the second the Red Sea Rift, the third divided the African plate lengthwise and formed the East African Rift that spread southwards.

The development of the East African Rift had repercussions on the landscape and climate throughout East Africa. The bulging of the African plate raised the average ground level to almost 1,600 metres above sea level in some areas. At the summit, the ground broke into the enormous crevasse that is the Rift, several kilometres wide and several hundred metres deep in some areas. This long gash and the elevated areas around it formed a kind of climatic barrier that separated the western hills from the eastern hills, divided the landscape into tropical forest and savanna, but also influenced the animal species that evolved in these new landscapes.

At the bottom of this gash, where the lithospheric plate is very thin and fragile, several fractures opened up allowing the underlying magma to rise up and evolve into different volcanic massifs, including famous peaks such as Ol Doyno Lengai or Kilimanjaro. Further south, the Rift divided into two parallel branches, the western branch and the eastern branch. The Virunga chain

6.1
Relief model drawn from satellite imagery showing the Virunga Volcanoes. Nyiragongo and Nyamulagira are at the centre of the image.

6.2
Three dimensional scale model, showing the two active volcanoes around Goma.

of volcanoes is part of the western branch, sometimes referred to as the Albertine Rift.

The violent opening of the Rift was not always regular and large pieces of rock sometimes detached from the walls. The Rwenzori massif - also named the Mountains of the Moon by Ptolemy who believed they were the source of the Nile – was formed in this way. It is a 120km-long and 40km-wide massif that reaches 5,119 metres, and is covered in glaciers. It is the third highest summit in Africa and the highest point in the Democratic Republic of Congo (DRC). It is made up of metamorphic rocks from an ancient volcanic range (essentially granitic gneiss and Precambrian quartzite that come from the platform of the African continent). About 10 million years ago the bottom of the rift collapsed while its edges were raised. The Rwenzori massif was formed from rock that was fractured, compressed and then raised more than 3,000 metres by the tectonic forces that opened up the Rift. The intense erosion of this block caused a thick sedimentary layer to accumulate in the adjacent rift.

6.2   THE VIRUNGA CHAIN

To the north of Lake Kivu, which runs the length of the western branch of the East African Rift, fractures caused the rising up of large quantities of magma. Some of these fractures are orientated north-south in the axis of the rift, while others are orientated east-west, almost perpendicular to the rift. These different fractures and their intersections are the source of the Virunga chain of volcanoes.

The oldest volcanoes are in fact located to the east of the rift (Muhavura, Gahinga, Sabinyo), or on the actual edge of the rift (Karisimbi, Mikeno, Visoke). Nyiragongo and Nyamulagira are at the bottom of the rift. The chain is orientated in an east-west direction, which is roughly perpendicular to the rift. Nyiragongo and Nyamulagira volcanoes, located in the same axis as the rift and separating the waters of the Congo and Nile rivers, are the only two active volcanoes, while the other volcanoes are considered dormant, despite a brief eruption episode (lasting 24 hours) of Visoke in 1956.

The formation of the Virunga chain had important effects on the hydrographic network throughout the region. It is generally considered that, during the central Pleistocene period, waters from a series of primitive lakes (Tanganyika, Edward and Albert) flowed to the north and most probably fed the Nile. Then, during the upper Pleistocene, less than a million years ago, the establishment of the Virunga volcanoes created a large barrier across the rift. The waters accumulated behind this barrier, flooded the existing contours and thus formed today's Lake Kivu where the edges marked by numerous bays and peninsulas are evidence of the flooded valleys. The level of Lake Kivu, which has no outlet in the volcanic barrier, rose to more than 100 metres above the current level. Only 10,000 or 20,000 years ago, the lake overflowed to the south, cutting through the edge of the Rusizi, lowering its level by feeding today's Lake Tanganyika and from there the River Congo.

The almost permanent activity of these two volcanoes is a testament to the dynamism of the rift and the continued deformation of the African plate. This activity is characterised by an almost permanent presence of active lava lakes. Only five volcanoes in the world are thought to present such a phenomenon: Erebus in Antarctica, Kilauea in Hawaii, Erta'Ale in Ethiopia, Nyiragongo and Nyamulagira in DRC.

Despite their proximity (less than 13 km between the two craters), it is generally considered that Nyiragongo and Nyamulagira are two completely separate entities, with different magma reservoirs and different dynamisms.

6.3   NYAMULAGIRA AND NYIRAGONGO VOLCANOES: ERUPTIVE ACTIVITIES

6.3.1   NYAMULAGIRA

Nyamulagira volcano is one of the best examples of a Hawaiian or shield volcano. Reaching 3,053 metres, it covers a large surface area, is topped by a caldera more than 2 km in diameter, and repeated lateral eruptions that emit long lava flows have given it wide flanks on a gentle slope.

The eruptive activity of Nyamulagira is characterised by frequent lateral eruptions. A lava lake appeared in the active part of the caldera in 1921 and then emptied when the big eruption of Tshambene occurred (1938–1940), from which one of the flows formed the limit of the national park to the east of Sake. The bottom of the caldera then caved in, and has been slowly filled in by an accumulation of lava flows from different eruptions.

Since the lava lake emptied, it seems that the frequency of lateral eruptions has increased, generally interspersed with periods of calm, varying in length from one to four years. Each eruption forms one or more new cones, which line up on the fractures linked to the tectonic of this western branch of the East African Rift. These eruptions produce significant volumes of lava (50 to 80 million m^3), in the form of powerful flows 10 to 30 km long. For the most part, the flows remain inside the limits of the national park and present very little risk to inhabited areas. There are often complaints that these incandescent flows cause significant damage to the park's environment, and it is true that during every eruption large quantities of vegetation are burnt and replaced with fresh lava fields, while the wildlife of the region flees towards new

6.3
The lava flow from Nyamulagira in May 2004. The two volcanoes, Nyiragongo and Nyamulagira, although very close to each other, are considered separate in terms of their activity. The eruptions on Nyamulagira are significantly more frequent. Over the past three decades, they have rarely been separated by more than three or four years. They project large quantities of gas and scoria, or small lava particles that are carried in the wind and deposited at a considerable distance from the eruption. The volcanoes represent a major tourism attraction, but also a health hazard for people living nearby.

horizons. However, these volcanic eruptions are part of a natural cycle and a national park should also care about protecting its mineral riches. Despite its apparent brutality, the volcanic environment is extremely fragile.

6.3.2 Nyiragongo

Like its neighbour, Nyiragongo has a very fluid lava flow but it has a completely different morphology. Reaching 3,470 metres, it is made up of a principal cone with very steep slopes, flanked on the north and south by two large lateral cones. The steepness of its slopes is due to the accumulation of small overflows from a lava lake that must once have been located at the summit of the volcano. The summit today consists of a vast crater with an average diameter of 1,200 metres, and its depth varies according to the volcano's activity.

The volcano's activity, one of the most remarkable in the world, consists of a permanent, active, lava lake at the bottom of the crater. The Nyiragongo lava lake was officially discovered at the time of the crater's first description in 1928, but it most probably existed well before then since a gas plume as well as red glows had already been described. This lava lake disappeared in 1977 in the internal collapse of the crater that followed its first historic external eruption. The lake reappeared for several months in 1982 and rose considerably in the months that followed. It then congealed for 12 years before becoming active again in 1994–1995. During that time its level rose by almost 100 metres.

The January 2002 eruption caused the internal part of the crater to collapse again. A new lava lake reappeared at the bottom in November 2002. It is still very active at the time of this book going to press (December 2008).

6.3.2.1 Historic eruptions of Nyiragongo Volcano

It was believed for a long time that the eruptive activity of this volcano was limited to the lava lake inside the crater. Therefore, due to the richness of the volcanic soil, a dense population settled over a large area near the volcano and the town of Goma is growing every day (more than 500,000 inhabitants in January 2006).

Two recent eruptions have brutally called this theory into question:

January 1977
On January 10, 1977, a system of north-south fractures cut through the volcano little by little, accompanied by a series of earthquakes throughout the region. At the base of the volcano the lava was angrily expelled in different directions. In fact, the entire magma column that supported the active lava lake in the crater was injected into the fractures and then spurted out in heavy flows. The flow rate was very high and a veritable 'wave' of lava swept through the area, leaving traces up to three metres high on some tree trunks. In just a few minutes the first flows reached inhabited areas travelling, according to witnesses, at between 20 and 60 km/h. Roads, villages and farmland were quickly covered in lava. According to sources there were between 70 and 400 victims. The main eruption would have lasted 60 minutes at the most, during which 20 million m³ of lava would have issued forth - the largest basaltic output ever recorded. It is estimated that around 17 million m³ of lava came directly from a combination of the active lake and the lava column in the crater. When this lava disappeared, the entire system of terraces and shafts in the bottom of the crater suddenly collapsed, leaving large shafts 800 metres deep. Five years later, in June 1982, a new lava lake appeared at the bottom of the shaft and significant upward movement began.

6.4
The residential sector of Goma, destroyed by the lava flows after the eruption of Nyiragongo on January 17, 2002.

6.5
The crater and lava lake of Nyiragongo seen at night in May 2005.

When this activity ended in September, the lake's level had risen more than 400 metres in the crater. A new phase of activity began in July 1994 and ended in December 1995. The lava lake's level then rose another 90 metres.

January 2002
Following some precursory earthquakes, the fractures that first appeared in 1977 suddenly spread south, towards the town of Goma. In keeping with the 1977 scenario, the magma was injected into the fractures and lava flowed out from various points. The highest points were in Shaheru, around 2,600 metres high. The flows devastated the forest that covered the slopes of the volcano. Further down, the flows invaded villages at the foot of the volcano and cut across the Goma-Rutshuru road in several places. The last flows came from just above Goma airport. A third of the runway was destroyed, preventing aeroplane access to the terminal. A second flow joined up with the first and crossed the town of Goma from north to south, coming to a stop in Lake Kivu. Seventeen per cent of the town was destroyed, leaving some 120,000 people homeless in one of the most densely populated neighbourhoods. The administrative and commercial centre was destroyed wiping out nearly 80 per cent of the regional economy. The population evacuated the town. The flows that invaded it were fortunately slow-moving allowing those in danger to flee. There were only around 40 direct victims of the eruption.

6.3.2.2 Current activity of Nyiragongo volcano:

When the lava from the volcano cone drained out at the time of the eruption, the bottom of the crater dropped from -250 metres to around -700 metres. To the surprise of scientists, new eruptive activity rapidly appeared at the bottom of the crater, initially consisting of a few explosions that threw up scraps of lava and pressurised gas. These activities intensified and, as of November 2002, a new lava lake formed at the bottom of the crater. It gradually increased in size before intense magmatic activity raised its level dramatically between December 2004 and January 2005. The lake was still growing in size at the time of writing this chapter (June 2006), and eruptive activity inside the Nyiragongo crater was very intense.

This activity generates a large permanent plume over the volcano, made up of ash, water vapour and different volcanic gases. The two main components of the plume, sulphur dioxide (SO_2) and fluorine (HF), are regularly analysed and quantified. The quantities of SO_2 are the highest in the world for permanent volcanic activity. Measurements, made by satellite, record quantities of 12,000 to 50,000 tons of SO_2 per day. This is the equivalent of the pollution emitted every day by the whole of France. It has also been calculated that, between November 2002 and November 2004, Nyiragongo alone emitted as much sulphur as all the other volcanoes in the world put together. This has an important impact on the environment. In a four-kilometre radius around the crater, the vegetation is destroyed by acidic gases. Further away, these gases dissolve in the atmospheric humidity and give rise to acid rain. This also has a significant impact on the national park's forest, as well as on farmland to the west of the park, with the prevailing winds.

Fluorine also dissolves in water and re-emerges in the rain. In the areas around the volcano, a significant part of the population does not have access to drinking water and therefore resorts to using rain water. This water is heavily polluted by fluorine (more than 20 times World Health Organisation regulations) but it is still used by an estimated 50,000 people exposing them to severe fluoride risks.

The permanent magmatic activity, which causes the continuous convection of the magma between the deep layers and the surface, illustrated by the changing levels of the lava lake, supports the idea of a risk of lateral eruption. Any future eruption could have more of an effect on the town of Goma than the January 2002 eruption. Permanent monitoring of the volcano and its activity is needed now more than ever.

Jacques Durieux

6.6
Lava lake of Nyiragongo seen from the inside of the crater, at a depth of 300 metres, on May 27, 2005.

6.7
The eruptions of Nyamulagira (seen here in May 2004) and Nyiragongo constantly transform the surrounding landscapes. Parts of the forest are burned, but within 30 to 40 years, a new forest will reappear, maturing in 60 to 80 years, thereby creating a mosaic of forest blocs of different ages.

The history of volcanic activity in Virunga National Park

Jacques Durieux

6.3.3 THE IMPORTANCE AND THE IMPACT OF NYIRAGONGO AND NYAMULAGIRA VOLCANOES IN TERMS OF THE CONSERVATION OF LANDSCAPE AND BIOLOGICAL DIVERSITY

The significant activity of the two volcanoes located in the Southern Sector of Virunga National Park is one of the most remarkable aspects of the park. Their aesthetic and scientific values were among the reasons the park was awarded World Heritage Site status. The rapid occurrence - on a geological scale at least - of eruptions and lava flows is also the basis of unique forest developments. Delvingt *et al.* (1990) described and illustrated the different stages of forest colonisation on lava flows. In the region to the east of Tongo, a sclerophyllous forest develops in around 40 years and can grow into a mature forest dominated by *Bersama* and *Afrocrania* after 60 years.

On the one hand, since these forests are young and constantly being renewed, they have a poor selection of endemic species. On the other hand, like the hot spring water associated with volcanic activity, they are of unparalleled scientific value since they are a true living laboratory where scientists and biologists can constantly observe the colonisation of different life forms and the evolution of ecosystems. The conservation of the forests and geological landscapes on the slopes of Nyiragongo and Nyamulagira volcanoes should therefore be a priority for the ICCN. These landscapes and phenomena represent an important source of revenue thanks to the tourism they generate.

6.8
The eruption of Mt Kimanura in 1989 attracted many visitors, who had to be guided by ICCN rangers. Some visitors chose to spend the night close to the volcano to get the better views possible in the darkness, whilst maintaining a safe distance and ensuring that they were not downwind of the eruption.

The Goma Volcano Observatory

First established in 1986, the Goma Volcano Observatory (GVO) is a product of the Geophysics Department of the Natural Sciences Research Centre (CRSN) in Lwiro (Bukavu). The centre was looted on several occasions before 2002 and the loss of vital resources made it impossible to fulfil its role of monitoring and providing warnings. From the moment of the January 17, 2002 eruption, the international community mobilised to re-equip the GVO. Today, this Goma-based facility is supported by a European Union project as well as the United Nations inter-agency programme, Volcano Risk Reduction.

The GVO has multiple, complementary duties, including the monitoring of Virunga volcanoes, the prediction of eruptions, risk management, communication and education. Monitoring of the active volcanoes is carried out by a seismological network made up of seven automated stations located around Nyamulagira and Nyiragongo volcanoes, and linked by radio to the GVO so data is received in real time. A second network measures ground deformations and monitors more closely the active fractures from 1977 and 2002. Lastly, geochemical monitoring of the gas plume is carried out by taking soil samples for spectrometer readings (DOAS) or by acquiring satellite data. The data is processed at the GVO and a summary of findings is published every week.

Emphasis is placed on communication and education: Alert notices have been put up in Goma and in neighbouring communities, information meetings supported by specialised video footage are regularly organised. An education program on volcanoes and their risks has been initiated in all schools in the region.

Lastly, a contingency plan and evacuation plans have been put together thanks to collaboration between the GVO, provincial authorities, Goma authorities and the various humanitarian organisations in the region.

6.9
Nyiragongo volcano, active above the town of Goma.

120 Jacques Durieux

The Mazuku - natural gas chambers   Jacques Verschuren

It was in April 1958 that an exceptional discovery was made in Albert Park. A biologist was canvassing the northernmost part of the lava plains, a very wooded and impenetrable area, when he discovered large rocky ditches (more than 50 metres long), scattered with ancient blocks of lava. In these channels, he came across dozens of elephant, hippopotamus, lion and monkey carcasses, in varying stages of decomposition. The researcher entered into the channel – roped up for safety, but became euphoric before quickly losing consciousness. It emerged that the channels were filled with toxic gas responsible for felling the animals that happened to wander into them. He was rescued safe and sound. Later, many more channels (around 15) were discovered in the same area (at the junction of the South and Central sectors). These channels were near crystalline resurgences from the Molindi River, a main tributary of the Rutshuru (80 percent of its downstream flow). Somewhat rudimentary methods were used to take samples of the toxic gases and they were analysed in Bukavu. These gases were made up of more than 60 to 65 percent carbon dioxide, CO_2, without a trace of carbon monoxide, CO.

Experiments conducted using domestic animals (high metabolism vertebrates) showed the channels had an almost instant, but reversible, anaesthetic quality. This is followed by a lethal anoxia. The animals observed constituted a perfect sample of the local fauna: rodents, hyraxes, birds collapsing instantly, frogs, reptiles and a multitude of small invertebrates, particularly butterflies. The victims' stomachs were teeming with larval forms, apparently unaffected by the gas. A succession of dead animals was established: A dead wart hog attracts a hyena, which in turn attracts another hyena, then a jackal and all end up being overcome and dying one after the other. The bodies of pygmies, presumed to have been looking for ivory were also found.

One of the findings that was particularly surprising was the large number of dead baby elephants, while it seemed the adults were resistant. This phenomenon was later explained by the discovery that the upper layer of the noxious gas layer varies, being quite near the ground during the day, rising to several metres high at night.

It is evident that these gases are connected to local volcanism, though the process by which this occurs is still a mystery. The emissions are irregular, with no chronological links to the eruptions. Most of the gas chambers are located at around 1,200 metres in altitude, but higher channels were discovered between Goma and Sake (1,500 metres) and even in the mountains above 2,000 metres.

The phenomenon plays an important role in regulating the number of vertebrates on the plain. No 'adaptation' to this phenomenon has been evident, since almost sedentary hippopotamuses (Ondo Lake, Kibuga Lake) have proven as susceptible as migratory birds from the north such as stints and sandpipers.

More recently (1972), elephant remains were discovered in channels that were several hundred metres long. Globally, hundreds of proboscideans no doubt perished due to the gases.

In 1958, following the unforeseen eruption of Mount Mugogo, in a sector of long extinct volcanoes, several carcasses were found on the edges of the lava flows, in particular semi-endogenous species had risen to the surface (such as golden moles, Chrysochloris). The distinctive smell of the gases brought to mind a derivative of sulphur.

From a scientific point of view, this phenomenon requires further study. The only other comparable case occurs sporadically at Lake Nyos in Cameroon. But as a reassurance to tourists: the mazuku are located outside the areas accessible to visitors, and getting there requires several hours of hiking through almost impenetrable vegetation.

6.10
The skeletons of several elephants, killed in a Mazuku.

6.11
All vertebrates can fall victim to the gases produced by the Mazukus. Their corpses attract scavengers, such as this hyena, and they are in turn asphyxiated.

The history of volcanic activity in Virunga National Park

Jacques Verschuren, Jan Van Gysel & Marc Languy

7 The vegetation: 80 years of evolution

Jacques Verschuren, Jan Van Gysel & Marc Languy

The combination of bio-geographic location and varied relief, gives Virunga National Park an uncommonly wide range of habitats. These habitats are dynamic, responding to many influences, both natural and not so natural. Three factors are of particular importance: climate change, direct anthropic influences and natural or induced fluctuations in the populations of large herbivores, notably elephants and hippopotamuses.

The resulting evolution of habitats in some sectors of Virunga National Park has been dynamic, considering the very brief time scale, bio-geographically speaking, even if this might appear a relatively long period in human terms (70 years since the first documented records).

This chapter's first author, more an ecologist than a botanist, observed the evolution of habitats in Virunga National Park between 1948 and 1990, while the second author carried out intensive studies of the changes in the vegetation on the Rwindi-Rutshuru Plain between 1977 and 1982, comparing his findings with those of previous investigations. The third author has used a selection of photographs, of the same tracts of vegetation taken years apart, to monitor typical, often striking, examples of habitat change.

This chapter is divided into three complementary sections: the first covers change in the vegetation of the entire national park over the 70 years since records began; the second focuses on the plains of Lake Edward over a five-year period, charting the dynamics of vegetation change at the species level; the third shows photographs of several sites taken at intervals of between 20 and 50 years.

7.1 Large-scale habitat changes

Since 1930, a number of habitat changes and trends have been observed. These changes are reflected in:
- observations made on foot or from four-wheel-drive vehicles;
- aerial photographs, including those taken by the *Institut Géographique Militaire* (IGM) in 1958, and those taken from ICCN aeroplanes, as well as images taken more recently, in 2004–2005, by the WWF;
- photographs taken by the first author in 1988 of sites previously photographed at intervals between 1930 and 1957, images of which were then retaken in 2005 and 2006 by Marc Languy;
- SPOT satellite images of high resolution, covering all of Virunga National Park, taken in 2004 and 2005 as part and parcel of the WWF/ICCN project.

Closed habitats are not well represented among the subject matter of early photographs, such as those taken in the 1930s by G.F. de Witte, which otherwise provide good grounds for comparison. For open biotopes, work of Robyns (whose map appeared in 1948) remains an excellent reference. Sequences of dated photographs have been analysed in other works by the first author, in collaboration with the zoologist, Luhunu.[1]

This assessment of habitat change, in the park and adjacent areas, will – for the sake of convenience – proceed in a north-south direction.

7.1.1 Equatorial forest

A comparison of recent data with observations the explorer H.M. Stanley reported in the late 19th Century confirms that the limits of the great equatorial forest outside what is now the park as well as inside it, have remained little altered, in areas where the natural habitat has not been destroyed by man. Today's forested massifs within the park constitute relict islets; the rare continuity that exists between the montane forests of the Rwenzori and the great western forests is one of the unique aspects of Virunga National Park.

There are, in the extreme northern reaches of the park, vast *esobe* (dry grasslands), with many *Borassus* palm trees. That these grasslands appear to be stable is confirmed by the presence of savanna mammals, such as waterbucks and lions. North of the Puemba-Semliki confluence, outside the park, the forest degrades rapidly before giving way abruptly to vast savannas, probably induced by the explosive growth of local human populations. Thereafter, on Lake Albert's southern plain, which lies mostly in Uganda, the grasslands are totally degraded.

The Watalinga Enclave, once forested, has rapidly deteriorated with the demographic expansion that followed the construction of a bridge over the Sem-

7.1
One of the very earliest photographs of scenery in Virunga National Park. Such photographs have proved invaluable in enabling scientists to monitor changes in the vegetation of the park for nearly 70 years. Shown here: Mount Ilehe, photographed by Harroy in December 1937.

liki River in around 1957. Above the Watalinga Bridge, there are some impressive rapids and waterfalls.

Inside the park, the forest is largely stable, except in the south, in the zone of transition towards savanna, where it is invaded by numerous exotic *Cassia* trees.

7.1.2 The Rwenzori Mountains

The retreat of glaciers from the high summits has been well documented photographically. Noted as early as in 1937, the phenomenon was studied by de Heinzelin in 1957 and has progressed rapidly since, as shown on Page 33 of this book. A continuation of this study would be useful, as would a resumption of records charting the visibility of the high summits—generally hidden between 11:00 a.m. and 3:00 p.m., according to past observations spanning more than four decades. Montane Forest once covered the lower, outlying Mutsora massif. Intensive land clearing, followed by sporadic cultivation, has encouraged the growth of tall *Pennisetum* grass on the lower slopes of the massif, up to an altitude of 2,000 metres, with the result that only degraded patches of forest remain, complete with many colonising trees. The mid-altitude montane forest is now discontinuous, with some residual blocks of woodland found in the valleys. *Cyathea* tree ferns characterise this level.

The tree heather zone has remained largely intact on the Congolese slopes to the north and west; this is the most heavily clouded zone. In the past, local people seldom crossed the heather to the 'white stone' (snow) level, as there were few animals to hunt there.

There is on the Rwenzori massif no homogeneous *Hagenia abyssinica* level, such as is found on the extinct volcanoes in the western part of the Virunga Massif. The vegetation at the alpine level remains intact—if little explored. Indeed, with the exception of the well-trodden trail of the Butahu, there are still vast uncharted territories, barely touched even by the climbers of the Grunne mission more than 75 years ago. Aerial photographs, taken at intervals, are essential; these need to be taken early in the morning, and from fairly powerful aeroplanes, as many small single-motor planes cannot safely venture above 4,500 metres. The Haute Lume, south of the mountain massif, also remains *terra incognita*. No scientific party appears to have visited this area since the initial delimitation of the park's boundaries.

7.1.3 The Upper Semliki River between the Beni-Kasindi road crossing and the Lake Edward outlet

The contrast between the southern part of the forest, invaded by *Cassia,* and the monotonous savannas that extend all the way to Ishango is a brutal one. This is perhaps the park's least 'primeval' region, having formerly been occupied sporadically by populations trying to escape trypanosomiasis (sleeping sickness).

The gallery forests lining many of the rivers in the west are generally in a state of regression. There are some *Borassus* palms, but to the east, outside the park, large expanses under papaya cultivation alternate with vast stands of *Imperata* grasses. Vegetation on steep cliffs rising from the Semliki River, seems to be have remained intact, downstream of the excavation sites at which 'Ishango man' and 'Ishango bone' were discovered.

To the east of the Lake Edward's outlet stand the modest grass-covered hills of Bukuku, which dominate the plain of the Lubilya River mouth. This was once a vast stretch of marshland, often with papyrus, alternating with forested stands of *Acacia* sp. The sector was degraded, however, by a dense human presence, originating from Uganda.

The Ishango site, characterised by age-old ditches created by the passage of multitudes of hippopotamuses—now sadly absent—is still relatively intact. The plain to the north of the lake seems to have been spared the ravage of major change, despite repeated fires, periods of illegal human presence and some topographical modifications to the outlet, not least as the result of intense trampling by hippopotamuses.

The savannas between the west bank of the Semliki River, close to where it drains from Lake Edward, and Kyavinyonge, at one time purely gramineous (grass-like), have been covered, since 1988, by a multitude of xerophilic shrubs—a result of the near disappearance of elephants. The sprawling Kyavinyonge fishery remains an eyesore, particularly inasmuch as it is now surrounded by cultivated fields as well.

7.1.4 Mount Tshiaberimu

Mount Tshiaberimu was, as long ago as 1959, surrounded by cultivation. Yet the natural vegetation on the mountain and in the narrow Tumbwe Valley, has suffered remarkably little at the hands of these local farmers. At the higher altitudes, this massif boasts vast expanses of bamboo, out of which rise some towering *Podocarpus* trees. Sub-alpine vegetation is confined to a few hectares around the summits.

The 21 gorillas that were counted on Mount Tshiaberimu in 2006 differ from the Mountain Gorillas found on the Virunga Volcanoes and are a relict of a once far larger population of Eastern Lowland Gorillas. Until 1960, immense uninterrupted stands of bamboo still covered the crests of the graben, extending south of the massif to roughly the latitude of Kabasha, allowing gorilla groups to migrate from Mount Tshiaberimu over a distance of 50–75 km. Added proof of these movements, which had already been documented, came in the shape of a heinous act on the part of a biologist who, in 1937, 'in order to demonstrate the gorillas' presence', needlessly killed an animal at Alimbongo, not far from Lubero, outside the park, despite the fact that total protection had already been accorded to the subspecies. Soon after 1960, almost all of the bamboo, except for that on the Tshiaberimu massif, was cleared.

7.1.5 The shores of Lake Edward from the mouth of the Tumbwe River (Muko) to Kamande

The western shores of Lake Edward, precipitous in places, had been included in the park in order to guarantee continuity between the Northern and Central Sectors.

This ecological corridor is an absolute limit for some species, in particular for Topi antelopes, which do not occur north of the lake.

Fires on these slopes have long since destroyed the original vegetation, which is likely to have been dry forest. This has been replaced by a shrubby savanna. The photographic record shows that, until recently, when illegal cultivation began, this habitat was relatively stable.

Mud flats with grassy vegetation and some large trees once extended around the mouths of the rivers entering the lake near Mosenda and Kisaka. These habitats are now severely degraded. Indeed, around some river mouths, there are hundreds of fishermen's huts. The resulting habitat destruction has breached the ecological continuity. No settlements of this sort had been present before 1988. Until then, some Nande people, living in the hilly regions outside the park, had been authorised to come, once a week, to buy fish from Vitshumbi, or from collection points at Kisaka, Mosenda or elsewhere. Their presence on the lakeshore was of limited duration, as spelled out in a signed agreement (dating from 1958–1959, but destroyed in 1960). The agreement was respected and did not set a precedent for the settlements that followed.

At Lunyasenge, at the mouth of the Talya River, where the flats are much broader, there was – until 1988 – a splendid primary riparian forest, frequented by numerous Giant Forest Hogs and elephants. A unique 'pseudo rain forest' bordering the lake, this was probably the result of the local microclimate. Another, xerophilic forest – at Nyakakoma, east of the Rutshuru River mouth – was destroyed by an illegal fishery in 1967. The forest at Lunyasenge was destroyed much more recently, after having been plundered by poaching gangs.

Further south, around the bay at Kamande, where thickets of impenetrable ambach *(Aeschynomene sp.)* once lined the shore, the plains beyond were covered with a densely wooded savanna, kept in check by the area's huge elephant herds. In the late 1950s, more than 500 hippopotamuses lived in this bay. Vast stands of papyrus lined the mouth of the Lula River.

7.1.6   Lake Edward's southern plain

The botanical composition and evolution of the vegetation on the plains south of Lake Edward is the subject of a detailed analysis presented in the second part of this chapter. In this section, we shall concentrate on providing a broad overview, complete with information from the periphery of the study area, while limiting ourselves to the larger plant formations only.

7.1.6.1   Eastern slopes of the Mitumba Mountains

Dominating the plain, these slopes are covered with a vegetation whose make-up, determined as much by the impact of fires as by a human presence, varies considerably from place to place. Forested and shrubby zones alternate with open, grassy areas, amid extensive fields of bracken.

The extent of the large wooded triangle, disposed at an altitude of about 1,500 metres altitude on the eastern flank of the Mitumba Mountains, and clearly vis-

7.2
The vegetation of the park's montane forests and Afro-alpine zones has yet to be studied in detail. Climate change may be a determining factor at these elevations.

7.3
Elephants and, perhaps to a lesser extent, hippopotamuses, are the principal agents of changes on wooded savannas: they maintain the open spaces in which grassland can develop. Their declining numbers are the cause of much of the habitat change that has occurred over recent decades.

7.4
The grassy savanna of the Rwindi Plain has for many years been the focus of detailed monitoring and ecological research.

ible from the Rwindi camp, has remained unchanged for several decades.

Until 1975, there were – between Kabasha and Kamande, beneath the mountain's characteristic bicephalic peak (standing about 2,200 metres above sea level) – still vast areas of intact savannas. Today, cultivated fields are omnipresent, covering all the peaks and slopes of the Mitumba outside the National Park, which not so long ago were clothed in natural vegetation.

The shrubby vegetation that has lined the route skirting the Kabasha Escarpment since 1948 is showing signs of reforestation. The road serves as a fire-break and differences in the vegetation to either side of this thoroughfare can be very striking.

7.1.6.2 The plain between the Rutshuru River and the base of the Mitumba Mountains

The limits of the Rwindi River gallery forest, both above and below the bridge, do not appear to have suffered any change over recent decades. Fires come up against the forest, but do not penetrate it. The shore of Lake Edward between Kamande and the mouth of the Rwindi has been spectacularly modified by the formation, then disappearance, of a number of small satellite lakes, some of which may at various times, have been joined to the main lake.

Along the traditional routes between the Muhaha/Lula and the Rutshuru River, travelled not long ago by thousands of visitors, the vegetation did not vary much between 1948 and 1988. These were mainly areas of short grasses and were beyond the reach of fires. In the east, a discrete invasion of acacias became apparent, coinciding perhaps with a reduction in the number of *Euphorbia candelabrum* and *Capparis tomentosa* trees.

The distribution of, and clearings in, the thickets of *Euphorbia dawei* along the lower Rwindi, in particular near Vitshumbi, have hardly varied over recent decades. The supposedly detrimental role of nesting marabou storks and pelicans has been insignificant. Fires seldom penetrate the xerophilic forests. It seems doubtful that the climax of the southern lake plain was the *Euphorbia dawei* forest, as envisaged by Lebrun. Cornet d'Elzius and Lejoly have also expressed their reservations on this matter.

The boundaries of modest forested galleries with *Phoenix reclinata* along the middle Rutshuru show few significant modifications. As was the case several decades ago, this river's delta into the lake remains covered by large *Papyrus* formations, avoided by hippopotamuses but serving as refuges for stray elephants. The main branch of the watercourse flowed northward in 1958; today it flows westward.

Aerial observations show that the numerous circumvolutions of the middle Rutshuru have varied little over the years. Some stretches are partially or completely separated from the main course. There is a tendency for some side valleys to be transformed into simple lakes that no longer flow into the river except during seasonal flooding.

Along the middle Rutshuru, and between Kagnero and Nyamushengero in particular, the banks give way to immense white beaches, said to be due to erosion. Their origin has been the subject of numerous debates: are edaphic factors responsible, or can the phenomenon be attributed entirely to overgrazing by hippopotamuses? In 1934, when there were fewer hippos, these beaches were very well developed. In 1974, when the hippopotamus population was at its peak, the beaches tended to contract. Now that hippopotamus numbers have been greatly reduced, it will be interesting to see what becomes of these 'erosion beaches'.

Some cliffs east of the Rwindi Camp and elsewhere on the plain have shown exceptional stability between 1934 and 1983. This site, often used as a didactic instrument to demonstrate a possible aeolian erosion or runoff is therefore badly chosen. Along the Rutshuru River, massive mudslides took place in a few hours, requiring a change in the layout of the track along the river. Between 1973 and 1989, the vegetation that recolonised these scars reflected the normal progression; exotic shrubs that are fairly common on savannas, such as *Lantana* and especially *Opuntia*, propagated very slowly, despite ideal climatic and edaphic conditions. The elimination of *Opuntia* – a plant that, in other regions, has caused serious problems – has become a pressing challenge.

7.1.6.3 The plains east of the Rutshuru River

Before 1970, these savannas were intact, unaltered by human activity. Access to them, formerly very difficult, has since been facilitated by the installation of a ferry at Nyamushengero, a fishery at Nyakakoma and a laboratory at Lulimbi. This has proved to be a disaster for the flora, as well as for the fauna, of the park.

North of the Ngesho River, there are several large 'crest' lakes, such as Lake Kizi. These are covered with beds of floating water lettuce *(Pistia)* and provided refuge to 500 hippopotamuses. Unbridled elephant poaching has altered the habitat through allowing the invasion of countless small acacias, which have contributed to the drying of these lakes.

The spread of the acacias, while discernible in the early 1980s, did not become widely apparent until 1989. The receding grassy areas of this savanna harboured the largest concentrations of Buffaloes, Topis and Kobs in the National Park.

On the eastern part of the plain, the locations of the immense, discontinuous thickets of *Euphorbia dawei*, near Lake Kizi in particular, do not seem to have changed between 1948 and 1989.

The middle course of the Ishasha River, upstream from Lulimbi, was the setting of a large gallery forest dominated by *Croton* and *Pterygota*. These forested blocks have remained intact. Chimpanzees used to live in this closed habitat, along with the last of the park's transient elephant herds, which were not overly poached, migrating back and forth between here and the eastern (Ugandan) side of the river.

Downstream from Lulimbi, after flowing through a small wooded area, the Ishasha River makes its way through savanna to its mouth on Lake Edward's eastern shore. The course of the river has changed a lot since the park was created.

Upstream from Lulimbi, there are still extensive bare beaches between the river and the plateau. The limits of these beaches and of the shrubby outlying savannas changed little over the years. Likewise, clearings in the xerophilic forest near Lulimbi, on both the Congolese and the Ugandan sides of the river, have remained largely unchanged.

The human population based in and around the Lulimbi Camp has increased unsustainably since the camp's creation. Between 1983 and 1988, the growth of this population accelerated markedly, reaching a level that today is unacceptable for a national park. In areas once visited only occasionally by wandering Wakiga poachers, entire villages have been constructed. So far there has been no settlement along the Rwindi River, even in areas – near Rwindi Camp along its access trails, and on the plateau – that were formerly occupied by man. The human invasion of Lulimbi needs to be checked, if this region is to retain any semblance of belonging to a national park. Expulsions are essential. The fact that the

7.5
The ecotones (zones of transition) between the savanna and the forest present perfect sites from which to study the dynamics of the park's forest-savanna mosaics.

illegal settlers are not farmers has ensured that some gallery forest remains, and that local stands of *Rauwolfia vomitoria* have not been damaged. But some evictions are going to be necessary, all the same.

Until the 1960s, the savanna on the southeastern quarter of the plain, between the Ngesho and Kwenda Rivers, was grassy, boasting few tree species. The subsequent disappearance of elephants triggered an invasion of *Acacia* bushes, thickets of which now cover this once open habitat. The flats of the lower Kwenda River have, in the absence of invading acacias, remained open as before.

Before 1955, a splendid forest, almost a rain forest, existed outside the park, near the upper reaches of the Kwenda and Evi rivers. Clearing of this forest began in the 1950s, and today nothing is left of the last remaining fragments of this forest at Kisharo.

7.1.7 The Kasali Mountains and outliers. Complex transition zones between the Central and Southern Sectors

The Kasali massif constitutes a 'mountainous islet' at the heart of the park. Its western flanks rise gently, and until 1968, supported grassy areas alternating with uncommon forest galleries and marshes. In 1958, a fairly extensive area of humid forest still existed on the peak. The eastern portion of this forest in particular, exposed to the full force of monsoon winds, was a favoured haunt of chimpanzees.

In the east, the slopes of the Kasali massif and its southern foothills are much more abrupt. Formerly covered with a brushy savanna, these slopes also harboured some islets of closed forest with well-defined limits. After the elephants, abundant on these slopes before 1960, were exterminated, there was a general forest invasion. The original ecotones, zones of transition between the forest and savanna ecosystems, are in some areas still readily discernible.

In 1958, there was—south of the Kasali massif—a stark contrast between stretches that had been included in the park (areas of brush, later forested areas) and the unprotected areas (notably Bambu Road), which had been completely degraded.

Lower elevations to the east of the Kasali Mountains, such as Mount Ilehe, once almost bare of vegetation, are today covered with invading bushes. Again, this is a consequence of the disappearance of elephants. The top of this fairly precipitous hill commands a good view of the plains east of the Rutshuru River. Once almost exclusively grass-covered, this open habitat has since 1983 seen a generalised invasion by shrubs, notably of small *Acacia* spp., which have now replaced much of the open grassland.

The hilly terrain surrounding the thermal waters of May-ya-Moto has likewise become overgrown, as have 90 percent of the hot springs located beneath steep cliffs further upstream, which remain largely unexplored.

Molindi area is known, not only for its crystalline waters, but also for the lethal vents known as *mazuku*, from which clouds of toxic gas are released. The gas has long been a major cause of natural mortality among elephants and other animals. The locations of these deadly vents and of others both further south and on Mount Muvo do not appear to have changed over recent decades.

The narrow gallery forests of *Phoenix* palms found south of Mabenga are replaced upstream along the Rutshuru River by very much larger formations of these palms. The north-south road near Mabenga, formerly in open habitat, now passes beneath a shady vault of olive trees *(Olea* sp.*)*. West of the town of Rutshuru, there are several small blocks of forest, clustered around the sites of underground springs. The spring water seeps back down, however, through interstitial cracks in ancient lava flows. Beside remnants of the Rutshuru Hunting Domain, the land outside the park is now a vast sea of cultivation; in the south at altitudes of around 1,500 metres, coffee plantations now cover entire hillsides. What fragments remain of the Rutshuru River's gallery forests are badly degraded, except for the area around the waterfalls close to the town itself.

The aquatic vegetation of the plain has been studied by Lebrun, among others. *Pistia,* sought after and spread by hippopotamuses, has invaded every stretch of standing water. We did not see water hyacinth, *Eichornia crassipes,* at all over the period 1948–1989. Were this exotic plant to appear, it would have to compete with the *Pistia*. Efforts should be made to monitor the impact of the spread of such exotic plants, particularly now that the region's hippopotamuses have all but disappeared.

7.1.8 The active volcanoes

The make-up of the plant communities that have developed on different lava flows is astonishingly varied; so much that it is difficult to generalise, even with the help of expert studies, such as that of Lebrun (1947). Northwest of the lava plain, extensive montane forests, growing in non-volcanic soil were included in the park (Kamatembe). Where these forests stop abruptly at the park boundary, they are replaced by intensive cultivation around the Mokoto Lakes. Further south, within the park, Lake Magera (2,200 metres above sea level) once supported a small hippopotamuses population. South of the lava field, the impact of human populations, and of the Rwandan refugees in particular, is so great as to render a comparative analysis of vegetation change virtually meaningless, since there is nothing left of the original habitat. On Tshegera Island, in Lake Kivu, which lies in the park, some traces of a former human occupation have been located.

The original composition of the lower montane forest in this region was confined to relatively few tree species.

Where rainfall is high (at Kibati, for example), the disintegration of volcanic rocks is rapid. Banana plants can be seen emerging from the lava within just a few months after an eruption (Nyiragongo, 1977 and 2002). Rates of disintegration are slower in the north, where there is less rainfall.

On the Rumangabo outlier, within its loose soils and very old lavas, montane forest was once present at this climax altitude of 1,700 metres. Almost completely cleared before its inclusion in the park, this outlier has benefited from 70 years of protection, having been re-colonised, first by *Pennisetum* grasses and, later, by indigenous trees. The same is true of the Mutsora outlier, in the northern Sector of the park. Much of the vegetation on the lava plain is fire resistant.

7.1.9 THE EXTINCT VOLCANOES

An old secondary *Neoboutonia* forest near Kakomero provides a natural corridor between the extinct and active volcanoes. Elephants once made frequent use of this corridor, crossing the Goma-Rutshuru road after nightfall. The gorillas, though, never made this crossing to the active volcanoes, from which they are absent. Clearing of the roadsides for 'security reasons' (explained in Chapter 4) has somewhat reduced the effectiveness of this natural corridor.

The surprise eruption of the Mugogo Volcano in 1957 triggered a mass die-off of vegetation in an area spanning hundreds of hectares in bamboo forest. The bodies of many vertebrates, including rare and endogenous mammals, such as golden moles, *Chrysochloris* sp., were discovered among the debris. Some of the deaths may have been the result of discharges of toxic gas *(mazuku)*, following the eruption. However, the very distinct, omnipresent sulphurous odour evoked a different phenomenon.

At around 3,100 metres above sea level, there is—on the western extinct volcanoes—an important *Hagenia abyssinica* belt, dominating an understorey of umbellifers, *Peucedanum*. Pastoralists and their stock have repeatedly invaded sections of this forest, but few trees have been felled during such temporary incursions. Where cattle have been driven out, the vegetation has regenerated fairly quickly, so benefiting the gorillas. Illegal grazing by domestic bovids on the flanks of the Virunga Volcanoes has long marred conservation efforts. By contrast, no such problem exists on the slopes of the Rwenzori, where no cattle venture. A homogenous *Hagenia* belt is absent on the Rwenzori massif and on the extinct eastern volcanoes.

On balance, most of the habitats found in Virunga National Park are fairly stable, with the exception of those on the plains north and south of Lake Edward, which have been invaded by a succession of small acacias. The fact that grassy habitats are diminishing could impact negatively on grazers such as Topis and, to some extent, buffaloes.

7.6
Acacias are the species most sensitive to the presence, or absence, of large mammals, and—given the opportunity—will quickly re-colonise open spaces.

7.2 THE PLAINS SOUTH OF LAKE EDWARD: A DETAILED STUDY OF THEIR BOTANICAL COMPOSITION AND EVOLUTION

The Rwindi-Rutshuru Plain is of particular botanical interest. Here, an area exceeding 140,000 hectares has seen no direct human intervention for more than 80 years.

Particular characteristics of the climate and relief have triggered a remarkable response by the vegetation.

The main dynamic agent of vegetation change is the biomass of herbivores. Herbivore communities have varied greatly in composition and size over the 80-year period (see next chapter).

7.2.1 RAINFALL AS AN AGENT OF DISTRIBUTION OF THE VEGETATION

The dry climate of the plain contrasts markedly with that on neighbouring elevations. The pluviometric gradient is substantial. The isohyets are very close together, and within a few kilometres the vegetation changes from acacia-dotted savanna to rain forest. On the plain, changes in the vegetation from the driest place (Vitshumbi) to the wettest (the Virunga and the Rwenzori) are subtle, but become striking at the higher elevations.

Grasses are excellent rainfall indicators. In the same sand-clay soil, one can find *Hyparrhenia familiaris* and *Themeda triandra* on the Lulimbi plateau; *Cymbopogon afronardus* further south towards Kikere (point 8), and *Brachiaria eminii* near Kibirizi. Closer to the mountains, one finds *Beckeropsis uniseta* and *Pennisetum purpureum,* and further south, at the level of May-ya-Moto, the plain has been colonised by *Imperata cylindrica,* which signifies an increase in rainfall. Trees also indicate changes in the climate; the *Capparis tomentosa* thickets that, near Lulimbi, grow only in warthog burrows in termite mounds are omnipresent near May-ya-Moto (a possible consequence of overgrazing).

7.2.2 THE GREAT GEOMORPHOLOGIC ASSEMBLAGES

The plain is part of the Western African Rift system, dominated by great lakes. These constituted a single unit before the lifting of the volcanic Virunga Range in the Quaternary.

7.2.2.1 THE PLATEAUX
During the Quaternary, the filling of the East African Rift depression created a sub-horizontal plain in what had been a steep-sided valley. This fluvio-lacustrine fill consisted of clearly stratified unconsolidated rock; this was fractured by present-day tectonics, forming plateaux that slope gently westwards from the scarps that mark their eastern limits. The main plateaux, oriented east to west, are:
– The Ishasha Plateau, between the Ngesho parallel and Lulimbi Station
– The Lake Kizi Plateau
– The Rwindi Plateau

These partially eroded plateaux have produced characteristic badlands colonised by cactiform forests. The intermittent hydrological network has a bayonet configuration, following the fault lines of the plateaux, particularly to the south of Ishasha.

7.2.2.2 THE LAKE DEPOSITS
During the lake's recent recession phase, currents formed sand bars and ridges leaving separate clayey, salt lagoons. These lagoon ridges are clearly visible on aerial and satellite images as features of the landscape, and their vegetation is characteristic.

7.2.2.3 THE RUTSHURU RIVER DELTA
In the delta of the Rutshuru River, a complex structure of un-eroded clay-silt deposits disposed along a north-south axis, is readily apparent on aerial and satellite images. These are probably submarine deposits dating from the period during which lake water still covered the plain. The present-day Rutshuru River has carved a new delta to the east of these deposits.

7.2.2.4 THE LIMESTONE SLAB
During a warm climate phase, algal organisms solidified limestone, forming the massive slab south of what is now the Rwindi Station that constitutes the plain visible from the Vitshumbi road crossing. Broken up by tectonic movement, the fractures in this slab are apparent on both sides of the Vitshumbi road. These fractures have facilitated exploitation of the rock for road construction, and are the park's only source of building stone. The slab marks the starting point of the spectacular canyon on the Rwindi River south of the station; the valley opens out as soon as the slab is interrupted.

7.2.2.5 THE SOILS
The soils of the plain are paramount in determining the distribution of the vegetation. A detailed study of the soils of the Rwindi-Rutshuru Plain and their role in supporting ecosystems in this part of the park was conducted by Vanoverstraeten (1989).

Soils vary considerably as a function of the contrasting relief. The main types of soils are:
– The ferralitic soils of the Mitumba and Kasali foothills, which experience relatively high rainfall
– The sandy soils formed on the lagoon bars and dunes
– Clays that may be either alluvial (as found on the left bank of the Rutshuru River) or lacustrine (mostly saline and found on the shores of the lake)
– Also clayey, yet non-saline, soils that have resulted from decomposition of the limestone pavement (on the *Botriochloa* plains along the Vitshumbi road).

The mosaic of soils is very striking, both on the ground and in aerial views. Perhaps most striking is the alternation between clayey and sandy soils in the lagoon cordons.

The vegetation follows these variations very faithfully, in the absence of major variations in the climate.

7.2.3 METHODS OF BOTANICAL INVESTIGATION

Having outlined the main climatic and morphological parameters, we can now embark on a detailed breakdown of the plain's vegetation assemblages. To this end, it is worth pointing out that methods of floral analysis have varied widely over time.

Lebrun (1947) used the qualitative method of plant

associations, based on the principle that plants gather together in identifiable associations. No geographical locations are given for the numerous vegetation samples he collected, which renders comparison with present-day samples difficult.

Robyns (1948) compiled an indispensable inventory of the park's flora, at a time when systematics was not yet fully understood.

Cornet d'Elzius (1964) used a purely phenological method of description based on the appearance of the vegetation (very short grass, shrubby savanna, and so on). He did map his observations, but as distribution records for individual species are not reflected in his cartography, it is difficult to use his map in comparison with the present-day situation.

Delvingt (1978) used the same method as Lebrun to describe areas frequented by hippopotamuses.

Verschuren (1986, 1993) made use of photographs showing views of the same places at different times, often decades apart. This approach has been helpful in detecting major changes in forest boundaries and in chronicling the effects of human impact (see Section 3 of this chapter).

Van Gysel (1982) used a strictly quantitative method of locating samples with the help of aerial photography.

For each observation point, a presence-absence sampling of the main herbaceous plants provides for a statistically analysable evaluation of the vegetation cover. This method was tested on areas of herb covers of different heights and of compositions both species-rich and species-poor. Repeat sampling (as many as 80 times) in each area ensured high statistical reliability. The size of each sampling surface was designed to ensure maximum species representation over a minimal area (one-fifth of a square metre).

Each species singled out for study met one, or more, of the following criteria:
- useful in differentiating vegetal associations
- important in the biomass
- consumed by herbivores.

In all, 27 species, or groups of species, were used to characterise the vegetation. Observation sites were selected using a 4-km-by-4-km grid, superimposed over a mosaic of aerial photographs dating from 1960. The points were given concrete markers in 1980, but many of these have since disappeared. All are recorded, however, on the isohyet map. Rangers have taken to using the numbers of these points as landmarks, instead of place names, which are often much more vague. It is not uncommon to hear rangers planning to meet each other 'at point 13', which is more precise than 'two kilometres north of the Nyakakoma fishery'.

Sampling was also done elsewhere in the park, but the 94 sites ensured that Virunga National Park's southern plain was systematically covered.

The advantage of this survey method was that it allowed samples to be compared quickly. This in turn meant that changes in composition could be speedily detected. In 1996, 17 of the points were re-measured. The same methods were then used to study changes in the make-up of the herbaceous vegetation. The Congolese operators from Lulimbi were able to demonstrate that they still knew their flora perfectly, down even to the vegetative stages of the grasses.

7.2.4 Description of the vegetation and its dynamics

This description will proceed from east to west (from Uganda to the Mitumba).

7.2.4.1 The Ishasha Valley

This narrow valley, limited by the Ishasha Plateau, is important for the region's fauna. In a 'bed' of fossil bones close to point seven in the valley (Kata-gata), a molar from an extinct species of elephant was discovered in 1980. Other bones found here have yet to be identified.

The gallery forest in the valley consists mainly of *Turea nilotica* and *Euclea schimperi*. At Lulimbi, the forest is made up exclusively of *Rauwolfia caffra,* which is exceptional, as this tree is rare in the region. These forests have suffered in being used as sources of firewood by rangers from the patrol posts nearby. Improved fireplaces, more efficient than traditional ones, would reduce this impact. On the exposed banks near the mouth of the Ishasha River there are dense stands of *Euphorbia dawei,* a cactiform tree occurring on the drier and more eroded parts of the plain. This tree was proposed as the climax species by Lebrun, at a time when botanists were convinced that for each climate zone, there existed an arborescent formation marking the end stage in the evolution of plant associations. Later, working in Burundi, Lebrun observed that colonisation by these cactiform formations was linked to eroded terrain, which is also the case in Virunga National Park.

Considering the heterogeneity of edaphic conditions, and the rainfall gradient, there is no single tree formation that is able to cover the entire plain. In the absence of fires, and with an animal load compatible with vegetation production, open forests of *Acacia* spp. would probably extend over most of the plateaus. As it is, the species of *Acacia* predominant in any particular area (mainly *A. sieberiana* and *A. gerardii*) depends on the soil type.

7.2.4.2 The Ishasha Plateau

Of tectonic origin, the Ishasha Plateau is 60 km long. From it, an east-facing escarpment drops down to the Ishasha River. The plateau loses itself in the north, near the mouth of the Ishasha River.

On the northern part of the plateau, sandy areas (probably old lake lagoon cordons) covered with an over-grazed savanna of *Sporobolus pyramidalis* and *Hyparrhenia familiaris* are juxtaposed with salty clay areas under *Craterostigma* and *Sporobolus spicatus*. Thickets of *Capparis tomentosa* form only where aardvarks and warthogs made holes in large termite mounds. Such termite mounds need free water in order to develop, unlike the smaller mounds made by termites that consume herbaceous organic matter, which are ubiquitous on the savanna. The *Capparis* thickets are associated with the larger termite mounds found in the saline zones around Lulimbi.

Further south, these saline areas disappear, giving way to a uniform *Sporobolus* savanna in which the grasses *Themeda triandra* and *Heteropogon contortus* appear. Areas of deep sand are colonised by the large *Hyparrhenia dissoluta.* Here too, large termite mounds are responsible for the development of thickets. In aerial views, these thickets appear as circles roughly five metres in diameter. On closer examination, it is obvious that each thicket has become established in a hole in a termite mound.

Experiments carried out in 1979 with the aim of protecting, and measuring the productivity of, the *Sporobolus pyramidalis–Hyparrhenia familiaris* savanna revealed a significant enrichment in both *Chloris gayana* and *Cenchrus ciliaris*. Delvingt has suggested this was probably the result of heavy overgrazing by a combination of hippopotamuses (on the northern part of the plateau), antelopes (Topis and Kobs), and numerous herds of buffaloes. It seems likely that a reduction in the grazing load would trigger enrichment and spur development among savanna species. The experiment lasted one year only, and has not been pursued further.

The southern part of the plateau is covered with a *Cymbopogon afronardus* savanna, a reflection of the increased rainfall. This savanna was much enriched by *Acacia gerardii* during the 1980s. To the south, meanwhile, *Acacia sieberiana* and *A. hockii* gradually enriched the savanna. The dominant grasses follow the rainfall gradient. *Cymbopogon*, accordingly, is gradually replaced by *Imperata* and *Beckeropsis*. The herbaceous vegetation changed little between 1980 and 1995, except near the lake and around Lulimbi, where enrichment in short grasses *(Chrysochloa* and *Crategostigma)* was probably triggered by the large concentrations of hippopotamuses then seeking refuge here from persecution in other areas.

7.2.4.3 Between the Ishasha Plateau and the Kizi Plateau

The eroded sections of the Ishasha Plateau boast the largest *Euphorbia dawei* forest in the park. Constituting what is known as the Nyirafunzo Depression, these eroded areas have a particularly rich and uncommon flora. Interspersed among the forests are fragments of the plateau covered with a *Sporobolus pyramidalis* and *Heteropogon contortus* savanna. Eroded areas that are unforested sport a discontinuous short grassland. In the Kizi fault, near Nyirafunzo Lagoon, in the north, there is an effervescent, strongly alkaline spring, but this is not thermal in origin. The breach between the two plateaux is narrower along the Ngesho River. The vegetation here has not visibly evolved over recent years.

7.2.4.4 The Kizi Plateau

This plateau is named after Lake Kizi, now dry. Sand banks, alternating with salty, clayey beaches, mark the shoreline of the dry lake in all but the area close to the Mitumba Massif. The lagoon cordons, such as those of Nyirafunzo and Kamande Bay, have progressively dried up with the retreat of the lake and the lifting of the plateaux.

The most recent cordons, lying in formation, are covered with a grassland of *Panicum repens*. The lagoons that do still contain water have an aquatic vegetation of *Pistia stratiotes* and *Lemna* spp.

The more recently dried cordons have deep, sandy soils, rich in calcareous material (mollusc shells) and non-saline. Their vegetation is similar to that found on the plateaus *(Sporobolus* and *Hyparrhenia)*.

In the dried lagoons there are patchy *Sporobolus consimilis* reed beds, surrounded by expanses of short grassland featuring *Sporobolus sanguinei* and *Sporobolus spicatus*.

The deep, sandy cordons located near the Rutshuru ferry crossing, on the right bank of the river, favour colonisation by *Hyparrhenia dissoluta* and *Acacia sieberiana*. A discontinuous *Chrysophyla peloorientalis* grassland has become established in hardened soils along the Kizi, in what was probably an old lagoon. Tectonic forces have dramatically broken up the plateau, leaving the occasional dendritic stream, complete with fringing thickets, to follow the resulting rifts. Grazing pressure on some areas (along the east bank of the Rutshuru River, in particular) has been intense, given the load of hippopotamuses.

Towards Ngesho, where there is higher rainfall and less grazing pressure, the short grassland disappears. It is replaced, first by a *Heteropogon* savanna, and then – between Ngesho and May-ya-Evi – by a *Cymbopogon* savanna, in a sequence reflecting that seen on the Ishasha Plateau, albeit unfolding in a different direction.

An *Imperata cylindrica* savanna dominates the block between the May-ya-Evi and the southern boundary of the park. Overgrazing near the Rutshuru River and the pools of Luilango has caused *Capparis tomentosa* to proliferate. *Imperata* is generally associated with abandoned fields (ruderal plant), but its presence in this case is explained by the higher rainfall, taking into account the altitude and the proximity of the Kasali Mountains. The disappearance of *Ophyoglossum* (possibly a seasonal occurrence) from the lagoon cordons is the only plant loss observed between 1980 and 1996.

7.2.4.5 Depositional features of the Rutshuru River deltas

The Rutshuru River has a fossil delta in which linear north-south depositional features are evident in banks combining fine mica sands with zones of compact clay. The terraces west of the Rutshuru River are subject to heavy grazing by hippopotamuses, which by feeding on the low herbs and pulling up most of the tufted grasses have created a discontinuous, low-cropped grassland. The area's relatively non-porous, clay soda soil is in any case not conducive to maintaining a thick grassy carpet.

The grassy patches look strange from the air, forming circles around the scattered *Capparis* bushes. The flat terrain probably accounts for this in allowing wind action and lateral erosion to carry the fine sand to the low points where the thickets grow. Concentric depositions of sand form in patches about five centimetres deep. Production rates are very low, with the result that hippopotamuses must wander over distances of five kilometres or more, in order to find their food (Delvingt, 1978).

Samples collected in 1995 show no marked differences from those collected in 1980. It seems likely, however, that diminution in the load of herbivores will allow the vegetation here to become richer in low savanna species. In conditions that are virtually indistinguishable from those at Mweya in Uganda, the vegetation would then resemble that found at the crossing of the Vitshumbi road: a low *Botriochloa* savanna studded with *Euphorbia calycina*.

7.2.4.6 The plain between Rwindi and Rutshuru

Beginning at the bridge on the Rwindi River, this plain is limited in the south by the Kasali Mountains and extends along the entire length of the Vitshumbi road. Towards the lake, relict lagoon cordons form sand banks colonised mainly by *Hyparrhenia dissoluta* and *Chloris gayana*.

A limestone slab subtends this plain. Fractured by tectonic movements, this slab has a strange crosspiece, visible from the air, with sand cordons oriented east-west and tectonic fractures north-south. Except for the delineating thickets, the fractures bear little relation to the vegetation. The slab continues to the south of the Rwindi River, where it is colonised by taller vegetation (80 cm) of *Indigofera* sp., generally ignored by the herbivores. Photographs taken from the Vitshumbi road crossing show no evolution since 1960 in the vegetation of this plain.

7.2.4.7 The Rwindi Plateau
Rising from the Rwindi-Rutshuru Plain, the Rwindi Plateau is in fact an upheaval of this plain, which explains why the limestone slab is visible along its entire length, from the bridge over the Rwindi to the lake. The soils are the same as those of the plain except for a big sandy 'dune' near the station, east of the river, which underlies the only growth of *Aristida* on the plain. This species is characteristic of dry habitats and sandy substrates.

There is an extension of the plateau on the other side of the Rwindi. Towards the northern end of the plateau, a few areas with vegetation similar to that of the Rutshuru are heavily grazed by the hippopotamuses from the Rwindi. Here, the vegetation is subject to the same considerations seen at the Rutshuru, particularly with respect to the reduction in the number of large herbivores. Otherwise, no perceptible change in the vegetation on the plateau has been observed between 1980 and 1996.

7.2.4.8 The Rwindi-Lula Plain
This plain, extending south of the Rwindi station to the mouth of the Rwindi River, is well known among tourists, since it attracts nearly all the herbivores and predators of the valley and lies close to the hotel. Draining the plain are the the Muhaha River (which has its source near Kibirizi), the Lula (which comes down from the Mitumba Mountains), and the Rwindi (emanating from the Mitumba mountains). Its sandy lagoon cordons in the north aside, the Rwindi-Lula Plain is one of short grasslands on clay terrain – not unlike the grasslands of Kizi and of the northern Lulimbi Plateau. The salinity of the soil and the vegetation to which this gives rise, may explain the frequent presence of large numbers of antelopes and other herbivores. Samples taken in 1996 show no major variation from those collected in 1980. The grazing load had remained little altered, given the protective function of the nearby station.

7.2.4.9 The Kibirizi Plain
The Kibirizi Plain is a southerly extension of the Rwindi-Lula Plain. The herbaceous component of the wooded savanna on the clay-silt soils north of Kibirizi, between the Kasali and the Mitumba Mountains, is the richest in the park, supporting a mixture of *Brachiaria, Themeda,* and *Panicum maximum,* shaded by *Acacia sieberiana*. Fires are an annual occurrence in this area. Given the tall vegetation and the fact that this plain is close to the park boundary, one might expect poaching to be heavier than has so far been the case. The animal load seems to have remained below the plain's capacity of forage production, a rare situation inside the park. The vegetation has responded accordingly, showing an increase in productive species, coupled with a consistent regeneration of the acacia cover. This pattern undoubtedly stems in part from the reduction in the number of elephants.

7.2.4.10 The foothills of the Mitumba Mountains
The rainfall gradient is reflected in the vegetation of the Mitumba foothills with the appearance (sequentially, from the plain to the summits) of:
– *Cymbopogon afronardus*
– *Panicum maximum*
– *Pennisetum purpureum*

Areas disturbed by man are often covered with an *Imperata cylindrica* savanna. Here too, re-growth of *Acacia sieberiana* is widespread, for the same reasons as on the Kibirizi Plain.

7.2.5 Conclusions on the vegetation of the plain south of Lake Edward

The formations of the vegetation are closely related to the soils and the rainfall gradient. Changes in the vegetation over the span of 80-year observation period are essentially linked to the evolution of the animal biomass.

The reduction in elephant numbers, in particular, has spurred a marked regeneration of the area's acacia forest.

The drastic reduction in the number of hippopotamuses will dramatically alter the make-up of the herbaceous carpet along the Rutshuru River. The anticipated changes were not yet visible in 1996.

It is important that the park's management's capacity to go on making quantitative observations of the vegetation at regular intervals should be maintained, as it is only through this method that effective long-term monitoring can be achieved.

7.7
Location of sites for which there are periodic photo sequences. The sites followed by J. Verschuren (S01 to S31) are identified by the same numbers as those used in his 1986 and 1993 publications. Other sites proposed here include sites S32 to S37, the sites of the legal fisheries (P01 to P03) and the illegal fisheries (P04 to P13), as well as the ICCN stations (C01 to C03).

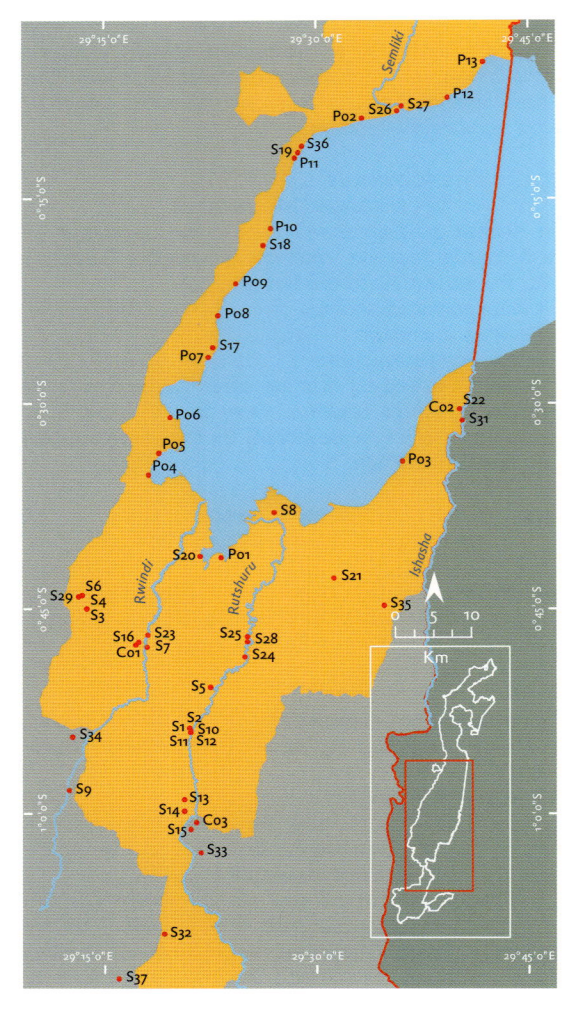

7.3 Monitoring habitat change by comparing periodic photographs

7.3.1 The principle

One of the best methods of detecting and documenting changes in a habitat is to draw comparisons from periodic photographs. The principle consists of photographing, from a given point, a representative portion of a habitat that is liable to natural or man-induced modifications, and then – at different times – to re-take the same photograph from the same point, at the same angle, and using the same focus. The interval between any two such photographs would depend on the type of changes at work, and on the dynamics of the vegetation. It is seldom necessary, for such photographs to be taken at intervals of less than five years. A gap of ten years is probably satisfactory, at least for natural changes.

Comparing photographs enables changes to be detected and documented. In the case of aerial photos taken vertically, it is also possible to quantify changes (to calculate areas of deforestation or reafforestation, for example). In most cases, however, the photographs are taken either on the ground or in oblique view from an aeroplane, meaning that quantification is not possible.

The monitoring exercise if of more value if one can, at the same time, document the probable causes of the changes that are detected. An obvious example in Virunga National Park is the fluctuation of elephant and hippopotamus populations, as we shall see later; another possible example is climate change.

7.3.2 The sources available for Virunga National Park

The oldest photographs taken in Virunga National Park that are usable within this framework are those of G. F. de Witte, which date back to the 1930s. The sites where these photos were taken are sometimes difficult to locate since, since at the time, G. F. de Witte had no plans for their subsequent use in comparative analyses. Thereafter, aerial photos taken by the *Institut Géographique Militaire* in 1958–59 would provide a great number of reference points – and an excellent basis for study, as these covered nearly the entire park. At the same time, J. Verschuren was taking photographs that would later prove invaluable. Indeed, it is the work of J. Verschuren (1986) that forms the basis of this monitoring system. In that work, Verschuren reviews seven sites, all in the Central Sector of Virunga National Park, between May-ya-Moto and the Rwindi River. A reference document by the same author, published seven years later (Verschuren, 1993), completed the work in two ways. From a temporal point of view, many photographs from 1988 were added to the series; while on a geographical perspective, the sites were extended southward (Bugina), eastward (Ishasha), and northward (Ishango) to cover 31 points. More photos were, at the same time, made available for other sites. More recently, M. Languy and P. Banza have re-taken most of the photographs of these sites and have also contributed a large set of aerial photos, aimed essentially at monitoring the fisheries and areas of the park under invasion.

In the present discussion, we refer to a map from the publication by Verschuren (1993), reworked and enhanced, and giving the approximate co-ordinates for each of the 31 sites, but to which we propose to add some new sites (Figure 7.7).

In this book, only eight sites are illustrated, all showing changes that are clearly visible, and whose probable causes are identifiable. Photographs for all of the sites will be presented in another publication. For clarity, the numbers of the sites detailed below correspond to the numbers in Verschuren (1993), and the authors strongly urge others to use the same references.

These eight sites were also selected with a view to alerting managers to changes that are occurring, and to highlighting their respective causes. It should be pointed out, moreover, that it can be just as important to document an absence of change, bearing in mind that this kind of work is very long term.

Just as the photos taken by de Witte more than 70 years ago are still in use, there is every reason to believe that such photos, as well as those taken during our own time, will still be examined a century or more from now. Thus, if a habitat shows strong variation between 1934 and, let's say, 2040, the future biologist will probably be interested to know what the situation was in 2008 or 2020. To be able to let this biologist know that in 2008 the habitat had not varied from that of 1934 is clearly of great value.

7.3.3 CASE STUDIES

7.3.3.1 SITE 1
RUTSHURU VALLEY SEEN FROM MOUNT ILEHE LOOKING NORTHEAST

The modifications are spectacular. In 1934, the Rutshuru River had only a border of gallery forest, while the plain to the east of it was a grassy savanna. Already in 1959, reforestation is visible. This is accentuated in 1983 and is confirmed again in 2005 (even though the photograph is taken from a slightly different angle). The plain has been invaded by a multitude of small acacias and the open habitat is almost completely closed. Variations in the climate are not the cause; this modification is entirely the result of the reduction, then virtual extirpation, of the area's elephant population, a phenomenon that took place in successive waves, between 1960 and 1970 and after 1985. The fauna of open habitats (Uganda Kob, for example) has almost completely disappeared.

7.3.3.2 SITE 4
THE WESTERN PLAIN OF THE RWINDI RIVER SEEN FROM KABASHA

Initially (1934), this plain was dotted with numerous trees, in particular *Albizzia* spp. and large acacias. Elephant numbers had, at that time, begun to increase sharply. Photographs taken in the 1950s show a near-total deforestation on account of the increase in the population of these large proboscidians. The habitat was then completely open (1959). From 1983, a typical reforestation of small acacias began. This has since continued apace, as shown in the 2005 photo. At the 'open plain' stage, the fauna of this habitat (Kobs) increased considerably. Elephants were thus responsible, ultimately, for successive modifications.

G.F. de Witte (1934)

J. Verschuren (1959)

J. Verschuren (1983)

P. Banza (2005)

G.F. de Witte (1934)

J. Verschuren (1959)

J. Verschuren (1983)

P. Banza (2005)

The vegetation – 80 years of evolution

7.3.3.3 SITE 15
VIEW TOWARDS THE CENTRE OF THE KASALI MOUNTAINS, LOOKING EAST FROM ABOUT ONE KILOMETRE SOUTH OF MABENGA

The photo by Gilliard, taken in May 1939, clearly shows that the eastern flanks of the Kasali massif were covered with an important savanna, dotted with acacias. Only the high summits were wooded. A large forested triangle, widely overflowing two of the valleys, was especially prominent. Successive invasions by bushes, then trees, are visible from 1988, and have continued ever since. In the 1988 photograph, as in the 2005 one, it is still possible to guess at the presence of the forested triangle, whose older, taller forest has become much less evident, however, now that the surrounding savanna has closed. Here too, the cause of the reforestation can be attributed directly to the disappearance of elephants, which by eating a great many of the young plants would have kept the savanna open.

7.3.3.4 SITE 16
CREST OF THE MITUMBA MOUNTAINS SOUTH OF THE KABASHA ESCARPMENT, SEEN FROM NEAR THE RWINDI CAMP

From 1956 to 1988, there is little obvious change on the mountains. In 1956, the slopes of the Mitumbas are entirely deforested, except for the gallery forests at the bottom of the valley. In 1988, the situation is broadly similar, although it is possible, on the slopes, to perceive some very sparse wooded areas. The 2005 photo, by contrast, clearly shows a generalised invasion of shrubs and trees, probably acacias. The formation is not yet closed, however, and it will be interesting, if this photograph can be re-taken in a few years' time, to see what happens. It is probable that the same phenomenon noted for the preceding sites will occur. Later still, after the hoped-for revival of elephant numbers, it might become necessary to follow a potential reversal of this trend.

Jacques Verschuren, Jan Van Gysel & Marc Languy

7.3.3.5 SITE 18
ROCKY PEAK ON THE WEST SHORE OF LAKE EDWARD, NEAR MOSENDA

There is very little change observable between 1959 and 1988, except for a light shrubby re-colonisation on the summit. The flanks are covered with bushes and the valley bottom has a well-developed gallery forest. The bushes extend down to the lake shore. The 2005 photograph shows a flagrant change. Nearly all of the peak is deforested and burned: nothing remains of the gallery forest, bar a few isolated patches here and there, and the flanks have been almost entirely stripped to make room for cultivated fields that reach down to the lake shore. This is the result of an invasion associated with the neighbouring fisheries, and shows the impact that the spread of illegal cultivation is having on the western shore of Lake Edward.

7.3.3.6 SITE 19
THE LOCALITY KISAKA, ON THE WEST SHORE OF LAKE EDWARD

In 1957, just as in 1988, the hillsides were mostly bare, except in the valley bottoms and at the foot of the escarpment, which were covered with well-developed forest vegetation. This part of the park was then well protected and neither fisheries nor agriculture existed there. The 2005 photograph shows a radical change, with a near-complete loss of the forest cover in the bottom of the valleys as well as at the foot of the escarpment, where an illegal fishery has moved in. All the hills have been burned and most have been cleared for agriculture. Unless or until protection can be restored, allowing these invaded areas to recuperate, the development of severe erosion can be expected to dominate future monitoring photographs.

The vegetation–80 years of evolution

7.3.3.7 SITE 20
MWIGA BAY

Preservation of the shores around Mwiga Bay, and especially of the bare saline areas, is apparent between 1958 and 1988. The same can be said, in a general way, of other areas in the park, notably along the Rutshuru River, where there are also numerous bare saline areas. However, the photo of Mwiga Bay, taken in March 2005, shows an invasion of vegetation at the expense of the bare areas, which have contracted markedly. This could be the first documentation of a major vegetation change resulting from the quasi-disappearance of hippopotamuses, which were abundant in this area but which have now become extremely rare.

7.3.3.8 SITE 23
THE RWINDI VALLEY TO THE LEVEL OF THE CAMP, RUTSHURU-KABASHA ROAD

Little change can be noted between the photographs taken in 1958 and in 1988, other than a slight development of the gallery forest. A close examination of the photo taken in March 2006, however, shows an invasion of shrubs that is very obvious in different places: while the area to the south of the Rwindi camp (on the right in the photo) and on both sides of the bend formed by the road was entirely bare (grassy savanna) in 1958 and 1988, it has – by 2006 – clearly been invaded. Furthermore, the 2006 photo shows reforestation east of the river (top) and, more precisely, to the right of the right-angle turn in the road. This trend is evident in most formerly bare areas abutting on the gallery forest, for such areas are being progressively invaded by acacias. The changes are just as visible when aerial photos taken in 1958 are compared with SPOT satellite images made in 2004. The perspective in both sets of images is perfectly vertical, making detailed comparison easier. The resolution (five metres) of the satellite image shown here is such that individual acacias, each plainly visible, can be identified, whereas none of these trees was present in the areas marked by crosses in the 1958 image.

J. Verschuren (April 1988)

M. Languy (July 2004)

7.3.4 Conclusions

The study of these few sites has revealed important changes in the vegetation over relatively short periods. For the savanna sites on firm ground and on the flanks of the Mitumba Mountains, the overall impression is of a fairly rapid reforestation, essentially of acacias. The role of climate change as a possible contributing factor can be set aside, since no important changes in climatic data are noted (see the introduction of this book). Inside the park, the major influence, unquestionably, has been the number of elephants present at any given time. Site 4 is the example *par excellence* of this. The impact of hippopotamuses on the vegetation has long been debated. So far, the recent sharp declines in the hippopotamus population appear to have affected only limited areas of the savanna close to water, but a far wider impact is expected to become apparent in the years to come. So, the re-taking of site photographs on the plains at the southern end of Lake Edward is essential in allowing potential further variations to be monitored.

The evolution of habitats outside the park is strongly driven by anthropic pressures, as will be demonstrated in Chapter 9. In a relatively new trend, however, evident since the end of the 1990s, human impact is now also a major factor in several of the areas inside the park. The photographs of 'littoral' plains, such as those at Kisaka and Mosenda, with growing areas occupied by fishing villages (see Chapter 13), provide striking examples of this. Such human impacts warrant particular attention, and the authors recommend that all future sites from which periodic photos are taken should include these points of pressure.

Periodic photos of permanently closed habitats within the park (lava plain and rain forest), though difficult to analyse, generally show little or no variation.

---

1 Ideally, the sites in question should have permanent markers. Yet this has not always been the case, save at Garamba. In the absence of such markers, it is often difficult to determine the precise locations of old sites.

Marc Languy

8 Dynamics of the large mammal populations

Marc Languy

The first chapters of this book retrace the rich history of Virunga National Park. This history has had a considerable impact on the park's animal populations, which are of essential nature conservation value and are a primary tourism target. The large mammals have been the subject of various censuses and studies over the past 50 years. The results of these investigations allow us to trace the evolution of populations of the largest species and to compile a report with a conclusion remarkably similar to that reached by J. Verschuren in 1969 (Chapter 3), 'situation very grave but not hopeless'.

8.1 History of large mammal censuses in Albert National Park / Virunga National Park

In accordance with its initial remit to allow and encourage research, the national park has received numerous expeditions and visits by researchers, international as well as national, over the past 80 years. The detail of the ensuing studies is treated in Chapter 5. Consequently, in this chapter we shall discuss only the main large mammal censuses.

8.1.1 Count C. Cornet d'Elzius 1959–1960

The titanic and detailed works of Count Cornet d'Elzius, senior warden at Rwindi at the end of the 1950s, represent, without any doubt, the most detailed censuses that have ever been done in Virunga National Park. This work is based on thousands of days of foot patrols between 1958 and 1960 and resulted in the collection of 69,603 'contacts' and a total of 976,952 observations of animals. The study covered the park's Central Sector, including the western shore of Lake Edward and all of the Rwindi-Rutshuru Plain to Ishasha. The results of this impressive work, for which one must give credit not only to the author but also to the dozens of rangers who collected the data, were published relatively recently (Cornet d'Elzius, 1996). They deal with 30 species of large mammals.

8.1.2 Bourlière and Verschuren 1960

One of the key references in ecological studies of the mammals of Virunga National Park is the publication by Bourlière and Verschuren (1960) on the ecology of the ungulates (hoofed animals). This work does not comprise censuses as such, but a set of detailed observations on population density, structure and dynamics for all the ungulates of the park, covering all sectors. The study provides precious indications of population size and/or relative abundance for many species, as well as on their distribution, which allow us to make meaningful comparisons half a century later.

8.1.3 Delvingt 1974

Within the framework of his doctoral thesis on the ecology of hippopotamuses, Delvingt (1978) carried out a very detailed census of the hippopotamuses of Virunga National Park in 1974. Most importantly, he compared the results of ground and aerial population counting methods, producing data from which correction coefficients for aerial counts can be calculated, taking into account the lengths of time the hippopotamuses spend

8.1
Virunga National Park is world famous for its exceptional concentrations of hippopotamuses.

submerged, the turbidity of the water and the surface visibility of Lake Edward.

8.1.4 Mertens 1981

Mertens (1983), who was working at the time in Virunga National Park under a programme financed by the Belgian AGCD *(Administration Générale de la Coopération au Développement),* conducted several aerial censuses in 1980 and 1981 of the main ungulates in the savannas south and north of Lake Edward (six species).

8.1.5 Verschuren 1988

Though no longer residing permanently at Virunga National Park at this time, J. Verschuren made a prolonged visit to the park in April-May 1988 which enabled him to collect various data and to publish several studies for the IZCN *(Institut Zairois pour la Conservation de la Nature)* and the AGCD. This included a report on the scientific and technical problems of Virunga National Park (Verschuren, 1988) and an update on the evolution of the habitats and large fauna (Verschuren, 1993).

8.1.6 Mackie 1989 and Aveling 1990

As part of the Kivu Programme financed by the European Union, a new aerial census of the hippopotamus populations was carried out by Mackie (1989), followed by a census of buffaloes and elephants by Aveling (1990).

8.1.7 Languy 1994

Under the World Wide Fund for Nature (WWF) project in support of the IZCN, an aerial census of the hippopotamuses covering the entire Congolese part of Lake Edward, interior ponds and rivers (as well as a count of the fishermen's dugouts on the lake) was conducted in July 1994, in collaboration with Fraser Smith of the WWF Garamba project (Languy, 1994).

8.1.8 Various authors, treating mountain gorilla populations (1986, 1989, 2003)

Harcourt *et al.* (1983) and Vedder and Aveling (1986) reported the results of Mountain Gorilla censuses during the 1980s, before security deteriorated to the point where no counting was possible for nearly 15 years. In 2003, some organisations and projects, including the International Gorilla Conservation Programme (AWF/FFI/WWF), the Wildlife Conservation Society (WCS) and the Institute for Tropical Forest Conservation (ITFC) undertook a census of the gorillas in the Virunga Mountains, including the Mikeno sub-Sector in Virunga National Park, while the WCS undertook a census of other mammals and birds.

8.1.9 de Merode and colleagues 2003–2005

In partnership with the WWF, the Frankfurt Zoological Society (FZS) and the Zoological Society of London (ZSL), E. de Merode, accompanied by experts from the ICCN and conservation projects, carried out three censuses in Virunga National Park. The first census, in March 2003, covered all of the savannas north of Lake Edward and extended to 12 species, including crocodiles (as well as illegally introduced cows, goats and sheep). This inventory was completed in November 2003 with the inclusion of the savannas south of the lake. Finally, a new census, this time of hippopotamuses only, was performed in August 2005 and covered the entire range of the species inside Virunga National Park.

8.1.10 Kujirakwinja and colleagues 2006

This latest census was carried out by the WCS and the ICCN from 9 to 12 June 2006 (Kujirakwinja *et al.,* 2006). The area covered is very similar to that of the preceding inventory but did not include the western shore of Lake Edward or the northern part of the park's Southern Sector. The sampling effort on this area was 16 percent. During this aerial census, ten species were counted, including the main ungulates but the hippopotamus.

8.2 Comparative results and trends from 1959 to 2006

8.2.1 Summary tables

8.2.1.1 The main savanna ungulate species

Four inventories were carried out in a sufficiently rigorous and complete way to permit the monitoring of the numbers of five savanna species in Virunga National Park: Buffalo *Syncerus caffer,* Elephant *Loxodonta africana,* Thomas's Kob (also called Uganda Kob) *Kobus kob thomasi,* Defassa Waterbuck *Kobus defassa* and Topi *Damaliscus korrigum.*

The first inventory, conducted by patrols on the ground, was that of Cornet d'Elzius (1996) in 1958–1960 and involved 30 species. It was followed 22 years later by one conducted by Mertens (1983) on five species, and another a further 22 years later again (2003) by de Merode *et al.* (2003–2005), conducted this time in part by aerial flyovers and covering essentially eight species (as well as domestic mammals). The latest inventory, undertaken by the WCS in June 2006 was very similar to the 2003 one as to the areas covered and the methods used. These last two censuses were not absolute. They used the transect method and total numbers were obtained by extrapolation. This technique gives an estimated total, while providing for a 95 percent confidence interval (see Chapter 12 for details).

It must be noted that these inventories concern only the savannas north and south of Lake Edward. Given that kob, waterbuck and topi are confined to savannas, the numbers obtained for these species correspond with numbers for the whole of Virunga National Park. For buffalo and elephant, on the other hand, these numbers refer to only a fraction of the total population, since these two species also frequent forests.

Table 8.2 and Figure 8.3 give the results of the 1959, 1981, 2003 and 2006 censuses, and show the individual numbers for the Semliki Plain and for the area south of Lake Edward. The 1959 data, collected and published by Mertens (1983), were revised by Verschuren (1993) and it is the latter that are published here.

The data from 1959, 1981, 2003 and 2006 clearly show a vertiginous drop in numbers for four species out of five. They show a loss of 86 percent for buffalo, 90 percent for elephants, 83 percent for Defassa

8.2 Evolution of numbers of five species of mammals in the Congolese plains of Lake Edward

	1959	1981	2003	South 2006	1959	1981	2003	North 2006	1959	1981	2003	Total 2006
Buffalo	23,678	8,916	2,240	3,748	4,629	799	52	74	28,307	9,715	2,292	3,822
Elephant	2,889	621	265	298	536	130	21	50	3,425	751	286	348
Thomas's Kob	10,731	9,750	11,588	12,399	487	550	533	583	11,218	10,300	12,121	12,982
Waterbuck	1,531	570	169	368	692	210	42	6	2,223	780	211	374
Topi	5,939	3,460	855	1,353	0	0	0	0	5,939	3,460	855	1,353

8.3 Evolution of numbers of five species of ungulates in the Congolese plains of Lake Edward

- Buffalo
- Elephant
- Thomas's Kob
- Defassa Waterbuck
- Topi

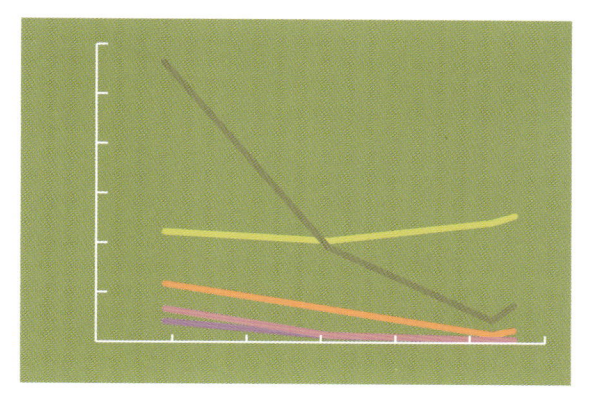

8.4 *Elephants are among the key species to be regularly censused.*

Waterbuck and 77 percent for topi. Although the decline of buffalo seems the most dramatic from a quantitative point of view, all four species have been decimated. All are taken either for their meat or their ivory, and there is no doubt that poaching is the most severe direct cause of decline.

Thomas's Kobs are the exception. Their numbers remain stable, both from 1959 to 1981 and from 1981 to 2006. It is very probable that this species, by far the smallest of the five, is not a favoured target of poachers, who prefer to concentrate their efforts on larger game of greater commercial value. It is probable, however, that if poaching were to continue at this unbridled rate, with the increasing scarcity of the large ungulates, Thomas's Kob would rapidly become a poaching target and suffer the same fate as the other species.

8.2.1.2 Elephant

As mentioned above, the population of more than 3,000 individuals, censused in the savanna in 1959, was only a fraction of the total population. Outside the savannas, Verschuren (1993) notes that the species was virtually omnipresent in Virunga National Park, being found in especially large numbers in the Kasali Mountains and along the lower and middle Semliki, in tropical forest. It was also regular, but in smaller numbers, on the extinct volcanoes (up to 2,000 metres) and was seen in the Nyamulagira crater. It seems that the elephant was never abundant in a natural state on the active volcanoes or on the Congolese part of the Rwenzori Mountains.

On the basis of systematic information and patrol reports, Verschuren (1993) estimated that in 1960 the total number of elephants in Virunga National Park was 8,000. Taking into account the rare elephants that survive in forest (very uncommon on the extinct volcanoes, no recent observations on the active volcanoes and absent from Rwenzori), the present total number for Virunga National Park is certainly not above 400 individuals, which means the population has plummeted to around five per cent of the numbers reported in the 1960s.

8.2.1.3 Eastern Lowland Gorilla

In Virunga National Park, the Eastern Lowland Gorilla *Gorilla beringei graueri* is unknown except on Mount Tshiaberimu, northwest of Lake Edward. The first estimate to be made for the Eastern Lowland Gorilla is by Cornet d'Elzius (1996) who mentions a group of 31 animals, discovered on 21 July 1958. In the following years, other authors give only fairly rough estimates, generally of 30 to 40 individuals at most, but no census ever exceeded 31. In 1986, Aveling (1986) found some groups totalling 25 to 30 individuals. Very recently, proper censuses have produced 19 individuals in 2002, 21 in 2006 and 20 in 2008.

8.5
Baby mountain gorilla, in the Mikeno Sector.

Marc Languy

8.6
Evolution of Mountain Gorilla numbers in the Virunga Mountains

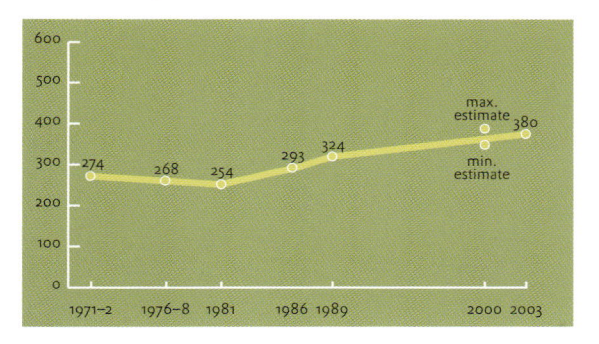

8.7
The Warthog is abundant in the plains south of Lake Edward but rare in the north.

8.2.1.4 MOUNTAIN GORILLA

The Mountain Gorilla populations of the Virunga Mountains have been the subject of much research and several censuses over the past 35 years (Harcourt et al., 1983; Weber and Vedder, 1983; Vedder, 1986; Sholley, 1991).

Over the past 15 years, the region has experienced great political instability, and it was impossible to carry out censuses between 1989 (when the population was estimated at 324 individuals) and 2003, date of the latest census (see Chapter 12). This census shows an increase in the total population of 17 percent compared with 1989, which is the equivalent of an annual growth rate of just over one per cent. Figure 8.6 shows the evolution in the number of gorillas for the entire population on the extinct volcanoes, including those in Rwandan and Ugandan territory (M. Gray, pers. comm.).

8.2.1.5 CHIMPANZEE

There is very little information on the past and present numbers of chimpanzees, *Pan troglodytes schweinfurthii,* in Virunga National Park. The park is known for harbouring many small populations, scattered throughout each of the sectors. For the Central Sector, Cornet d'Elzius (1996) estimates that in 1960 the population comprised between 200 and 230 individuals, mostly spread out over the escarpment west of Lake Edward near the Lunyasenge River. In the Southern Sector, a population of about 50 was habituated to tourism at the end of the 1980s. The population at Tongo has visibly suffered from the presence of refugee camps in 1994–1996 and from the vast illegal charcoal production operations, under way since 2003 and still rampant in 2009 (Chapter 14). Although the chimpanzees are still confirmed to be at Tongo, it is not possible to estimate the size of the present population, or to quantify the evolution of its numbers. The species has always been absent from the extinct volcanoes.

There is no historical data on the number of chimpanzees in the Northern Sector of Virunga National Park, although their presence is very well established, as Verschuren (1972) indicates, and is confirmed by ICCN patrols. The best indication of changes in the chimpanzee population comes from a distribution map dating back to 1958–1960, produced by Verschuren (1972). This map was re-made in 2006 by the WWF project, based on many surveys and from the results of patrols in the field by the ICCN and by the project's field trips. Comparison of the 1960 and 2006 maps gives the following results:

- In the Southern Sector, nearly all the observation sites known in 1960 still have chimpanzees in 2006: flank of Mount Nyiragongo, Maroba (north of Mushebele, with 26 individuals), Rumoka volcano near Tongo, area extending from Katwa and Mulalamule to the upper Rutshuru. A population is also noted outside the park at Muhungezi (18 individuals). The 1960 site northeast of Kitsimbani volcano has not been visited by ICCN patrols for many years; it is thus impossible to determine whether chimpanzees still occur in this area.
- In the Central Sector, all of the 1960 sites still have chimpanzees: Mabenga area at May-ya-Moto, Kasali Mountains, upper Ishasha and Lunyasenge. Moreover, chimpanzees have been reported recently at Kitiriba (eight individuals) as well as between Talia and Kamandi (nine individuals). The latter population and that at Lunyasenge are, however, seriously threatened by farming and human invasions of the land along the western side of Lake Edward.
- Finally, in the Northern Sector, the observation sites also agree with the situation in 1960, with an apparent absence from the northern shore of Lake Edward at Nyaleke but presence further north at the foot of the Rwenzori (Balegha, Kikingi and Bamundjoma) and in the Semliki forest (Bahatsa, Lesse).

In conclusion, it seems that the general distribution of chimpanzees has not changed much since 1960 and that the great majority of sites – if not all of them – still harbour some chimpanzees. The population at Tongo and those of the western shore of Lake Edward are, however, very probably decreasing, and may even be threatened with disappearance, because of *makala* production and illegal extension of fields, respectively.

8.2.1.6 WARTHOG

In 1960, Cornet d'Elzius (1996) estimated that the population of warthogs *Phacochoerus aethiopicus* on the plains south of Lake Edward numbered at least 1,732 with a possible maximum of 2,700. In 1971, d'Huart (1971) evaluated the population in the same area to be 3,874, while Mertens (1983) estimated it at 1,200 animals in 1981. Finally, the aerial surveys of de Merode (2003) give an estimate of 445 individuals in 2003 and those of Kujirakwinja et al. (2006) estimated 694 in 2006. It is noted that, contrary to other data used in this chapter, the numbers of warthog are essentially based on estimates and must therefore be considered with greater prudence.

Dynamics of the large mammal populations

8.8
Estimate of the warthog population in Virunga National Park from 1960 to 2006

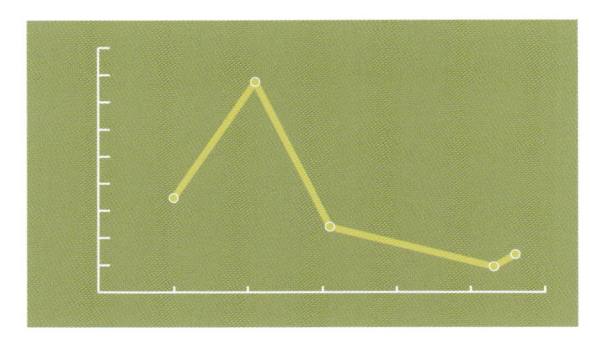

8.9
The hippopotamuses of Virunga National Park have often been censused.

However, the downward trend in numbers cannot be doubted and is another reflection of the significant drop in animal populations in Virunga National Park (Figure 8.8). The drop in the 2000s can be attributed to different factors, including a reduction of Aardvarks (whose burrows are regularly occupied by warthogs), disturbance resulting from a significant increase in human activities in Virunga National Park (traffic, access to the fisheries, military activities, farming) and poaching with snares.

8.2.1.7 HIPPOPOTAMUS

Of all the species living in Virunga National Park, the hippopotamus *Hippopotamus amphibius* is certainly the one that has been the subject of the greatest number of censuses. Even though methods used have sometimes differed, seven inventories made between 1959 and 2005 have ensured very good monitoring of the hippopotamus populations for the entire park. Table 8.10 summarises the inventories that have focused particularly on the hippopotamus.

The very detailed study by Delvingt (1974) included a comparison of results from ground censuses with those obtained by flyovers. His observations and other comparative studies enabled the author to establish correction coefficients that can be applied to aerial counts. These coefficients take into account the turbidity of the water and the average length of time the animals dive, in order to estimate the proportion of

8.10 Principal censuses of hippopotamuses in Virunga National Park

Authors	Date	Method
Bourlière and Verschuren	1959–60	Ground censuses
Delvingt	1974	Ground and aerial censuses
Mertens	May–June 1981	Aerial censuses
Mackie	1989	Aerial censuses
Languy	July 1994	Aerial censuses
de Merode	March–November 2003	Aerial censuses
de Merode	August 2005	Aerial censuses

8.11 Numbers of hippopotamuses in Virunga National Park from 1959 to 2005

	1959	1974	1981	1989	1994	2003	2005
Semliki River	8,811	3,852	2,325	995	141	34	50
Lake Edward	7,804	9,638	7,769	7,019	4,011	892	683
Rwindi River	1,300	1,278	920	2,324	1,314	78	35
Ishasha River	100	335	462	467	400	141	61
Rutshuru River	7,340	10,262	7,337	9,121	4,417	164	58
Interior ponds	1,175	3,813	2,282	2,949	566	0	0
TOTAL	26,530	29,178	21,095	22,875	10,849	1,309	887

8.12 Evolution of hippopotamus numbers in ViNP

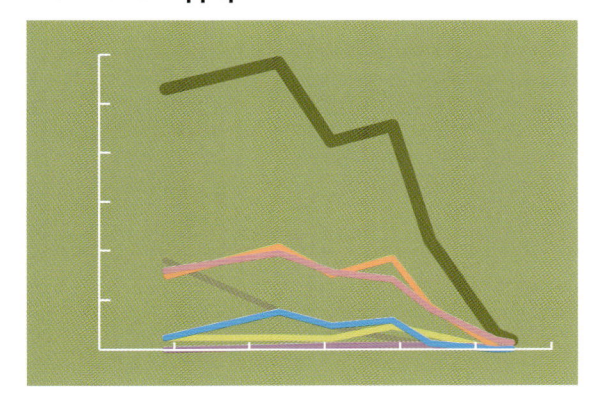

— Semliki River
— Lake Edward
— Rwindi River
— Ishasha River
— Rutshuru River
— Interior ponds
— Total

hippopotamuses that are submerged and therefore go unseen during an aerial passage. All of the authors who have carried out subsequent aerial counts have followed this method and applied correction factors to their censuses. This results in a fairly reliable estimate of numbers. Languy (1994) also shows how different observers obtain remarkably similar results, and that there is therefore a very small margin of error due to the 'quality' of the observer. Table 8.11 and Figure 8.12 summarise the results of these censuses. Since all of the authors censused identical sectors and published these results by sector, it is possible to follow not only the total evolution of numbers but also the evolution for each of these sectors. The evolution of the hippopotamus population shows a period of growth as a result of substantial protection efforts, permitting the population to peak at almost 30,000 individuals in the mid-1970s. This population was then the largest in the world, representing no less than 20 per cent of the world population of the species. Starting in the 1990s, the population experienced a dramatic decline and was decimated to the point where there were no more than 1,000 individuals in 2005. This represents a loss of 97 per cent of the population in 30 years.

Beyond the negative impact on tourism and on the maintenance of short grasslands, the quasi-disappearance of this mega-herbivore is having a dramatic impact on the trophic chain of Lake Edward. A hippopotamus can consume 30 to 50 kg of grass per night, which is expelled as excrement in the lake or in the rivers that empty into it. Before the population decline, the animals delivered more than 600 tonnes per day of essential nutrients to the plankton of the lake and therefore to the fish that feed on it. For some years now, many fishermen – and consumers of fish – have complained about the diminishing number of tilapia and their decreasing size.

8.2.1.8 OKAPI

The Okapi *Okapia johnstoni* was not discovered until 1901 because of its very secretive habits. It was discovered *'in sylvis fluvio Semliki adjacentibus'*, that is, in the

**8.13–8.16
Evolution of the distribution
of hippopotamuses in Virunga National Park**

Each point represents 20 individuals.
The placement of the points is approximate.

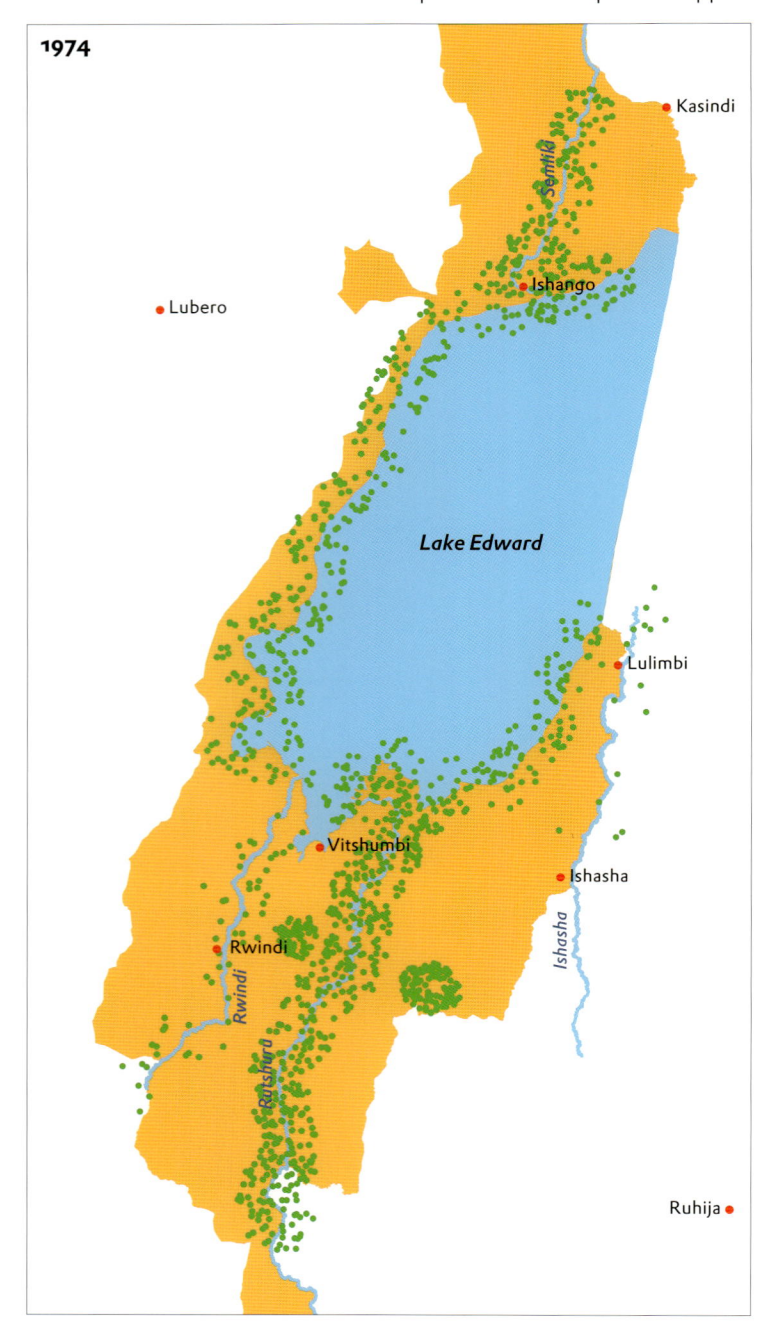

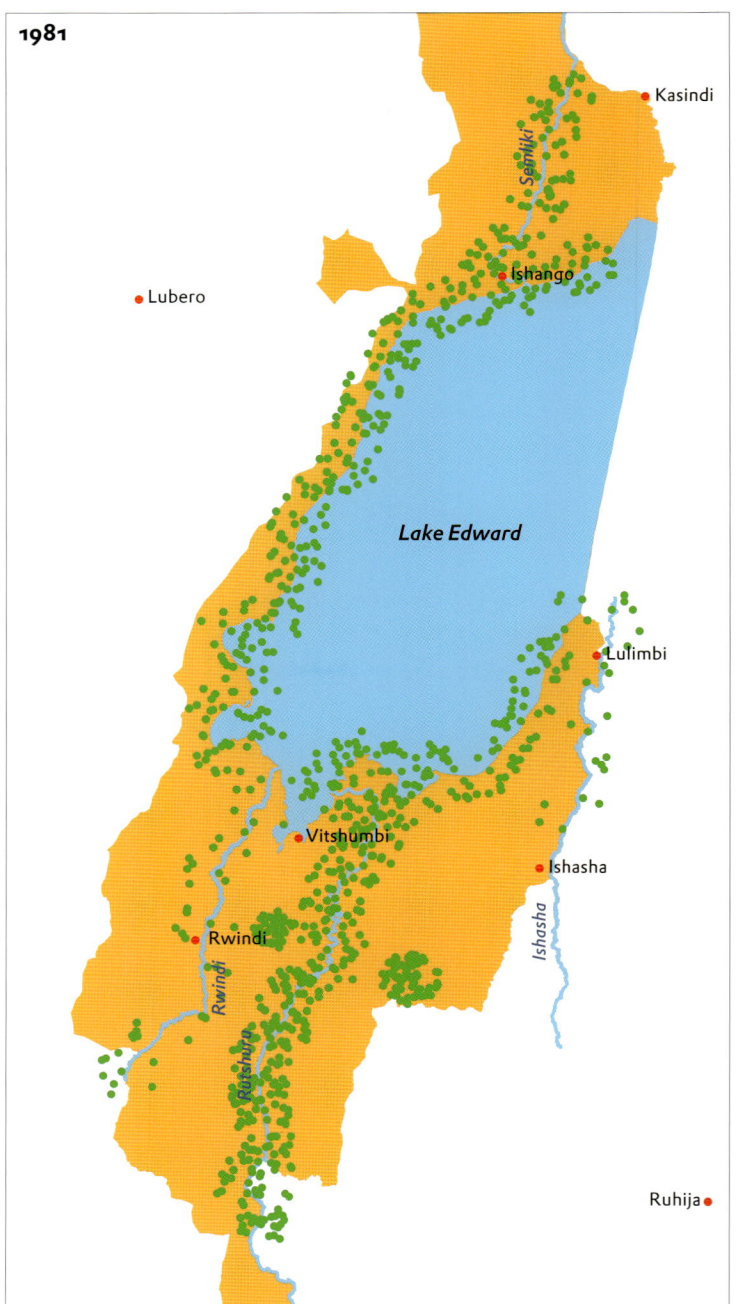

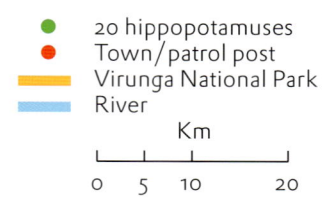

- 20 hippopotamuses
- Town/patrol post
- Virunga National Park
- River

Marc Languy

| 1994 | 2005 |

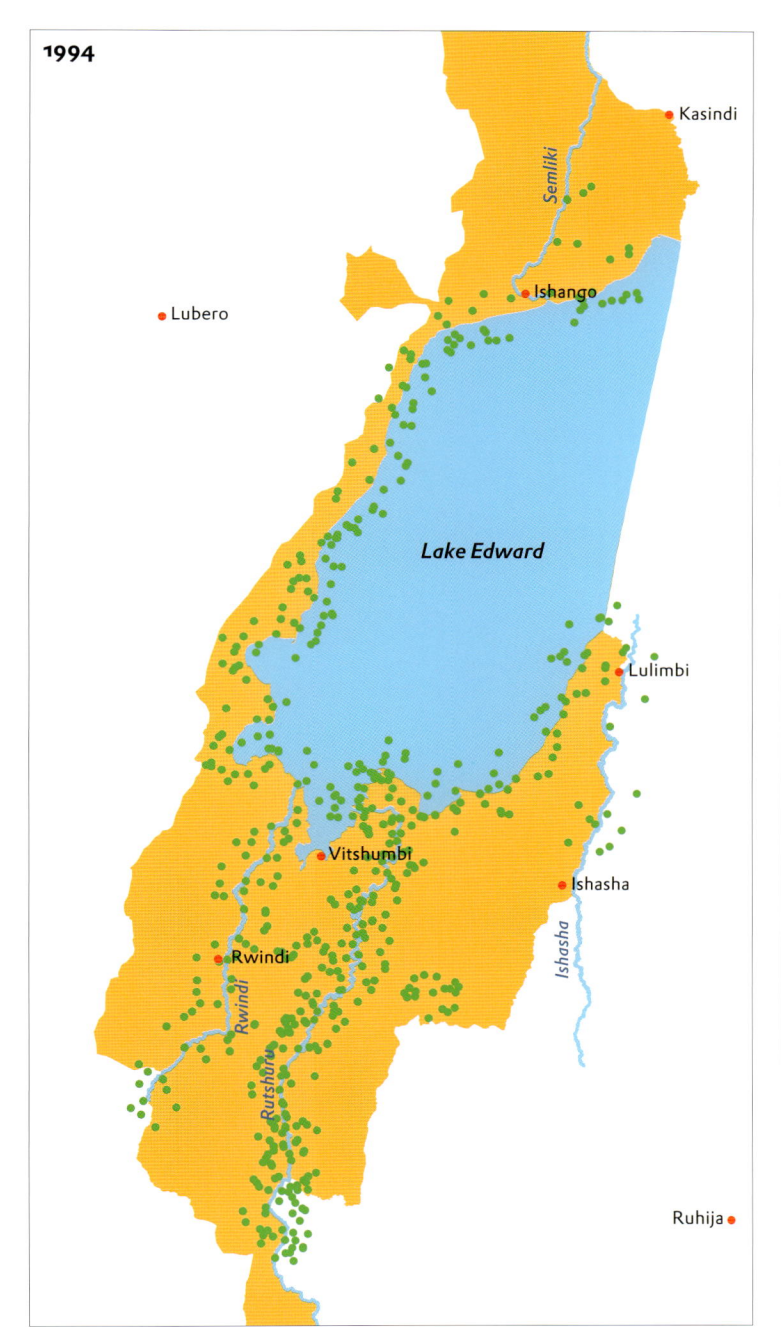

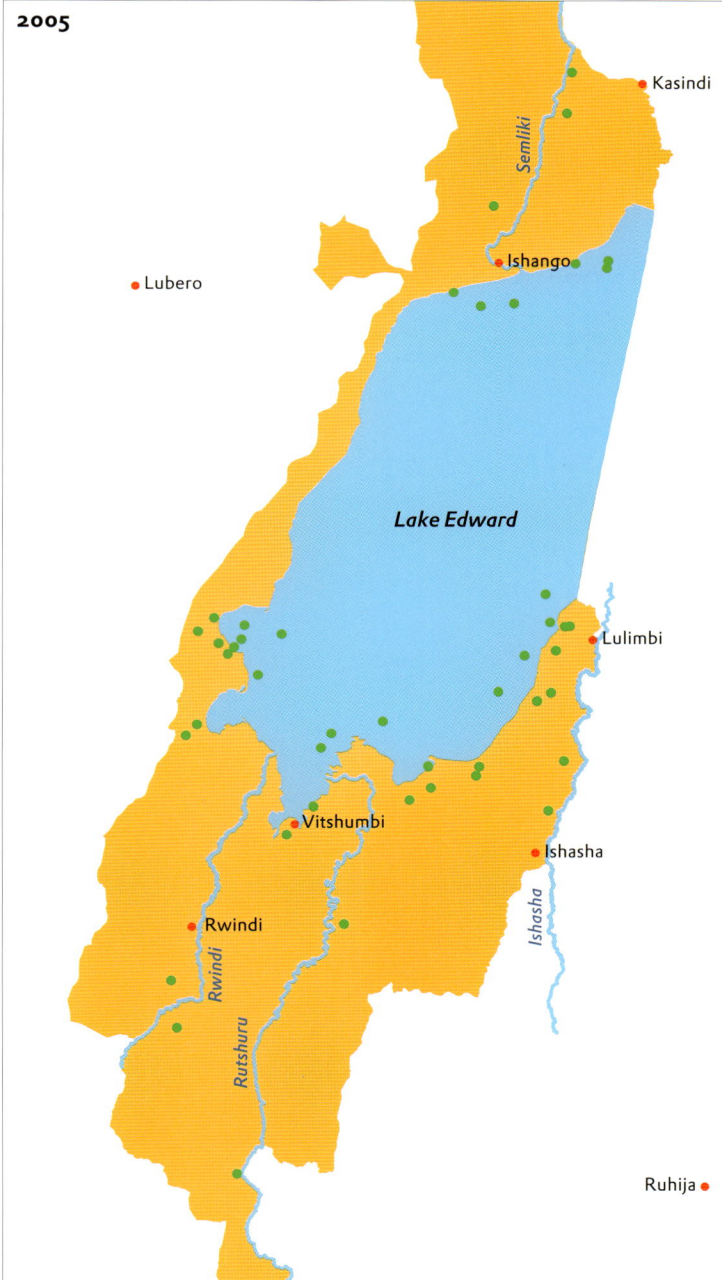

Dynamics of the large mammal populations

Semliki forest, probably near Mundala (Sclater, 1901), thus within the national park before its creation. There has never been a census of this species inside the park, but Bourlière and Verschuren (1960) mapped Okapi records in the 1950s. Other than an unconfirmed observation in 1988, near the Biangolo River, to our knowledge there were no observations of Okapi after 1959 until its rediscovery in May 2006 during an investigative mission organised by the WWF and the ICCN, with the support of Gilman International Conservation. In eight days, a total of 17 signs of the species' presence were recorded between the Makoyobo and Lesse rivers, and there is every indication that a viable population still survives in Virunga National Park, which should be confirmed by conducting inventories further north, including on the east bank of the Semliki River (see Chapter 12).

8.2.1.9 Lion

The lion *Panthera leo* is a difficult species to census. Hubert (1947) estimated that, in 1931, the number of lions in the Rwindi-Rutshuru plain was 250 individuals, a number that fell to 100 ten years later (Bourlière and Verschuren, 1960). For his part, Cornet d'Elzius (1996), on the basis of a large number of records of the species, estimated the population south of Lake Edward in 1960 to be between, at the very least, 105 individuals and, at the most, 200. It is generally accepted, though not demonstrated by actual censuses, that the number of lions substantially increased in the 1970s as a result of conservation efforts in favour of the species and its prey. A significant indication is that in 1965, a tourist had one chance in three of seeing the species, while its observation was almost guaranteed between 1970 and 1990. The size of the population in 2006 is unknown but the species is still seen regularly in the region of the Rwindi and on the Lulimbi plateau. A lone male is still reported in the Northern Sector (in the savannas of the upper Semliki), and it is possible that a small population survives there. The WCS recently set up a lion monitoring programme, using radio transmitter collars in Queen Elizabeth National Park and in the Ishasha area. This should allow park managers to determine whether there are movements between the two protected areas, which could guarantee the survival of the species in the savanna ecosystem south of Lake Edward.

8.2.1.10 Other species

The Cheetah *Acinonyx jubatus* was mentioned once by K. Curry-Lindahl (1961) between the Rwindi and Nyabugenda, but the conditions in which this very fleeting observation was made do not permit inclusion of the species on the park's list of mammals. An intensive search in the days following this observation was unsuccessful. There have been no other records for this species.

Although this chapter deals with mammals, it is important to comment on the expansion and distribution of the crocodile *Crocodilus niloticus* in Virunga National Park. The species inhabited the lower Semliki to below the falls located between the Vieux-Beni bridge and the confluence of the Lusilube and Semliki Rivers (Delvingt, 1990). It was in June 1986 that the first crocodile was observed near Ishango (Verschuren et al., 1989). The species rapidly spread and was noted on the western shore of Lake Edward near Mosenda in 1989, and reached the Rutshuru, near Nyamushengero, in mid-1991. Finally, several individuals were observed in 2005 downstream from May-ya-Moto.

8.17
Evolution of biomass of ungulates in the savannas, southern part of ViNP (1959–2006) (metric tons)

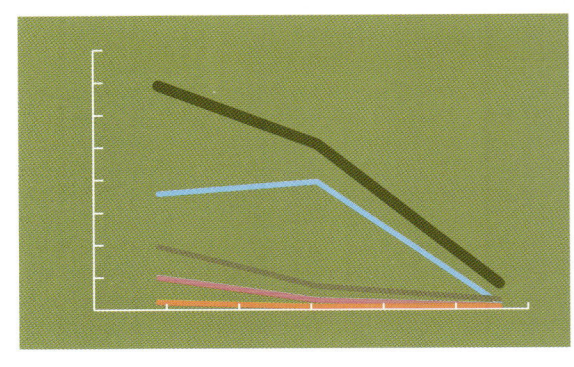

- Hippopotamus
- Buffalo
- Elephant
- Others
- Total

8.2.2 Evolution of the biomass in the plain south of Lake Edward

Virunga National Park has long been famous for the exceptional, world-record-breaking density of its biomass. Bourlière and Verschuren (1960) showed that the biomass of ungulates per km² of the savannas in the park was more than four times greater than that found in the Serengeti.

In 1959, the total weight of the main ungulates in the savannas south of Lake Edward (1,250 km²) was 34,523 tonnes, compared with 25,567 tonnes in 1981 and 4,043 tonnes in 2006 (Figure 8.17).

The dramatic decline over the past 20 years is particularly shocking. Some scientists in the 1980s suggested that the biomass density (27,619 kg/km²) was abnormally high (essentially because of the substantial number of hippopotamuses that then represented three-quarters of the total animal biomass) and that this phenomenon could have a negative impact on entire ecosystems through overgrazing and trampling. This proposal has always been controversial. In 2006, however, the debate is no longer relevant since the biomass has dropped by 90 per cent to 3,235 kg/km². It is quite probable that the present situation is causing a modification of habitats in favour of a re-colonisation by acacias and other trees and shrubs to the detriment of the grassy savannas, a situation where continued monitoring will be important.

8.3 Observed causes of decline

The main cause of the dramatic decline of all the large mammals in Virunga National Park is unquestionably the unprecedented level of poaching in the park over the past 15 years. Indeed, there have been no epidemics and no unusual natural mortality reported for the past two decades.

The fluctuations of elephant numbers mirror particularly well the efforts made in the anti-poaching war and the varying support given to the park authority, now the ICCN. It was during the 1960s that elephant numbers noticeably declined due to ivory poaching, with the disappearance of the biggest tuskers. Beginning in 1969, the park authority was re-established in the shape of the *Institut de la Conservation de la Nature du Congo* (ICNC) and the anti-poaching fight proved effective with elephant numbers stabilising, or even increasing, until 1978–1979. Due to a decline in support for the then park authority, the IZCN, the anti-poaching battle began to lose ground and a second drop in elephant numbers was recorded until around 1985. Between 1985 and 1991, a new stability or a slight increase in numbers was noted. From the early 1990s, on the contrary, the degradation of social and political conditions in the country brought with it a substantial relaxation of nature conservation efforts, and a new decrease in elephant numbers.

As far as the hippopotamus is concerned, the population's decimation is a direct consequence of intense poaching in the park over 15 years. As many as 12,000 individuals, representing more than half of the population present in 1989, were slaughtered in just five years, between 1989 and 1994. Therefore, this intensive poaching was not related to either the Rwandan refugee crisis or to the 1996 and 1998 wars. Starting in 1992, military camps were set up in Virunga National Park, and several patrols by the IZCN clearly exposed the poaching activities of the military inside the park. At this time, there were already large batches of hippopotamus canine teeth on sale in Goma (nearly 10,000 teeth in all) representing at least 2,500 individuals. In 2003 the discovery of groups of more than 30 hippopotamuses killed on the Rutshuru, whose canines had been removed, but which had not been butchered, also indicated unequivocally a substantial trade in hippopotamus ivory. In the early 2000s, the price of hippopotamus ivory was between three and five US dollars a kilo in the park and five to ten US dollars a kilo in Goma. Yet the hippopotamus is also poached for its meat, which is sold in many villages bordering the park, as well as in some of the illegal fisheries. At the beginning of the 2000s, poaching was so intense that the price of hippopotamus meat did not exceed one US dollar per kilo. In 2002 along the west coast of Lake Edward a whole hippopotamus would sell for 50 US dollars.

8.4 CONCLUSIONS

For the last ten years, at least, there has been considerable tangible proof of the direct involvement in poaching by men in uniform belonging to both the rebel groups and the regular army. As soon as a return to peace is completely established and unification of the government and the army fully carried out, it will be crucial to tackle this problem and to be sure that all the military factions leave the park. A corollary to this is the urgent need to build up the capacity of the ICCN (training, logistics and administration). This reinforcement, already initiated by various partners of the ICCN at the beginning of the 2000s, has been under way since 2005, with training sessions supported by different partners under the aegis of the FZS, and significant support from the European Union via the ZSL and the WWF.

It is important to emphasise the catastrophic plight of the large mammal populations of Virunga National Park. As the pillars of wildlife viewing tourism, they are key to the long-term existence of the park, and are thus an essential source of revenue for the local and national economies. It is imperative that urgent measures be energetically applied to remedy this situation. Yet grave though the situation is, it is by no means hopeless. No species has disappeared from Virunga National Park since its creation, with the sole exception of the African Wild Dog *Lycaon pictus,* exterminated from the park in the middle of the 1950s as in many other parts of Africa. As long as essential habitats can be preserved, poaching eradicated and support for the ICCN increased by the Congolese Government and external partners, there is every reason to believe that the park's animal populations could all be restored to their former levels. The stabilisation of numbers observed between 2003 and 2006, amounting even to a slight increase for some species, offers a glimmer of real hope in a context of emerging, if cautious, optimism.

One of the last aspects to consider in this framework is the key role played by neighbouring protected areas, in particular Queen Elizabeth National Park (QENP), immediately adjacent to the Central Sector of Virunga National Park. The historical differential fluctuations of elephant populations in Queen Elizabeth and Virunga National Parks, sometimes very rapid, cannot be explained by birth rate alone, but crucially point rather to migratory movements between these two protected areas (A. Plumptre, pers. comm.). We can expect that various animals, and notably elephants, having found refuge in Queen Elizabeth National Park, will return to Virunga National Park once poaching has been stopped.

Marc Languy, Carlos de Wasseige, Baudouin Desclée, Grégory Duveiller Bogdan & Stéphane Laime

9 Changes in land use on the periphery of Virunga National Park

*Marc Languy,
Carlos de Wasseige,
Baudouin Desclée,
Grégory Duveiller
Bogdan
&
Stéphane Laime*

When Albert National Park was created, more than 80 years ago, population pressure was already making itself felt, but it was fairly limited. 'Glass bowl' protection was the mode of conservation at this time, possible because human activities outside of the protected area had very little impact on the interior. There were, in addition, adjacent lands that played the role of buffer zones.

Now population pressure has intensified massively. This, coupled with increased agricultural activity, has forced a change in the conservation game. The number of people living on the immediate periphery of the park has exploded, virgin land has been invaded, ecological corridors cut and the demand for fuel and construction wood has increased tenfold.

The authors of this chapter examine and document the evolution of the anthropic pressures bordering the park over the course of time. The aim is to help future managers identify the priority areas for action, which should be integrated into wider park management. Some of these priority areas include the setting up of new buffer zones, the protection of the Rutshuru Hunting Domain, the promotion of alternatives to deforestation in the north of the park and the supply of firewood for the rapidly expanding cities.

9.1 Introduction

In 1925 vast areas on the periphery of the national park were neither inhabited nor farmed, neighbouring towns held a few thousand or, at most, a few tens of thousands of inhabitants, and some did not even exist. Agricultural pressure and urban pressure for firewood or construction timber were extremely limited and did not worry park managers. But since then there has been a demographic explosion in this region and the human landscape around the periphery of the park has dramatically changed. As a result, pressures on the park have multiplied to a point where they now constitute one of the foremost priorities for conservationists and, in parallel, for the authorities in charge of territorial planning.

It is vital we understand these pressures on the land inside and outside the park so we can identify the best ways to manage them. It is equally important to understand the dynamics of these pressures, so this chapter retraces the evolution of land use and its effects on Virunga National Park. The documentation of land use today will also allow the manager of tomorrow to understand the forces at work.

This chapter does not attempt to draw a complete map of all the human pressures on the park. Human population growth, societal structures, macro-economic mechanisms and even political stakes are only a few of the determining factors. We concentrate on land use, especially the recent rocketing of subsistence farming. Other chapters in this book tackle other pressures such as fishing, wood, politics and poaching.

Finally, there is very little quantified data available on some fundamental parameters such as population statistics, revenues or socio-economic activities, either past or present. Nevertheless, the changes observed are so significant that precise statistics are not necessary to document their impact.

9.2 The great population changes in the Democratic Republic of Congo

Since the creation of the national park, the world population has risen from 2 to 6.5 billion people. The greatest part of this increase has taken place in developing countries. The population of the Democratic Republic of Congo (DRC) is now 60,000,000 while in 1950 it was around a fifth of the size, at 12,184,000. The population growth rate calculated on the past five years is 2.79 per cent per year and, contrary to most other African countries, it is not likely to decrease before 2015. Thus, when Virunga National Park reaches its hundredth anniversary in 2025, the projected population of DRC will be 103,000,000-strong.

The average person in DRC is 16.3 year-old today, compared with 18.1 year-old in 1950. Thus, in 2008 DRC

9.1
In many places in the park, in particular in the Southern Sector, the two main land uses, farming and conservation, clash abruptly.

9.3 Residents adjacent to the park: evolution of pressure points

This section reviews the examination that Verschuren (1993) made of the progression of human populations towards the park since its creation and brings his findings up-to-date. Mentioned here are the sections for which there are witnesses or information documenting the *evolution*. At the end of the section, we bring together and summarise the pressures on park boundaries.

At the time the national park was established, human population density was low in the region of North Kivu, except for a few localities. Correspondingly, the creation of the park caused few land problems, except in some parts of the Northern Sector concerning fishermen and some stretches of the Beni area. Most of the people from the plains were evacuated before the park was created because of trypanosomiasis (sleeping sickness).

Until the beginning of the 1960s, the human population remained relatively stable or increased only slightly. But since the 1960s, a population explosion has been observed around the edges of the park.

9.3.1 Northern Sector

9.3.1.1 The Djuma-Nyaleke-Mavivi region

As aerial photographs from 1957 show, the region between Oisha (25 km north of Beni) and the Djuma River (forming the western boundary of Virunga National Park) was then entirely covered by forest. Today, this area of around 50,000 ha is almost entirely under cultivation. Landsat images from the 1980s clearly indicate that not all the forest degradation is recent but, at present, the 'agricultural front' reaches the Djuma River, and thus the immediate boundary of Virunga National Park. As illustrated in Section 9.4.3, the last wooded areas in the north of this sector were cut down recently (2000–2005).

Further south, between Mavivi and Nyaleke (7 km east of Beni), an examination of satellite images shows that this region was already deforested in the 1980s, but that the forest inside Virunga National Park was still intact in 2000. SPOT images from 2005 indicate, however, an invasion of more than 6,000 ha inside the park. A more detailed analysis reveals that in less than five years the rate of deforestation here reached 8.5 per cent per year (Figure 9.4). The forest was cut over an area of nearly 3,000 ha. This invasion ended in December 2005 when the area was entirely vacated, but the scars remain and it will take many decades for a natural habitat to be re-established.

9.3.1.2 From the bridge at Vieux-Beni to Mwenda

Vast areas of intact evergreen rainforest existed outside the park in this area in 1960. The situation persisted as it was until 1989, but with significant local deforestation outside the park. Verschuren noted hardly any population pressure up to the foot of the Rwenzori massif in 1989.

By 2005, however, fields had reached the park limits along an 8 km stretch immediately to the north-west of Mwenda. Further west, some 1,000 ha of forest still remain outside the park, but this forest is being cleared rapidly and fields will probably run along all the boundaries of the park within 5 to 10 years.

9.2
The village of Kanyabayonga, on the escarpment of the Mitumba Mountains, north-west of the Rwindi, is one of numerous pressure points in full development on the boundary of the park.

9.3
Farming is the principal activity of people in North Kivu.

ranked among the 10 countries in the world with the youngest population.

A rise in population does not necessarily imply a similar rise in pressures on natural resources since it depends on the level of development, the means of production, the agricultural productivity and the services provided by the state, such as energy supply. Unfortunately, in the case of the DRC, the socio-economic situation has consistently worsened since the 1980s. Food production per capita has fallen 14 per cent since 1985. Less than half of the population has access to drinking water, 29 per cent in rural areas. Only 35 per cent of children attend primary school and a meagre 12 per cent attend secondary school. These few indicators, among others, reflect the disengagement of the state over the past 20 years, leading to a situation where many people have no access to services.

In these conditions, the impact on natural resources is particularly severe. Many inhabitants have no alternative but to return to the forests – including those of the national parks – to satisfy their need for essential protein and energy, as will be seen in the following chapters.

While this is the case at the national level, the situation is more serious in North Kivu. The Rwandan refugee crisis of 1994–1996 and the population displacements provoked by the two wars in the 1990s damaged the economic security of many people and accelerated the explosion of urban populations, as will be seen in Section 9.5.

Marc Languy, Carlos de Wasseige, Baudouin Desclée, Grégory Duveiller Bogdan & Stéphane Laime

April 2000 *February 2005*

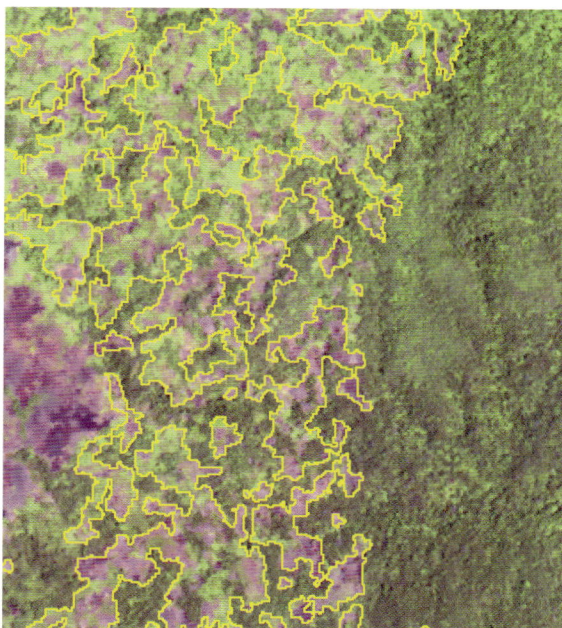

9.4
Deforestation in the area of Nyaleke-Mavivi inside the park (see Figure 9.13 for the location). On the February 2005 image, the forest appears in dark green. All the violet-purple polygons outlined in yellow are cleared areas that were still wooded in 2000.

9.3.1.3 FROM THE INTERSECTION OF THE BENI ROAD TO MOUNT TSHIABERIMU

Population pressure was already heavy along the boundaries in 1958, and today this is one of the sectors where the number of people living immediately adjacent to the park is highest.

Pressures were such that at the beginning of the 2000s, during an unfavourable political climate, 700 ha of forest on Mount Tshiaberimu inside Virunga National Park were cleared to make room for fields. The agricultural invasion of the park was rectified in 2003 and the section of Mount Tshiaberimu that lies inside the park's boundaries is again protected. However, the corridor linking the mountain to the shore of Lake Edward was still occupied at the end of 2008.

9.3.2 THE CENTRAL SECTOR

9.3.2.1 NEAR TSANZERWA-NYAMILIMA AND ALONG THE ISHASHA ROAD

The road north of the Kwenda River, crossing savannas abundant with game, was opened around 1957. A series of roads perpendicular to the main north-south axis come close to the park boundary. This new road had an enormous impact on the area as it brought tens of thousands of farmers and cattle breeders. This allowed the northern part of the Rutshuru Hunting Domain (*Domaine de Chasse de Rutshuru, DCR*) to be heavily invaded in 1989. Examination of satellite images shows that in 2006, *all* of the DCR north of Nyamilima is cultivated, people having settled in the immediate proximity of the valleys of the Ishasha and Kasoso rivers, both limits of the national park.

The National Park's boundary is locally defined by the upper eastern edge of the valley of the Kasoso River. This area is one of the most critical in the park. Before 1958–1959, all of the areas outside the park located north of Kisharo were made up of nearly intact immense savannas.

At the middle the Evi River, the forest of Kisharo (a stand of semi-evergreen forest in an area of the plain with fairly high rainfall) was originally proposed as an extension to the park from 1957–1958. Even then part of the forest had already been transformed into colonists' plantations, but today practically nothing remains of this forest.

9.3.2.2 WEST SHORE OF LAKE EDWARD AND CRESTS OF THE MITUMBA MOUNTAINS TO THE INTERSECTION WITH KABASHA

Population pressure, already strong in 1960, became intense nearly everywhere. A series of tracks take off perpendicularly to the crest road linking Kabasha, Lubero, Butembo and Beni, reaching the borders of the park. At the end of these tracks, large villages are found, often perched overlooking Lake Edward. The park boundary is made up of a series of straight lines joining the summits. The entire western part of these summits is cultivated. A typical example is a mountain dominating the plain west of the Lula River, very visible from the Rwindi and recognisable by its cleft summits. The croplands that dominate the crests overhanging the lake can be seen just about everywhere from the lake and the plain, which was not the case in the past.

9.3.3 SOUTHERN SECTOR

Verschuren (1993) gives figures on the number of inhabitants around the Southern Sector of Virunga National Park, noting that this region was always very populated. According to a report from the senior warden at Rumangabo, dated 19 August 1987, the number of people in each community was the following:
– Bakumu   30,360
– Bahunde   179,860
– Bahahi-Mokoto   173,976
– Bwisha   264,120
– Bwito   186,684

Changes in land use on the periphery of Virunga National Park

9.5
The village of Kibirizi, on the western flank of the Central Sector, has undergone significant growth over the past ten years. The village did not exist at the beginning of the 1960s.

9.6
Pressure on the land is very strong along the Mikeno sector, and park boundaries there have been sharply defined for more than half a century.

In total there were about 700,000 inhabitants surrounding the Southern Sector in 1987. It is difficult to obtain figures nearly 20 years later. Without taking into account the recent large-scale migrations from Ituri and the Maniema caused by the wars, but the natural growth rate of 2.7 per cent per year means that, after 20 years, the population should have reached at least 1,200,000. The town of Goma alone had more than 550,000 inhabitants in 2006.

9.3.3.1 Near Bugina-Tongo
In 1958, the Mabenga-Tongo track, lying south of the strip that links the Central Sector to the Southern Sector, traversed a nearly uninhabited region as far as Bugina. The expanses between Lake Ondo and Lake Kibuga, the sources of the Molindi River and the mazukus ('evil winds') on the one hand, and the foot of the mountain on the other, were unoccupied then. Human occupation on the mountain flanks was rare. Until 1961, villagers were settled only near the crests, and only an inconspicuous invasion of the slopes was to be seen. Now all the land outside the park is under cultivation, whether in the plain or on the slopes.

9.3.3.2 From Kibirizi to Tongo
This is a region of the park where the changes as a result of population pressures are the most striking. Before 1960, the cultivated areas were far from the park. Now thousands of hectares of the plain bordering Mount Kasali are cultivated by new people who have cleared virtually all the land. An example that is particularly representative of this phenomenon is illustrated in Section 9.4.

9.3.3.3 From Tongo to Sake
At Tongo, the areas at the foot of the escarpment (the triangle) are occupied only very locally. Farmers have not yet occupied the lava fields outside the park, nor several of the forest blocks important for the chimpanzees. However, these forests are the source of large-scale, if not industrial, charcoal production supervised by military personnel. An extension of the park in this region had been intended and it is still conceivable; the new boundary would be located near the foot of the escarpment.

In the area between Butambira and Sake, people progressively settled all the way up to the park's boundaries, but respected its forests, until around 1998. But from this year on, and mainly starting in 2002, at least 60,000 displaced people took over more than 10,000 ha of the forest (Dziedzic, 2005).

9.3.3.4 Sake-Mugunga-Kibati
Population pressures increased in the vicinity of Goma. All arable land outside the park is now occupied. Systematic cutting of firewood on the old lava fields outside the park is still happening. This activity is directly related to the extremely rapid expansion that Goma underwent at the end of the 1990s (see Section 9.5 for a particularly striking illustration).

9.3.3.5 Kibati-Kakomero
Contrary to Verschuren's indication (1993), the main road does correspond to the Virunga National Park boundary. The lava fields on the edge of the park are generally not farmed, except in the extreme north of this segment. They are not suitable for agriculture, but are vast areas of pastureland with very few settlements due to a lack of permanent water. This area constitutes good terrain for management as a buffer zone where wood production tests should be encouraged, in collaboration with the Ministry of Environment.

9.3.3.6 From Kakomero to the level of the Rutshuru airfield
The park boundary lies about five km west of the main road. Most of this area is an old lava field heavily wooded but hardly occupied, nor usable, except for wood-cutting. This tract constitutes a *de facto* protected area for the park, to be managed as a buffer zone. East of the road, by contrast, fields and plantations are everywhere and there is no available space left.

9.3.3.7 Boundary along the Congolese dormant volcanoes from the Rwandan border (south of Kibati) to the Ugandan border

From time immemorial, population pressures have always been strong along the boundaries of the dormant volcanoes.

On the Congolese side, the park includes vast stretches of middle altitude forests. The boundaries of the park here are at a much lower altitude than in Rwanda or Uganda. Strong population pressure is exerted at Kibumba, where a strip of forest connects the extinct volcanoes and the active volcanoes (Mwaro corridor, see Chapter 4). This forest has been respected by the people, but there is no natural habitat left on the periphery of the Congolese dormant volcanoes sector.

9.3.3.8 Synthesis of pressures on the ViNP boundaries

We have conducted a detailed analysis of the six SPOT satellite images covering the limits of the park in order to determine the immediate farming pressures on the *boundaries* of Virunga National Park. These images, taken in 2004 and 2005, have a spatial resolution of five metres in black and white and of ten metres in colour, which permits the identification of land use, and notably the detection of all the fields and other human occupation. The intensity of pressure on the park's boundaries has been classified into three categories:

– Low pressure: when natural habitats border the boundaries outside the park up to at least two kilometres, leaving a band of land that is not permanently occupied by man.
– Medium pressure: when the park's boundaries are not bordered by fields but these are found less than two kilometres away and/or when degrading human activity is carried out in the adjoining area.
– Strong pressure: when farmed fields are directly adjacent to the park boundaries and/or fields are inside the park.

Figure 9.8 shows these three levels of land-use pressures. The percentages for the categories of low, medium and high pressure are 33 per cent, 21 per cent and 46 per cent respectively.

It must be pointed out however that more than half of the low category areas benefit from the international character of the park's boundary, which is bordered by other protected areas on the other side of the frontier (Volcanoes National Park, Queen Elizabeth National Park, Rwenzori Mountains National Park and Semuliki National Park) or by the Ugandan part of Lake Edward.

Aside from the boundaries shared with other protected areas, the lowest pressure is exerted in the northern end of the park and to the west of the Rutshuru Hunting Domain. Between Rutshuru and Rumangabo, the immediate pressure is also weak because of the lava fields that border the park which are geologically recent and not suitable for agricultural purposes.

These 'green' zones must urgently be considered for durable natural resource management projects. They would benefit from a legal statute enforcing this type of management.

However, nearly the entire western flank of the park suffers strong or medium pressure, as well as the region adjacent to Goma of course.

9.4 The advance of the farming front towards Virunga National Park

As outlined above, most of the park was bordered, at its creation, by entirely natural areas without any form of agricultural activity. This situation still largely prevailed in 1959, as attested by the aerial photos from 1958–1959. The population pressures described above have, however, driven people to extend the cultivated areas. In many places, a 'farming front' has progressed right up to the park boundaries. To illustrate this, we look at three sites from the Northern and Central Sectors of Virunga National Park.

9.4.1 The case of the Rutshuru Hunting Domain

The first illustration of the impact of population explosion on land use around the periphery of Virunga National Park is demonstrated by the Rutshuru Hunting Domain, created in 1946. Its management and borders were modified several times until the promulgation of Law No. 00024 of 14 February 1974, which placed it under the management of the *Institut Zaïrois pour la Conservation de la Nature* (IZCN), a precursor to today's *Institut Congolais pour la Conservation de la Nature* (ICCN).

The analysis of agricultural expansion was based on aerial photos taken in 1959, Landsat images from 1987, and SPOT images from 2004. The results are represented in Figures 9.9 and 9.10. The study area forms a rectangle of 163,000 ha encompassing the Hunting Domain and adjacent areas. An area of 157,000 ha was free of clouds and cloud shadows on the three dates the images were acquired.

In this study area, land that was subjected to agricultural activity totalled slightly less than 14,000 ha in 1959. It increased to nearly 81,000 ha in 1987, and to 92,000 ha in 2004.

9.7
The Kilolirwe region, north of Sake on the western flank of the Southern Sector, has always been occupied by large cattle farms. At the beginning of the 2000s, however, the interior of the park was invaded by several tens of thousands of people. On this photograph, the park boundary follows the row of eucalyptus forming a diagonal line. The upper left of the photo shows meadows outside the park that have been established for several decades. The bottom right of the photo shows the area inside the park which has been illegally farmed since 2002.

9.8
The categories of intensity of pressures on park boundaries (see text).

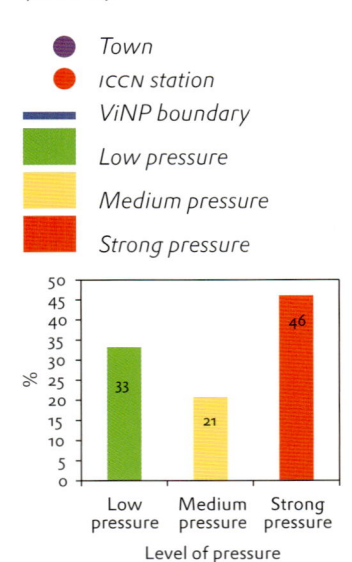

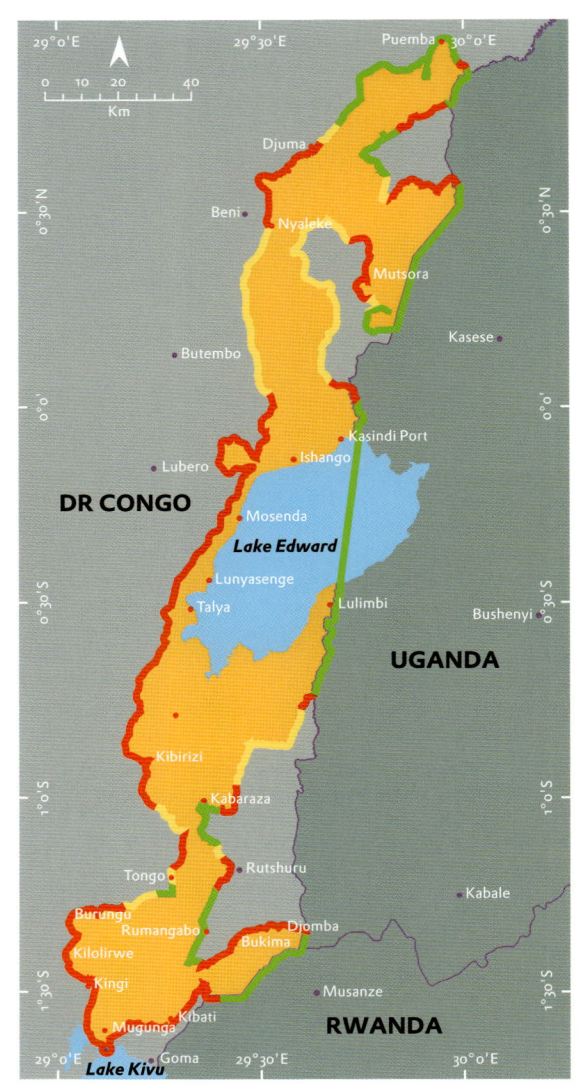

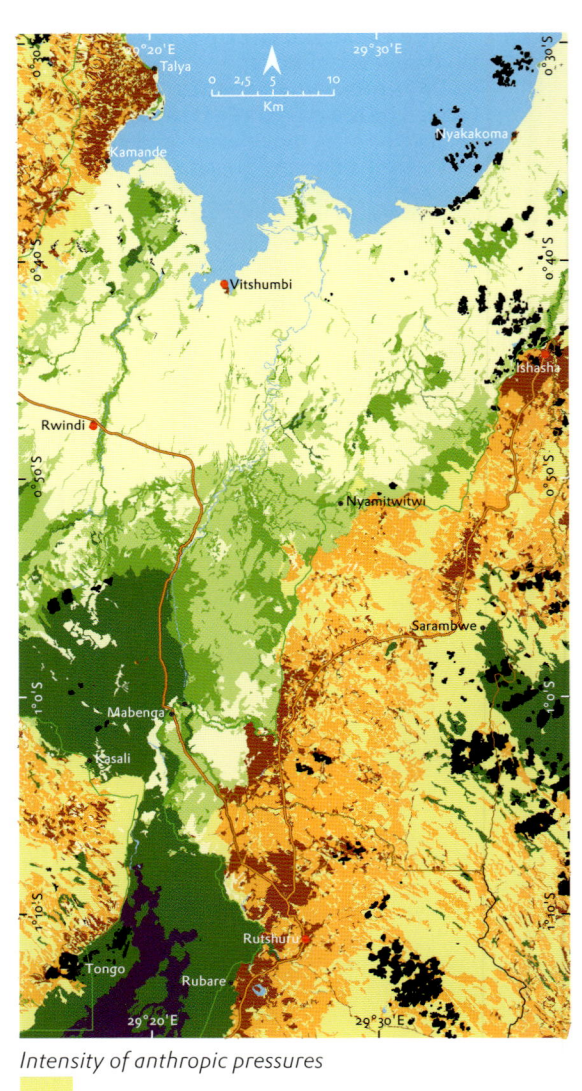

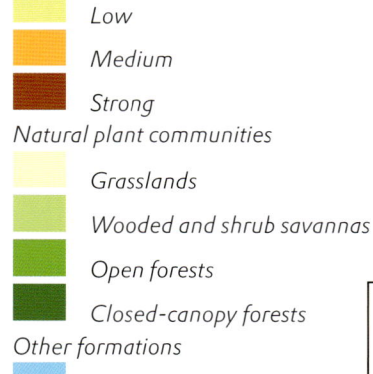

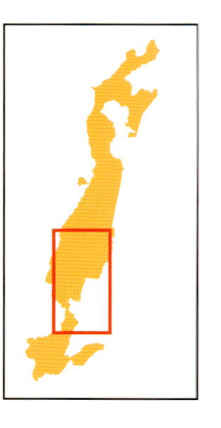

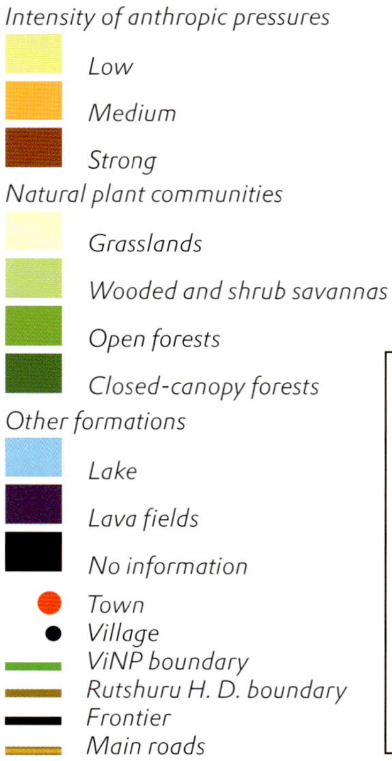

- Town
- ICCN station
- ViNP boundary
- Low pressure
- Medium pressure
- Strong pressure

In 1959, it was necessary to travel 17 km from Mabenga in the direction of Rutshuru before encountering the agricultural front that was radiating out from the town. It took only nine km to reach it in 1978, and seven km in 1994 (Languy, 1994).

The 2004 satellite images show that since Languy's observations in 1994, the agricultural front has stabilised along the main road. This pause probably reflects the combined effect of increasing impoverishment of the soil (becoming more and more sandy and salty) and lower rainfall (result of the relief causing a *foehn* effect) as one approaches the park from Kiwanja. However, the invasions continued further north and south of the road, where fields now reach the boundaries of the park formed by the May-ya-Kwenda and Rutshuru rivers, respectively. This expansion is, of course, in parallel with that of the towns of Kiwanja and Rutshuru, as shown in the following section of this chapter.

9.4.2 THE AREA LYING BETWEEN THE RWINDI AND RWERE RIVERS.

This area constitutes one of the sites in the comparative photos (see Chapters 7 and 23) established by J. Ver-

9.9
Land use and intensity of anthropic pressures in the western part of the Rutshuru Hunting Domain

Intensity of anthropic pressures
- Low
- Medium
- Strong

Natural plant communities
- Grasslands
- Wooded and shrub savannas
- Open forests
- Closed-canopy forests

Other formations
- Lake
- Lava fields
- No information
- Town
- Village
- ViNP boundary
- Rutshuru H. D. boundary
- Frontier
- Main roads

158 — Marc Languy, Carlos de Wasseige, Baudouin Desclée, Grégory Duveiller Bogdan & Stéphane Laime

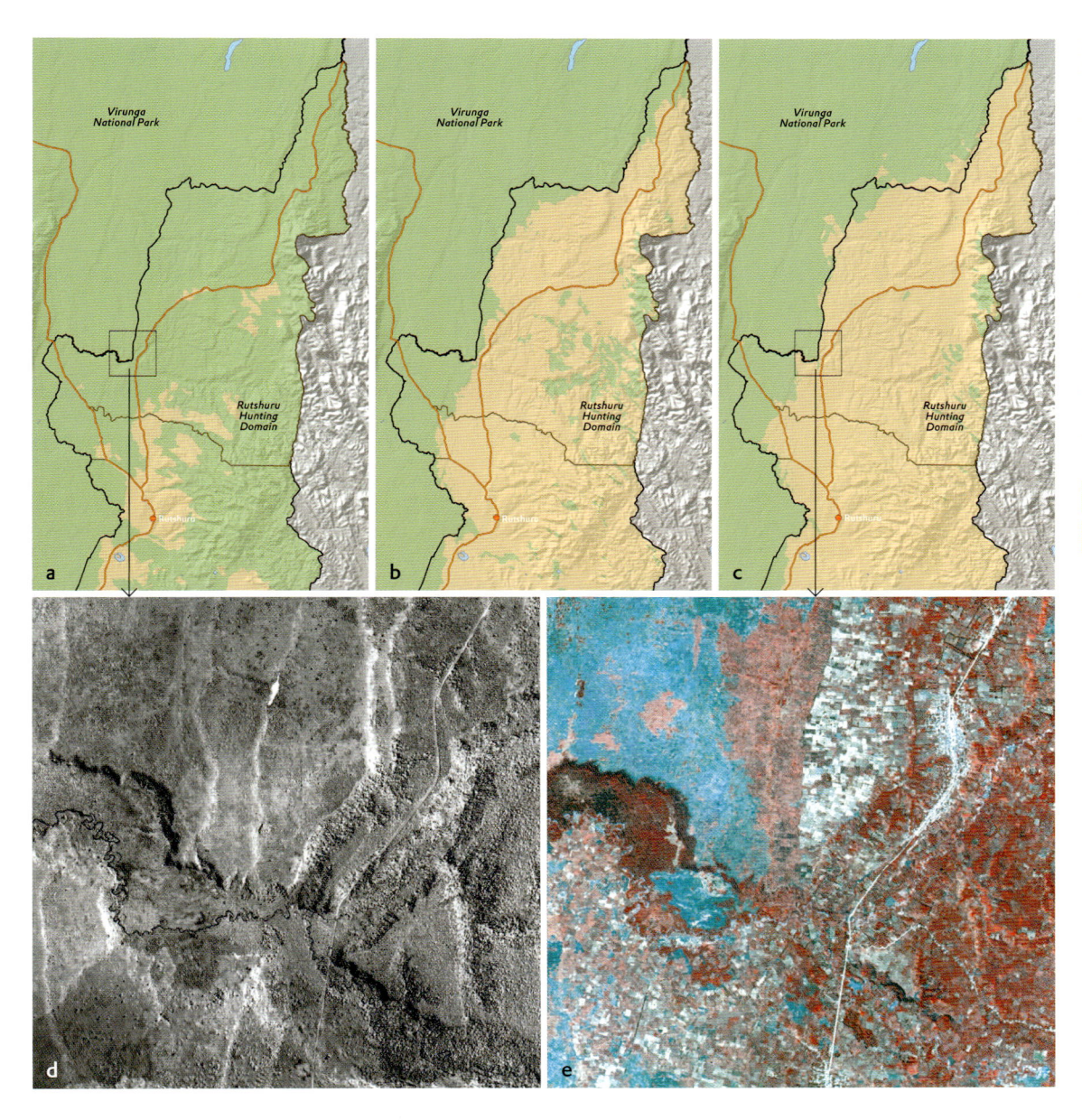

9.10
Extent of the cultivated areas in the Rutshuru Hunting Domain in (a) 1959, (b) 1986 and (c) 2004. The three images above show progressive invasion of the Rutshuru Hunting Domain and the advance of farm fields towards the park. The two images below are extracts of (d) an aerial photo from 1959 and (e) a spot image of the same area in 2004, showing details of the area near to the May-ya-Kwenda River. The agricultural front is very visible in the centre of the upper part of the satellite image.

- Natural habitats
- Cultivated areas

schuren. Here we have the 1959, 1988 and 2006 photos, as well as a SPOT image from July 2004. The confluent of the Rwindi and Rwere is located about 18 km south of Rwindi Camp. Starting from the junction of these two rivers, in the south, the park boundary corresponds to the Rwere, the Rwindi 'leaving' the park (Figure 9.11).

The 1959 photographs show that the whole expanse between the Rwindi and the Rwere (outside the park) was then uninhabited and covered with natural sclerophyllous forest and wooded savanna.

In April 1988, the entire area west of the Rwindi and between the two rivers was invaded by agricultural fields. There is no longer a tree standing. The area to the east of the Rwere is intact, however, and natural reforestation occurred between 1958 and 1988, very probably because of the reduction in the number of elephants.

In March 2006, progress of the invasion can be seen. The stretch between the rivers is farmed but the area east of the Rwere still remains intact. One will notice, however, a return of tree cover in the cultivated area, but this is because of tree farming or orchard plantations.

It is thus possible to observe that there is no longer any 'free' space between the national park and the cultivated areas of more than 3,000 ha. This land was totally uninhabited when the park was created, up to at least the beginning of the 1960s.

9.4.3 The lower Semliki

Between 2000 and 2005, we analysed the deforestation in the lower Semliki region, within and outside the park. The results are illustrated by a detailed map that shows the area bordering the Djuma River (Figure 9.12) and the region studied (Figure 9.13).

Figure 9.12 clearly shows that the region around the park at the level of the Djuma River was already very degraded in 2000 but that it still had some visible for-

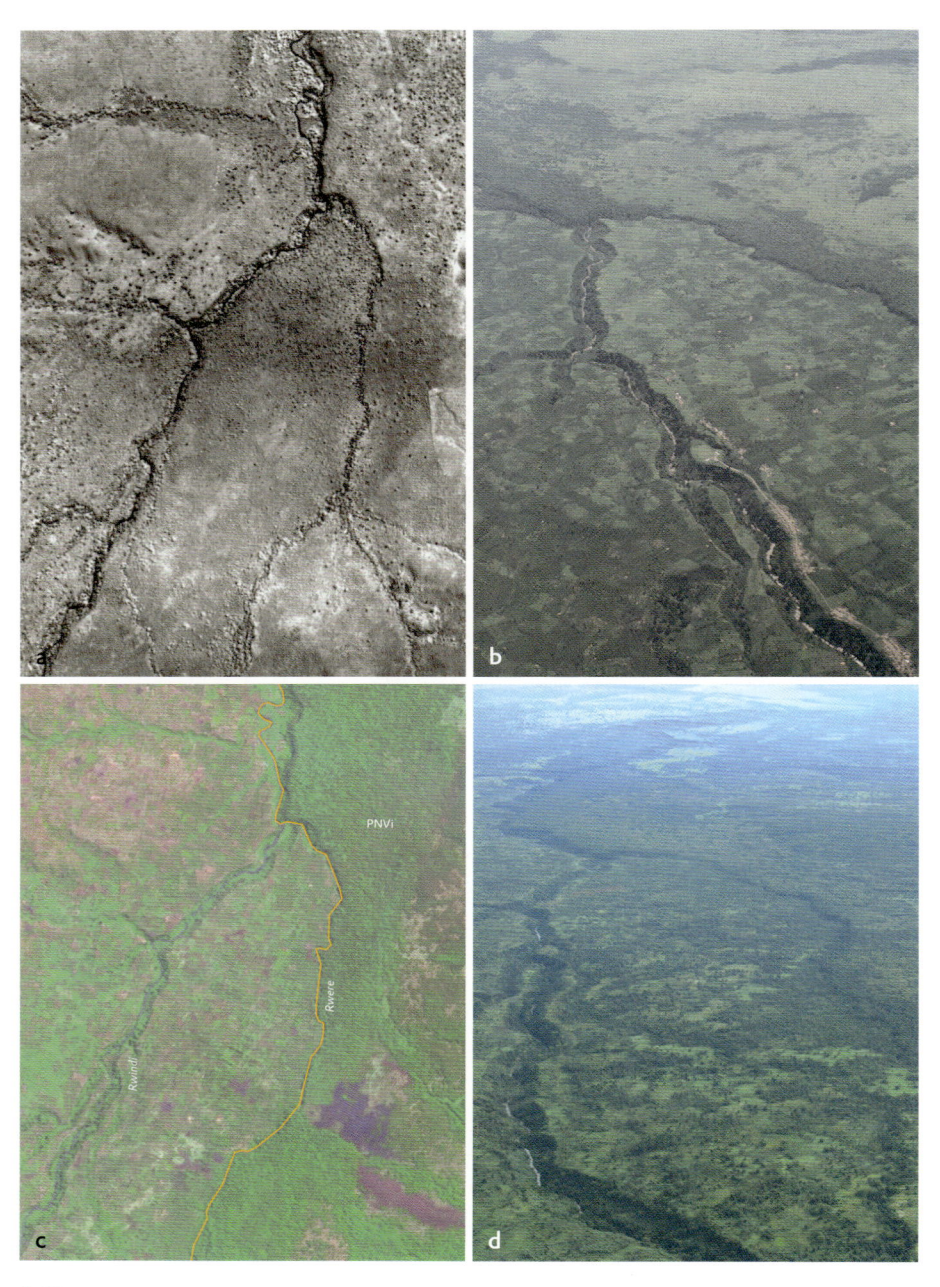

9.11
The area lying between the Rwindi and Rwere rivers. Photographs taken in (a) 1959, (b) 1988 and (d) 2006 and (c) an extract from a satellite image made in 2004.

On this map, four areas of natural forest bordering the park can be identified:
- The area between the Semliki River and the Rwenzori Massif, along a line from Nyaleke to Mwenda (3,580 ha),
- The area immediately north of Djuma, west of this river (6,800 ha),
- The area right on the border in the extreme north of the park (8,200 ha),
- The area east of Virunga National Park and west of Kamango, between the Mbau-Watalinga road and the Rwenzori – centred on 00° 39' N, 29° 47' E – (about 11,000 ha).

A plan for land use is urgently needed in these four areas, including an investigation into whether community forests, founded on the new Congolese forestry code and managed by the resident communities, would be an option for ensuring a more durable management of some of these forest patches. In fact, maintenance of the forest cover between this part of the park and the Ituri Forest, located in the west, would permit the preservation of an ecological corridor between these two areas.

There are no longer many areas where the natural habitat is still intact in the immediate periphery of the park. As well as areas of the Northern Sector identified above, there remain the extreme west of the Rutshuru Hunting Domain (see Section 9.4.1) and the eastern flank of the park between Kakomero and Rubare (Section 9.3.3). These different areas (represented in Figure 9.14) constitute the last buffer zones of the park, although none had been managed as such. This kind of management must be developed as part of the park's management plan in the years to come.

9.5 THE EXPANSION OF LARGE CITIES AND NEIGHBOURING VILLAGES

Two urban centres located on the periphery of the park were studied to document their expansion, Goma and Kiwanja/Rutshuru. In these two cases, the aerial photos taken in 1959, Landsat images in 1987 and 2001, and SPOT images from 2004 were used to determine the area occupied by the cities.

9.5.1 Goma

In 1959, Goma was a very small town. It grew progressively until the beginning of the 1990s. Then, the Rwandan refugee crisis and the two wars caused a large influx of people who were either following the economic prosperity engendered by the humanitarian operations, or fleeing the insecure areas in Maniema, Masisi, and even Ituri. In 1959, Goma covered an area of 482 ha. In 1987, the town's surface area nearly tripled, extending to 1,375 ha. In 2004, the town occupied 3,508 ha, seven times more than a half century earlier, with a population estimated at 550,000 people (Figure 9.14).

9.5.2 Rutshuru/Kiwanja

The expansion of Rutshuru, and then of Kiwanja, is just as spectacular. In 1937 Rutshuru had between 5,000 and 6,000 inhabitants (Verschuren, 1993). This grew to 50,000 in 1991 (Harroy, 1987; Verschuren, 1993) – bringing together the inhabitants of both

est patches. Although there were some patches left in 2005, several blocks of up to 10 ha have been cleared. At this rate, it can be estimated that in less than 10 years there will not be a single forest patch left next to the park. The average rate of deforestation in the Djuma region was 1.5 per cent per year.

The general map of deforestation in the lower Semliki region (Figure 9.13) illustrates two important elements. One is that all the forests bordering Virunga National Park suffer from clearing that increases as one approaches the park, and the other is that the rate of deforestation varies quite strongly from one region to another.

Over the entire area represented in Figure 9.13, the loss of forest between 2000 and 2005 is 3,500 ha, including some 2,000 ha along a strip of 10 km outside the park.

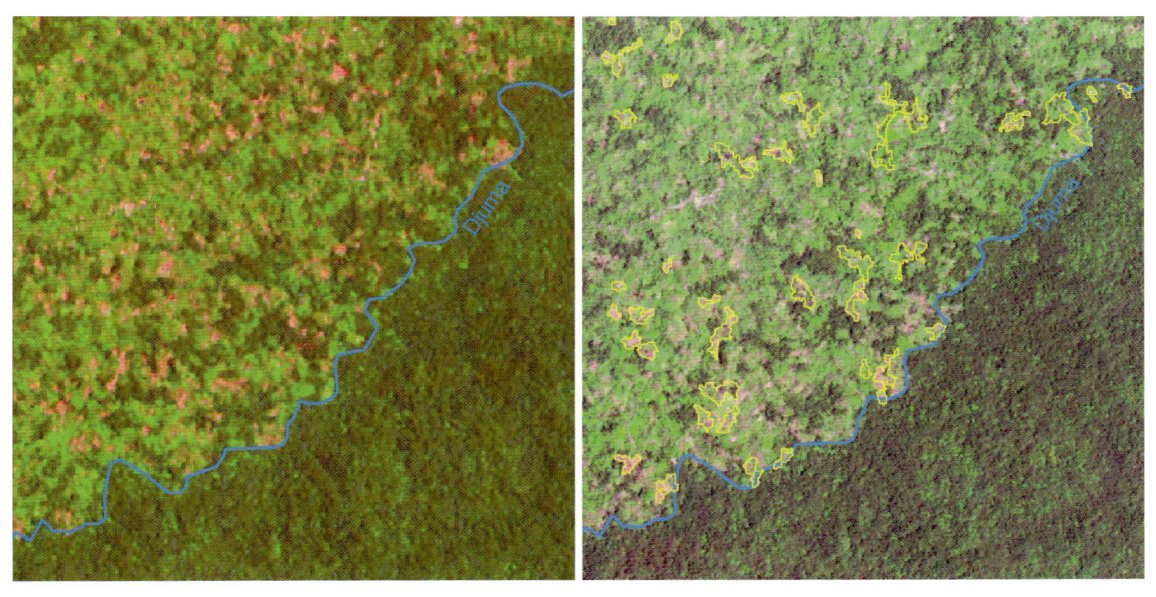

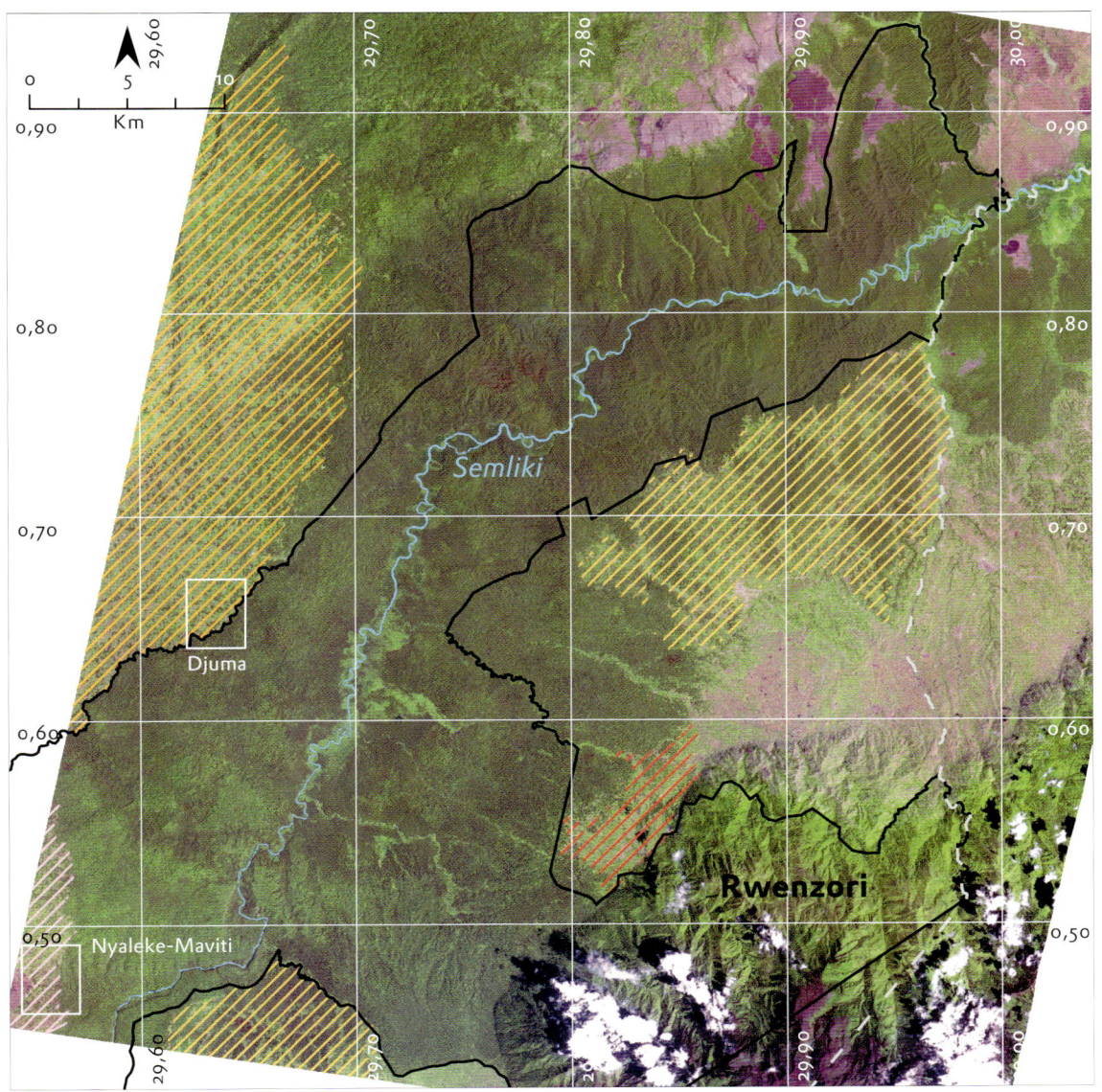

9.12
Deforestation in the area of Djuma, between April 2000 and February 2005. The river forms the boundary between the park (below right) and the non-protected area (above left). In 2000, the area outside the park was already broken up but many forest patches (marked off by yellow on the 2005 image) were cleared between 2000 and 2005.

9.13
Map of deforestation in the lower Semliki between 2000 and 2005.

Legend
Annual rate of deforestation
- 0,5 %
- 1,5 %
- 2,0 %
- 8,5 %

Extracts of satellite images
- Closed forest
- Savanna
- Agricultural complex
- Cloud and shadow

Changes in land use on the periphery of Virunga National Park

9.14
Expansion of the town of Goma from 1959 to 2004 on the basis of the study of (a) aerial photographs taken in 1959, (b) of a Landsat image from 1987 and (c) a SPOT image from 2004.

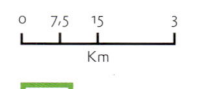

1959	Extent of Goma
1987	
2005	
	International frontier

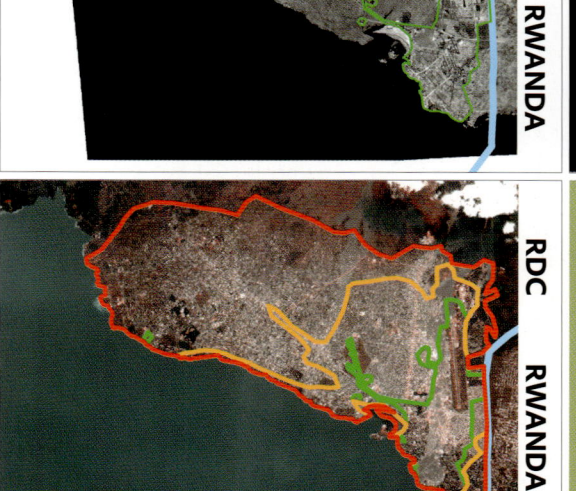

Rutshuru and Kiwanja. Kiwanja was created around the middle of the 1950s, when the road to Ishasha was opened. In 2008, the populations of Rutshuru and Kiwanja are estimated to be 15,000 and 40,000 inhabitants respectively.

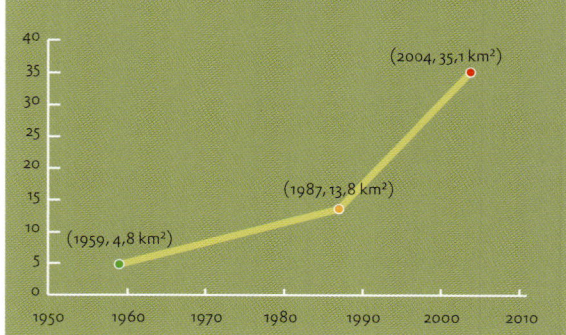

9.6 Conclusions

This chapter documents the enormous expansion of human activity, essentially agricultural, bordering the park since it was created. The aerial photos from 1958–59 are very important because they show that vast expanses along nearly the entire periphery of the park (with the exception of the Mikeno sector) were uninhabited and unfarmed. This is particularly visible along the whole western flank of the park and the Rutshuru Hunting Domain.

Examination of satellite images and the results of surveillance patrols have identified those parts of the park's boundaries that are most subjected to heavy pressures and therefore the areas where boundary demarcation work must be a priority. The western shore of Lake Edward and of the Kilolirwe region are two 'hot spots' where all efforts to recover the integrity of the park must be concentrated.

The same examinations located areas along the immediate boundary of the park that are still natural and so able to play the role of buffer zones, such as the lava fields on the eastern flank of the Southern Sector between Kakomero and Rubare; the western part of the Rutshuru Hunting Domain; the Tongo triangle; the forests of Djuma and Mwenda and, lastly, the western part of the Watalinga enclave. These are areas where durable forest management initiatives should be undertaken with, if possible, controlled wood extraction.

Examination of the expansion of Goma, along with other towns and large villages on the edge of the park, shows that the Congolese Government must act quickly to contain these developments (for example, by reopening areas of strong agricultural potential further to the west, combined with improving security). The Government must manage the growing towns' increasing demands for energy and make sure that the

9.15
Aerial view of Kiwanja, March 26, 2006.

9.16
SPOT image showing the extension of Rutshuru and Kiwanja in July 2004.

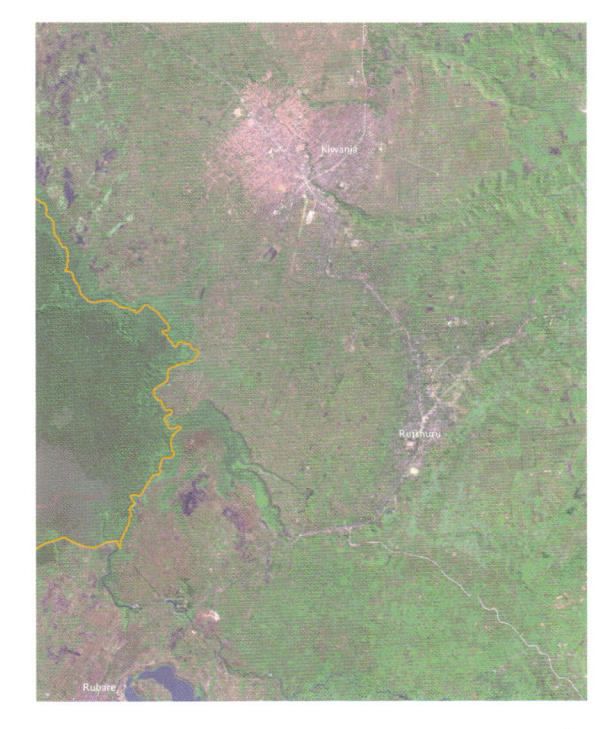

park provides economic and ecological resources without being exploited illegally and unsustainably, threatening the future existence of this national treasure and world heritage site.

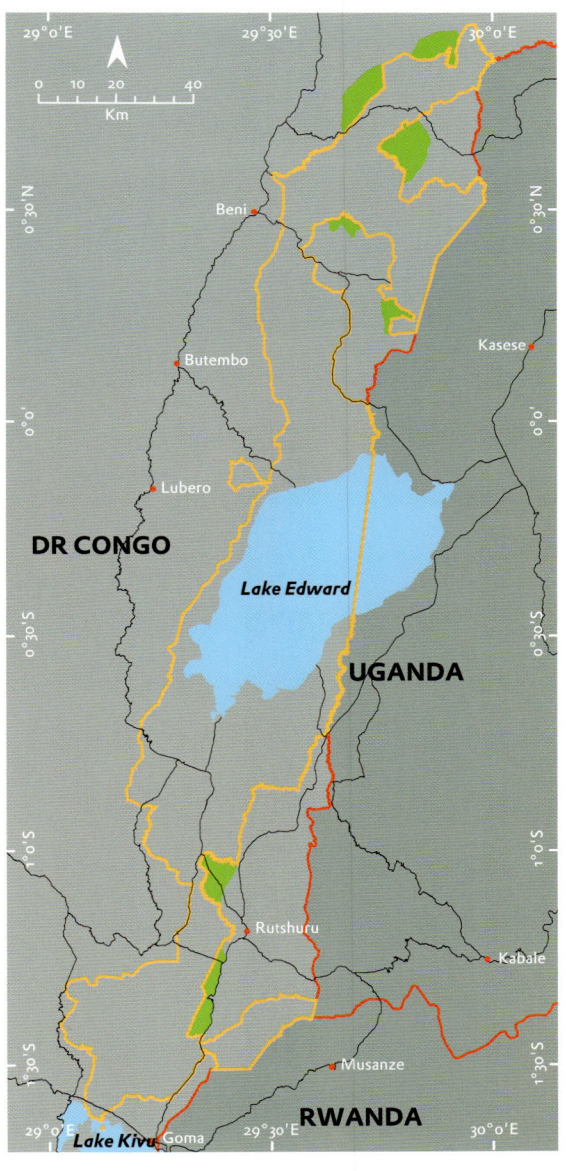

9.17
Areas of remaining natural habitat bordering Virunga National Park in 2006. These areas could serve as buffer zones to be managed jointly by the ICCN and resident communities.

Legend
― Roads
― International frontier
― ViNP boundary
▇ Buffer zone

Changes in land use on the periphery of Virunga National Park

163

Frédéric Kasonia & Norbert Mushenzi

10 History of the COPEVI: the use and management of Lake Edward

Frédéric Kasonia & Norbert Mushenzi

Lake Edward, because of its geographical position and physical characteristics, is among the most productive lakes in the world. The section of the Lake that is within the DRC is classified as part of Virunga National Park. Thus, the management of access rights to the Lakes resources by the park authorities is a necessity, and needs to be handled with considerable care. As such, it is a constant challenge. As a solution, the national and local authorities, advocated for the organisation of the fishermen into a co-operative (see Chapter 1 for the general context). This co-operative was structured to ensure the management and control of fishing while optimising production, as much for the well-being of the people as for the protection of the park.

The past years, characterised by a loss of control by the national authorities, has deeply disrupted the management structures described in this chapter. The preoccupying growth of established fisheries and the creation of illegal new fisheries are described in Chapter 13; this chapter discusses the co-operative itself.

10.1 The COPEVI, the former COPILE

The COPEVI (*Coopérative des Pêcheries des Virunga*, Co-operative for Virunga's Fisheries) was established as a private enterprise, originally named the COPILE (*Coopérative des Pêcheurs Indigènes du Lac Edouard*, Co-operative of the local fishermen of Lake Edward) to administer the fishing rights of the following chiefdoms: the Baswagha, Batangi and Bamate chiefdoms and the Bapere sector in the territory of Lubero; the chiefdoms of Bashu, the Watalinga and the Beni-Mbau and the Rwenzori chiefdoms in the territory of Beni; the chiefdoms of Bwito, Bwisha in the territory of Rutshuru and, lastly, the chiefdom of Bukumu in the territory of Nyiragongo-Goma. The COPILE represented the people of these chiefdoms and was created in 1949 in compensation for the lands conceded to the IPNCB (*Institut des Parcs Nationaux du Congo Belge*), following the recommendations of the committee of inquiry (*commission d'enquête*) of Louis De Waersegger in 1947 (see Chapter 1). The COPILE had its headquarters at Vitshumbi, in Rutshuru territory, and originally supervised the two state fishing companies in Vitshumbi and in Kyavinyonge, also created in 1949.

The people who directly benefit from the fishing are those from Beni, Lubero, Rutshuru, Nyiragongo and Goma and, indirectly, the populations of other provinces of the Democratic Republic of Congo.

The COPILE was created to exploit fish on Lake Edward, and was given the exclusive right to do so. These rights were assigned through the acts of cession of indigenous rights, in compensation for cessation of lands by the people when the national park was created. Authority within the COPEVI could be exercised either individually or by delegation, as provided for in Article 5 of its bylaws.

The eleven chiefs of the chiefdoms and sectors were automatically nominated as co-operative members. At the time, the *Commissaire de District* was the president of the board of directors of the COPILE, and the vice-chairmanship went to one of the co-operative chiefs elected by the General Assembly along with five other members, one per chiefdom or sector. Each Administrator would automatically be nominated as technical advisor to the COPILE in his respective territory and would attend meetings in an advisory capacity.

The name of the Lake Edward fisheries co-operative has been changed many times over the past fifty years.

The following are some of the changes that have been made:

COPILE: *Coopérative des Pêches Indigènes du Lac Edouard,* at its creation in 1949.

COPILA: *Coopérative des Pêcheries Industrielles du Lac Amin,* a change of name in 1972 and 1973, when Lake Edward became Lake Idi Amin.

10.1 *A product typical of Vitshumbi is the famous* makayabo, *dried and salted tilapia.*

10.2
Sunset on Lake Edward.

ONP: *Office National de Pêche,* transformed into a state company by nationalisation on 17 April 1975, and the headquarters and office of the directorate general established in Kinkole (Kinshasa), answerable to the Presidency of the Republic until 1976, date on which the ONP put an end to the fishing activities on Lake Edward.

COPILA: The co-operative members of the former COPILA, who were now nationalised into a state enterprise ONP, saw in it an abuse of confidence on the part of the *Office National de Pêche* and demanded the restoration of their fishing rights, which led to reinstating the former name COPILA in 1978.

COPEVI I: *Coopérative des Pêcheries de Vitshumbi.*

After the recuperation of their fishing rights on Lake Amin, when this president was no longer in power in Uganda, the company became the *Co-opérative des Pêcheries de Vitshumbi,* named after the location of its headquarters in 1970.

COPEVI II: *Coopérative des Pêcheries des Virunga.*

This brusque change of name occurred after the theft of the company seal, when the change of name and a new stamp were made in 1983 to avoid fraud.

10.2 THE OBJECTIVES OF THE COPEVI CO-OPERATIVE

The COPEVI was, from the beginning, a company with commercial and social objectives; according to its statutes, its primary objectives were as follows:
– Fishing, treatment and commercialisation of fish and their by-products from Lake Edward.
– The removal, sale and providing of fish to the people of the territories whose chiefdoms and sectors are members of the cooperative at the most competitive prices possible.

Formerly, the colonial administration had created a state-run fishery at Vitshumbi around 1943. At the request of the Governor of the Province of Costermansville (now Bukavu), it was handed over to the indigenous fisheries that were being created. The colonial administration also authorised the establishment of a fish storage facility in Lunyasenge to provide the local population with animal proteins. This storage facility could handle only the temporary transit of fish and was not permitted to build installations, and certainly not permanent ones.

10.3 PRODUCTION AND MEANS OF FISHING

Traditional fishing was practised long before the park and the COPEVI had come into existence. Early settlers in the region used a type of net called a *ngana* in shallow waters. The more daring of these traditional fishermen would gather three or four floating banana tree trunks to fish in deeper waters. Now and then, a small dugout would be used. When the COPILE came into existence, the fishermen undertook 'industrial' fishing and began using the appropriate fishing equipment that included 32 fishing boats from the Vitshumbi and Kyavinyonge fisheries, using an experienced, qualified and conscientious work force. At this point, the co-operative was flourishing. The production statistics for the years 1949 to 1964, presented below, testify to this success:

Year	Production
1949	2,000 tons
1950	1,773 tons
1951	1,227 tons
1952	2,062 tons
1953	2,836 tons
1954	3,241 tons
1955	3,003 tons
1956	3,101 tons
1957	2,200 tons
1958	4,211 tons
1959	5,000 tons
1960	6,010 tons
1961	5,443 tons
1962	4,000 tons
1963	3,200 tons
1964	3,900 tons

During these prosperous times, the COPEVI had a monopoly on all fishing activities. The cooperative became an industrial fishing operation with a sophisticated monitoring programme collecting regular data (platform scales, weight register, etc).

The 1964 rebellion marked the beginning of a progressive decline of the COPEVI and a dramatic drop in production from an average yearly production of 3,325 tons (cf. report by Dr. Seundi, March 1999), mainly because of the growing number of unrecorded dugouts whose fishing efforts escaped the COPEVI and its record keeping.

There was a modification of the COPEVI's bylaws in 1983, when the sub-regional Commission for the COPEVI turned over the presidency of the co-operative to Mwami Buunda Birere, a co-operative member, and the vice-presidency to Mwami Atshongya Mutombo Kasereka, while the sub-regional Commissioner thus became a simple member of the Council of Administration. In 1984 the COPEVI formally resigned its monopoly and turned over part of its fishing rights to individual fishermen, effectively taking them on as sub-contractors of the COPEVI with the idea of increasing production. This initiated a period of liberalised fishing, which significantly increased production, but was largely unstructured and unplanned, leading to a progressive breakdown of fisheries management around Lake Edward.

In principle, fishing quotas were regulated. The Government fixed the number of dugouts legally authorised in the Vitshumbi and Kyavinyonge fisheries based on the recommendations of a panel of experts. A key personality here was J.M. Vakily, an expert who, after conducting a scientific study on Lake Edward funded by the European Commission, established the number of dugouts to be 700 units, distributed as follows:
- Vitshumbi fishery 400 dugouts
- Kyavinyonge fishery 213 dugouts
- Nyakakoma private fishery 87 dugouts

Totalling 700 authorised dugouts along the Congolese shore.

The quota of 700 dugouts was based on this very detailed study to ensure that fishing yields did not exceed the productive capacity of the lake. But according to Vakily (1989) even at the time of study, fishing pressure was already unsustainable.

In July 1994, Languy (1994) counted more than 1,100 dugouts on Lake Edward. Today, because of 'pirate' fishermen and clandestine fisheries, that number is over 2,000 and includes motorised and non-motorised dugouts, the latter usually fishing in the spawning grounds using nets with prohibited mesh sizes. The COPEVI is powerless to stop them.

10.4 THE CASE OF THE NYAKAKOMA FISHERY

The Nyakakoma fishery was established in 1965 for local residents who had handed over their rights to lands included in Albert National Park. Astonish-

10.3
All of the fisheries on Lake Edward in the Congo are located inside the park which is managed by the ICCN.

10.4
All of the fishing practised on Lake Edward in 2008 is artisanal.

10.5
ICCN guards questioning fishermen found working in one of the spawning grounds, areas where fishing is prohibited in order to protect fish reproduction.

ingly, the fishery has become the private property of Mwami René Ndeze, of the Bwisha chiefdom in Rutshuru territory. The Mwami took possession of Nyakakoma shortly after replacing his father, Mwami Daniel Ndeze, who became a political refugee in Belgium during the post-independence period. This form of privatisation by the late president Mobutu was not uncommon between 1965 and 1968, but went against the co-operative principles on which the lake had been managed and undermined collective fishing rights of local residents.

Nyakakoma fishery was closed for a time during this period by the central government and placed under the supervision and conduct of the directorate of the Vitshumbi COPEVI, as was the Kyavinyonge station.

10.5 The establishment of the COPEVI Directorate at Kyavinyonge

In 1999, with the rebellion that took place and, in particular, the creation of two RCDS (RCD/KML separated from RCD/Goma), the Vitshumbi directorate remained in the south working with co-operative member chiefdoms including Rutshuru and Nyiragongo-Goma territories administered by the RCD/Goma, while the Kyavinyonge directorate was created by those of the Beni and Lubero territories which were administered by the RCD/KML. The COPEVI directorate of Kyavinyonge was thus created during the conflict period and was administratively independent of the Vitshumbi directorate from 28 June 2000. The same decree that created the COPEVI directorate of Kyavinyonge (No. 01/023/CAB/CP-NK/2000) also established three other COPEVI stations dependant on this new directorate, namely the stations at Kasaka, Lunyasenge and Kasindi. They are still operational today, in spite of the fact that ICCN was not associated with these decisions.

This was a clear breach of process given that the territory where the fisheries operate is an integral part of the national park and comes clearly under the responsibility of ICCN.

10.6 Administrative organisation of the COPEVI-Kyavinyonge Directorate

The following governance structure oversees the management of the Kyavinyonge COPEVI.
- A General Assembly of which the governor of the province of North Kivu and the bishops of the dioceses of Butembo-Beni and Goma are honorary members.
- Counsel of Administration according to the bylaws of the co-operative.
- An executive committee of three accountants to oversee the financial affairs of the enterprise.
- An Executive Director to administer daily activities.

10.7 The question of services associated to the fisheries

The various interventions of state actors together with those of religious groups has precipitated the growth trade and increased immigration. The breakdown of law and order during the conflict years has further exacerbated a chaotic and unstructured growth of Kyavinyonge. However, the gradual restoration of the rule of law has begun to have an impact. In 2003, for example, the population had reached 22,000 inhabitants at Kyavinyonge; in 2005, it was down to 18,000, a substantial reduction attributable to the energetic banning of subsistence agriculture inside the park.

No further census has been conducted since the beginning of 2005. However, the number of boats officially authorised by the Central Government in

Kinshasa is still fixed at 700 dugouts on the Congolese shore (see above). To these legal dugouts must be added 100 dugouts at Kisaka port, authorised without the agreement of the ICCN during the wartime period, as well as 100 completely illegal dugouts at the Lunyasenge and Kasindi port stations, 'authorised' by Decree No. 01/023 of 28 June 2000 mentioned above, again without ICCN approval.

10.8 Recommendations

In order to remedy the intractable problem of multiple fisheries on Lake Edward, within Virunga National Park, a number of issues need to be addressed:
- Reinstate the COPEVI with its power to allocate collective fishing rights on Lake Edward, rights that were violated by actors such as the illegal taxi-dugout operators.
- Assign legal powers to COPEVI that would guarantee the protection of its rights on the lake as well as in the fishing concessions.
- Improve the conditions for regular fishermen by allocating two legal fisheries with drinking water and adequate sanitary conditions.
- Suppress all pirate fisheries and re-establish patrol posts at the park boundaries
- Reduce all activities that have little to do with the fishing activity except where they can clearly be shown to have a zero or positive impact on the management of the park.
- Encourage the authorities at the various decision-making levels and the ICCN to help the COPEVI restructure itself and rebuild its infrastructure, notably by recognising the collective convention bearing on all of Lake Edward.

10.6
One of the numerous spawning grounds of the lake; these are very shallow places sheltered by vegetation.

11.1
The ascent of Rwenzori is one of the main tourist attractions of Virunga National Park.

Annette Lanjouw, Marc Languy & Jacques Verschuren

11 Virunga National Park: a jewel for tourism in the Democratic Republic of Congo and in the Great Lakes region

Annette Lanjouw, Marc Languy & Jacques Verschuren

Prior to the political crisis in the Democratic Republic of Congo, Virunga National Park represented the major source of tourism revenue for North Kivu, the country and for the whole region. It was the most visited park and it was endowed with some of the most developed tourist facilities. The park includes a particularly rich and varied diversity of ecosystems, the mountains of the Rwenzori range, expanses of savanna that extend to Queen Elizabeth National Park in Uganda, a network of lakes and rivers, vast surfaces of marshes and wetlands, low, middle and high altitude forests, active volcanoes and the extinct volcanic Virunga range. The forests and savannas of Virunga National Park still harbour an enormous variety of mammals, although their populations are now much reduced (see Chapter 8), including elephant, hippopotamus, lion, hyena, leopard, buffalo, mountain gorilla and chimpanzees, colobus and other monkeys, and suids including Forest Hog, Giant Forest Hog and Warthog.

The description of Virunga National Park and its biological diversity in the first Chapter of this book outlines the exceptional value of this Park as the richest protected area on the African continent with more than 200 species of mammals and over 700 species of birds (more than double the number of bird species in western Europe) and a large number of endemic and threatened species.

While there were periods when tourism flourished, such as in the 1950s and again in the 1970s, tourism never developed to its full potential. This was largely because of the political and economical context, but the park nevertheless remained an important tourist destination. At a national level, Virunga National Park was the main focus for the growth of a tourism industry, and also provided the main source of revenue for ICCN; it effectively funded activities in all the national parks of the country, more than seven percent of the total surface area of the country.

This Chapter was written to document the enormous tourism potential of Virunga National Park. A brief historical review recalls the golden age for tourism in the park, and then suggests avenues for a tourism revival in Virunga National Park.

11.1 Tourism in Albert/Virunga National Park before 1980: valuable lessons that can still be used

With the exception of the occasional gorilla visit, tourism has essentially died in Kivu as a significant economic activity. This was not always the case, and its development remains essential for the future of Virunga National Park. Along with conservation and scientific research, tourism remains one of the three fundamental objectives for ICCN.

Before 1960, 90 percent of visits were concentrated at the Rwindi Lodge. The first lodge, established between 1930 and 1935, was fairly basic and was managed by the Danly and later by Jean Ballegeer. It was composed of a dozen huts, a small restaurant and the houses of the hotel manager and the senior warden. In 1958 and 1959, the camp and its dependencies were improved by the senior warden, Cornet d'Elzius: more pavilions were built, others upgraded, garages and *lazarets* were built, and camps established for the rangers. Major alterations took place in 1970, on the occasion of the visit of King Baudouin and Queen Fabiola. More accomodation, some almost luxurious, were built during this period.

Until 1960, and even slightly later, the Rwindi Lodge was administered directly by the *Institut des Parc Nationaux.* Later, it was attached to a variety of private or public organisations *(Commissariat d'Etat au Tourisme),* which was highly contested. A strongly held view at the time was that the Park's authorities should always be maintained at a site that is situated within the boundaries of a protected area. The lodge was built so that it could be maintained to a high standard: water came from the Rwindi River; sewers and septic tanks were built, and generators were installed to provide visitors with a high degree of comfort. Lights went out at 10pm. Supplies came mainly from Goma. Bread was baked locally, while fresh food and meat came from the area around Butembo, often referred to as the fertile crest, or from Kibati, where large numbers of labourers offered fresh vegetables and strawberries to the passers by. In contrast to Lulimbi, the negative impact of this camp on the National Park was almost nil, as there was no attempt to plant crops on the soils surrounding the station.

Generally, visitors never stayed for more than 48 hours and on the tracks which did not require a four-wheel drive vehicle. Visitors had to be accompanied by a guide. Unlike Kagera National Park, that was managed almost like a hunting reserve, no game was offered to the tourists for consumption. Instead, fresh fish, especially tilapias, was provided from the fishing camp of Vitshumbi.

11.2
The welcome tourists receive (as here, at Mabenga), even before they reach stations in the park proper, is seen as important.

Visitor numbers for Rwindi Lodge in an average year, 1957, were:

	Total	Paying
January	371	304
March	403	374
April	456	356
September	310	252
October	279	225
November	456	385

The park was accessible throughout the year. In spite of the muddy roads, the rainy season, with its clear skies, provided the best conditions for photographs.

Other figures, still for 1957:

Resident paying visitors	1,982
Resident non-paying visitors	597
Non-resident paying visitors	1,807
Non-resident non-paying visitors	193

Tourist numbers remained stable until June 1960. In total, they amounted to 4,000–5,000 adult male visitors per year (tourist statistics were not registered for women and children at that time!). Strict hygiene standards were enforced around the station, although wildlife was able to roam freely within the compound: tourists would be woken up by elephants passing within five metres of their bed or when a hippopotamus splattered dung on the walls of their pavilions. Ballegeer recorded how elephants sometimes 'illegally' grabbed banana bunches stored in an outbuilding; monitor lizards ate the eggs in the chicken coop; and a good clear-up of the surroundings was guaranteed by hyenas and jackals.

The tracks were not paved, intentionally, to avoid altering the landscape as in the South African Parks like Kruger. Vehicles would occasionally get stuck in the mud, but that was very rarely a cause for complaint. Tourists were not allowed to leave their vehicle except in specific spots and off-road driving was forbidden. Excursions to see the large numbers of animals on the Rwindi plains mainly took place at dawn, although many of them were also visible after 3pm.

Before 1960, three circuits were available for visitors, each 40–60 kilometres long:

- the Kamande track, which was progressively abandoned because of the difficulty of crossing the ford of the Muwe-Lula;
- the Rwindi track, which had limited wildlife (except for large buffalo herds), but with dramatic *Euphorbia* forests. At Vitshumbi, it joined one end of the Rutshuru track;
- the Rutshuru track, the most spectacular, with exceptional hippopotamus populations. Later, the Kibirizi circuit and a track south of the Katanda were also accessible.

In 1972, when the ferry on the Rutshuru was installed, a number of tracks were established in the eastern part of the plain, towards the illegal fishing camp of Nyakakoma and Lulimbi. Previously, this region had been difficult to access, because of a dangerous diversion of the Ishasha road. This area remains very sensitive and care will have to be exercised in developing tourism in such a site. Nocturnal visits on the tracks were prohibited.

- At the time, the guides, whose distinct uniforms were different from those of the rangers, spoke French, and no English. Many visitors, before 1960, could speak some *lingala* or *kiswahili*. The station had a senior tourist guide who played a critical role in welcoming and helping visitors.
- Before and after 1960, a number of cultural monuments existed that were regularly visited:
  - statue of King Albert at the foot of Kabasha, and
  - Danly's plate on the escarpment;
  - monument of *May ya Moto* honouring the courage of the rangers who died fighting poachers.
  - Tourism had a positive impact on the conservation of the site. The frequency of poaching incidents was significantly less in tourism areas, as guides and tourists provided a form of constant surveillance.
  - a clinic *(lazaret)* existed in Rwindi and was available to staff and to all visitors. In 1957, the officer on duty handled over one thousand cases.

Although the Rwindi Lodge constituted the main site (with 90 percent of visitors going there), other sectors were open to tourism. It does not seem appropriate to open up other parts of the southern plain to tourists, in particular the right bank of the Rutshuru, or the middle Semliki, between Kasindi and the forest.

11.1.1 Ishango and the upper Semliki

This is one of the most inspiring natural sites of Africa. Ishango is one of the sources of the White Nile and wildlife was abundant, including elephants, hippos and waterbuck, while myriads of aquatic birds make it an extremely attractive wildlife area. ICCN built three self-catering houses for visitors in the 1950s, which were rehabilitated in 2008.

Whereas Ishango was extremely appealing for many visitors, it had its limitations. The tracks were limited and tended to end abruptly. The tracks in the Bukuku Mountains and towards the Lubilya delta were not good for observing wildlife. Nevertheless, visitors would swim in the waters of the outflow of the Semliki

11.3
One of the early tourist pamphlets distributed in the then newly independent Congo-Kinshasa.

River (strictly speaking, this was illegal, but largely tolerated). More recently, the arrival of crocodiles in the area prohibits swimming. Fly fishing was authorised in the 1980s.

11.1.2 The Rwenzori

Walking and mountaineering expeditions could be organised through ICCN and the Mutwanga Hotel. It took three to four days of climbing from the park's entrance (at 1,700 metres above sea level) to reach the base of the glaciers at 4,400 metres. Well-equipped refuges were installed at Kalonge, at Mahangu and at Kiondo, as well as a shelter near the glaciers. While requiring good physical fitness, the ascent did not demand any climbing ability. With a permit, tourists could try to reach Point Margerita, the third highest peak in Africa at 5,119 metres.

No official contact existed between park authorities of the Democratic Republic of Congo (western Rwenzori) and those of the contiguous Ugandan park. The situation was similar in the other sectors (Ishasha).

11.1.3 Active volcanoes

Active volcanoes were always accessible to tourists who were not discouraged by the long marches on the lava. They had to replace their sandals upon their return, because of the rapid wear due to the sharp lava. Mount Nyamulagira (3,056 metres) is reputed to be

Virunga National Park: a jewel for tourism in the Democratic Republic of Congo and in the Great Lakes region

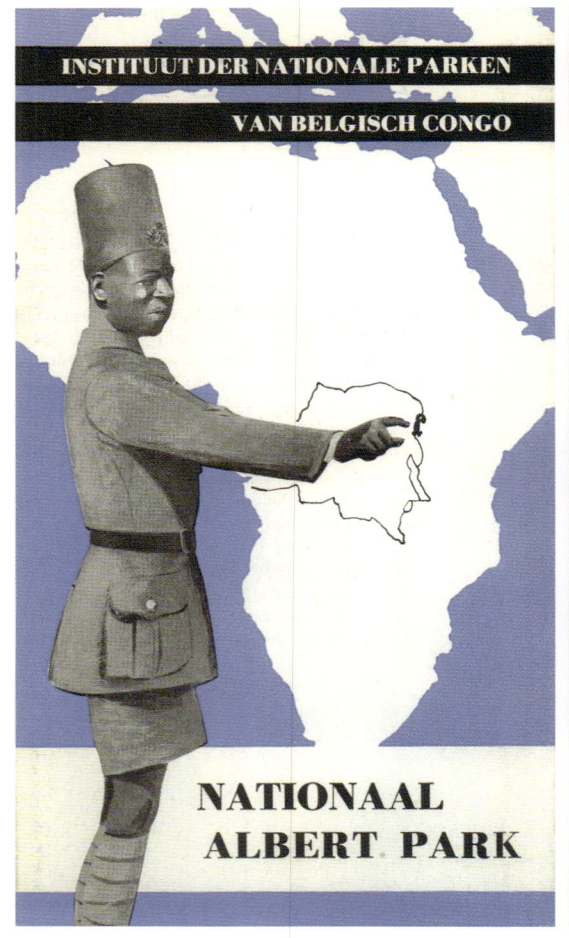

11.4
During the colonial period, tourism was accorded a high profile; promotional pamphlets were available in French, Dutch and English.

an easy climb but Mount Nyiragongo (3,470 metres) has steeper slopes. Several refuges had been installed at various altitudes, and a simple shelter existed in the crater of Mount Nyamulagira. A metal refuge, constructed on the orders of the ICCN Executive Director at the time (1973), was rather an eyesore on the eastern slopes of Mount Nyiragongo. As on the Rwenzori massif, climbers needed porters for their equipment. These men were characterised as the Sherpas of central Africa, although, disliking the cold, they sometimes preferred to descend in the evening and return early, at dawn.

Previously, Nyiragongo's climbing route was by the north via the extinct Baruta volcano; nowadays the climb is direct, starting from the Goma-Rutshuru road. The sight, especially at night, of satellite volcanoes like Kitsimbani, attracted numerous tourists who would venture onto the lava fields.

North of the active volcanoes, at Tongo, a chimpanzee population was known to exist, but had yet to be habituated for tourism.

11.1.4 Extinct volcanoes

In the early years, tourism was entirely banned from the Mikeno sub-Sector. However, after G. Schaller discovered ways of approaching the gorillas in 1960s, permission was progressively granted. At present, most visits are made through Rwanda or Uganda. The Congolese area, at Djomba or Bukima, little visited for the moment, remains a marvellous resource for the field naturalist.

11.1.5 Tourism after 30 June 1960

In the 1960s, during the conflict years, tourism stopped abruptly and visitor numbers were low until 1969. This year marked the beginning of an economic revival in the Congo and the re-organisation of the *Institut des Parcs Nationaux*. The Rwindi Hotel was taken over by the *Commissariat Général au Tourisme*, directed by Njoli M'Balanga, who set up some excellent initiatives. At the beginning of 1970, on the occasion of the official visit of King Baudouin, many pavilions were built near the River Rwindi. A real revival of tourism was evident until the end of the 1970s. The status of the guides was upgraded and they were provided with a visitor centre. Despite some insecurity, which would worsen over time, visitors progressively returned to Virunga. One strange event of the early 1970s was the arrival of all the Miss Belgium candidates at the invitation of the Congolese state. In 1975, part of the general assembly of the IUCN took place in the Rwindi Hotel. Later, many official visitors came to the Rwindi Hotel, such as Prince Bernhard of the Netherlands. Congo's head of state came on several occasions by helicopter. He had a strong affection for Rwindi, and it was there that he decided to create four new national parks, 'we would like 12 to 15 percent of our country be established as national parks... impeccably managed, like Rwindi'.

Annette Lanjouw, Marc Languy & Jacques Verschuren

The deterioration, caused by the troubles resulting from the rebellions, was slow at first, but reached its climax at the end of the 1990s. Rwindi Camp was pillaged on numerous occasions and transformed into a playground for armed individuals, dressed in colourful uniforms.

11.2 Typology of the tourist activities for Virunga National Park

This section examines, from north to south, a series of tourist activities that could be considered in Virunga National Park, with a brief proposal for possible extensions. This description of activities would obviously only be valid when considered in a stable context. Recommendations for the preparation, context and strategy necessary for these activities are set out at the end of the chapter.

11.2.1 Northern Sector

Three major attractions are immediately achievable in this sector:
- Ascent of the Rwenzori massif. Several shelters are still in place along the trajectory that begins at Mutsora station and leads to the highest point in the country. This route crosses a particularly diversified ensemble of forested strata, of tropical forest to the Afro-alpine level, passing through montane forest, tree heather forest, giant *Lobelia* and *Senecio* assemblages. Since the re-opening of the Rwenzori Mountains National Park in Uganda in 2002, it is possible to imagine the creation of a circuit where one could, for example, ascend from Fort Portal on the Ugandan side, come down to Mutsora in the Democratic Republic of Congo and continue towards the Tshiaberimu and Ishango before joining Queen Elizabeth National Park in Uganda via Ishasha.
- Mount Tshiaberimu: this is the only site in Virunga National Park where the Eastern Lowland Gorilla occurs and is the only place in the Democratic Republic of Congo, with Kahuzi-Biega, where the animals are habituated to visitors. Like all the mountains of Virunga National Park, Mount Tshiaberimu also offers a very rich and particular avifauna, with many bird species endemic to the Albertine Rift. This site is also very attractive for other forms of tourism, such as hiking for sport and scenic tourism (breathtaking views over Lake Edward).
- Ishango: a site of exceptional scenic beauty, with large concentrations of aquatic birds, crocodiles and also mammals (elephants, buffaloes, hippopotamus, lions, etc.). Large mammals were somewhat less numerous than south of the lake.

Other activities that can be imagined in this sector include treks in the great Semliki forest, perhaps on the trail of okapis, bongos and the many smaller primate species; cultural activities with the pygmies of Mwenda and the descent of the Semliki River by canoe (and on foot in some places).

An important opportunity for tourism development in the Northern Sector is the linking of visits to Mount Hoyo to the north of Beni. This site, which has always attracted the curiosity of visitors, has just been re-established by ICCN who have named a resident warden. The site was always managed by ICCN as an extension to Virunga National Park. Finally, there are visits to the exceptionally beautiful Sinda Valley – north of the present park boundary – that ought to be considered.

11.2.2 Central Sector

This sector is characterised by the famous Rwindi-Rutshuru Plains which has some of the most dramatic wildlife viewing opportunities in Africa.
- The savanna trails (Rwindi trail, Rutshuru trail, etc.) are now largely abandoned, but enabled the viewing of most of the mammals of Virunga National Park. Abundance of lions and elephants made this area a major attraction.
- A visit to Vitshumbi and a stop at the end of the morning gives an opportunity to taste the famous *mustiko* – tilapia – of Lake Edward grilled the local way.
- Mwiga Bay and other neighbouring areas, as well as many parts of the Rutshuru River provide an opportunity to see large concentrations of hippopotamus.
- Fishing on the Rutshuru, if correctly regulated and controlled, offers potentially important tourism (including local tourism, notably for people based in Goma or Butembo).

11.2.3 Eastern Sector

At present, the trails east of the Rutshuru offer few wildlife-watching opportunities. Certainly they do not offer the abundance of wildlife seen on the Rwindi Plain. Therefore, in the short term, the re-opening of tourist trails in this sector should not be considered a priority, but could become important in the longer term, particularly if tourism numbers were to increase dramatically in the Rwindi-Rutshuru area, and if a transboundary tourism with Queen Elizabeth National Park were to develop.

11.2.4 Southern Sector

Several spectacular attractions are available in the Southern Sector. The visit to the chimpanzees in the Tongo area was hugely successful in the early 1990s: chimpanzee populations still exist here, but their habituation to visitors has to be re-established. Tongo provides a fascinating complement to the Mountain Gorillas of the Mikeno sector and the lowland gorillas of Tshiaberimu, making Virunga the only park in the world in which three taxa of Great Apes can be observed.

Climbing the active volcanoes, Mount Nyiragongo (one day round trip or two days with a night in the shelters near the summit) and Mount Nyamulagira (ideally three days), is an activity that is unique in the world. Even during the relatively frequent eruptions, visits near to the eruption sites can be easily organised, and aerial fly-overs are a possibility.

The extinct volcanoes offer exceptional trekking opportunities and could be developed. Some volcanoes, such as the Mikeno, require advanced mountaineering experience, while others are merely a stiff walk.

The visit to the Mountain Gorillas in the Mikeno sub-Sector obviously is Virunga's greatest tourism attraction

11.5
The reception building at Rwindi in August 1989. Tourists were invited to register their observations. In one typical example, a three-day visit yielded a tally including 12 lions, one hyeana, a leopard and a serval cat.

11.6
This luxury tourist camp on the Rutshuru, established by Albert Ngezayo-Prigogine in 1993, was unfortunately destroyed a few years later.

Annette Lanjouw, Marc Languy & Jacques Verschuren

and can be done from Djomba or Bukima. Visiting the habituated gorillas provides a unique experience that is the main focus of most of the visits to Virunga Massif, but other attractions exist as well. For example, visitors exploring the montane forests of the Southern Sector can see the Golden Monkey *Cercopithecus kandti* and other endemic primates, and there is also a very rare avifauna, prised by international ornithologists.

11.3 Virunga National Park in the global tourism market place

The enormous concentrations of hippopotamus populations in the Central Sector, the mountain gorillas, the active and extinct volcanoes, and the Rwenzori Massif in the north, have all contributed to Virunga National Park's international reputation. This was recognised in 1979 when the park was designated a World Heritage Site by UNESCO.

Before the wars began in the 1990s, the park attracted several thousand tourists a year. Between 1990 and 1995, nearly 2000 people visited the gorillas annually. These visitors were very strictly controlled; only three families of gorillas were habituated and only six people a day were allowed to visit those families. The 2000 visitors represent an average 'occupancy rate' of 30 percent (18 permits to visit the gorillas available per day at that time). Occupancy rate of visitors to the gorillas rose regularly to over 80 percent during the high season (summer holidays and Christmas). The average revenue from all of the visitors to Virunga National Park was 150,000 US dollars per month. These funds contributed directly to the financing of the *Institut Congolais pour la Conservation de la Nature* (ICCN) and the operating costs of all the parks in the country.

The Democratic Republic of Congo would never qualify as a leisurely destination like Kenya, Tanzania and South Africa. Rather, it has always been featured amongst the more adventurous destinations. The lack of adequate infrastructures, the poor state of its roads, the less than welcoming behaviour of its immigration officers and police, corruption and enormous distances to travel, as well as the ensuing costs, all serve to discourage the average tourist. The country had a reputation that frightened most foreigners, and especially the older travellers who are financially better off. Nevertheless, the country has an enormous tourist potential. In 1998, an important workshop and a debate on tourist development took place to prepare some strategies and priorities for the revival of tourism in the country. The priorities focused on the behaviour and activities of the police, the military, immigration and other authorities, in order that corruption and aggressiveness toward tourists could be reduced. The global effects of the events of 11 September 2001 have certainly yielded even greater obstacles to the development of tourism in DRC.

Once political stability has returned, tourism will undoubtedly be one of the first economic activities to resume in the region. Enormous pressures will then be brought to bear on the park authorities, and on the fragile primate populations, given their potential as an important source of revenue. Without a solid, well-structured strategy, and a programme of tourism development, tourist resources run a great risk of being over exploited and thus badly exploited. The fragile basis for this burgeoning tourism industry could quickly become threatened. Two-thirds of the habitat of the Virunga volcanoes lies within the Democratic Republic of Congo, along with about half the Virunga mountain gorillas. Over-exploitation or destruction of the ecosystem in the Mikeno sub-Sector would have a very harmful effect on the entire region.

11.4 Recent statistics on visits to the gorillas in the Virunga massif

In 2003, the International Gorilla Conservation Programme commissioned a study on the tourist market in the Virunga Massif and at Bwindi (Uganda), providing useful information on the present and potential economic value of gorilla tourism. It also provided the basis for a marketing strategy at government and international levels. The study also enabled an evaluation of the distribution of benefits between local, national and international communities. The study is based on statistical and financial data of gorilla tourism in the four parks that harbour mountain gorillas: Volcanoes National Park in Rwanda, Mgahinga and Bwindi National Parks in Uganda, and the Southern Sector of Virunga National Park in the Democratic Republic of Congo. Most of the data were collected between 2000 and 2002, a period when gorilla tourism in the Democratic Republic of Congo was closed. The data are based on visits to seven groups of gorillas. Now that tourism can resume in the Democratic Republic of Congo and all the groups habituated to tourists are available, the figures are higher, but the observed trends can be generalised.

International visitors represent 84 percent of all visitors to the four parks studied (including Virunga National Park) and 81 percent come from three distinct geographical areas: North America (20 percent); Europe (42 percent) and Australia, New Zealand and Japan (19 percent).

At that time, a permit to visit the gorillas costed 250 US dollars. The permits alone brought in nearly three million dollars each year for the seven groups of gorillas, with an occupancy rate of 30 percent.

This occupancy rate was low, because of the problems that the region has experienced and because not all of the gorilla groups were open to tourism. With an increase in prices, but especially in the occupancy rate, potential revenues are now well above five million US dollars per year.

Of course, there is substantial indirect expenditure related to a gorilla visit: transportation, hotels guides, porters, meals, artisanal goods, and so on. These supplementary benefits represent nearly five million US dollars each year (again, a total for the four parks) and 3.1 million US dollars in the form of taxes.

As shown in graph 11.8, the groups that represent the greatest financial flow are the direct revenues at the national level (24 percent) and taxation (20 percent). International travel represents 13 percent of the profits. As one might expect, local revenues represent the smallest component of the income. This is the most urgent area to look at, and is a priority both in terms of ICCN's development and for the local communities, who must benefit from a larger part of the profits generated by gorilla tourism.

Precise impact studies of development and integrated conservation projects around the protected areas in Uganda, have clearly shown that the greatest

Annette Lanjouw, Marc Languy & Jacques Verschuren

11.7
A close view of the Mountain Gorillas at Djomba or Bukima, the high point of many a tourist's stay, is still the park's biggest single attraction. It is important, if difficult at times, to adhere to the strict rule specifying that a minimum distance of seven metres be maintained between the gorillas and the tourists, in order to minimise the risk of disease transmission.

11.8
Distribution of revenues generated by gorilla tourism.

- International revenues
- National revenues (non-local)
- Park entries
- Impact of the national revenue (non-local)
- Impact of government taxes
- Local revenues
- Impact of local revenue

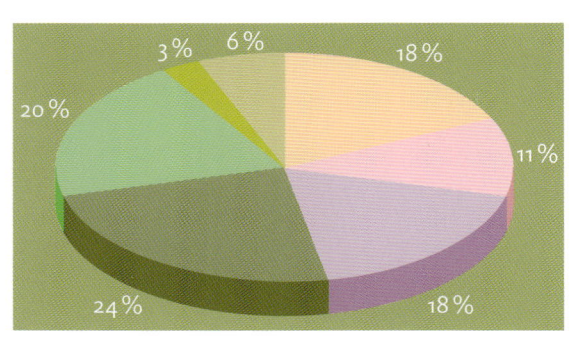

beneficial impact comes when revenue is equitably divided and goes directly to local communities. It has a greater impact than either agroforestry activities or sharing access to natural resources.

With an occupancy rate of 30–40 percent and a price per visit fixed at 250 US dollars or more, only visits to the gorillas in the Democratic Republic of Congo could induce benefits of more than 20 million US dollars per year.

One critical condition, in addition to rebuilding a tourism industry with all that this entails, is for the ICCN to regain control of the management of the revenues that it attracts, rather than the usage that was established during the rebellion.

11.5 Strengths and weaknesses of tourism: the Virunga National Park experience

It is important to take a critical look at the strengths and weaknesses of tourism, locally and in general, in order to identify what actions need to be taken. We present here the principal strengths and weaknesses:

11.5.1 Objectives of tourism

- Financial investment permitting conservation and development (reduction of poverty, infrastructure development);
- Cultural exchanges and development of relationships between countries and people;
- Environmental education and behavioural changes in tourists through their appreciation of Africa's natural and cultural wealth.
- Improved awareness, thereby facilitating and stimulating investment in DRC.

11.5.2 Strengths of tourism in general

- The biggest industry in the world;
- The industry with the greatest growth in number and revenues, in the world;
- An industry that can link the objectives of development (reduction of poverty) with economic objectives (individual, public, private) and conservation (financial support of the natural sites visited).

11.5.3 Strengths of tourism in Virunga National Park

- Enormous potential because of the exceptional assets
- Landscapes, geography, fauna and diversity;
- Historical experience on which the ICCN can rely;
- ICCN is relatively well-organised, with a competent, friendly and motivated staff, capable of accompanying the tourists and promoting the country;
- Language: the majority of the guides speak French and many have learned English
- Congolese culture is varied, attractive and interesting;
- Development of tourist infrastructure in the region, from which the local populations also benefit (roads, airstrips, lodges, and so on);
- The ICCN which could be relatively well financed (consider the busy periods when tourism revenues were coming to the country), capable of conducting good park management programmes, with the possibility of working harmoniously with neighbouring communities;
- A deterrent for poaching and other illicit activities in frequently visited sites.

11.5.4 Weaknesses of tourism in general

- High risk, an industry that is strongly influenced by external factors (economy of the host country, foreign policy of the country of origin, global and regional security);
- Strong dependence on the image of the country visited, sometimes distorted by the media;
- Continuous need for promotion at an international level, because of a highly competitive industry, constantly in competition with fashionable, new destinations.

11.5.5 Weaknesses of tourism in Virunga National Park

- Infrastructures regularly destroyed during periods of armed conflict;
- Present image of the Democratic Republic of Congo, and in particular of its eastern region, still very negative;
- Lack of a development strategy for tourism, the 'destination' and the desired image for this destination, and identifying visitors who should be attracted primarily to this destination;
- No local population involvement and very little benefit to them. This makes it easy to politicise the park and for politicians to exploit the negative attitudes of populations towards the park (the only value perceived by the populations is the 'unexploited land');
- Interest of other authorities (immigration, police and others) to exploit visitors: corruption, imposition of mythical 'taxes' (for example, for photography), costly visas, and so on;
- Anarchic development of tourist infrastructures in the region and too few links between private investments, plans of ICCN and local populations;
- Absence of a tourism officer with the ability to develop tourism, with adequate means;
- Marked weakness of national tourism.

11.6 Lessons learned and recommendations for re-development of tourism in ViNP

This section considers the various initiatives that need to be undertaken in order to relaunch tourism in Virunga National Park and to ensure its long-term development in a way that secures maximum benefit for the protected area, while at the same time providing benefits for local populations and for the economy, both local and national. As the basis for preparing a tourism development strategy, these recommendations are grouped under broad themes, which inevitably overlap.

11.6.1 Institutional and strategic considerations

- Establish a clear strategy for tourism development. This should include the following aspects: categories of visitors to be targeted and their countries of origin; types of infrastructure to be developed (and those to be avoided!); tourist information strategies; capacities and knowledge to be developed in building competent human resources; the revenue sharing system; new sites to be developed. In the planning of prospective tourist facilities, the close collaboration of ICCN Headquarters is of course essential in ensuring that there is harmony between sites, which might then be developed in coordination with other institutions, such as the National Tourism Office.
- Integrate the tourism development strategy with national development strategies to ensure that common priorities can be identified: the need for new airstrips or roads, for example, or for the introduction of tax reduction and other incentives that will encourage private investment in the tourism sector.
- Coordinate the implementation of the tourism development strategy with the operations of other national authorities (including immigration and the police) so as to ensure that the issue of visas, arrival formalities and onward travel in the country is made agreeable and easy for tourists.
- Develop and apply a fair and effective programme under which revenues and benefits accruing from tourism can be shared with those communities living adjacent to the park. The aim of such a programme should be to maximise tourism's potential role in poverty reduction, while giving local people a vested interest in wanting to support the park through tolerating and protecting its fauna, through contributing directly to the conservation effort (the anti-poaching battle included) and through welcoming tourists to the area. The expertise of the ICCN's various partners at Virunga National Park should be solicited directly in formulating such a programme.
- Involve community representatives and local authorities in some aspects of the management of park tourism resources, so their priorities, opinions and needs can be taken into account in helping to broaden and consolidate the conservation effort, while boosting tourism.
- Clearly identify the necessarily separate mandates of the responsible authorities vis-à-vis tourism and nature conservation (protected areas) in order to ensure that the parks do not become just tools for tourism. Since, for the moment at least, the attractions the Democratic Republic of Congo can offer to tourists are its fauna and its landscapes, these resources must be managed strictly to guarantee their survival in the long term. Unregulated tourism can jeopardise entire habitats, having negative impacts on the fauna and the flora as well as on the quality of the natural resources (water, for example). Over-development can rapidly undermine the entire base of a country's tourism industry. It is therefore imperative that the authority put in charge of tourist development should not have priority over the authority in charge of nature conservation and the management of protected areas.
- Make sure that all tour guides are given adequate training and can respond even to specialist tourist needs, including the demand for ecotourism.
- Envisage the creation, within the ICCN Directorate-General, of the post of a technical counsellor charged with tourism in Virunga National Park and in other protected areas within the Democratic Republic of Congo.
- Resolve the question of how tourism revenues are to be shared as a matter of urgency. *Direct* revenues must benefit the rangers and wardens, as well as the neighbouring communities. Other players, notably instruments of the State, will benefit automatically from the indirect revenues that, in economic terms, are even more significant (as demonstrated in Section 4, above).

11.6.2 Diversification of tourist activities

It is important to ensure that tourism's primary target – the Mounatin Gorilla, say, in the case of Virunga national Park – does not remain the only attraction. Only through diversifying the attractions of an area can one maximise the lengths of time individual tourists will wish to stay, while guarding against the risk of disappointment should the primary focus of interest be unavailable or become the subject of excessive demand.

- Concentrate on developing a uniquely varied appeal for each destination. The more striking selling points a destination has, the more likely it is to gain acceptance as a destination of choice. To this end, it may be necessary to look beyond an area's administrative boundaries in devising itineraries that take in the attractions of neighbouring regions as well. A circuit that includes such activities as hiking the Rwenzori, savanna game drives, river walks, lake cruises in boats, shoreline birdwatching in marshlands and on beaches, and climbing volcanoes, as well as an opportunity to observe the region's celebrated Great Apes, could command an almost unresistible appeal, for example.
- Make provision for ecotourism and ornithological outings, both now increasingly in demand, and for backpacking tourism, all with access to well managed camping facilities inside the park. Today's adventurous young people with their Spartan equipment will be the rich tourists of the future. In the much longer term, and in exceptional cases only, the re-introduction of hunting tourism in specially designated zones might even be considered.
- Organise, under strict supervision, boating excursions on Lake Edward, making these conditional, however, on avoiding the vicinity of Ishango. Explore avenues for adventure tourism, such as a descent of the Semliki River in a dugout to reach the 'eighth wonder of the world', the spectacular gorges of the Sinda.
- Consider the potential of exploiting one, or more, of the transects between Kasindi and Ishango

11.9
The hotel at Rwindi in 1993. Located in the heart of the park's Central Sector, the Rwindi is the perfect base for tourists wanting to see the mammals of the Rwindi-Rutshuru Plain, and is also a 'required' stop on the North Kivu tourist circuits.

that afford spectacular views of the banks and cliffs of the Semliki, in particular the pre-1960 excavations that yielded the Ishango man and the Ishango bone. Where the track from Kasindi crosses the Equator, new signposts should be erected, so that tourists—for whom the old posts, long since removed, were a popular photographic backdrop—can again enjoy the novelty of being photographed on Latitude Zero.
- Climbing of the major volcanoes can itself be the object of many a tourist visit, as distinct from viewing the gorillas. Some mountaineering skill is called for on the ascent of Mikeno, but on Karisimbi, the highest of the volcanoes, an exhilarating walk is all that is required.
- Habituation of the chimpanzees at Tongo should be resumed. Then perhaps, a 'Great Apes certificate' could be developed, for issue to proud visitors who get to see both the gorillas and the chimpanzees.
- Modalities for visiting Mount Tshiaberimu should be established, and the tourism options there (trekking, ornithology, adventure tourism) should be developed and diversified.

11.6.3 Transborder prospects

Ecosystems that are shared with neighbouring countries can—with the necessary collaboration and coordination—benefit greatly from a combination of their respective tourist attractions. Areas that stand to benefit especially from cross-border tourism cooperation include:
- The Rwenzori Mountains, where a loop circuit involving a climb up the Ugandan side and down the Congolese side, with the option of continuing on to Mutsora, Tshiaberimu and Ishango, would be a very appealing proposition.
- The savannas of Rwindi and Ishasha and adjacent areas of Uganda's Queen Elizabeth National Park could—together—make for an outstanding cross-border game-viewing experience.
- The Virunga Volcanoes, where a mountaineering circuit based on climbing all six extinct volcanoes would make an unforgettable experience—and one that, to date, few people have managed to achieve. Only with the help of cross-border tourism cooperation would it be possible, on one expedition, to gain access to all six of these summits. The option of going on to scale the two active volcanoes as well would give climbers the rare distinction of 'bagging' all eight of the major summits in the Virunga Massif.

11.6.4 Promotion

Without advertising, there can be no tourism, not even wildlife viewing. Before 1960, the Institute published a great many promotional booklets and brochures, including a series of leaflets in French, Dutch and English. Pamphlets in Kiswahili and Lingala, were produced in 1970. The *Institut des Parcs Nationaux du Congo Belge* (IPNCB) had also published hundreds of post cards and other advertising vehicles. In a travel guide compiled by *Touring Club du Congo,* Albert National Park is discussed at length in three languages. A comprehensive guide book covering Virunga National Park was subsequently published by Delvingt, Mankoto and Lejoly (Delvingt *et al.,* 1990).

Listed below are some concrete proposals for local, national and international promotion of tourism in Virunga National Park.

- Compile and publish a new tourism map of Virunga National Park.
- Produce a tourism pamphlet in French and English on the attractions of the park.
- Invite tour operators and journalists, once security has been reinstated, to visit the park to convince them of the feasibility of organised tourism.
- With the same security proviso, invite teams from one or more of television's major natural history and wildlife documentary film units with an established global viewership (National Geographic, BBC, and Ushuaïa are good examples) to acquaint potential tourists with the wonders of Virunga National Park.
- Launch a national publicity campaign, and one directed at expatriates as well, before targeting selected countries abroad, concentrating on established travel markets in the United States, Canada, the United Kingdom, Germany, France, Belgium, Switzerland, Scandinavia, South Africa, Asia and Australasia.
- Considering staging a grand event at which a major declaration can be made, re-launching tourism in the Democratic Republic of Congo.
- Involve embassies of the Democratic Republic of Congo in new efforts to give all prospective visitors a particularly warm reception, since trips effectively begin with such formalities. It is important, moreover, that consular personnel, whose knowledge of the parks may be limited, should have a ready supply of promotional leaflets.
- Develop a strategy that will enhance domestic tourism, which has always been weak. Such a strategy should be tailored to the needs and means of the Congolese tourist, and should include local tourism, notably for school parties.

11.6.5 Technical aspects, logistics and infrastructure

- In general, only serious investments in infrastructure development, catering for the needs of particular tourist groups, should be countenanced, putting an end to anarchic developments.
- Long-standing problems surrounding the ownership and management of the Rwindi Hotel, in particular, should be resolved, and the necessary funding secured for its rehabilitation.
- Rehabilitation of the shelters on the Rwenzori and Nyiragongo Mountains and of the trails leading up to them, in progress since the end of 2005, should be completed without further delay.
- The thatched hut at Vitshumbi should be rehabilitated, giving tourists a rest stop away from the centre of the fishery, where they can partake of refreshments or light meals.
- There should be a pamphlet, available to all, clearly explaining the sites to be visited, the charges, and the rules of conduct.
- Control over the production, issuing and management of visitor permits should be handed back entirely to the ICCN, as part of a professional system under which a larger number of stakeholders can be integrated.
- All informal 'taxes' or 'permits' for photography and other routine tourist activities should be abolished.

11.7 In conclusion

The various sections making up this chapter have amply demonstrated, were any proof required, that Virunga National Park harbours enormous tourism potential. Recent experience has shown that, even with relatively little infrastructure, the returns gained from what limited tourism there is can generate significant revenue for the local and the national economy, and of course for both the park and the ICCN.

The lingering insecurity problem notwithstanding, the health of the tourism sector has far-reaching implications, not just for the ICCN, but for the country as a whole. The development of a thorough and well thought-out tourism strategy, integrating all facts of the tourism sector, is therefore a priority for the ICCN and for all its partners.

A diversification of attractions, aimed at extending the visits of tourists after they have seen the Mountain Gorillas, is an important element of such a strategy. In this respect, Virunga National Park can easily rival even the best known tourist destinations on the African continent today. The recoveries seen in neighbouring countries, such as Uganda and Rwanda, which – in the 1980s and 1990s respectively – suffered acute security problems and political crises, allow us to be reasonably optimistic about a rapid return of tourism to Virunga National Park.

---

* Only non-resident tourists were considered 'true' tourists. Most of the park's resident visitors were Belgian colonists, often from as far away as Léopoldville.

Emmanuel de Merode, Andy Plumptre, Maryke Gray, Alastair McNeilage, Katie Fawcett & Marc Languy

12 The status of Virunga's large mammals

Emmanuel de Merode, Andy Plumptre, Maryke Gray, Alastair McNeilage, Katie Fawcett & Marc Languy

In this chapter, the authors present data on the current status of Virunga National Park's large mammals. As outlined in Chapters 2 and 8, the park has held the world's highest recorded mammal biomass, species diversity and endemism. Years of political instability and armed conflict caused a crash in animal numbers, to the extent that many conservationists have a real concern for the imminent total or local extinctions of certain species. Efforts to document the distribution and abundance of large mammals vary considerably according to habitat. Certain species, such as the Great Apes or hippopotamus, have been the subject of detailed studies. This is why the chapter examines forest and savanna species separately. It is the result of recent (between 2003 and 2006) and intense efforts by ICCN and its partners to document the status of the park's mammals following a period of armed conflict.

12.1 Introduction

This chapter examines the large mammals of the park. Biologically, a large mammal is defined as any mammal species whose adult average weight is over 1 kilogramme (Diamond, 1975). But, for the purposes of this chapter, we have been restricted by survey constraints to those species above 15 kilogrammes. This is a standard approach used in most ecological studies in savanna environments (Norton-Griffiths, 1976; Dublin, 1991; Sinclair, 1995). In forest habitats, we examine a much more diverse assemblage of mammals, but with the exception of the Great Apes, we limit ourselves to animal presence, and not abundance, which is extremely difficult to measure.

The population trends of large mammals were examined in Chapter 8, and so the subject will not be treated in any great detail this chapter. However, it becomes difficult to discuss the results of presented in this chapter without reference to past data.

Over the historical record, ungulate mammal populations have always been the most significant in terms of numbers and biomass. Lake Edward and its river networks was once home to the greatest hippo population on the planet. Today, this population has been decimated, a fact which lead to the hippopotamus species being registered as vulnerable on the IUCN Red List for the first time in 2006. Similar patterns have been recorded for other key species, including the elephant and buffalo. Thus, it is important to emphasise that the estimates on the current status of these populations should not be considered a stable state. On the contrary, Virunga National Park, and in particular its savanna habitats, should be viewed in an extreme state of dynamic disequilibrium, as often found in marginal, semi-arid environments characterised by unpredictable rainfall (Walker, 1981; Dublin et al., 1990; Sinclair & Arcese, 1995).

Savanna habitats are classified into various broad zones: The Upper Semliki, made up of dense gallery forests is situated in the north. The more open *Cymbopogon* savannas are situated further south, with more marshy habitats in the east in the delta region of the Lubilya. The savanna mosaics to the south of Lake Edward, made up primarily of low grass savannas dominated by *Sporobolus* and *Cynodon*. In this part of the park, there is also a more arboreal savanna, dominated by euphorbias. Gallery forests are also very present along the Rwindi and the Rutshuru Rivers. There is an unusual small savanna of a few hundred hectares in the extreme north of the park, but as far as we know, there are no species of large mammals in this isolated habitat.

The forested habitats of Virunga National Park cover five zones: the Virunga Volcanoes (the Mikeno Sector) and the active volcanoes (the Nyamulagira

12.1
Lions are among the rare large mammal species for which we have no accurate population data.

12.2
Areas within Virunga National Park that were censused for large mammals in 2006.

- • Town
- Southern Transects
- Northern Transects
- Central and East Transects
- International Boundary
- Virunga National Park

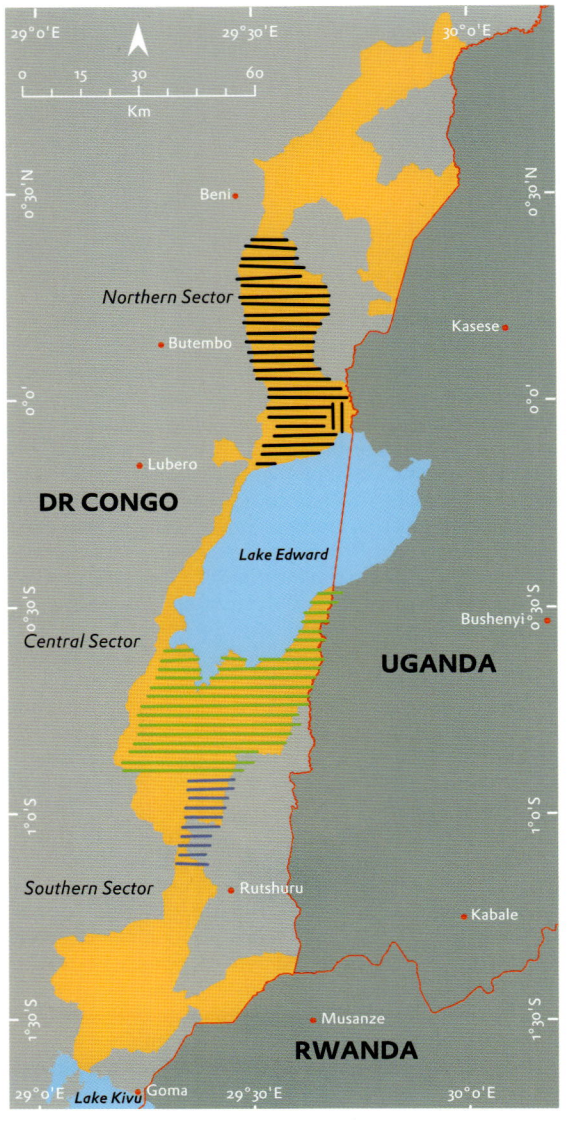

the Virunga Volcanoes. The exact taxonomic status of the Bwindi gorillas is the subject of some controversy (Grub *et al.*, 2003).

In Virunga, the mountain gorillas can be found in the ecosystem of the Virunga Volcanoes, which covers the border between eastern Democratic Republic of Congo, Rwanda and Uganda, between 1°20' and 1°35' south and 29°20' and 29°40' east. The mountain gorilla is the largest of living primates, one of the most well studied, and sadly, the most threatened. Approximately 380 Mountain Gorillas survive in the Virunga Volcanoes (about 450 square kilometres, between altitudes of 1850 metres and 4507 metres), where they are protected within three national parks, Virunga National Park in the Democratic Republic of Congo, *Parc des Volcans* in Rwanda and Mgahinga Gorilla National Park in Uganda.

The chimpanzee, *Pan troglodytes*, can also be placed among the many endangered species of African large mammals, which is categorised as endangered on the IUCN Red List (Hilton-Taylor 2002). Three sub-species of common chimpanzees are currently recognised: the western chimpanzee *Pan troglodytes versus*, the central chimpanzee *Pan troglodytes troglodytes*, and the eastern chimpanzee, *Pan troglodytes schweinfurthii*. The eastern chimpanzee dominates in most of eastern and north eastern Democratic Republic of Congo, including Virunga National Park, and its range stretches as far as the western forests of Burundi, Rwanda, Tanzania and Uganda. The sub-species is classified as vulnerable, based on IUCN criteria, as are all members of the species. This is based on the massive decline of population numbers across the continent, as a result of hunting and habitat loss (Plumptre *et al.*, 2003).

12.2 METHODS

12.2.1 SAVANNA SURVEYS

The current status of large savanna mammals is largely based on recent aerial survey results. The most recent general counts, that established estimates for the entire assemblage of large mammals was carried out in August and November 2003 (Mushenzi *et al.*, 2003), and more recently in June 2006 (Kujirakwinja *et al.*, 2006). The results for almost all species are remarkably similar on both counts, with a slight, but non-significant, increase in the number of buffaloes and other species. These surveys both covered the savannas from the northern slopes of Nyamulagira, in the south, to the Beni-Kasindi road in the north (Figure 12.2). We also present the most recent hippo counts carried out in August 2005 on the shores of Lake Edward, and along its primary rivers (the Rwindi, Rutshuru, Ishasha and Semliki rivers).

The 2003 and 2006 counts followed the standard method for systematic reconnaissance flights (Norton-Griffiths, 1976). An experimental category aircraft, a Glastar (in 2003) and a Cessna 182 (in 2006) were used, each carrying four crew. The counting method for 2003 was as follows: the pilot's role, other than flying the aircraft, is to read out altitude data and flight positions to the observers. The front-seat observer, to the right of the pilot, records observations on the vegetation and human presence. The two rear-seat observers count wildlife on either side of the aircraft. The observers can only count the animals that are contained within a strip width on the ground. These strip widths represent a sample size

Sector), together with Tongo in the south, the forest bloc of Tshiaberimu and the low altitude tropical forest in the northern sector of the park. Other small forest outcrops also exist, especially along river courses, but their combined area is significantly less that that of these five blocks. The vegetation of four of the five blocs is made up primarily of montane forests, while the fifth bloc, the low altitude forest, is in the park's northern lowlands (at around 700 metres). The species that live in these forests are dependant on their specific environment, and for convenience, we describe each species in turn, based on their broad habitats.

In this chapter, gorillas and chimpanzees are treated individually because of their iconic significance for Virunga National Park. Gorillas as a genus are believed to be one of the most threatened of all African primates. Currently two species, and four sub-species of gorillas, including the Mountain Gorilla, *Gorilla beringei beringei*, are generally recognised (Grub *et al.*, 2003). Within the Mountain Gorilla subspecies, there are two populations, both of which are critically in danger of extinction, one in the Bwindi Impenetrable Forest National Park in Uganda, and the other in the massif of

Estimating and comparing animal counts from systematic reconnaissance flights (Norton-Griffiths 1976)

Equation 1
Jolly's method 2 for estimating animal densities from systematic reconnaissance flight data: the ratio method for unequal sample points (from Inamdar 1996).

$$D = \frac{\sum y}{\sum a}$$

$$SE\,D = \sqrt{(1 - (\sum a)/A)} \times \frac{n}{\sum a} \times \sqrt{\left(\frac{1}{n(n-1)}\right)(\sum y^2 + D^2 \sum a^2 - 2D \sum ay)}$$

$$Y = A \times D$$

$$SE\,Y = A \times SE\,D$$

Where:
- y = the number of animals in a given sample unit
- a = the area of a given sample unit
- A = the total area of the region being surveyed
- n = the number of units sampled
- D = the estimate of mean density
- $SE\,D$ = the standard error of the estimated mean density
- Y = the estimate of total numbers in the region size A
- $SE\,Y$ = the standard error of the estimate of total numbers

Equation 2
Test for the significance of the difference between two sample estimates

$$t = \frac{(Y1 - Y2)}{\sqrt{VarY1 + VarY2}}$$

- $Y1$ = population estimate at the beginning of the study period
- $Y2$ = population estimate at the end of the study period
- $VarY1$ = variance of the estimate at the beginning of the study period
- $VarY2$ = variance of the estimate at the end of the study period
- t = t-value (if t is greater than 1.96, estimates differ significantly at the 0.05 level)

---

just under 10% of the total survey area. Observers are able to locate the strip widths by limiting their observations to the area that is defined by rods attached to the wing struts of the aircraft. The flight path of the aircraft follows north-south parallels, at five kilometer intervals. A GPS is used to provided accurate navigation along the transects, and an extremely precise altitude is maintained with the help of a radar altimetre. Thus, when the aircraft maintains an altitude of exactly 350 feet above ground level, observers can be sure of restricting their observations to an area 500 meters wide.

The 2006 survey used the same method, adopting more or less the same altitude, but used transects that were oriented east-west. The northern part of the southern sector and the western shores of Lake Edward were not flown, because it was well established that no large mammals existed in those areas. The sampling intensity was of 17.6% for the northern sector and 14.4% for the southern and central sectors (Kujirakwinja, 2006).

The calibration of strip widths is done by repeatedly flying across visual markers placed at twenty meter intervals, at altitudes that vary between 140 and 400 feet. During each passage, observers count and record the number of markers, which makes it possible to establish the exact relationship between altitude and strip width. A regression coefficient enables us to extrapolate the sample count to an overall population estimate with 95% confidence intervals, for a given species using a standard statistical analysis (Norton-Griffiths, 1976).

The hippo count in 2005 was not carried out using transects, but rather as a total count of all individuals counted along the shores of the lake and along the main rivers. Given the substantial reduction in the areas occupied by hippopotamus pods, the area that needed to be surveyed was substantially reduced. As a consequence the whole survey could be completed in less than a day (28 August 2005). The average altitude was 300 to 400 feet above ground level and the average speed was 183 km/h. The survey flight crew was made up of one pilot and two observers counting independently. To

12.3
There are approximately 720 Mountain Gorillas left in the world. Half of this population lives in the Virunga Volcanoes, shared between Rwanda, DRC and Uganda.

include observer omissions, caused by diving hippos that could not be seen, a correction coefficient of 1.25 was applied. The coefficient is based on the studies of Delvingt (1974), and provides for a much more accurate estimate of the park's hippo population.

12.2.2 Forest Surveys

Most studies of the mammals of the forested areas have been inventories, to determine species lists and distributions for the park. These usually involved the collecting of specimens for identification in the Africa Museum at Tervuren (De Witte, 1938) or elsewhere (Chifundera *et al.*, 2003; Owiunji *et al.*, 2005). Very few surveys have been made in the forests to quantify the abundance of large mammals, with the exception of Mountain Gorillas.

12.2.2.1 Detailed study in Rwanda
Surveys were made in Rwanda which measured the population density of ungulates in the Virunga Volcanoes massif (Plumptre, 1991; Plumptre and Harris, 1995; Plumptre *et al.*, 1997). These studies used permanent plots and transects to count dung of the ungulates and convert the dung counts to population density estimates.

12.2.2.2 Distribution of mammals in Virunga Volcanoes
Some more recent surveys have looked at the relative abundance of large mammals in the Virunga Volcanoes (Owiunji *et al.*, 2005). These used sightings of mammals from a Mountain Gorilla census undertaken in 2003 (Gray *et al.*, 2005) and complemented with dung counts undertaken in January 2004. Reconnaissance walks were made throughout the massif as part of the gorilla survey looking for fresh gorilla trail. Sightings of large mammals were recorded while on these reconnaissance walks until fresh gorilla trail was located. The distance walked was measured with a hipchain and thread which allows the calculation of encounter rates for the animals. The reconnaissance walks aimed to enter every 500m x 500m block area throughout the massif. Sightings of any large mammal and dung observations for elephants, Giant Forest Hogs and buffaloes were recorded.

For the second survey in January 2004 the focus was primarily on collecting biodiversity data on birds and plants of the massif. However, to complement the mammal data from the gorilla census dung counts were made in 10 m radius circular plots along reconnaissance trails and the dung observations of elephants, buffaloes, bushbucks and Black-fronted Duikers recorded.

12.2.2.3 Mountain Gorillas
As mentioned above, the 2003 Mountain Gorilla survey used irregular reconnaissance tracks throughout the Virunga massif, to search for recent gorilla tracks. When a track of no more than 5 to 7 days was found, it would be followed until the nests were found. From the nesting site, the number of gorillas, together with their age and sex was estimated by counting the individual nests, by measuring faecal matter, and by collecting hair samples. By locating and dating each of the nests and the gorilla tracks, and by marking discovered nesting sites it was possible to avoid double counting.

In order to establish an estimate for the total population, a correction coefficient was applied to the nest count estimates. This takes account of the fact that infants of less than one year are very difficult to detect in the nest counts. It also accounts for the small groups and solitary males that were not detected in the survey. The coefficient is based on data from a large sample of very well documented habituated gorilla groups. These habituated groups were also surveyed using the same nest counting techniques. The correction coefficient is based on the number of infants that are known from the habituated groups, but which were not counted in the nest survey. The same approach is used for small groups and solitary males that were never detected during the survey, but which are known to exist.

Emmanuel de Merode, Andy Plumptre, Maryke Gray, Alastair McNeilage, Katie Fawcett & Marc Languy

12.5 Results of a sampled aerial census in June 2006 over Virunga National Park (Kujirakwinja et al., 2006)

	Northern Sector		Central Sector		Total
	Estimate	95% Confidence interval	Estimate	95% Confidence interval	
Buffalo	74	12–208	3,748	811–6,685	3,822
Elephant	50	8–153	298	145–470	348
Kob	583	94–1,172	12,399	6,654–18,144	12,982
Waterbuck	6	1–17	368	130–600	374
Topi	0	0	1,353	408–2298	1,353
Warthogs	0	0	694	347–1,041	694
Baboons	0	0	737	104–1,389	737

12.2.2.4 Chimpanzees

Chimpanzee distributions established in 2006 were obtained using two complementary methods. The first set of data is drawn from ICCN recorded field observations, based on either biological monitoring, or patrol monitoring, or even opportunistic observations. A second source of information was based on a questionnaire carried out on rangers and wardens in each of the main stations of the park, using a detailed map of the park. Only observations from a reliable source, and that were made no more than a year before the inquiry, were taken into consideration. Added to that were the observations made by WWF staff during the numerous field visits in the National Park.

12.2.2.5 Okapis and other mammals of the lower Semliki

From 20 to 27 May 2006 a joint team made up of ICCN rangers, WWF and Gilman International Conservation staff carried out a wildlife survey on the west bank of the middle Semliki River, in the area of the Makoyobo, Lesse and Abia rivers, through to Abatupi river. The primary reason for carrying out his survey was to establish whether the okapi *Okapia johnstoni*, still ranged within Virunga National Park. The last documented observation of okapis in Virunga had been in 1959. Another species of interest was the bongo *Tragelaphus eurycerus*.

In the Abia sector, an inventory was carried out based on reconnaissance surveys using wildlife or human tracks in the forest. Tracks were used along a path of least resistance. In the Makoyobo and Lesse sectors, straight line transects were used. These transects were rigidly maintained, irrespective of the terrain. Transects were cut in the forest by teams of trackers with machetes, accompanied by a biologist who guided the team using a compass. For both the straight line transects and the reconnaissance surveys, indicators of animal presence were searched within a strip width of 10 meters either side of the transect. Each animal indicator (faecal matter, foot print, traces of eaten vegetation, vocalisation or direct observation) was systematically recorded and the location registered using a GPS. All indicators of human presence (abandoned camps, tracks, cartridges etc.) were recorded in the same manner. Several team members were Bambuti Pygmies, working for the GIC project in the Okapi Faunal Reserve. These team members were extremely experienced in identifying wildlife tracks, because it was their daily work in the Okapi Reserve.

12.3 Results

12.3.1 Savanna mammals

Table 12.5 presents the estimates for several species in the Northern and Central Sectors of Virunga National Park in June 2006. The results for each of the two sectors were analysed separately, and an overall total for the whole park was estimated.

The distribution of species is very similar to those of previous counts. The largest concentration of mammals is around Lulimbi. Elephants are particularly concentrated in the eastern part of the Central sector, even if a large population of over 60 elephants was counted in the Kabaraza area, both in 2003 and 2006. The same population was also seen on several *ad hoc* flights over the central sector between 2003 and 2006 (de Merode pers. com.). Although hippo numbers were counted in the 2003 aerial survey, they are not included in this table, because much more accurate results are available from the hippo total count carried out in 2005. The results of the hippo total count are presented in Table 12.6.

12.3.2 Forest Mammals

12.3.2.1 The Virunga Volcanoes

The Virunga Volcanoes range from around 1800 to 4507 m altitude above sea level and contain several montane habitats including montane forest, bamboo, *Hagenia-Hypericum* woodland, giant *Senecio* and giant *Lobelia* and alpine moorland. The Mountain Gorilla, *Gorilla beringei beringei*, is probably the best known mammal in this massif and has been intensively studied at the Karisoke Research Station, funded by the Dian Fossey Gorilla Fund International (Robbins, Sicotte and Stewart, 2003). Other primates occur here but have been little studied. Of particular interest for conservation is the golden monkey, *Cercopithecus kandti*, which is almost completely confined to the Virunga Volcanoes (a very small population occurs in the remnants of Gishwati forest to the south in Rwanda). The Blue Monkey, *C. mitis*, and the L'Hoest's Monkey, *C. lhoesti*, also occur here. Other large mammals include the elephant, *Loxodonta africana*, buffalo, *Syncerus caffer*, Bushbuck, *Tragelaphus scriptus*, Black-fronted Duiker, *Cephalophus nigrifrons*, and Giant Forest Hog, *Hylochoerus meinertzhageni*. The Yellow-backed Duiker, *C. silvicultor*, used to occur in the massif but has not been seen for many years. Of the large predators, Spotted

12.4

Chimpanzees are well represented throughout the park, but in small numbers, except in the Mikeno Sector. Their populations are believed to be at around 300.

12.7 *Topis are common on the Lulimbi plateaux, to the south of the lake. In contrast, they are completely absent in the north.*

12.6 Hippopotamus populations in Virunga National Park in August 2005

Sector	Numbers observed	Estimate*
N shore of Lake Edward	17	21
Upper Semliki	40	50
W shore of Lake Edward	0	0
S & E shores of Lake Edward	530	662
Ishasha River	49	61
Rutshuru River	46	58
Rwindi River	28	35
Inland waters and marshes	0	0
Total	**710**	**887**

*Correction coefficient 1.25

Hyaenas, *Crocuta crocuta*, are still seen with feral dogs, *Canis domestica*.

12.3.2.2 Nyiragongo-Nyamulagira

The active volcanoes of Nyiragongo (3469 m) and Nyamulagira (3055 m) have been less well studied and the vegetation is in a dynamic flux because of the activity of the volcanoes. There is a series of successional vegetation types from bare lava rock to lichen encrustations, mosses and then xerophytic vegetation that can withstand long periods without water because it drains away quickly through the lava rock. Where there has not been relatively recent lava flows, montane forest occurs up to sub-alpine habitat at 3470 m on Nyiragongo. Primate species include chimpanzees, *Pan troglodytes*, Blue Monkey, Red-tailed Monkey, *Cercopithecus ascanius*, and Olive Baboon, *Papio anubis*. Elephant, Bushbuck and Bushpigs, *Potamochoerus larvatus*, are the main ungulate species recorded and leopard, *Pantherus pardus*, and Golden Cat, *Felis aurata*, are the two large predators although there have been no recent sightings to our knowledge.

12.3.2.3 Tshiaberimu

This block of montane forest on the western Rift escarpment is small (60–77 km^2) and lies between 2450–3117 m altitude. Despite its small size, 276 plant species and 106 animal species have been recorded from this forested area (Chifundera *et al.*, 2003) including Grauer's Gorilla, *Gorilla beringei graueri*, Owl-faced Monkey, *C. hamlyni*, Blue Monkey, L'Hoest's Monkey, Guereza, *Colobus guereza*, buffalo, elephant and Black-fronted Duiker. The montane forest is linked by a narrow corridor, 'Mulango-wa-nyama' with the rest of the Virunga Park at lower altitudes in the Rift valley.

12.3.2.4 The Rwenzori massif

Virunga National Park links the lowland forest which stretches to the Ituri with the glaciers on the peaks of the Rwenzori Massif. As such this region spans an altitude of 680–5,100 metres and includes lowland forest, medium-altitude forest, montane forest, bamboo, heather forest, subalpine, and alpine habitats. Among the mammal species that have been recorded here are the chimpanzee, blue monkey and Angolan Colobus, *Colobus angolensis*. The ungulates include elephant, Black-fronted Duiker, Rwenzori duiker, *Cephalophus rubidus*, and Bushpig. Leopards and Golden Cats probably occur here as they are recorded from the Rwenzori massif in Uganda.

12.3.2.5 The Semliki Forest

The 2006 inventories detected 17 different signs of recent Okapi presence: 12 in the Lesse area and five in the Makayabo area. No signs were found in the Abia region, which is the area where the most signs of human presence were found. Bongos were also found in this region. An interesting find is the White-bellied Duiker *Cephalophus leucogaster*, a species never seen before in the park. Six signs of this species were found in the park. Other duikers were also found, including *Cephalophus weynsi*, *C. dorsalis*, *C. sylvicultor et Philantomba monticola*. Lastly, *Cercopithecus mona denti* was also found. There was no recent evidence of elephants. Bushbucks, Yellow back Duikers, hippos, leopards and Golden Cats are also known to occur in the Semliki Valley.

Table 12.9 presents the number of mammal species in each of the five forest blocs in Virunga National Park. Included in the table is the total number of mammal species, the number of species that are endemic to the Albertine Rift, and the number of threatened species (vulnerable or in danger of extinction).

The results show a much higher number of species in the Virunga Volcanoes, but this may be because the area has been much more intensively surveyed for smaller mammals over many years (rodents, shrews and bats).

12.3.3 The distribution of mammals in the extinct volcanoes of Virunga

Data from the mammal surveys in the Virunga Volcanoes can easily be transposed to a map that also depicts the relative abundance of those species. It was not possible to get actual densities because of the errors caused by certain key variables, such as the differential decomposition rates of faecal matter according to season and habitat (Owiunji et al., 2005; Plumptre & Harris, 1995; Plumptre, 1991). Data drawn from faecal matter are on Figure 12.10 and those based on direct observations are in Figure 12.11.

12.3.3.1 Mountain gorillas

The Mountain Gorilla population in the Virunga Volcanoes has been the subject of many surveys and studies over the past 35 years (Harcourt et al., 1983; Weber and Vedder, 1986; Sholley, 1991). Nevertheless, over the past 15 years, the region was badly affected by political instability and armed conflict. As a consequence, there was no comprehensive census undertaken between 1989 (when the total population was estimated at 324 individuals) and 2003. The 2003 census was the last one to be carried out (Gray et al., 2005) and it estimates the population at 380 individuals. This represents a population increase of 17% since 1989, or an average annual growth of 1.15%. Significantly larger groups were found during the most recent census, especially among the habituated groups.

12.9 Number of species, number of endemic species to the Albertine Rift, and number of threatened species in Virunga National Park

Forest Block	No. of species	No. of endemic species	No. of threatened species
Virunga Volcanoes	86	18	6
Nyiragongo-Nyamulagira	26	4	5
Tshiaberimu	21	1	2
Rwenzori	44	7	3
Semliki	47	1	4

12.3.3.2 Chimpanzees

Chimpanzee were noted in a number of areas between 2005 and 2006 (Figure 12.15). These sites are as follows: in the Southern Sector, on the flanks of the Nyiragongo, Maroba, the Rumoka volcano near Tongo, the area between Katwa and Mulalamule on the upper Rutshuru river, together with a site outside the park at Muhungezi. In the Central Sector, in the area between Mabenga and May ya Moto, the Kasali Mountains, the upper Ishasha and Lunyasenge, Kitibira and Talia-Kamandi. In the Northern Sector, in the foothills of the Rwwenzori Mountains (Balegha, Kikingi and Bamundjoma) and the Semliki Forest (Bahatsa and Lesse).

12.8
Buffaloes are one of five species of ungulates who have been regularly censused. Almost 4000 were counted in June 2006.

12.10 *Relative Faecal abundance in each sector studied*

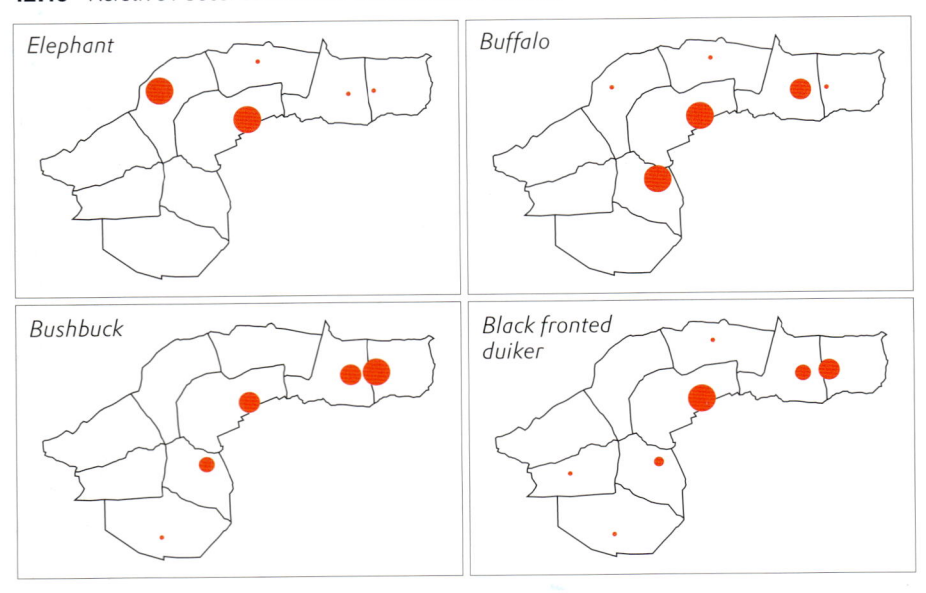

Elephant | Buffalo

Bushbuck | Black fronted duiker

12.11 *Relative abundance based on the gorilla census in 2003*

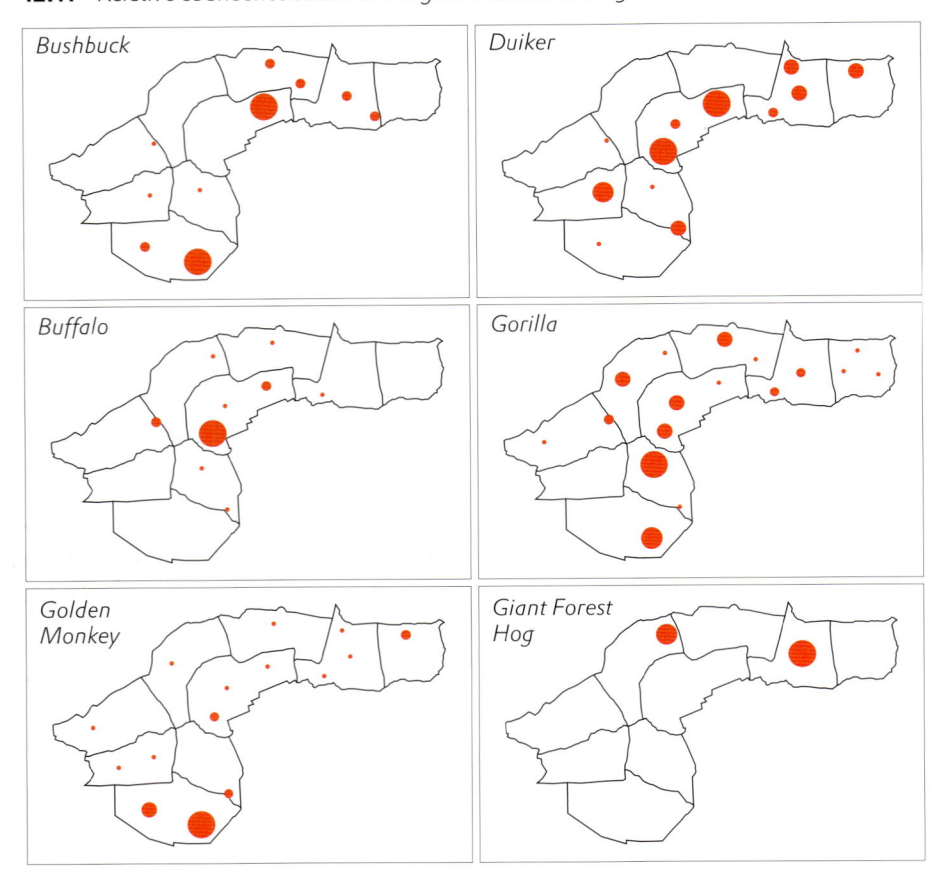

Bushbuck | Duiker

Buffalo | Gorilla

Golden Monkey | Giant Forest Hog

Note that more sectors were defined during this study, hence the presence of more than one spot in a single sector.

12.12 *A male silverback defends his group in the Mikeno Sector*

12.4 DISCUSSION

12.4.1 SAVANNA MAMMALS

The savanna surveys undertaken in recent years need to be assessed from a historical perspective, taking into account the dynamic nature of population changes at the moment. Chapter 8 clearly showed that many populations of savanna species have effectively been decimated. The changes that have affected these populations, and their movements into the adjacent Queen Elizabeth National Park, nevertheless indicate that the populations could one day be re-established, once the rule of law is brought back to Virunga National Park. The savanna mammal surveys must remain a priority and need to be carried out at least at three year intervals.

It is encouraging that not a single species has been lost from Virunga National Park in recent years. The only exception is the African Wild Dog *Lycaon pictus*, which was wiped out from Virunga National Park, and throughout the region, in the 1950s. In the long term, when prey populations have been re-established, the re-introduction of wild dogs could be considered.

12.4.2 FOREST MAMMALS

It is clear that the forested areas of the Virunga Park have had few studies of wildlife in comparison with the savanna areas. The reality is that most studies were designed to establish inventories of species and even these have primarily catalogued the large mammals and not included the smaller species. The only site with some additional information on relative abundance of species is the Virunga Volcanoes region (Mikeno sector of Virunga Park and adjacent parks in Uganda and Rwanda). There is therefore an obvious need to undertake some surveys in these other areas of the park to obtain quantitative baseline estimates of species numbers. Baseline numbers are important because of the

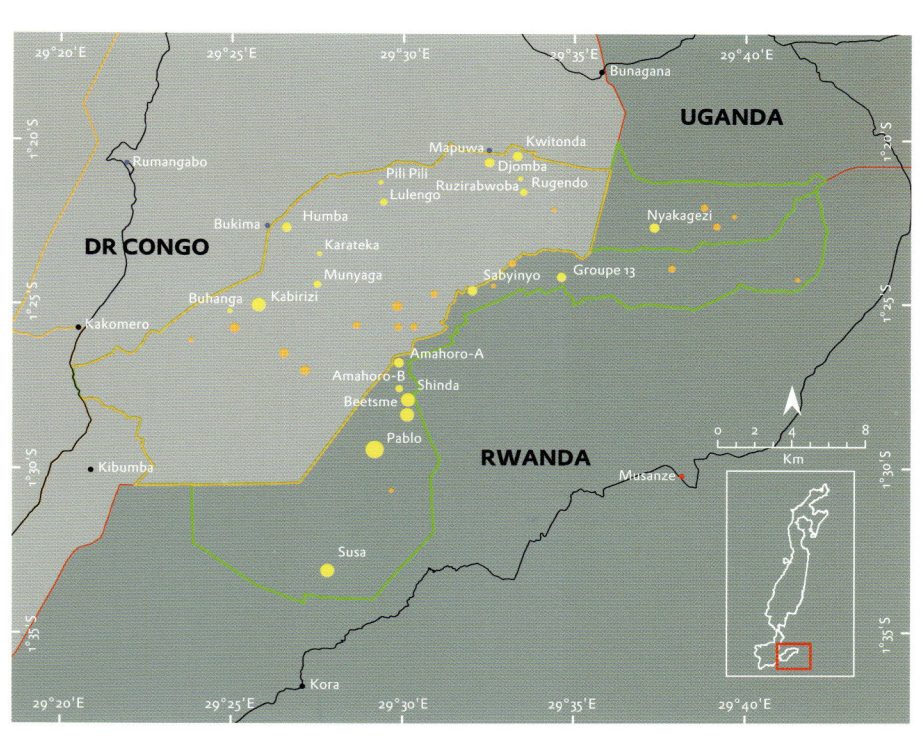

12.13
Group distribution during the mountain gorilla census in the Virunga in 2003

Group size
Non-habituated groups
- • 1
- • 3–7
- • 8–11

Habituated groups
- • 1
- • 1–6
- • 6–14
- • 14–35
- • 35–52
- ● Town
- ● Village
- ● Patrol post
- ▬ International boundary
- ▬ Road
- ▬ Park Boundary
- ▬ Virunga Massif

12.14
The hippopotamus is the best censused species in the park. It represents an important tourist attraction, but also a keystone species in the savannas of Virunga National Park

The status of Virunga's large mammals

12.15
Distribution of chimpanzee observations in ViNP in 2005

- ● Town
- ● Chimpanzees
- ▬ Roads
- ▬ International boundary
- ▯ ViNP

Virunga Volcanoes, around Mt Visoke and on the saddle between Mts Visoke and Sabinyo, together with a few groups in the south and east of the ecosystem. In contrast to the previous censuses, no groups were found on Mt Mikeno. The growth experienced by the population since 1989 was almost entirely concentrated in one section of the ecosystem – the area to the south of Mt Visoke and on the flanks of Mt Karisimbi in Rwanda. This is the area covered by the Karisoke research groups, and the Susa group that was habituated for tourism (Figure 12.13). The distribution of Mountain Gorillas was found to be negatively correlated with human disturbance.

12.4.4 CHIMPANZEES

The figure for chimpanzee populations in Virunga National Park in the 1960s suggest that the majority of these animals were located in small forest blocks completely isolated from each other and from the main populations in the lowland tropical forests of the Congo Basin. As such, they were considered to be relict populations. Chapter 8 compares the results of 1960 with those of 2005–2006 and shows that most of these populations still exist in their former range, even if some of these have diminished and are now severely threatened.

There is no accurate estimate for the number of chimpanzees in Virunga National Park, because of the years of insecurity which made a thorough survey effectively impossible. However, recent data collected by the WWF/PEVi project, in collaboration with ICCN, suggest that the population is around 100 in the Southern Sector, and is well over 100 in the Central Sector. In 1960, the population estimate for the Central Sector was between 200 and 230 individuals (Chapter 8). It is impossible to give an estimate for the Northern Sector, but if chimpanzees are present in most of the northern forests, their populations would have to be at least 150. Thus, the minimum number of chimpanzees in Virunga National Park is at least 300.

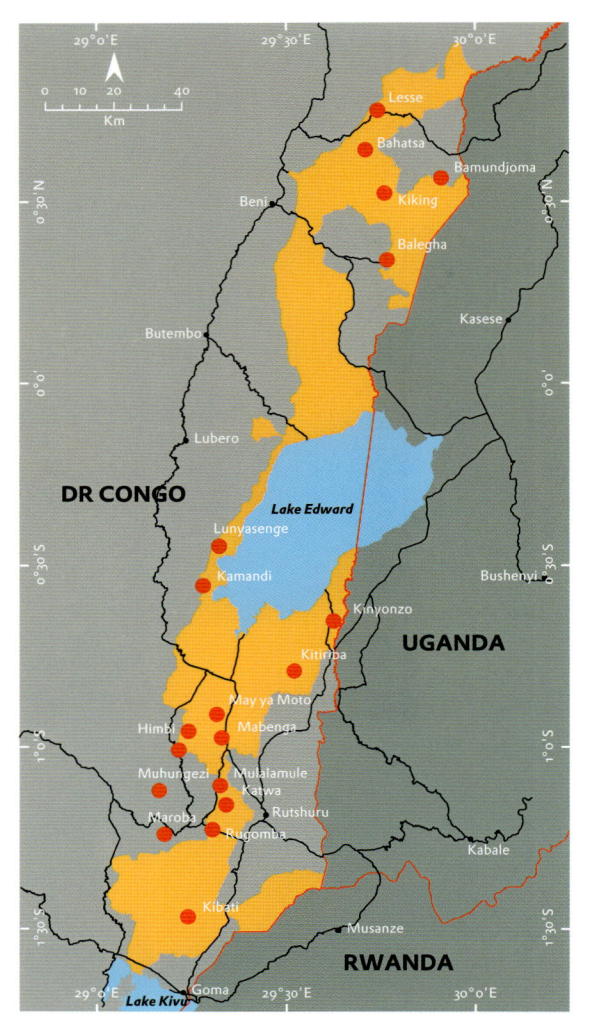

need to monitor animal populations over time and assess whether they are increasing or decreasing.

The rediscovery in May 2006 of the okapi and the bongo in the middle Semliki, after vanishing for almost half a century, was a very encouraging event that reminds us that, for all the destruction of the park's wildlife populations, all is not lost. Furthermore, the recent discovery of a species of duiker new to the park, also demonstrates that the park has many unknown species and populations that have yet to be discovered, given that the northern forests have not benefited from a comprehensive survey since the 1960s. A real priority in the coming years will be to establish more accurate baseline data in the northern forests, especially on iconic species such as okapi and bongo, so that effective wildlife conservation can take place both inside and immediately outside the park. Rumors of the presence of the okapi on the east bank of the Semliki should also be verified. If this were to be shown to be correct, it would not be beyond reasonable doubt for a small okapi population to be present in Uganda, in the Semuliki National Park.

12.4.3 MOUNTAIN GORILLAS

As in previous counts, the majority of the gorillas counted in 2003 were found in the central part of the

12.5 CONCLUSIONS

The sampling methods used provide the best trade-off between survey effort and accuracy of results. A total count of all species would be prohibitively expensive to implement. Where a comparison is possible, the aerial surveys seem to corroborate well with ground surveys and *ad hoc* observations.

On the one hand, these results provide an accurate assessment of the status of the wildlife resources of the park after seven years of political instability and civil war. On the other hand, it provides us with a benchmark, against which it will be possible to measure the success of the various large scale rehabilitation initiatives in Virunga National Park. The survey method has been standardised, making it possible to measure change over time and space. Thus, it also becomes possible to compare the status of Virunga's populations with those of sites like Garamba National Park, where similar methods were used. It is clear that these data, and those of subsequent surveys, will provide a fundamental tool for evaluating the rehabilitation programme for Virunga National Park. As such, it would certainly be useful to repeat the surveys on an annual basis, for the duration of the project, in order to provide regular assessments, and help wardens and project managers to understand where the main priority areas and issues are and to help them improve their project strategies.

It has not been possible to collect much information on the forest species in Virunga. The data presented in this chapter are largely restricted to species presence and distribution. It is probable that the decline of forest species has not been as great as that of the savanna species, largely because they are much harder for hunters to find. However, the widespread destruction of their habitats is likely to make them increasingly vulnerable, to the extent that local extinctions are likely in the near future. Surveying the forests of the middle Semliki, between the Watalinga Bridge and the Puemba River is without a doubt one of the greatest survey priorities for Virunga National Park.

The gorillas and chimpanzees of Virunga National Park have always been quite restricted in their range, and are both vulnerable, according to their IUCN Red List classifications. These populations face many threats, including poaching, habitat loss, armed conflicts and civil unrest, and disease. If the status of the chimpanzee populations remains largely uncertain, there is no question of doubt that the two subspecies of gorilla, in Mikeno and Tshiaberimu, have increased in recent years. The increase in the population of Mountain Gorillas, despite many years of intense political insecurity, is a remarkable conservation achievement; it is a tribute to the phenomenal dedication and devotion of its park staff and their partners, and of the collaboration of three countries to achieve a transboundary approach to conservation. Nevertheless, a detailed examination of the results does show that the success is largely concentrated in a very restricted part of the ecosystem. In some areas, the population has only remained stable. In others, it has even diminished. In many ways, we could have expected a much greater population growth, given that habitat availability is not a constraint. The typical growth for such a population could reasonably be expected the reach 3%. Some gorillas have fallen victim to the insecurity. Other disturbances, both of the habitat as well as the population itself, may have stifled the population growth.

12.16
There is an important population of lions in the central sector of the park, but it is extremely difficult to get an accurate estimate of their numbers.

Marc Languy & Déo Kujirakwinja

13 The pressure of legal and illegal fisheries on Virunga National Park

Marc Languy & Déo Kujirakwinja

As shown in Chapter 10, it is a major challenge to keep a balance between fish exploitation and conservation on Lake Edward, whose Congolese waters lie entirely within Virunga National Park. This chapter examines the uncontrolled expansion of legal and illegal fisheries along the shores of Lake Edward, including the recent development of numerous, larger fisheries on the western shore of the lake. These fisheries were established during the wars, without any legal framework or validation by State authorities or the ICCN in Kinshasa, and they are linked to significant illegal agricultural activities. Lastly, we discuss the impact of this on the sustainable exploitation of the lake's fish stock and the urgent need for the ICCN to reorganise the fisheries of Lake Edward.

13.1 History of the Virunga National Park fisheries

Verschuren (1993) gives a succinct but very instructive summary of the history of the Lake Edward fisheries and Chapter 10 provides additional information. It is useful to repeat the essential points here.

When Albert National Park was created, various small traditional fisheries were in existence along the Rutshuru River and the shores of Lake Edward.

Analysis of the 1934 maps and of old documents show that the fisheries were really very small: Vitshumbi, in particular, was no more than a few huts with 92 fishermen when the park was created (Harroy, 1987).

Then the authorities evacuated the local fisheries because of the high prevalence of trypanosomiasis (sleeping sickness). In the early 1940s fishing activities were re-established as co-operatives so the local communities could extract an economic benefit from the lake.

An important element of this was the fixing of a maximum number of dugout boats (pirogues) that can be registered and, therefore, allowed to exercise fishing rights on the lake – a process carried out in accordance with the ICCN. This limit set was 700, however since then there have been a spate of questionable local decrees (issued during the time of the rebellion, without the agreement of Kinshasa or the ICCN leadership) attempting to increase that number (see Chapter 10).

The legal status of the fisheries remains unchanged in 2008. Two fisheries, Vitshumbi and Kyavinyonge, are recognised inside Virunga National Park and a third, Nyakakoma, is tolerated.

Since the mid-1990s, however, various illegal fisheries have appeared, essentially along the western shore of the lake. Neither the ICCN nor the central government has authorised (the national park being governed by national laws) their installation: They are thus illegal. Unfortunately, they were allowed to take root during the war years as at this time the ICCN was unable to exercise the control to have them dismantled.

Some of the fisheries, such as the one at Birwa, east of Vitshumbi, have been evacuated now, but many others still persist (see section 13.3 below).

It is important to emphasise here that all of the Congolese waters of Lake Edward are part of Virunga National Park and that, because of this, any fishery established is *de facto* on the territory of the park, and therefore comes under the authority of the ICCN. In contrast, the Ugandan part of the lake is not a part of Queen Elizabeth National Park.

13.1
An artisanal dugout returning from fishing.

13.2 *Growth of Vitshumbi between 1959 and 2005.*

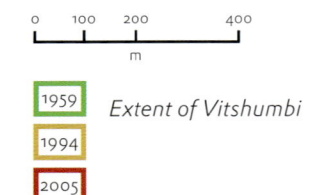

0 100 200 400
m

- 1959 — Extent of Vitshumbi
- 1994
- 2005

13.2 EVOLUTION OF THE LEGAL FISHERIES IN VIRUNGA NATIONAL PARK

With the population growth that the region has experienced and the weakening of the various authorities governing fishing activities, the three legal fisheries of Virunga National Park have not stopped growing, in area as well as in number of inhabitants.

13.2.1 VITSHUMBI

When the park was established, there were 92 fishermen based in Vitshumbi (Harroy, 1987). By 1989 the number of inhabitants was estimated to be 10,000 (Verschuren, 1993). By the year 2000 Mugangu (2001) estimated 24,000 inhabitants. However this figure does not appear to be the result of any study and later studies indicate it is an overestimate.

It is difficult to know the precise number of inhabitants in Vitshumbi today. It is, on the other hand, easier to document the expansion of the fisheries on the basis of the number of habitations with the help of aerial photographs, which were taken in 1959, 1988, 1994 and 2005. It is, in fact, possible to identify administrative or collective buildings (schools, churches, and so on) as well as houses and even latrines.

The study on Vitshumbi's growth is summarised in Figure 13.2. The 1948 maps show only a handful of huts in existence at that time, and the aerial photos of 1959 indicate about a hundred habitations spread over 19 hectares. In 1988, no less than 700 habitations can be counted. There is slow growth from 1988 to 1994 (41 additional habitations, but they occupied a small area and were confined to the east of the main road); by 1994 the fishery occupied an area of 38.7 hectares. Then followed a large increase with 350 new habitations built in a new area covering nearly 30 hectares, west of the road, where the housing density is much lower than in the 'old' quarters. The total area of Vitshumbi was 70.2 hectares in November 2005, having more than tripled since 1959.

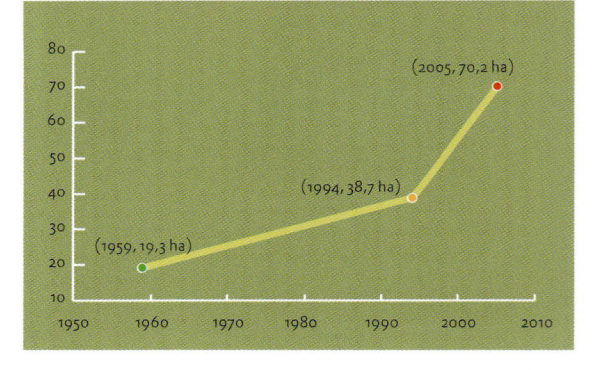

A detailed analysis of recent aerial photographs shows that Vitshumbi had 747 habitations in July 1994 with an increase to 1,324 by November 2005. On the basis of inquiries and visits in the field, it is estimated that there are on average 10 people per building, which gives an approximate total of 13,240 inhabitants in Vitshumbi.

A close-up photograph of these new neighbourhoods, taken in April 2006, shows the new houses that have appeared within the space of six months (Figure 13.3). The very new corrugated roofs and the many houses under construction are unambiguous evidence of Vitshumbi's growth.

Regulations limit the number of dugouts based at Vitshumbi to 400, with seven people per dugout, so in theory, the number of people working legally at Vitshumbi should be 2,800 - a fifth of the actual number of people living there.

Since its establishment, Vitshumbi has also seen the sort of infrastructure appear that make it more of a village than a fishery. A list of the principal infrastructures and their date of creation is shown in Table 13.4.

13.3
Close-up aerial view of the new neighbourhood in Vitshumbi, April 2006. One can see the spatial layout and a preponderance of corrugated tin roofs.

13.4 Principal Infrastructures at Vitshumbi in 2006

Category	Date of installation	Construction	Type of construction
COOPEVI	1949	1949	blocks
Catholic Church	1949	1973	stones
Catholic Primary School	1949	1949	blocks and stones
CBCA Church	1954	1984	bricks
Bureau Localité	1967	1945	blocks
Kimbanguiste Church	1968	1973	adobe blocks
Jehovah's Witness	1975	1980	adobe blocks
Brothers of Jesus	1982	1992	adobe blocks
Kyaviboko Primary School	1982	2005	bricks
Bwera Primary School	1985	1985	adobe blocks
12th CADECO church	1987		wood
CEBCE	1978	1987	adobe blocks
Community Health Centre	1949	1997	cement blocks
8th CEPAC	1974	1987	adobe blocks
Fishing Institute	2000	2003	adobe blocks
Kyaviboko Institute	1995		adobe blocks
EPHIGO	1997		wood
New Apostolic Church	1984	1984	adobe blocks
Bwera Institute	1999	1999	adobe blocks
37th CADECO	1973	1983	adobe blocks
Adventist Church	1974	1975	adobe blocks
FEPACO	1985	1992	adobe blocks
Vitshumbi Institute	1976	1976	adobe blocks
COOPEVI Health Centre	1949	1949	cement blocks
Fishermen's Committee	1984	1988	adobe blocks

13.5
Evolution of the size of Kyavinyonge, from 1959 to 2006.

Extent of Kyavinyonge
- 1959
- 1989
- 2006

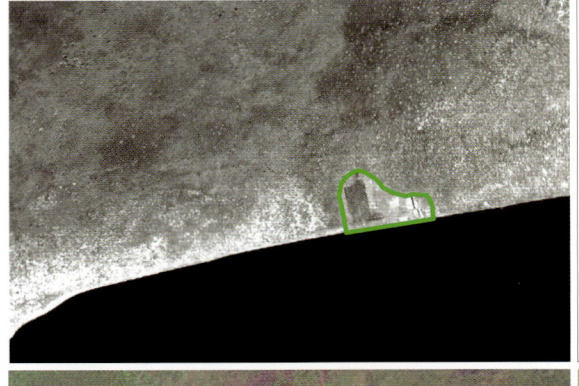

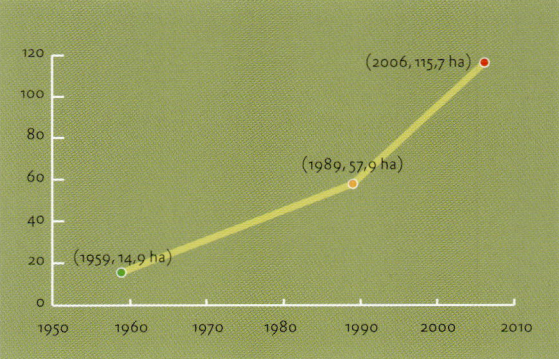

13.2.2 Kyavinyonge

The Kyavinyonge fishery was established in 1949. The 1948 maps clearly show that no form of habitation existed before this period.

Aerial photographs taken in 1959 demonstrate that there were 65 buildings on the site that year. An aerial photo taken by C. Aveling in 1989 is clear enough to be able to count the habitations which number 1,735. Verschuren (1993) estimated in 1989 that there were between 5,000 (personal estimates) and 6,000 (Norbert Mushenzi, *pers com*) inhabitants, but these figures are probably an underestimate.

An aerial photograph, taken in March 2006, very clearly shows the geographic expansion of the fishery. By this date, 2,008 buildings can be counted which tallies exactly with the censuses made by the COPEVI in 2003 and 2005 (22,000 and 18,000 inhabitants respectively). On the basis of verification in the field and comparison of photographs, it has been possible to determine the geographic growth of Kyavinyonge; from 15 hectares in 1959, to 58 hectares in 1989 and doubling 17 years later to 116 hectares in 2006 (seven times its size in 1959). This expansion is summarised in Figure 13.5.

13.2.3 Nyakakoma

At the time of its creation in 1967, Nyakakoma was no more than a very small fishery, essentially used as an embarkation and landing point for dugouts so the catch could be taken to Ishasha village.

Like the other fisheries, Nyakakoma has undergone a significant expansion, but one that is proportionally greater than either Vitshumbi or Kyavinyonge.

The 1958 aerial photographs confirm that there were no habitations and Nyakakoma did not exist at this time (Figure 13.6). The comparative examination of two aerial photos from July 1994 and March 2006 make it possible to estimate the number of habitations at 122 and 329 respectively, with a corresponding 1,830 and 4,935 inhabitants. This means the fishery tripled in size within the space of 12 years. The structures of the habitations are a little different; the buildings are larger than at Vitshumbi and Kyavinyonge and they are walled so that they can house several families. Typically, each building houses around fifteen people. Between 1994 and 2006, the area of Nyakakoma increased from 14 hectares to 35.5 hectares.

The collective infrastructures of Nyakakoma also multiplied, as shown by the list presented here (Table 13.7).

The study of the growth of the legal fisheries in Virunga National Park thus shows the population explosion of these three areas, inside the park. In 1959, a total of 165 habitations were counted. In 1994, 2,600 were counted, and by 2006 this figure was 3,661. In less than 50 years the number of habitations in the three fisheries has multiplied by 20.

13.3 The emergence of illegal fisheries on Lake Edward

As explained above and in Chapter 10, three fisheries are recognised on Lake Edward. Other fisheries have also appeared, notably at the end of the 1990s and in the early 2000s. It is necessary, however, to differentiate the fisheries that are based at the ICCN Patrol Posts from those that have been entirely created on new sites.

The illegal expansion of fishing activity dates from the beginning of the 1990s and it began at the Patrol Posts. The economic crisis, coupled with the ICCN's decreasing ability to intervene and the negligible salaries received by ICCN staff, rendered law enforcement

13.7 Principal Infrastructures at Nyakakoma and date of installation

Category	Date of installation	Construction	Type of construction
Ndeze Fishery	1964	1964	blocks and wood
Fishermens' Committee	1984	1985	wood
Bureau de la Localité	1964		
Maritime Services Office	1982		
Environment	1986		
Police (ex. Gendarmerie)	1986		
DGM/DSR (ex. SNIP)	1987		
Nyakakoma Health Centre	1997	1997	wood
Primary School	1967	1967	wood
Secondary School	1999	1999	wood
Kindergarten	2001	2005	wood
CEBCE Church	1964–65	1988	wood
Mosque	1965	1970	wood
Catholic Church	1965	1992	blocks
CBCA	1966	1991	wood
FEPACO	1986	1997	wood
8th CEPAC	1980	1994	wood
Brothers of Jesus	1982	1997	wood
Kimbanguiste Church	1994	1997	wood
34th CADAF	2001	2001	wood
Jesus Only Saviour Church	2002	2005	wood
Adventist Church			wood
Jehovah's Witness			wood

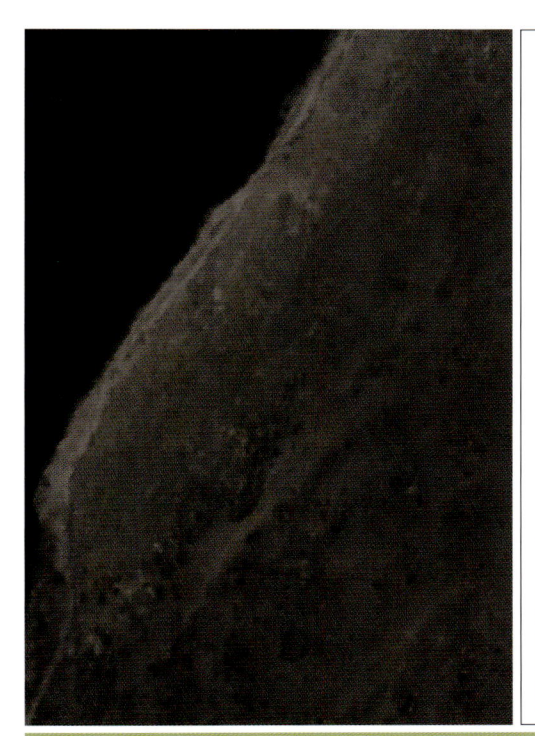

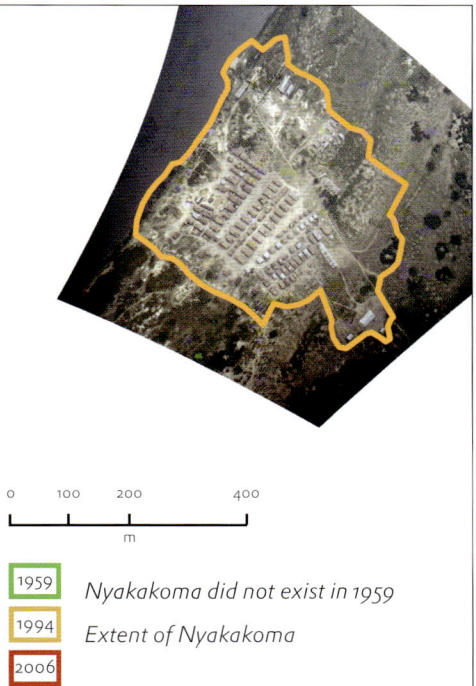

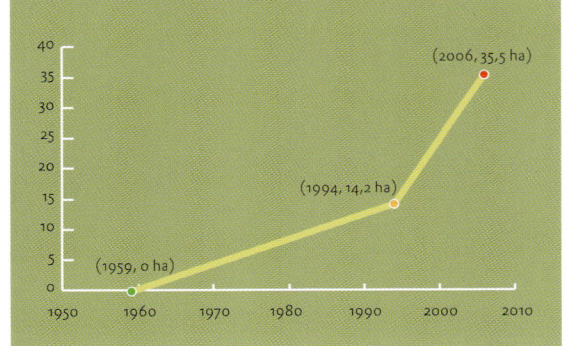

by the rangers very difficult, creating an incentive to allow fishing activity for direct or indirect revenues. This has led directly to an increase in the number of dugouts operating from Patrol Posts.

A World Wide Fund for Nature (WWF) aerial fly-over in 1994 showed different docking places dotted about, usually for groups of three to five dugouts, and one or more rudimentary huts of reeds and branches and signs of illegal encampments, but no solid structures, and extremely limited in number and size.

Fly-overs in March and June 2006, together with aerial photographs and the study of satellite images from July 2004 and January 2005, allow a more precise assessment of the present extent of illegal fisheries working in the Congolese area of Lake Edward.

Altogether ten well-structured illegal fisheries were noted, in addition to a more diffuse collection of habitations mostly around Kamande Bay.

The appearance of cultivated fields around the fisheries is a new and major development. It is clear that in the total absence of law enforcement for nearly 10 years, little by little, the local populations living in the fishing villages have been able to develop their agricultural activities. It is possible the fishermen's wives are cultivating the land in order to meet the needs of the household, but it is also very probable, at least around Kamande, that it is an expansion of the agricultural activity of the villages located on the summit of the escarpment of the Mitumba Mountains. This explanation makes sense as the escarpment runs along the western shore of the lake and largely corresponds to the boundary of Virunga National Park. Figures 13.8 and 13.9 show the location of these ten illegal fisheries

13.6
Evolution of the size of Nyakakoma, from 1959 to 2006.

The pressure of legal and illegal fisheries on Virunga National Park

13.8
Aerial photos of the ten illegal fisheries on Lake Edward.

13.9
Location of illegal fisheries on Lake Edward.

● illegal fishery
● legal fishery

along the shores of Lake Edward, as well as an aerial view of each of them (taken in March or June 2006).

Along the entire western shore, the illegal fisheries have resulted in 2,785 illegal habitations (Table 13.10). The various habitations scattered in the cultivated areas should also be added to this figure. Analysis of the satellite images shows that in 2005 over 11,685 hectares were under cultivation along the western shore of Lake Edward. The most invaded area is around Kamande Bay and Talya.

13.4 THE EVOLUTION OF FISHING ON LAKE EDWARD

In parallel with the burgeoning fisheries, the number of fishing dugouts working on the lake has also increased.

Since the COPILE *(Coopérative des Pêches Indigènes du Lac Edouard)* and then the COPEVI *(Coopérative des Pêcheries de Vitshumbi)* were founded, all dugouts have had to be registered. Following a detailed study on the fishing potential of Lake Edward, in 1989 the central government authorised a total of 700 dugouts with Vitshumbi, Kyavinyonge and Nyakakoma as home ports (see Chapter 10).

This quota of 700 dugouts was fixed, on the basis of a thorough scientific study, to ensure that the fishing effort would not be greater than the lake's reproductive capacity. The fishing potential for the whole of Lake Edward is between 15,000 and 16,000 tons per year, of which 3,000 to 4,000 tons are from the Ugandan side. At the end of the 1980s Vakily (1989) calculated that the fishing effort on the Congolese side represented between 10,000 and 11,000 tons per year, which was already close to the maximum capacity.

Mackie (1989) counted no less than 285 dugouts just in the area between the mouths of the Rwindi and Ishasha Rivers. From this it would be sensible to assume the authorised quota of dugouts was being ignored.

In 1994 a census of dugouts on Lake Edward, conducted by the WWF and the ICCN (Languy, 1994), confirmed the number of dugouts greatly exceeded the quota established under the COPEVI-ICCN convention, with a minimum of 1,100 dugouts counted.

In 2005 the COPEVI itself estimated a total of 2,000 dugouts (motorised and otherwise). This figure included the small craft of the fisheries and the clandestine 'pirate' fishermen.

As well as ignoring dugout limits, the fisheries and fishermen have also exceeded geographical limits. Practically all of the aerial fly-overs carried out since 1994 (and there have been dozens) have shown men fishing within the spawning grounds of most of the bays seriously damaging the reproductive potential of the fish. This information was already partly represented in the report of the General Assembly of the COPEVI in April 2000, which confirmed that the fishermen did not respect the ruling to keep a minimum distance of 500 metres from the shore and 1,000 metres from the spawning grounds.

Finally, direct observations in the field and from the COPEVI have recorded the use of prohibited nets (with too small a mesh). This is corroborated by the observation of very small tilapia on the market.

13.10 Illegal fisheries along Lake Edward in 2006

Fishery	Position	Habitations	Inhabitants
Kasindi Port	00°05'05"S; 29°41'50"E	327	3,270
Mahiha	00°07'40"S; 29°39'19"E	700	700
Muramba	00°11'40"S; 29°28'45"E	202	2,020
Kisaka	00°17'11"S; 29°26'49"E	429	4,290
Mosenda	00°21'25"S; 29°24'25"E	198	1,980
Katundu	00°23'36"S; 29°23'05"E	139	1,390
Lunyasenge	00°26'40"S; 29°22'22"E	546	5,460
Talya	00°31'02"S; 29°19'40"E	406	4,060
Baie de Kamande	00°33'38"S; 29°18'51"E	60	600
Kamande	00°35'13"S; 29°18'08"E	408	4,080
Total		**2,785**	**27,850**

13.5 Impact of various ongoing pressures on the productivity of Lake Edward

Until now no one has repeated Vakily's study (1989) and so it is difficult to demonstrate scientifically the impact of the different illegal activities recorded since the beginning of the 1990s, and particularly since 1996.

Four important elements lead us, however, to a depressing conclusion:
– Tripling of the desirable number of dugouts: **overfishing (increased mortality).**
– Fishing in spawning grounds: **reduction of the birth rate,** or more precisely of the survival rate, of the fry.
– Fishing with a net mesh that is too small: **reduction in fish reproduction.**
– Decimation of the hippopotamus population: important loss of input in the trophic chain of the lake, leading to a **reduction of productivity.**

These four elements, taken as a whole, can only lead to an appreciable reduction in the annual fish production of Lake Edward, to the direct detriment of the local economy. In fact, this situation has already been confirmed by numerous fishermen.

Another effect of the increasing number of people inhabiting fisheries, especially along the western shore, is a net increase in human traffic in the area, rendering the control work of the ICCN practically impossible. Furthermore, it is feared that, with the tens of thousands of people now living inside the park, on the edge of Lake Edward, the demand and trade in bush meat will also experience a net increase.

13.6 Conclusions and recommendations

Study of the development of the Lake Edward fisheries clearly shows that the present situation has moved dangerously far away from what was intended by the legislator when the fisheries were established.

Far from helping to maintain traditional fishing rights in a cooperative way and to sustainable development of the resident communities in the park, the fishing industry, as currently practised, has led to the anarchic development of infrastructures and to conflicts between the COPEVI, the private fishermen, and the administrative and military authorities. Above all, there has been an unsustainable exploitation of the fish stock. It is unlikely to disappear entirely but, in the short term, this means the lake can no longer supply the 15,000 to 16,000 tons of fish local communities need.

There is an urgent need for a new convention to be established between the ICCN and the COPEVI, and for an end to the toleration of the unregistered 'pirate' fishermen.

Above all, it is essential to complete the dismantling of the illegal fisheries along the western shore, so that good fishing management can begin to be practised for the benefit of everyone, while preserving the primary objective of protecting the exceptional natural habitats of Virunga National Park. To begin with, all agricultural activity must be prevented by the central governing body, with support from the local administrative authorities, respecting the orders of the Ministry of Environment (circular letter no. 1283/CAB/MIN/ECN-EF/2005 and press release of 20 June 2005 issuing an injunction to all farmers to stop cultivating the park).

As well as abiding by national law, it is important to reinforce control of the types of nets used and to reinforce respect for the spawning grounds which must be physically demarcated with buoys and ropes.

Lastly, let us remember the importance of allowing the hippopotamus population to re-establish itself in Virunga National Park, particularly in regard to producing the 'fuel' needed for good healthy fish reproduction.

14.1
The production of Makala (charcoal), though illegal, has become an enormous commercial enterprise, particularly in the Southern Sector of the park. Here, smoke from makala kilns can be seen emerging from the forest at the foot of Nyiragongo Volcano in April 2006.

Marc Languy, Samuel Boendi Lihamba & Walter Dziedzic

14 The supply of wood in the areas adjacent to Virunga National Park

*Marc Languy,
Samuel Boendi
Lihamba
&
Walter Dziedzic*

Chapter 9 showed that the region around Virunga National Park has experienced a particularly marked population explosion. When more than three million people live less than a day's walk from the park, the supply of firewood, charcoal (known as *makala* in Kiswahili), and lumber for construction represents one of the most acute threats for the forests of Virunga National Park (see also Chapter 19). Indeed, electricity and alternative energy sources are extremely limited; in fact they are simply non-existent in the vast majority of neighbouring villages. With such a demand, and taking into account the very limited capacities of the ICCN to enforce the law, the national park has been – and remains in 2008 – the only source of wood for energy. This chapter studies the current demands for wood and looks at how to quantify the deficit and identify the possible solutions. It also draws on more than 15 years of experience with the WWF's agroforestry programme and from the experiences of private and community plantations.

14.1 Study on the consumption of wood for heating and construction in the town of Goma

14.1.1 Context

In 1987, the WWF programme in Virunga National Park conducted a study into the firewood crisis around the park. As a result of its findings, WWF developed an important agroforestry programme and support for communities and individuals around Virunga National Park wishing to plant trees for private or commercial use.

Following the upheavals of the 1990s and the population explosion in the main cities, the WWF programme examined wood consumption in the town of Goma. This documented the pressure exerted on Virunga National Park, and also extrapolated the size of such a demand around the park, in both small and large communities. This study was conducted in November 2005 by students of the *Centre Universitaire de Goma,* supervised and supported by the WWF Environmental Programme. Wood consumption in Goma was evaluated using three independent approaches to guarantee the validity of the results.

The first, fairly basic, approach consisted of extrapolating the town's wood consumption by using standard data on consumption per person or per family in the region (on the basis of various previous enquiries carried out by other NGOs), multiplied by the number of people or households in Goma.

The second approach was to undertake a detailed survey of 4,067 households (including hotels, hospitals, orphanages, forges and other key consumption points in Goma) in order to quantify, not only their consumption of wood and charcoal, but also the expenses related to purchasing wood for energy.

Thirdly, a study was made of the different supply centres in the town that provided *makala* and construction wood, in order to learn how much wood was imported into town each day.

14.1.2 Results

In 2007, the town of Goma had 550,000 inhabitants with an average household size of nine people. Only three percent of households had access to semi-reliable electricity and only occasionally needed to supply themselves with wood.

A staggering 97 percent use charcoal almost exclusively as their source of energy: 55 percent buy large sacks of *makala* and 42 percent buy charcoal in retail quantities, a day at a time (Table 14.3).

The different calculations, by consumption of sacks per month or by calculating daily expenditures in char-

14.2
Firewood collecting is a daily chore for numerous women and children. Many get their supplies directly from the park, as can be seen here, in the Mwaro corridor between the Mikeno and the Nyamulagira sub-Sectors.

14.3 Consumption of makala by households in the town of Goma

Category of consumption	Average consumption per household per year (35-kg sacks of *makala* per year)	Number of households sampled	Percentage of the total of households interviewed	Extrapolation of the number of households for the whole town of Goma	Total consumption per year (35-kg sacks of charcoal per year)
Purchase per sack of 35 kg	21	2,218	55 %	33,300	699,300
Purchase from retail	24	1,700	42 %	25,540	612,960
Electricity		149	3 %	2,260	0
Total		**4,067**	**100 %**	**61,100**	**1,312,260**

14.4 Consumption of wood and charcoal by different types of consumers in Goma

Type of consumer	Cubic metres of wood per month	Sacks of charcoal per month sampled	Number of units	Consumption per year in sack (1 stère = 2 sacs) equivalents (1 cubic metre = 2 sacks)
Brickworks	25	–	5	3,000
Forges	–	2	15	360
Poultry houses	18	4	17	8,160
Orphanages	20	12	5	3,120
Medical buildings	–	1	4	48
Restaurants and cafes	–	7	13	1,092
Hotels	–	36	25	10,800
Bakeries	84	–	8	16,128
Total	–	–	**92**	**42,708**

Marc Languy, Samuel Boendi Lihamba & Walter Dziedzic

coal equivalents, give reasonably close results: an average household uses 24 sacks (of 35 kg each) of charcoal per year. This is the equivalent of 840 kg of charcoal or the equivalent of 5,040 kg of wood (six kilos of wood produce one kilo of charcoal). This result is reasonably close to that achieved by calculating an average consumption of 2 kg per person per day, or 6,570 kg of wood per household per year.

Consumers, other than households, use around 28,300 sacks of charcoal per year (about 1,500 tons). The quantity of construction wood (rafters and planks) used in Goma is equivalent to 60,000 eucalyptus trees per year (Table 14.4).

The combined data shows that Goma's total consumption is 1,355,000 sacks of charcoal per year, the equivalent of 47,425 tons each year.

It should be noted that enquiries near selling points show that, for about two-thirds of the buildings surveyed, 704,000 sacks are imported to Goma each year. Considering that around a quarter of the supply is direct and does not pass through the sales depots, the calculated quantity is very similar to those obtained above: 1,400,000 sacks per year.

14.2 THE PRODUCTION REQUIREMENTS THAT WOULD MEET THE DEMAND FOR WOOD IN GOMA AND OTHER CITIES

In 2006, annual consumption in Goma of 47,425 tons of charcoal per year corresponds to around 285,500 tons of wood, or to a volume of wood equal to 476,000 cubic metres. A number of studies of productivity in Africa, and notably in the Albertine Rift region, show that a well-managed eucalyptus plantation benefiting from improved silviculture techniques can sustain production of between 20 and 25 cubic metres of wood per hectare, per year.

Therefore, a state-run plantation of between 19,000 and 24,000 hectares would be sufficient to supply the town of Goma. It is certain that, even if such an area were available, such an enterprise, including the management of over 50 millions trees, would represent a real challenge.

It is more difficult to extrapolate the needs in wood for energy for all of the neighbouring populations of Virunga National Park. The referendum of November 2005 noted that a total of 2,350,000 people were registered as voters in the province of North Kivu (CEI, *Commission Electorale Indépendente, Nord-Kivu,* 2005). By adding young people below the age of 18, by accounting for a percentage of unregistered adults, and considering the population's spatial distribution along the park boundaries, it is possible to estimate that the number of people living within a day's walk of Virunga National Park is nearly 3,500,000. Given that energy sources, other than wood, throughout North Kivu are practically as limited as in Goma, these 3,500,000 people represent the consumption of 1,780,000 tons of wood or 3,000,000 cubic metres of wood annually. Such a demand – if should be met by wood from eucalyptus – would require a total plantation area of around 125 to 150,000 hectares. It is obvious that a significant proportion of wood is coming from private plantations in gardens and fields, as well as from some forests to the west of the park. However, there are already small-scale commercial plantations around the northern part of Virunga National Park. This simple calculation sufficiently demonstrates the need to attack head-on the energy and wood crisis in the region of Virunga National Park.

14.3 SUPPLY SOURCES

As far as Goma and the rest of the Southern Sector of the park are concerned, neither the enquiry teams nor officials from the various conservation projects active in Virunga National Park were able to identify *a single* charcoal production site outside of the national park.

Moreover, not a single day passes without numerous active *makala* kilns being seen in the Nyamulagira and Tongo Sectors of Virunga National Park, easily identifiable by the plumes of white smoke they emit. These kilns are also very visible from the Rumangabo station, during aerial fly-overs, and even on SPOT satellite images. Indeed, in July 2004 two spot images covering the Southern Sector, from Goma to Tongo, revealed no less than 75 active *makala* kilns, corresponding to a production of 2,100 sacks for that particular day alone.

In common with Goma, it would seem that the towns of Rutshuru, Kiwanja, Butembo and especially Beni (but to a lesser extent Lubero) obtain nearly their entire supply of charcoal from Virunga National Park. However, the Butembo-Beni area has several eucalyptus plantations. These eucalyptus trees are mainly used as construction lumber but also for charcoal production, so that pressure on the park, though strong, is a little less than in the Southern Sector.

14.4 THE IMPACT OF ILLEGAL PRODUCTION OF CHARCOAL AND THE POSSIBLE RESPONSES

The problem of illegal exploitation within Virunga National Park of wood destined mostly for charcoal production must be addressed soon. Indeed, even leaving aside the illegal nature of the production of charcoal within the park, a harvest of the order of three million cubic metres of wood per year directly threatens the forests of the park and numerous animal species. For instance, the production of *makala* in the Tongo area

14.5
Firewood is also gathered as a commercial product, destined for sale in Goma or to lime makers. Here, indigenous trees felled in the dry forests of the Southern Sector have been reduced to a truck-load of firewood.

14.6
Many of the trees felled in the park are transformed on the spot into charcoal, commonly called Makala *in Kiswahili.*

14.7
Detail from a spot satellite image taken in July 2004 just a few kilometres north of Nyiragongo, showing a cluster of active makala kilns.

is such that vast expanses of the open sclerophyllous forest that harbours chimpanzees are burned (sometimes involuntarily by uncontrolled fires originating in the kilns), thus reducing the habitat available to the species. Moreover, chimpanzees are highly sensitive to disturbances. So intensive is the traffic taking place at Tongo and elsewhere on the flanks of Nyiragongo and Nyamulagira that its highly detrimental impact on chimpanzee populations is threatening the tourism that generates revenues for the park and for the human populations that live around it.

It is difficult to quantify the precise impact of such exploitation in terms of deforestation or degradation of the forest. The impact of supplying Goma alone (to the extent of 500,000 cubic metres annually) corresponds to several hundreds of thousands of trees felled every year. Considering the low growth rate on these lands (the soil is much less fertile than in the extinct volcanoes sector), this level of exploitation is clearly not sustainable.

The problem is relatively simple: a balance between the production and use of wood must be found outside the park. Whereas the problem thus expressed is relatively simple, its solutions by contrast are far from clear, given the complexity of the context. A variety of approaches must be adopted, in a parallel and complementary way. Intervention is needed in three main areas. First, the use of wood at the household level must be limited as much as possible. Secondly, production outside the park must be increased. Finally, the illegal traffic originating in the park must be curbed.

14.4.1 REDUCE THE CONSUMPTION OF WOOD AND CHARCOAL

The first line of intervention is to reduce the demand for charcoal and firewood.

As things stand, it would be possible to reduce the cutting of wood for the production of charcoal by simply improving the energy efficiency of this transformation. Most *makala* producers have engaged haphazardly in this occupation, as a result of the laxity, over recent years, in the application of the law, and they do not use the best burning techniques. The typical production yield is of the order of 10 to 15 percent, whereas this could reach 20 to 25 percent with adequate techniques (and as much as 30 percent in industrial ovens), which would provide for a substantial reduction of tree-felling.

Intervention is also possible at another link in the chain: the consumption of charcoal in households. Here it is essential to disseminate widely information on improved portable stoves. Numerous experiments have been conducted in the field across the region surrounding Virunga National Park, particularly during the Rwandan refugee crisis and its aftermath when several humanitarian NGOs introduced improved stoves. For such stoves to gain the widest possible use, three elements are essential: the production of portable stoves whose improved characteristics are compatible with people's habits; the popularisation of these stoves; and then – wherever possible – the introduction of a system of subsidies that makes the stoves affordable for household users.

A third approach to reducing the consumption of charcoal and firewood is the promotion of alternative sources of energy. The most evident alternative energy source in the region is electricity. There are numerous obstacles to a sustained and dependable supply of electricity to Goma and other urban centres, but none of these obstacles is insurmountable, particularly when the enormous potential of the hydrological network (and of other available energy sources) to produce electricity is taken into consideration. Essentially, what is needed is political will at the national level to develop and implement a comprehensive strategy for the energy sector in the Democratic Republic of Congo, one that recognises the true value of wood, which is gravely under-valued at present through a failure to take into account the negative ecological and economic impacts of the massive destruction of the country's forests, both within and outside protected areas.

In addition to electricity, other potential sources of energy must be studied: the methane of Lake Kivu is an obvious alternative; it raises considerable technical and political challenges, but its immense energy potential is beyond doubt. Other towns in the region use oil and paraffin with great success, so much so that these fuels, or their derivatives, are the primary source of energy for low-income households.

Here too, with the right political will, it might be possible, say, to ease taxation on the sale of this type of fuel for domestic use, and so encourage the conversion from charcoal to petroleum derivatives.

14.4.2 INCREASE WOOD PRODUCTION OUTSIDE THE PARK

After reducing the demand, the second line of intervention would be to increase the amount of wood on offer. To this end also, many experiments have been attempted in the region, often with a degree of success, but with a limited quantitative impact.

For Virunga National Park, the most pertinent

14.8
One of the many tree nurseries established under the ICCN/WWF programme that is promoting the growing of trees in plantations around the fringes of the park.

experiment so far is one undertaken as part of the WWF programme around the park, which is supported by numerous private and community plantations and, more recently, by the initiatives of those plantations with commercial goals as well.

14.4.2.1 Individual subsistence plantations and community plantations

In 1987, in response to the ligneous resource needs of local populations, the PEVi (Virunga Environmental Programme) launched an ambitious reafforestation programme, as part of its existing environmental education campaign. This programme had many objectives, but the community plantations (through schools, churches and associations) were the main focus of intervention, especially during the first ten years of the programme. At the same time, meetings to raise public awareness of the value of trees and the benefits of plantations and agro-forestry were developed and held in many villages. Millions of seedlings were produced and sold at a modest price (but never given away free, so as to avoid creating the impression that a seedling has no value).

Between 1988 and 2005, a total of 9,109,000 trees were planted in the immediate periphery of Virunga National Park, as shown in Figure 14.9.

The project also conducted a survey on the rate of seedling survival in a total of 162 private plantations and 139 community plantations. This survey shows that 50 percent of the seedlings survive after the first 12 months, which are by far the most critical (there is, so to speak, no mortality after one year). The survival rate varies, however, according to the type of plantation, as shown in Figure 14.11. The survival rate in private plantations is much better (73 percent) than in community plantations (45 percent). This observation is not surprising in that people habitually take better care of their own property than they do of community property. The results for the community plantations nevertheless remain very positive, not least because this system generally frees up relatively large areas, thus engendering a greater quantitative impact in terms of volume produced.

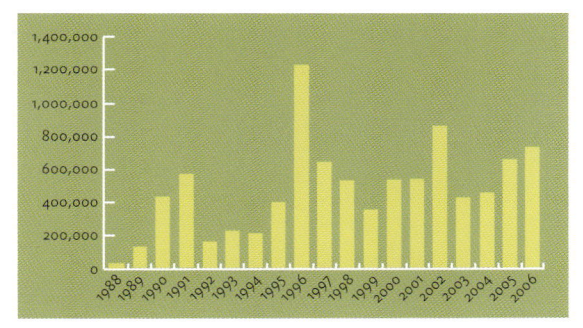

In all, no fewer than 60 different tree species were tested and popularised. Preferred species varied widely from one region to another, if only on account of differing climatic and edaphic conditions. Nevertheless, *Eucalyptus* and *Grevillea* remain the preferred trees for the production of sticks and construction wood. The third most favoured tree is *Acacia mearnsi*, singled out for its calorific and fertilizing qualities, its tannins, and its resistance to termite attacks, which renders it adequate as construction wood. Eucalyptus is preferred by local people because it is fast growing, which makes it possible, in the Rutshuru region, to start using the tree as construction 'stick' from its third year on. After three to four years, the stick, 10 to 12 metres long, can be sold for one US dollar (USD). Thus, a peasant who owns a field of one hectare can easily collect USD 700 at the first harvest and still expect later profits, given the high regeneration capacity of these trees.

Grevillea is somewhat less favoured because of its slower growth, but it was widely used in the past, par-

14.9
Graph showing the numbers of seedlings produced with the help of the WWF PEVi programme from 1988 to 2006. The fluctuations in production are related to security constraints in the field, which have hindered some interventions and slowed down production activities. The year 1992 saw a noticeable drop following the sudden withdrawal of donor agencies in November 1991. The year 1996 was a record year thanks to a reinforcement of PEVi reforestation activities by the wwf, with significant inputs forthcoming from both the UNHCR and the LWF, which found it useful to work with the PEVi in responding to the habitat destruction wrought by the massive influx of Rwandan refugees in July 1994. Thereafter, the outbreak of the war prevented the creation of many plantations. In 1999, security was gradually restored, allowing the project to intensify its activities from 2000.

The supply of wood in the areas adjacent to Virunga National Park

14.10
More than nine million seedlings have been produced since 1988, both in the project's tree nurseries and in the numerous local nurseries that are today managed entirely by local communities with technical support from the WWF.

14.11
Survival rates of seedlings (% after one year) in three different sectors of the WWF project.

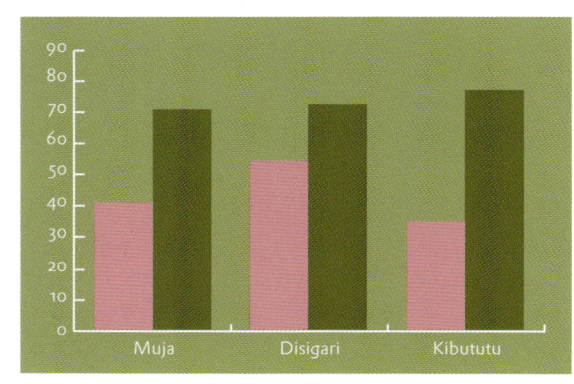

■ Community
■ Private

ticularly in association with coffee plantations. Finally, another widespread tree is *Markhamia*, which produces a good wood for making tools.

14.4.2.2 Green belts

Among community plantations, one type warrants particular attention. It is represented by those plantations bordering Virunga National Park that act as green belts. We refer here to real plantations, and not to trees planted in one or two lines along the park boundaries, where their main function is to provide and obvious physical delimitation between park land and agricultural areas.

An example of this type is the Rubare plantation that was started in 1992, and which now totals 70 hectares. The project facilitated an agreement between the ICCN and the peripheral communities under which, on the one hand, the ICCN grants the use (but not ownership) of land in the park's buffer zone to local people, while, on the other hand, these people plant, develop and maintain a plantation there and may farm within it during the trees' first years of growth. Moreover, benefits generated trough sustainable management of the plantation also go to the local communities. Having attained a particularly high yield (because of climate, soil quality and technical support), this plantation now produces a volume of more than 30 cubic metres per hectare per year, which represents a commercial value of more than USD 15,000/year.

14.4.2.3 Private commercial plantations

In recent years, as private initiatives have emerged for the production of sticks, many people have approached the project and its nurseries to establish small or middle-sized plantations. Besides seedlings, it is mostly technical advice these people seek to conduct such enterprises. Thus, the project encouraged interested people to form an organised body, which they did, calling their network *Réseau des Planteurs pour la Valorisation de l'Arbre,* or RPVA (literally 'network of planters for the economic promotion of trees'). This facilitates organisation of external support and related logistics, while enabling the peasant farmers to exchange their experiences and to achieve economies of scale by joining forces at various steps in the production chain – as in sharing the cost of transporting the cut timber, for example.

Such initiatives have considerable potential: as indicated above, recently conducted field studies show that one hectare of a eucalyptus plantation can yield construction sticks worth a minimum of USD 700 at the first harvest. Moreover, such a network provides for better marketing, which in turn makes selling the products easier.

Where the partners of the ICCN can most effectively intervene is in the organisation of training sessions. The first such session, organised by the WWF, found the strongest needs to be of a technical nature (establishment and management of a nursery, sustainable management of a plantation, and identification of exploitation perimeters), although administrative needs (accounting, book-keeping, network organisation, benefit evaluation, and investments to be made) were also identified as a priority.

Finally, what distinguishes this kind of network from standard community and association initiatives is that each individual remains the sole owner of his production and his benefits, which means that he is not exposed to some of the risks inherent in community production.

14.4.2.4 State-controlled plantations

Although private, village, or community plantations can contribute significantly to the supply of both energy and construction wood, the volume of the need, which will continue to increase in line with the demographic projections for the region (see Chapter 9), is such that establishment of large-scale plantations must be envisaged. Such plantations have already been created in other African states, so we can learn from their accumulated experiences. Large-scale plantations have some obvious advantages: they can be located on vacant land at a greater distance from the park, since the transport costs become more economical for such large volumes, which in turn can make a far greater contribution to the overall supply. On the other hand, very efficient management is called for, if such a system is to work. A badly run system, or one that is abandoned, will substantially reduce the productivity of the plantations. In the Democratic Republic of Congo there are state institutions (such as the *Service National de Reboisement*) and other organisations that could take charge of such initiatives.

The first step would be to identify potential areas of production, and to integrate these into an overall land-use plan, ensuring from the outset that there is no conflict over land allocation. A high-performance institutional mechanism should then be developed. This should have clear lines of responsibility and precise rules and regulations, as well as an awareness of the different interested parties, both state and private, so that administrative bottlenecks can be avoided, while ensuring that tax breaks and other incentives can be brought into play. Finally, a large programme devoted to supervision and training should be set up – probably with the support of external technical partners and subsidised by donors.

14.4.3 Dismantle the illegal charcoal traffic coming from the park

The surveys described above and the fact that there is no fuelwood production site (for the southern part of Virunga National Park at least) outside the park mean that more than 1,300,000 sacks of charcoal are removed from the Southern Sector of Virunga National Park each year. This figure is a minimum; it is public knowledge that hundreds of people cross the Rwandan border every day with loads of charcoal, since Rwanda has prohibited all cutting of wood and has therefore greatly reduced its own local supplies of *makala,* thereby encouraging the trans-border traffic from the Democratic Republic of Congo.

To supply *makala* in such enormous quantities is possible only through a very well-organised network spanning production, transport and sale. At the foot of Mount Nyiragongo there is a veritable *makala* 'highway', a wide trail along which the sacks are taken out with the full knowledge of numerous State agents who line the route, collecting various 'taxes' from the carriers.

It is necessary to bear in mind the importance of the money flow this illegal traffic engenders. The average selling price of a sack of *makala* in Goma is USD 15. The commercial value of all the charcoal sold in Goma totals around USD 20 million annually, which explains why so many people are implicated in the traffic. The asking price inside the park is about USD 3, and along the Rutshuru-Goma axis this increases to around USD 8. Even if the officials of the various institutions take only, say, five percent of this sum, their take would still represent an astronomical amount of money. A coherent policy must be adopted in order to put an end to this plague.

As a first step, the authorities should be made aware of the dire ecological and economic impacts associated with the illegal production of charcoal: deforestation, disturbance of river regimes, loss of tourism potential, and loss of revenue for both the central government and the local authorities because no part of the process is controlled or subjected to any national laws or regulations. Finally, wood must be assigned its real value. It is grossly under-valued at present, and the forests of the park are being sold off for a pittance, with not so much as a passing thought's being given to the need to find alternative solutions.

Another urgent priority for regulators and law makers is to develop a range of incentive measures that will boost the production of wood outside the park (and, as seen above, encourage the implementation of more efficient fuel conversion techniques, along with the use of alternative energies), while dissuading people from maintaining the illegal traffic from the park. Incentive measures might include allocating land and commercial state-controlled plantations, setting up micro-credits, de-taxing sustainable wood production outside the park, and developing reforestation projects in partnership with institutions and NGOs, Congolese or international, that could be financed, in part at least, by donors. The deterrent aspects would have to include simple measures geared to ensuring that national and international laws protecting Virunga National Park are respected. With a new government in place and a restructured ICCN, it is possible to introduce the necessary strong, clear measures against the various forms of corruption and complaisance that are driving the traffic in *makala*. For too long, this illegal traffic has been allowing just a few isolated individuals to accumulate vast profits at the expense of the wellbeing of the entire wider community.

It is essential, finally, that the ICCN be given the means to execute in full the mandate the Congolese Government has entrusted to it. In being allowed to implement its policies, the Institute will contribute significantly to rebuilding the rule of law in the Democratic Republic of Congo as a whole. To begin with, the required resources must include a reinforcement of the ICCN's technical capacity (training in different fields, including strict application of the law in keeping with the legal framework protecting the national park and its resources), the development of information and monitoring systems, and the resolution of conflicts. The logistical means of the ICCN (movements, communications and patrolling gear) will also need to be reinforced. It is essential that ICCN staff, like all staff of the Congolese State, should receive a fair monthly salary that is paid on time, so the risk of their turning to other, irregular activities to fund their basic needs can be averted.

15.1
An ICCN agent examines a dart used to immobilize a gorilla before treatment. Some aspects of the management of Virunga National Park require particular expertise, such as veterinary care for the mountain gorillas. Such skills, as well as other more fundamental ones, must be reinforced after the many years of crisis during which training was almost absent.

Henri Paul Eloma Ikoleki & Jean-Pierre d'Huart

15 The capacity of the *Institut Congolais pour la Conservation de la Nature* to manage Virunga National Park

Henri Paul Eloma Ikoleki & Jean-Pierre d'Huart

Formerly autonomous, Albert National Park was, at the time of the creation of other protected areas in the country, put under the aegis of a central management authority. Does the *Institut Congolais pour la Conservation de la Nature* (ICCN) have the capacity to manage the park well today? Are the legal texts, men, material and financial means adequate for its effective management? What are the deficiencies and how can the situation be remedied to create optimal management of the park? It is in the spirit of offering part of the answers to these questions that this chapter has been written.

15.1 The legal framework: the ICCN's strengths and weaknesses

As explained in more detail in Chapter 1, the management of Albert National Park – an autonomous institution at its creation in 1925 – was put under the authority of the *Institut des Parcs Nationaux du Congo Belge* by the Decree of 26 November 1934. This institute is the ancestor of the *Institut Congolais pour la Conservation de la Nature* (ICCN). It successively took the name of *Institut des Parcs Nationaux du Congo* (Ordonnance-loi no. 67-514 of 1 December 1967), *Institut de la Conservation de la Nature du Congo* (Ordonnance-loi no. 69-041 of 22 August 1969) and *Institut Zaïrois pour la Conservation de la Nature* (Law no. 75-023 of 22 July 1975).

These trust institutions of Virunga National Park had the same mission, which included the three following objectives:
- To ensure the protection of the fauna and flora in the integrally or near-integrally protected nature reserves,
- To promote scientific research and tourism in accordance with the fundamental principles of nature conservation,
- To manage so-called 'trapping stations' established inside or outside the reserves.

This specific mission will invariably be passed on to each institution that succeeds the *Institut des Parcs Nationaux du Congo Belge*. It guarantees full protection to integrally protected nature reserves, as stipulated by Ordonnance-loi no. 69-041 relating to nature conservation. According to Articles 1 and 2, Albert National Park, which is an 'integrally protected nature reserve', is an environment which '... presents a special interest and which it is important to preserve from any intervention capable of modifying its appearance, its composition or its evolution'.

To guarantee the effectiveness of this conservation mission, the legislation by means of Law no. 75-023, provides for the *Institut Zaïrois pour la Conservation de la Nature* (IZCN now ICCN) to hire wardens *(Officiers de Police Judiciaires)* and rangers. The latter have the following responsibilities (Article 12):
- To foresee, seek out and report to the wardens any violation of the present law (law creating the IZCN), any violation of the hunting and fishing legislation and any violation of the forestry legislation,
- To identify or, failing this, to bodily apprehend and take before the competent authority any person caught violating the law on nature reserves and hunting reserves, as well as those found in possession of items that prove their guilt, in particular, weapons, instruments, papers, plants, animals, skins or trophies;

15.2
After about 15 years without any training in the application of laws, this type of training for the rangers must be a priority for the improvement of the ICCN's skills in managing the national park. Here, rangers are being trained at Ishango in 2006, thanks to support from the Frankfurt Zoological Society.

15.3
Ranger Ndabananye at Rumangabo in 1956. Men are the most important element of the park's management capacities. Their pride in belonging to the Institut Congolais pour la Conservation de la Nature *is an asset that is important to the protected area.*

– To prevent any person from suppressing material traces of infractions.

With time, this policing mission will come up against growing opposition on the part of the local population living in or around the protected areas. In effect:
– Among the constant objectives consigned to the institute that is responsible for Virunga National Park, over the past 80 years, there has never been any precaution to ensure that the overall policy on conservation of ecosystems and biodiversity is in harmony with socio-economic development of the populations living next to the protected area;
– The indigenous rights to hunting, fishing, gathering and wood-cutting, as well as salt extraction, granted to the populations living in or around the park since the decree of 1925, were suppressed by *Ordonnance* no. 3 of 6 January 1944, having been exchanged for financial compensation or for the allocation of equivalent rights outside the park;
– The concept of integral protection of nature, which the integrally protected nature reserves were subject to, is difficult to apply and enforce because of population pressures and demands on this land (Maldague *et al.*, 1997);
– Since the United Nations Conference on Environment and Development (Rio de Janeiro, 1992), the concept of conservation is linked to that of development, and defined as 'everything that humanity plans and achieves to ease its impact on the natural environment to satisfy its real needs, while allowing the perpetuation of this environment in functional balance' (Maldague *et al.*, 1997).

The reasons evoked above and many others mean that, legally, the *Institut Congolais pour la Conservation de la Nature* (ICCN) has neither the mandate, nor the personnel or the technical capacity to officially implicate itself in the socio-economic development of the residents of protected areas and, therefore, of Virunga National Park. Most of the interventions in favour of neighbouring villages have been generated by the NGO partners of the ICCN, without sufficiently benefiting from an active and visible contribution by the Institute.

This legal void renders Virunga National Park difficult to manage: a clear ICCN mandate is absent and this, together with its modest means, keep the Institute from contributing to solutions to the deplorable living conditions of the growing numbers of neighbouring populations (even if this mandate is attributed to other state institutions). The overall context of the region, characterised by repeated socio-political crises, recently brought to light some effects of this shortcoming. The park's neighbours have the impression that the ICCN grants more importance to the animals than to people.

It is therefore urgent that from now on the governing powers consider reviewing the ICCN mandate and apply, in addition to its traditional mandate, objectives *contributing to* the promotion of socio-economic development of populations living around the protected areas – without its becoming an objective in itself.

The period of post-conflict reconstruction, associated with the input of foreign capital, is particularly suitable for equipping the country with legal instruments that favour the active input of the ICCN.

Following the enactment of the new laws on Forestry, Mining and Investment Codes, it is fortunate that a new law on Nature Conservation, taking into account the human and landscape dimensions of conservation, has been in preparation since the end of 2005. An institutional review of the ICCN has just been concluded and will allow the new law to adapt the ICCN's mandate and capabilities to the new requirements of conservation.

The ICCN should not experience any difficulty in taking on this option. It is already dealing with the needs of the populations by its implication in community conservation, an approach adopted in the framework of various local projects.

The state must acknowledge this new dimension of the ICCN and grant it the adequate means to play a significant role in this field. This would recognise the place that the environment sector can hold, alongside the other sectors of national life (mines, forests, education, security infrastructures, etc.) in the pursuit of the Millennium Development Goals.

Recognition at the national level is very important because, at this stage in the Democratic Republic of Congo development, efforts of the Government remain focused on maintenance of interior peace and security, unification of the army and growth of economic production. As a consequence, the emphasis has been on mines, energy and security. This includes the creation of military integration and training centres—called *Centres de brassage*—within the park boundaries, such as the one at Nyaleke, the establishment of road blocks, the encampment of armed troops inside the park boundaries, the requisitioning of park property and material goods by the Provincial Administration, and largely excludes environmental considerations.

The Strategic Document on the Reduction of Poverty (*Le Document Stratégique sur la Réduction de la Pauvreté,* DSRP) cites bad governance, degraded social values, dilapidation of infrastructures and shortage of financial means, as common causes of poverty (Ministry of Planning of the Democratic Republic of the Congo, 2002). If the destruction of the natural environment is taken into account in only a very marginal way in the national DSRP, it is, on the other hand, cited as one of the principal sources of poverty in the DSRP of North Kivu. At the same time, experts are calling attention to the fact that 'changes taking place in the ecosystems continue to worsen and compromise the achievement of global development objectives' (United Nations, March 2005). This is true for the eastern part of the DRC, whose ecosystems and natural habitats have suffered from systematic pillaging and illegal exploitation by refugees and uncontrolled armed groups since the 1990s.

Within the context of reinforcing its capacities, it will be equally pertinent to reconsider which authority the ICCN should come under. While at present the ICCN is under the ministries of Environment and the Portfolio, it might be justified to imagine, given its nature and its special mandate, the ICCN to be an institution directly under the Presidency of the Republic, as was the case in the past. An increase in the autonomy and means of the Institute would contribute to reinforcing its development and its own capacities.

15.2 HUMAN RESOURCES: ICCN STRENGTHS AND WEAKNESSES

As for the *quality* of ICCN personnel of Virunga National Park, park senior wardens (just like those in the ICCN in general) are trained, for the most part, at the *Ecole de Faune de Garoua,* in Cameroon. This school offers the classic panoply of courses in biology, ornithology, veterinary techniques, animal ecology, botany, management techniques, national parks, zoological gardens, mechanics and civil engineering, fire arms, economics of the fauna, legislation and administration, first aid and sport; as such, it certainly ensures a good basic training for senior wardens. It remains weak, however, in various fields including those of planning and public relations with the local populations.

Quite recently, some ICCN agents have also been sent to the College of African Wildlife Management in Mweka, Tanzania. In this school, in addition to biology and paramilitary tactics, courses are offered in improving relations with local populations such as population and conservation, tourism and tour-guiding (certificate level), environmental impact studies, community conservation, communication techniques for conservation and formulating proposals (diploma level). But so far, very few of the ICCN wardens have studied there.

Rangers are trained on the job, and this method has many strengths. The supervision of the young by older, experienced rangers, and the transmission of an *esprit de corps* and loyalty to the institution are important values to revive. Until the 1980s, ICCN rangers set the standard in Africa for their level of professionalism and training, but 20 years of near abandonment of ICCN by the state has changed the situation. Instructors from South Africa were shocked by the low levels of competence of Congolese students, even in simple tasks such as handling a classic firearm or basic anti-poaching techniques. The training organized for them in the Democratic Republic of the Congo (Lulimbi, Ishango, Garamba) and abroad (South African Wildlife College) has mostly emphasized paramilitary training and anti-poaching techniques. It is regrettable that the ICCN does not yet have an internal team of instructors that can guarantee continuous training of its rangers.

The lack of emphasis on the importance of contact with its resident populations stands out in this analysis of the ICCN agent-training programme. The training given at Mweka would be the most appropriate to fill this gap, but the language barrier is very real (courses are taught in English in Tanzania). Beyond this are many other aspects, such as statistical analysis, database management, use of computer and geographic tools, which are not well developed. All current efforts aimed at setting up a Geographic Information System for the protected areas (SYGIAP—*Système de Gestion d'Information pour les Aires Protégées*—, Monitoring of Illegal Killing of Elephants, Law Enforcement

15.4
Good quality supervision is essential to ensure continuous training of scientific and technical staff by self-training and through close collaboration with the partners.

15.5
Some rangers, supported by technical partners, intervene to rescue a young gorilla from a snare and to care for it.

km² in Virunga National Park compared with 27 km² in Salonga National Park, the most far-flung of the ICCN's sites. Of course, this comparison is only partially useful given the important differences in the context, the habitats (savanna/forest) and surveillance techniques. The quality of protection will depend on the quality of the men, on their motivation, on the equipment at their disposal, on the population pressure and on the political situation in the region — Kivu is the most densely populated region in the country and it has suffered the turmoil of the wars in the Great Lakes region for more than a decade.

The ideal number of personnel at Virunga National Park and their distribution in patrol posts or stations are being reconsidered in the institutional review, based on light of several factors, such as: the level of ICCN support to its personnel, exterior pressures, quality of leadership, level of training and equipment, type of terrain and ease of access, size of sectors and quality of communications, quality of the information network and availability of reinforcement personnel, etc.

It should be noted that, among these personnel, that should, in principle, be sufficient in Virunga National Park,* there are many men who have reached retirement age, and the youngest are now being confronted with financial and material problems that undermine their effectiveness. A thorough evaluation of their abilities and training needs is considered essential for a sound new start. The objective should be to energetically rejuvenate the personnel. Those who are ready for retirement — and there are many — should leave and those who remain should be trained. The recruitment and training should then follow, depending on available resources. In this way, Virunga National Park could, in about ten years, have a less numerous staff (and therefore a less heavy financial burden), but one composed of well-equipped agents, well-paid and well-trained.

15.3 THE INSTITUTIONAL FRAMEWORK

The management structure of Virunga National Park represents another challenge. For the past years, the organisation of the ICCN has made provision for intermediate structures between Virunga National Park and the National Headquarters in Kinshasa. Those concerned are the Provincial Directorate of North Kivu and the Liaison Office of the ICCN. These structures are both local representations of the ICCN and effective communication between Kinshasa and Virunga National Park, installed at Goma and responding directly to the National Headquarters.

In 1999, the rebel movement, the *Rassemblement Congolais pour la Démocratie* (RCD-Goma), established a new 'National Co-ordination' of ICCN, as the management body for Virunga National Park and of other protected areas located in the part of the country formerly under the control of the RCD-Goma, substituting itself at the ICCN Directorate. This structure, which has kept administrative and financial control of Virunga National Park from 1999 to 2004, took over the ICCN staff management. It then found itself in some dilemmas when differences between the National Co-ordination and the National Headquarters in Kinshasa appeared. Lacking competent technical staff, this unit committed a string of management errors that impacted on the park and its personnel.

Monitoring) and its associated training (for example by MIKE/CITES or by the *Université Catholique de Louvain*) were designed to meet these needs, but they require harmonisation with existing practices. The progressive establishment of a corps of ICCN instructors is needed to ensure continuous internal training for the future.

A workshop held in May 2004 brought together UNESCO, the ICCN and all its partners for the evaluation of the UNESCO/UNF Project in support of the five World Heritage Sites in DRC. Participants agreed that the reinforcement of personnel skills remained the top priority at the end of the crisis. They recommended the development of an overall training strategy at the ICCN linked to the results of the institutional review. Such a strategy would include different modules, a system of monitoring and a stable source of financing. Concerning the protected areas, the three major themes retained were the link between the protection and monitoring, the interactions with local communities and the resolution or prevention of conflicts.

Concerning the *quantitative aspect* of personnel, Table 15.6 compares the average number of hectares per ranger for each of the five World Heritage Sites (WHS) in the DRC (Virunga National Park/ViNP, Garamba National Park/GNP, Kahuzi-Biega National Park/KBNP, Salonga National Park/SNP and the Okapi Faunal Reserve /OFR) as well as for Maiko National Park/MNP.

Aware that maintenance of the integrity of a site (space and biodiversity) depends on many factors (political will, legal framework, general regional context, material and financial human resources), we will nevertheless consider the theoretical *number of hectares watched over per ranger* (ratio between the surface and the number of rangers on the site) in order to assess the management capacity of the site. The data for establishing this ratio are accessible and reliable.

From these data it can be seen that Virunga National Park must be the ICCN's best-protected site. In fact, it turns out that there is, on average, one ranger per 12

15.6 Area in km², number of rangers and average number of km² per ranger in six major protected areas in DRC

	Area (km²)	Number of rangers	Number of km²/ranger
ViPN	8,000	689	12
GNP	5,000	263	19
KBNP	6,000	125	48
MNP	10,000	71	141
OFR	13,726	85	161
SNP	36,000	134	269

15.7 *The management capacity of the ICCN depends equally on infrastructure: the patrol post after and before rehabilitation.*

At a time when the authority of the State is re-establishing itself, extending over the whole country, it is essential to restore an adequate decentralized structure under of the Provincial Authority in Goma, that holds a clear mandate and required capacity.

There is, in the World Heritage Sites and in other parks with several stations, a senior warden with the title of Chief Warden. He co-ordinates the activities of the wardens based in the park stations. He also heads the Site Co-ordination Committee (*Comité de Coordination du Site,* CoCoSi) as a representative of ICCN (see Chapter 20). This structure has led to confusion with the remit of the Provincial Director or Park Warden when the latter includes only one station. Although the function should be better defined, it is more appropriate to ensure liaison between the National Headquarters and the park, than that of Provincial Director or Head of the Liaison Bureau.

15.4 CONCLUSION

In conclusion, the law accords to the *Institut Congolais pour la Conservation de la Nature* the power of management of Virunga National Park. Given that it would be useful for the park to contribute to the development of the local population while fulfilling its original responsibilities (surveillance, monitoring, research, co-ordination, etc.), the current mandate of the ICCN is no longer appropriate in the social and economic context of the country, and its means are insufficient to fulfill the new mandate that should be conferred on it. It is important, however, to note that other services of the state are mandated to promote for economic development for the population, and that in this area, ICCN's mandate should be to improve neighbour relations rather than economic development in the proper sense. The warden and his technical assistants should be educated in co-management with the local population and brought up-to-date with information and communication technologies.

The biological importance and the very complex geographical, social and political context shape of Virunga National Park justify a sufficient and well-adapted personnel who have the required skills. The present personnel need training and rejuvenation to be able to respond to the challenges of tomorrow. In the framework of their conservation programmes, the ICCN and its partners must pay adequate attention to the training of ICCN personnel, at all levels, if the riches of the park are to be recovered and the park conserved for the next 80 years and beyond.

* Comparing the situation in Virunga National Park (one ranger/1,200 ha) with that of Kruger National Park in South Africa (one ranger/11,400 ha), there is clearly an enormous difference that must be tempered in light of the factors mentioned in this Chapter. In comparison, in 200 forested, protected areas in 37 developed countries, the average is one ranger/2,000 ha. (D. Zeller, International Rangers Federation, pers. comm.).

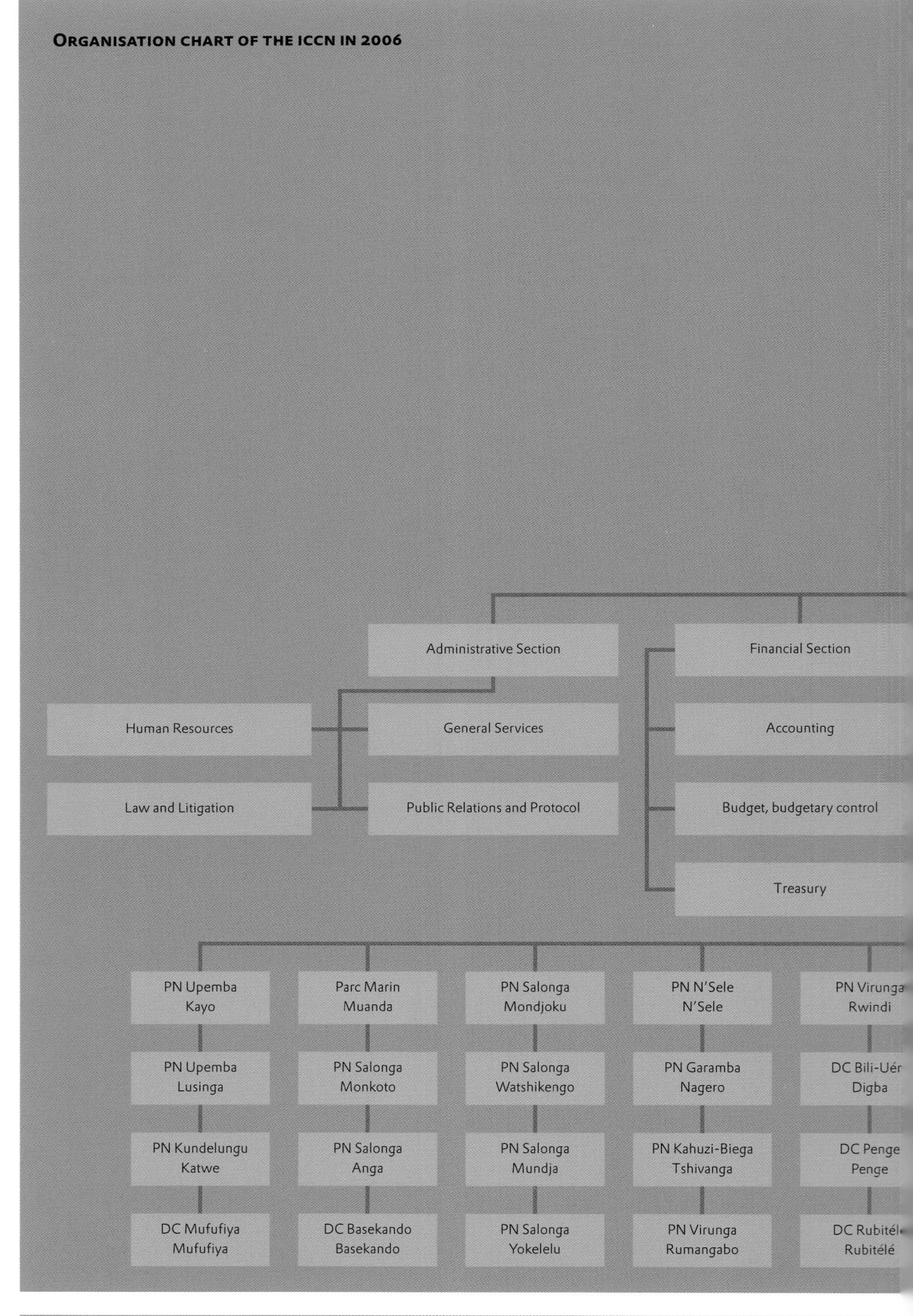

Organisation chart of the ICCN in 2006

	Administrative Section		Financial Section	
Human Resources	General Services		Accounting	
Law and Litigation	Public Relations and Protocol		Budget, budgetary control	
			Treasury	

PN Upemba Kayo	Parc Marin Muanda	PN Salonga Mondjoku	PN N'Sele N'Sele	PN Virunga Rwindi
PN Upemba Lusinga	PN Salonga Monkoto	PN Salonga Watshikengo	PN Garamba Nagero	DC Bili-Uér Digba
PN Kundelungu Katwe	PN Salonga Anga	PN Salonga Mundja	PN Kahuzi-Biega Tshivanga	DC Penge Penge
DC Mufufiya Mufufiya	DC Basekando Basekando	PN Salonga Yokelelu	PN Virunga Rumangabo	DC Rubitélé Rubitélé

Henri Paul Eloma Ikoleki & Jean-Pierre d'Huart

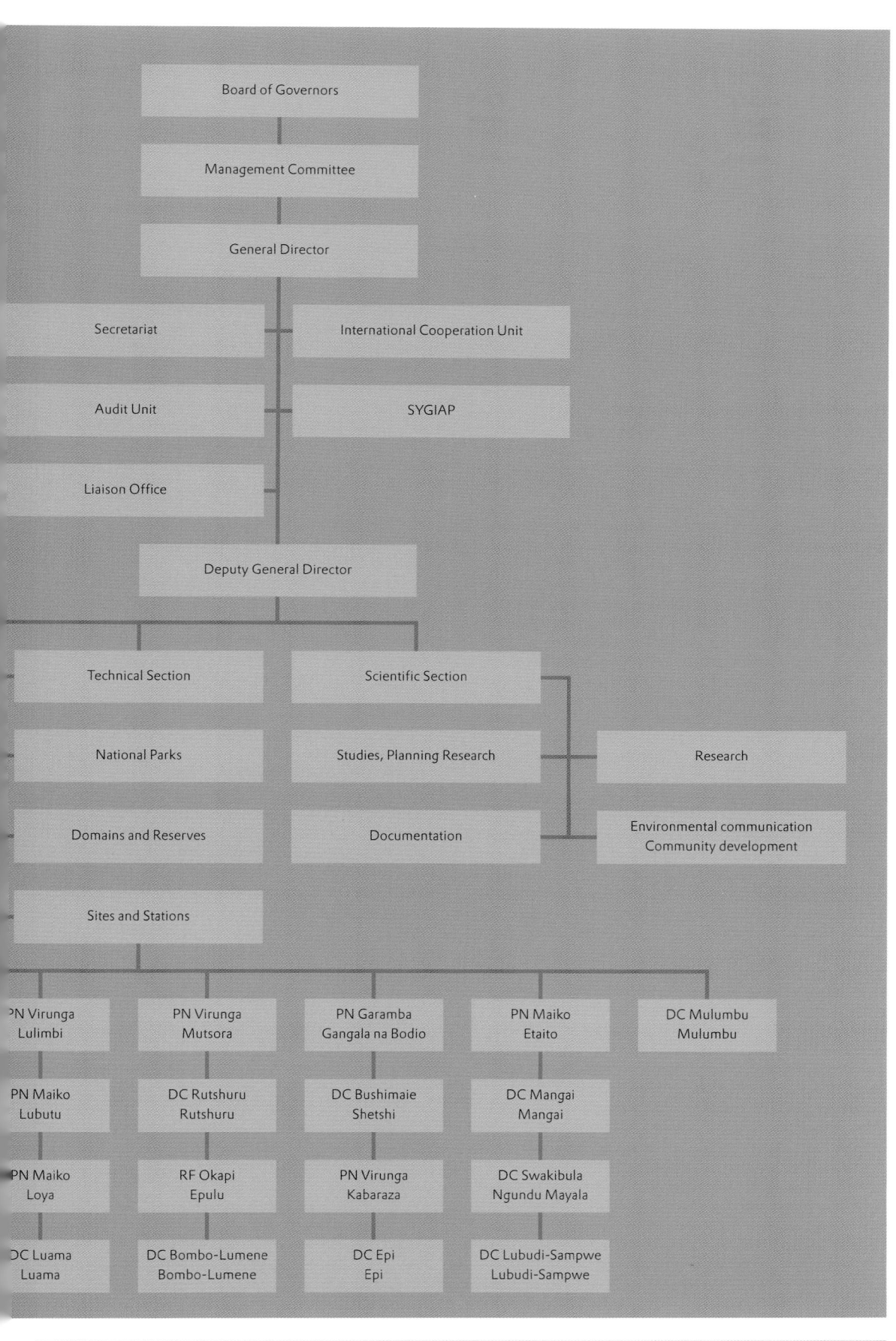

```
                        Board of Governors
                               |
                       Management Committee
                               |
                        General Director
                               |
  Secretariat  ————————————————+———————————————— International Cooperation Unit
                               |
  Audit Unit  ————————————————+———————————————— SYGIAP
                               |
  Liaison Office ——————————————+
                               |
                       Deputy General Director
                               |
         ——————————————————————+——————————————————————
         |                     |
  Technical Section     Scientific Section
         |                     |
  National Parks        Studies, Planning Research ———— Research
         |                     |
  Domains and Reserves   Documentation              Environmental communication
                                                    Community development
         |
  Sites and Stations
         |
  ———————+———————————+———————————+———————————+———————————
```

PN Virunga Lulimbi	PN Virunga Mutsora	PN Garamba Gangala na Bodio	PN Maiko Etaito	DC Mulumbu Mulumbu
PN Maiko Lubutu	DC Rutshuru Rutshuru	DC Bushimaie Shetshi	DC Mangai Mangai	
PN Maiko Loya	RF Okapi Epulu	PN Virunga Kabaraza	DC Swakibula Ngundu Mayala	
DC Luama Luama	DC Bombo-Lumene Bombo-Lumene	DC Epi Epi	DC Lubudi-Sampwe Lubudi-Sampwe	

The capacity of the *Institut Congolais pour la Conservation de la Nature* to manage Virunga National Park

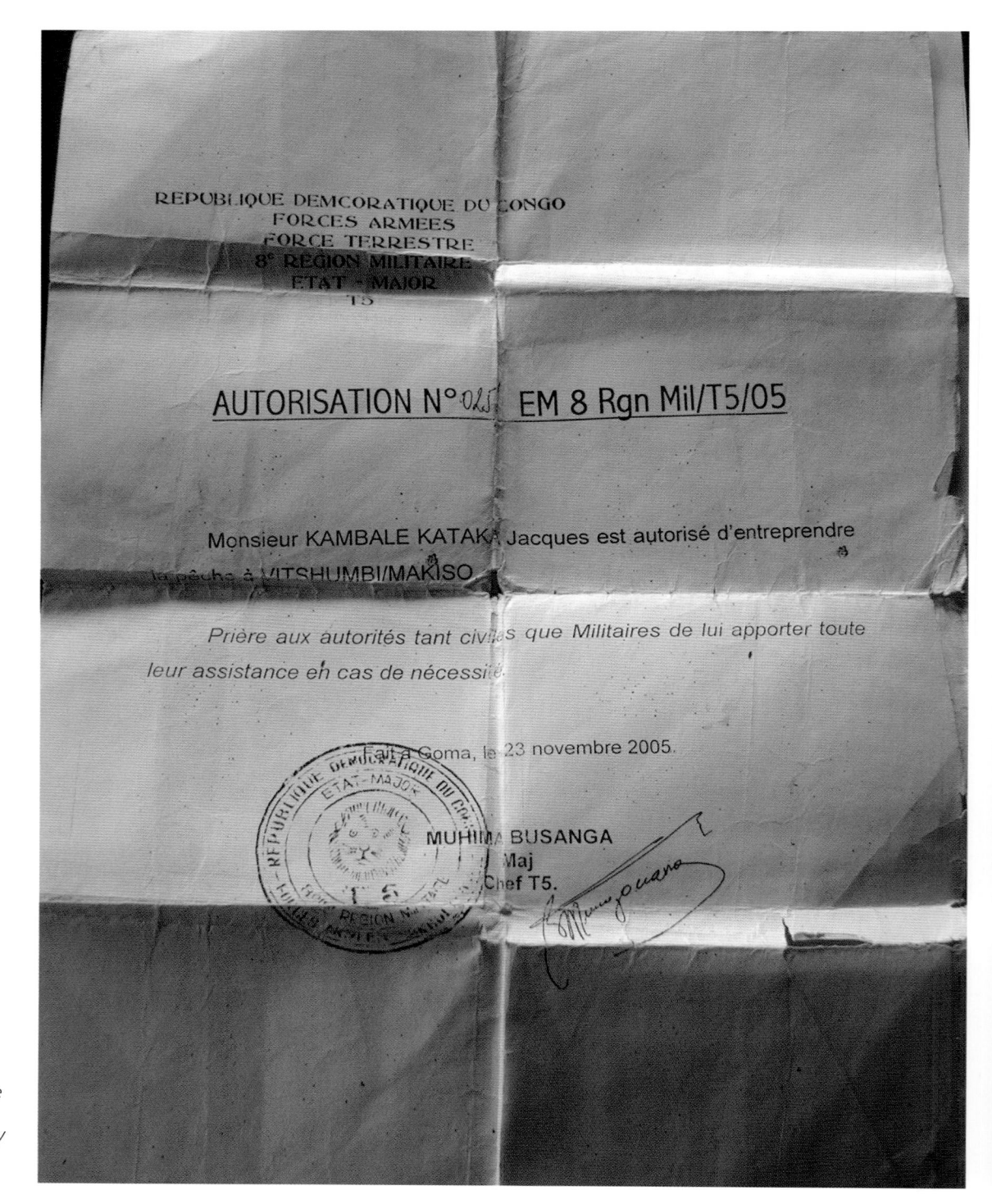

16.1
Conflict with other institutions in the management of the ICCN is a major challenge for conservation. Here, the military authorities have given permission to fish inside the park.

16 Political challenges to conservation in North Kivu

Ephrem Balole, Norbert Mushenzi & Emmanuel de Merode

The arrival *en masse* of Rwandan refugees (around 740,000) to the vicinity of Virunga National Park in July 1994 was the beginning of a major destabilisation of the protected area. Their stay not only brought serious harm to the park's natural resources, it was also the pretext for a cycle of wars that have gravely affected it. The war that broke out in September 1996 – in the eastern part of ex-Zaire, now the Democratic Republic of the Congo – brought with it even more detrimental effects, although officially it ended in June 2003. Details of this epoch are developed in Chapter 4.

In this chapter, the authors analyse the political context over the past ten years with respect to the conservation efforts of the ICCN. This analysis will focus on four main aspects of the political trend of Kivu, namely:
- geopolitical strategies at the regional level,
- loss of control of Kivu by Kinshasa,
- disintegration of the rule of law and political opportunism,
- inadequacy of the institutional framework.

The political dimension creates a very particular context where the classic tools of protected area management are not sufficient. It became indispensable for the managers of the protected areas to become familiar with political tools in order to deal with the complex problems of Virunga National Park. At the end of this chapter the authors analyse these tools and their efficiency.

16.1 Analysis of the political context of nature conservation in North Kivu

16.1.1 The sundry political tendencies in Kivu (both national and of Ugandan or Rwandan origin) and their position with regard to conservation and the national parks.

With the arrival of the Rwandan refugees in 1994, the lands adjacent to the park (mainly Masisi territory) suffered from conflicts between the indigenous populations who had organised into militias (Katuku, which constituted the nucleus of the Maï-Maï) and the Rwandophones: Hutu – later allied with the refugees – and Tutsi populations. There was a sizeable Tutsi migration from the Congo towards Rwanda. The presence of refugee camps contributed to the dissemination of small arms in the region, with the complicity of the Zairian ex-army who were incapable of stopping the conflict. These arms were used by a poaching network active throughout the entire park.

The 1996 war dispersed the refugees, many of them returning to Rwanda, whilst other groups, more dangerous because they were composed of Interahamwe and ex-FAR armed groups, settled in the Congolese forests, including Virunga National Park.

The 1998 war complicated the factions already present and rendered Virunga National Park even more fragile. It began with the RCD, ally of Rwanda and Uganda. But following internal quarrels of leadership, this movement rapidly split into RCD-Goma, ally of Rwanda, and RCD-ML, ally of Uganda (there was also the RCD-N). These first two movements had an important impact on the park, to the extent that the RCD-Goma controlled the southern part and the RCD-ML the northern part.

Alongside these two large rebel movements, several other armed groups exerted intense pressure on the natural resources of Virunga National Park. First of all, the Interahamwe and ex-FAR regrouped into two large ensembles: the *Armée de Libération du Rwanda* (ALIR) and the *Forces Démocratiques pour la Libération du Rwanda* (FDLR). But other Interahamwe organised into more or less independent groups and behaved more like highway bandits. These bands are mostly active in the southern part of the park that was formerly under the control of the RCD-Goma.

The Maï-Maï constitute the second militia group hostile to the national park. There were several groups of combatants without a command unit under the Maï-Maï denomination.[1] During the war, there were some Maï-Maï close to the Kinshasa Government, others were close to the RCD-Goma and therefore to Kigali, and still others close to the RCD-ML and thus close to

16.2
After loss of human life and the depletion of wildlife, the destruction of park infrastructure is one of the most negative consequences of political conflicts. Here, the buildings of the Rwindi in 2006.

Kampala. After the war, some of the Maï-Maï integrated into the FARDC while others continued independently, living like warlords in the territories under their control. The latter are mainly located in the north of Virunga National Park.

In the south, formerly under the control of the RCD-Goma, the militias formerly called 'Local Defence' still subsist. These are young people, generally Rwandophone, recruited to ensure the security of their villages against the recurrent attacks of the Interahamwe.

The other sizeable players are the regular armies of the three countries, Rwanda, Uganda and the Democratic Republic of Congo.

16.1.2 Politically, why is there a military presence in the park? Myths and realities

All the players have different motivations. Beyond political motives, economic reasons seem to take precedence. We shall examine the motives of each of them in detail.

From a political point of view, Rwanda has consistently played the security card to justify its interventions in the eastern part of the Democratic Republic of the Congo. Since Virunga National Park is the refuge of the Interahamwe, from which incursions to Rwanda are launched, the Rwandan Government thought it a good idea to organise its defence from the park.

As for the Interahamwe, the park constitutes an ideal space for several reasons. First of all, security: the forests become a real haven against surprise attacks. Then, survival: the park serves as a supply source for game and a retreat camp for pillaging of villages and attacks on the roads. Finally, through the network of smugglers, profits from the sale of natural resources exploited in the park are used to bring in arms and munitions. We emphasise that the park has become insecure for the local residents and for travellers.

For the Maï-Maï and other Congolese militia, the motives appear to be essentially economic. Lacking logistics, the Maï-Maï use the park to supply themselves with game. They encourage exploitation of the forest resources by the local residents (charcoal, firewood, timber and canes as supports for runner beans). The same motives are valid for 'local defence', which mobilises the militia to protect the fields conquered in Virunga National Park and the fisheries on Lake Edward. We point out that the uncontrolled bands of Maï-Maï spread desolation in and around the park just as much as the Interahamwe.

Just like the Interahamwe, the Ugandan NALU rebels are entrenched in the Rwenzori Mountains from where they launch attacks into Uganda, as well as into the Democratic Republic of Congo. Their motives are political and economic, notably the exploitation of mineral resources (coltan and gold) and wood.

Thus, Virunga National Park is at the crossroads of the different warring parties in the eastern part of the Democratic Republic of Congo. Because of its natural resources, the park itself has become the object of conflict.

16.2 Analysis of the main political problem of conservation in Virunga National Park

16.2.1 The regional geopolitical strategy and the recurrent wars

Distinct from the official political motives, the cycle of wars that Kivu has experienced since 1996 has been encouraged by implicit and unofficial motives on the part of Rwanda and Uganda. These motives are essentially of an economical and territorial nature. If the economic motives are completely implicit, the territorial motives are much more subtle and are conducted by manipulating the people.

Rwanda and Uganda have been cited as the primary culprits in the experts' reports to the United Nations on the illegal exploitation and the pillaging of natural resources in the Democratic Republic of Congo (United Nations 2003). Virunga lies at the heart of a process that saw the looting of Congo's natural resources, a process contributing to the deaths of over 5.4 million Congolese civilians (IRC 2008). As far as Virunga National Park is concerned, poaching was not directly attributable to the governments of Rwanda and Uganda. It is probable Congolese soldiers were also heavily involved in the pillaging of Virunga's natural resources. But in deciding to billet their troops in the park, these governments were facilitating a level of poaching rarely seen in any protected area in Africa. Moreover, in violation of the embargo decreed by CITES in 1990 on the trade in ivory, it is common to see ivory necklaces on sale in cities like Kigali and Kinshasa.

Concerning territorial issues, at the beginning of the 1998 war, the Rwandan leadership, on many occasions, made calls for a revision of the Berlin conference (Berlin II) on the grounds that the territorial division resulting from that meeting had slighted Rwanda. These claims revived distrust of the real intentions of the war of the RCD's allies. Indeed, a myth perpetuated in Rwanda holds that the Congolese area occupied by the Rwandophones, including the southern part of

Virunga National Park, belonged to the Kingdom of Rwanda before the division in Berlin. A similar myth suggests that the Semliki Plain, in the north of Virunga National Park, belongs instead to a tribe of the Banyarwanzururu, Ugandan pastoralists inhabiting the Rwenzori. These myths seem to make up the historical basis for the territorial claims of Rwanda and Uganda against the Democratic Republic of Congo.

But following the unfavourable international response to a re-negotiation of the Berlin meeting, territorial occupation continued in a more subtle way. Two cases can illustrate this as far as Virunga National Park is concerned: between April and June 2004, some 6,000 squatters, essentially of Rwandan origin, presented themselves as Congolese refugees and, protected by armed men, invaded and destroyed more than 1,500 ha of montane forest in the Mikeno sector, in the south of the park, in the sanctuary of the Mountain Gorillas. High level lobbying to the Rwandan government was required in order to make their men retreat from the park (see Chapters 18 and 24). In the same way, in the Karuruma sector in the northern Sector of Virunga National Park, a large group of over 300 families of Ugandan pastoralists settled within the boundaries supported by the Ugandan armed forces in 2000. Through intense negotiation by the ICCN, backed up by higher level diplomatic pressure, these families and their cattle (at least 5,000 head of cattle) returned to Uganda in March 2006.

16.2.2 Kinshasa's loss of control over Kivu

The 1996 war and especially the 1998 war, caused Kinshasa to lose administrative and economic control of Kivu. Control has been exercised by the RCD since that time.

From a political point of view, it was natural for the RCD to appoint its members to the various posts, including territory, security and intelligence, and the army. In spite of setting up a transitional government, political reunification has produced a lot of serious problems that have made the transition process extremely challenging and bitter.

Economically speaking, within the state financial institutions and state-owned enterprises in the territory under its control, the RCD put in place a structure designed to ensure financing for itself. This structure is Co-ordination, comprising a co-ordinator, his assistant and a person in charge of finances. At the level of the provincial directorates, this structure appointed directors at the head of the public enterprises.

In all these public administrations, the RCD's structures invariably had to contend with intense staff conflicts. Indeed, the revenues allocation scheme seriously impaired the functioning of these enterprises. The co-ordinators and provincial directors were expected to ensure maximum revenues from the state financial institutions for the rebellion and did not pay much attention to the social conditions of the workers or to the maintenance of infrastructure and equipment. After the reunification of the country, these structures have persisted in the state enterprises, constituting a serious obstacle to the economic reunification of the country.

16.3 Impact on Virunga National Park

The operations of the Co-ordination created a schism in park management. The southern Sector, under the control of the RCD, operated under the authority of the RCD, while the northern Sector worked under the authority of the ICCN/Kinshasa. The management of the site was disjointed. The ICCN's partners, as well as the Virunga National Park staff, suffered from the lack of professionalism of the Co-ordination, which pretended to be an indispensable administrative interface. Without tourist revenues, and with no possibility of generating other income sources, it is not hard to understand the financial difficulties at the heart of the problem.

16.3.1. Disintegration of the rule of law and political opportunism

The disintegration of the state is a result of the bad management of public property that had prevailed for a very long time in the Democratic Republic of Congo. For Virunga National Park, the problem can be seen at several levels: lack of policy in land use planning and rural development; non-payment of civil servants, mainly the soldiers; and lack of co-ordination of the activities lead by the State.

Virunga National Park is in a region of strong demographic growth, where a lack of policy in land use and rural development resulted in a concentration of agricultural populations around the park, increasingly dependent on scanty scraps of land. These populations were highly dependent on traditional agriculture that requires large amounts of land in order to be profitable. But, around the park, the area cultivated per household (0.5 ha on average at Kibumba)[2] has reduced to the point where currently there are peasants without land. Add to this the frequent movements of people fleeing the war; they settle in relatively secure villages around the park and end up living on resources taken from the park. The populations surrounding the park live in a state of worsening poverty following the collapse of the Congolese economy, the loss of jobs on the plantations and in factories around the park, and the loss of most non-agricultural jobs in the region.

Non-payment of the military personnel is without any doubt the principal exogenous puzzle that Virunga National Park is faced with at present. Since the soldiers do not have enough resources to live on, they are compelled to engage in the destruction of the park's resources at every level: animal poaching, exploitation of wood and other natural resources and violation of boundaries.

These factors are carefully exploited by the political opportunism that is gaining ground in the region. Politicians, stakeholders from the civil society or leaders of environmental NGOs, develop a double language concerning the park: on the one hand they use an 'ecological' approach and a level of co-operation with the park and its partners (donors) in order to receive benefits; and on the other hand, a 'militant' approach, such as trade-unionists, demanding rights for the park's populations, justifying even the overrunning of the park's boundaries. On this subject, Mugangu et al. (2000) distinguish between systematic violations organised by powerful networks whose objective is to take over agricultural space in the park and occasional violations

16.3
The wars and the activities of the different rebel groups, supported by different political factions, caused enormous damage to the park by allowing insecurity to rule and by preventing ICCN from acting.

16.4
The close collaboration between the ICCN and other key players helps to resolve political conflicts. Director Mushenzi receives equipment from the MONUC (United Nations Organisation Mission in the Democratic Republic of Congo) that will enable the ICCN to take back control of the park from the military.

carried out by the residents to survive. But all these violations have already had a significant impact in the park. One of the great ironies remains that, more often than not, the people who – given their political or social position – should be acting to save the park, are actually manipulating the population, and indeed turning it against the park. Concern for public property has yet to become a major preoccupation for some Congolese politicians. At about every level, personal interests take precedence.

16.3.2 The institutional crisis within the ICCN

The institutional crisis within the ICCN is illustrated by the lack of park management reform, by financial difficulties and by an inadequate institutional framework of co-operation with the partners.

Concerning reforms, Mugangu *et al.* (2004) lament the persistent insular management that prevailed for a long time in Virunga National Park, which did not favour development of the areas around the park. This management was effective as long as the state was strong enough to watch over the park and deal severely with offences (by police repression, fines, etc.). But the weaknesses of the state in the past decade have exposed the inefficiency of this management system, notably with the violations of park boundaries. In many countries, the new management strategy for protected areas is increasingly a concerted effort between the local populations, represented by legitimate structures, the managers of the site and the donors. This strategy has been introduced in Virunga National Park by the partners, including the Gorilla Organisation in the north, the IGCP and the WWF in the south.

As for the ICCN's financial difficulties, the non-payment of officials and rangers and the lack of equipment lead to staff demoralisation. Some officials are implicated in networks of illicit exploitation of the park's resources (wood-cutting, charcoal making, etc.). It must be noted, however, that UNESCO and different partners (IGCP, WWF, ZSL etc.) have, as much as possible, paid premiums to the guards, and important European Union support is underway. Overall, the present challenges of conservation exceed the capacities of human resources available at the ICCN, making training and recycling indispensable.

Finally, the inadequacy of the institutional staff is obvious: the partners experience difficulties related to unclear mechanisms and to the sheer number of representatives for any given site. In addition, for several years, the ICCN, like many public institutions, has worked under the auspices of a Co-ordination, a former structure of the RCD. Even though the Co-ordination was suppressed, the reality of a state with two hats has not yet completely disappeared: there is an ICCN hat, engaged in the protection and defence of park interests, and a public administration hat that is accommodating, even partisan during conflict arbitration. It is this balance between protecting the land and accommodating the people that sorely tests the rigorous application of the law.

16.4 Chances for viable solutions

16.4.1 The catastrophe scenario

A catastrophe scenario, the collapse of the park and its infrastructure, is a worst case scenario that unfortunately remains imaginable for Virunga National Park. In fact, the information available from various sources (MONUC, ICCN personnel, local and humanitarian NGOs) reveals that the armed bands destabilising the province still circulate in and around the park.[3] All that it takes is one 'favourable' opportunity for these elements to mobilise into a new eruption of violence. Any new violence compromises all the efforts already invested in restoring the integrity of the park and its natural resources. The threats are real: an increase in poaching, renewed violations of the park boundaries, weakening of the ICCN's capacity to control the park if their staffs are again forced to evacuate their posts, the inevitable pillaging and destruction. It is a familiar scene. This is what happened during the war.

It is not easy to find solutions to this catastrophe scenario. However, lessons have been learnt from the previous major crises in the Virunga National Park and the ICCN and its partners are already equipped with diplomatic and political lobbying skills. The guards have been retrained at the Ishango training centre and an elite corps has been formed, offering a chance to control some crises. However, in an acute crisis leading to large-scale violence inside and around the park, several measures could be taken:
- anticipate such violence by preparing a practical strategic framework that defines what needs to be done (evacuation of non-essential personnel, gathering at easily defendable key points, a communication protocol to be used during crisis periods, provision of necessary equipment, notably food rations, medical kits, etc)

16.5
The wars of 1996 and 1998 and the attacks in 2006 have almost entirely destroyed the infrastructure at Rwindi.

- political mobilisation: pleas to the MONUC and to political decision-making bodies, in particular the Governments and the armed forces. It is essential that political and diplomatic measures are carefully co-ordinated: a single message, clear and coherent, communicated by a designated spokesman.
- an international plea for the park. It is essential to have a clear communication strategy in place that can be quickly adapted to the context of the moment. A communication network should be established, in the media and at the political level, through a large number of personalities who sympathise with the cause of Virunga National Park and who would be aware of the role that they may be asked to play.

If these measures can be established in advance, it is certain that the ICCN and its partners will be able to minimise the impact of a political crisis on Virunga National Park.

16.4.2 THE NEED FOR A REGIONAL MECHANISM

The Democratic Republic of Congo has signed numerous regional co-operation agreements, some of which deal with trans-border conservation, integration of protected areas, tropical forests, the great apes, etc. All these agreements must be built on to reinforce the synergy of Virunga National Park, this will also minimise attacks that other signatory states could make on the country's natural resources. All of this requires the country to deploy diplomatic effort.

The political challenge remains one of the big problems that has to be addressed by the park's managers. Fortunately, several conservators already have some experience in this area and have successfully resolved challenges of this kind over the past decades.

1 In North Kivu there were at least three different categories: The Maï-Maï called *Kasindiens,* active in the territories of Butembo and Beni, the Maï-Maï called *Kifuafua,* active in the territories of Masisi, Walikale and Kalehe and the Maï-Maï called *Mongols,* Rwandophones active in the territories of Masisi, Rutshuru and Goma. The first two groups were allies of the Congolese Government and were seated in the transitional institutions while the latter, close to Rwanda, did not take part in the Maï-Maï component within the Government of Transition. There were also other bands commanded by Jackson who operated in the east of the park and were sometimes allied to the Interahamwe, and sometimes to the Ugandans.
2 Kivunda and Balole-Bwami: *La pression économique sur les ressources naturelles du PNVi* (Economic pressure on the natural resources of VINP), Unpublished report, PICG (International Gorilla Conservation Programme), 2001.
3 It is to be noted that, since the beginning of the process of transition in the Democratic Republic of Congo, the region of Virunga National Park has been the theatre of numerous armed conflicts, at times the regular army against the ex-armed branch of the RCD, at times the FARCD supported by the MONUC against Jackson's Congolese militias and the Interahamwe, sometimes brigades of the regular army... In addition, numerous armed attacks and pillaging in the villages and along the roads make Virunga National Park a dangerous place with very precarious security.

17.1
Water is a rare commodity in some of the park's neighbouring areas, but the park has a constant supply of drinking water.

Ephrem Balole & Samuel Boendi

17 Involving local communities in the protection of Virunga National Park

Ephrem Balole & Samuel Boendi

Virunga National Park faces greater encroachment from surrounding populations than any other protected area in the Democratic Republic of Congo (DRC). More than three million people live within a day's walk of the park and many of them depend on its natural resources. Clearly, these people can no longer be ignored.

Traditionally, the most viable means of protecting animal populations and plant life was felt to be fortress conservation, so long as there were enough resources for their survival within the protected area and there were connections with neighbouring protected areas. As shown in Chapter 19, the main threat to the park comes from the populations around its edges, for whom fishing is one of the only real benefits of living next to a protected area.

It is vital that these local communities begin to understand the ecological and economic importance of the park, and how it can benefit them both directly and indirectly. They should be involved in the management of the park, at least in the peripheral or buffer zone. In the future, the ICCN should develop a program allowing local people to make sustainable use of some of the park's resources. This chapter looks at the general framework for establishing a system of 'community conservation' for Virunga National Park, examining what has been achieved so far and proposing ways to take this forward.

17.1 Introduction: the theory behind the management of protected areas

17.1.1 Centralised management and total protection as dominant models

The creation in 1872 of Yellowstone National Park, in the United States, under a system of centralised management, inspired the creation of protected areas in many other countries. The focus at the time was on protection. Nineteenth century naturalists thought access to protected natural resources had to be controlled to successfully reduce the level of damage, particularly from people. They were concerned about the pressure human activity exerts on natural resources and the subsequent threat to the survival of some species. Given the Malthusian theory of demographic growth in the absence of an auto-regulating mechanism, these scientists feared that there would be an uncontrolled increase in human activity bringing about larger-scale damage and even a loss of biodiversity altogether.

There are other arguments in favour of this centralised approach which take a more philosophical view, based on 'the right of species to exist' (Nash, 1970). Nevertheless, the key factor supporting the centralised management of protected areas is the undeniable fact that the threat of human activity on biodiversity is so great that numerous species have already disappeared and others are at serious risk of extinction. This is the justification for excluding all human activity from protected areas.

Contrary to claims that national parks, particularly in Africa, were a construct of colonial leaders in a last desperate attempt to create 'Gardens of Eden' prior to independence (Brockington and Homewood, 1995), most African protected areas were created after independence in the 1970s and 1980s. However, Virunga National Park was created 35 years before independence, with the dual objectives of natural resource protection alongside scientific study.

Rather than concentrating on the separation of protected areas and people, an alternative attitude is to see protected areas as providing significant benefits to society (International Union for Conservation of Nature, 1994). These include access to natural resources, clean water, a healthy environment and eco-tourism. However, the economic and social costs that protected areas present for local populations can seem excessive. The frequent poverty of people who live in their vicinity is compounded by the lack of any alternative economic activity. This combination of pressures not only threatens conservation efforts, but places the very existence of the parks themselves. Efforts to find a solution to

17.2
A good community conservation program has a greater chance of success if it is run in parallel with an environmental education program. Hundreds of public awareness meetings were organised up to 1996 when fighting broke out. The meetings only started again in 2005 and then only very cautiously.

17.3
The WWF Environmental Program produces an environmental education magazine for the communities around Virunga National Park that is much appreciated by students and adults of all backgrounds. Over 50,000 copies are printed, and they are rapidly taken up. Here, a special issue commemorated the park's 80th anniversary in 2005.

this problem have prompted a need for alternative management models that combine a more effective means of nature conservation with the development needs of local populations. In 1992, the president of the IUCN stated that, 'we cannot rescue the parks without also rescuing the human communities, and we cannot rescue these communities without rescuing the parks' (Kamstra, 1992).

17.1.2 Decentralised management and controlled exploitation as alternative models

Following criticisms of the centralised management system for protected areas, alternative decentralised models have come to the fore, aimed at involving and developing the local population and increasing access to the benefits of nature conservation. This paradigm shift has taken place over the past twenty years. In that time, two main approaches have come to the fore: the participative approach and the neo-liberal approach.

17.1.3 The participative approach to the management of protected areas

Called community conservation (International Institute for Environment and Development, IIED, 1993) or populist conservation (Blaikie and Jeanrenaud, 1997), the basic idea is the same, that is to give local populations – otherwise marginalised in the management of protected areas – better access to the monetary and non-monetary benefits. The aim is to reduce the economic and social costs borne by the local populations and to for them to reap benefits through the sustainable use of natural resources.

The participative approach is founded on community development through self-help, self-reliance and self-management, with external support provided by park staff or the park's partners. People are given encouragement to use their skills and local resources to build up a community development micro-project. In return they get involved in safeguarding the protected area (for example, by reducing the pressure of human activity or by denouncing poachers). The first tests of this approach in Virunga National Park showed that the system worked to satisfy the social needs of the community, but did little to alleviate the more pressing problem of their poverty. With this in mind, the participative approach is increasingly being used to develop rural micro-enterprises and micro-finance around protected areas.

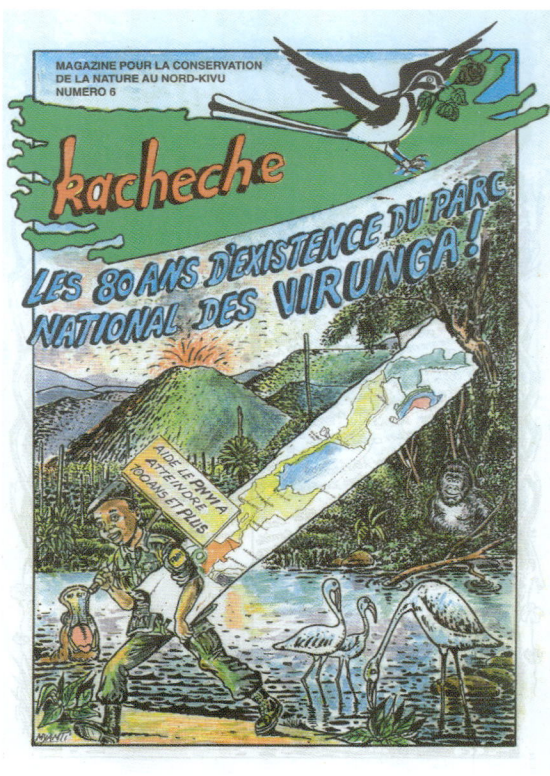

It is useful here to look at the investigations undertaken by the Institute for Tropical Forest Conservation on the successes and limitations of the Integrated Conservation and Development Programmes (ICDP). These show that such programs, if they are well run, have a very positive impact on people's attitudes towards protected areas, such as Bwindi. But changes in behaviour are more limited because of restricted economic choices. Consequently, it seems that public awareness and education programs alone have little impact because although local populations rally to the cause of the protected areas, a lack of alternatives means they are still forced to exploit them, often in an unsustainable way.

A further observation is that these kinds of development programs have little impact if there is not already a well-established link between the local communities and the protected area. For example, during the Kivu Programme begun in the late 1980s in Virunga National Park, it became clear that despite the numerous development initiatives that were effective in helping local populations, their link with the park and the ICCN was not strong enough to change their attitudes about the park.

17.1.4 The neo-liberal management of protected areas

Contrary to participative management, neo-liberal management takes an economic approach to environmental problems. It considers natural assets to be financial property, each with an 'opportunity cost' depending on an individual's economic preferences. The inference of this approach is that sustainable management of a protected area should be preferable to any other land use (Child, 1996). According to this analysis, the natural resource should generate high enough revenue to both maintain it and dissuade the use of it in any other way. This is a purely economic approach based on the value of using the resource and it does not necessarily take into account ecological imperatives. The local population's incentive to conserve the resource is, notably, the monetary benefit that it generates, which are transferred to it by the managers of the protected area.

This neo-liberal approach has been tested in the CAMPFIRE (Communal Areas Management Programme For Indigenous Resources) programs in Zimbabwe and in the ADMADE (Administrative Management Design) programs in Zambia. In both cases, part of the revenue generated by the protected areas is returned to the local communities for their development. In return each village forms a democratically-elected conservation committee in charge of duties such as animal inventories, anti-poaching activities, park-population conflict resolution and environmental education. Local people are also trained to support the anti-poaching struggle. In the CAMPFIRE system, 80 % of the revenue goes direct to the local communities who have established a transparent management system for deciding their financial priorities. A further 20 % goes to the district council to administer and manage the program (Metcalfe, 1995; Lyons, 1998).

17.2 The advantages of community involvement in the management of natural resources around the park

Virunga National Park is largely run on the principle of centralised management. The park's creation was made possible with the consent of local people, through payments of customary rights, land exchange or the voluntary resettling of communities. Despite this, some local communities still feel that land that was rightfully theirs was stolen from them and, furthermore, claim the creation of the park has led to a land shortage. Chapter 9 shows that the park was, from its creation until DRC's independence, surrounded by vast expanses of unoccupied land, but many local communities would still claim that they were unfairly evicted. It is important to redress these misunderstandings. One way is to include those populations who habitually invade the park's boundaries into its protection.

Congolese law confers on Virunga National Park the status of an 'Integrally Protected Reserve', forbidding local populations any access to its resources, with the important exception of the Lake Edward fisheries (Chapters 1, 10 and 13). In addition, the ICCN misunderstood local communities' sense of having lost their rightful land and relations between the two deteriorated. Currently conflicts are triggered by the following:

17.4–5
Support to the villages or private enterprises neighbouring the park is always targeted at activities that directly or indirectly affect conservation. Here a traditional pharmacy stocks wild plants gathered from the park and cultivated in a nursery.

- The *feeling* that land was appropriated without compensation,
- The demographic explosion of the population around the park,
- The lack of any economic alternatives to subsistence agriculture,
- Rangers' behaviour towards local populations on some occasions,
- Damage caused by park animals to the fields and goods of residents,
- The central management of revenues generated by the park, without benefit to the local populations.

17.3 The History of Community Conservation around Virunga National Park

17.3.1 The particular case of the Lake Edward fisheries

The agreement reached over 50 years ago between the ICCN and the chiefdoms around Lake Edward to create fisheries is an example of community conservation before the term was even coined. Both the fish and Lake Edward in general are part of the national park, and therefore fall under the jurisdiction of the ICCN.

It was a *de facto* community-based approach since the fishermen were organised into a co-operative. The agreement between the fishermen and the park authorities laid down how the resources should be managed and this was formalised by a contract. It also laid down how the resources should be technically managed, with a quota of dugouts and a maximum fishing effort authorised.

Even if this system of community conservation has come under considerable strain since the deterioration of the state in the late 1980s, it is often forgotten that this historic example worked satisfactorily for more than 40 years.

17.3.2 The case of more recent initiatives conducted by the ICCN and the WWF

More extensive community conservation began in Virunga National Park in 1987 with the *Programme Environmental autour des Virunga* (PEVi) initiated by the World Wide Fund for Nature (WWF) and the ICCN. This program was the result of a WWF study into why local people were exerting pressure on the park. The conclusion was that the people were driven by the need for firewood, building timber, garden poles for agricultural land and for food (hunting game for meat). The study also showed that the majority of people did not understand the park's ecological and economic benefits. The study's recommendations were to establish an environmental education program, coupled with an agro-forestry program (see Chapter 14) to make the population aware of the need to protect the park's natural resources.

The program was developed in three phases:
- Phase I (1987–1997): raising public awareness about the park's economic and ecological importance, particularly about the need to protect the mountain gorillas and Virunga National Park in general, coupled with an agro-forestry program.
- Phase II (1998–2001): integration of community activities concerned with sustainable management of the natural resources around the park.
- Phase III (2002 to the present): maintenance of the awareness program and of the management of natural resources around the park, but this time involving communities in the activities linked directly to park protection: rebuilding infrastructure and the community's ability to sustainably manage natural resources.

17.4 The organisation and functioning of community conservation in Virunga National Park

Community conservation in Virunga National Park is based on the principles of community development (see above). The PEVi works with associations made up of people living around the park's periphery. The associations' objectives have to be compatible with nature conservation. Some associations were established through this public awareness program although most were already operating prior to the programme's establishment. Collaboration with the associations goes through the following stages:
- Identification, with the population and the ICCN, of actions supporting habitats,
- Operating request made by the association addressed to both the ICCN and the WWF- PEVi. The request includes the association's legal operating documents (statutes and internal by-laws), as well as its achievements and work program,
- Identification, with the ICCN, of the most appropriate partner associations,
- Letter of Agreement. This is between the associations, the ICCN and the WWF (as technical partner) and specifies the responsibilities of each party as follows:
  - WWF- PEVi: technical support via supervision of technicians and punctual supply of materials,
  - Association: land and work premises, the participation of members in the work, completion of activities and joint decision-making about the resultant benefits (sale or reinvestment),
  - ICCN: participation in dialogue and in information exchange meetings concerning the poaching of resources in the park.
- Supervision of the association by WWF- PEVi technicians. This support covers both technical and cultural aspects, reforestation, structuring and management, as well as organisational development,
- Support for self-financing initiatives and economic sustainability.

A classic error in projects around a protected area is to try to solve all the social and economic problems of the resident populations. It is imperative not to spread limited human and logistical resources too thinly over poorly defined activities (and geographic areas). In the case of Virunga National Park, community conservation activities concentrate on two areas: 'The wood problem' in the widest possible sense including agro-forestry, reforestation, etc., and on public awareness, supporting NGOs and the associations in the field. The following are examples of activities:
- Forestry and agro-forestry: seedling production, identification of plantations, development of plantations, fitting out nursery sites, development of nurseries, charcoal-making technologies,

17.6
One of the nurseries set up to propagate plants gathered from the park for use by local communities.

17.7
A stream coming from Mount Tshiaberimu, in the Northern Sector of the park, provides clean water the whole year. The continuous provision of drinking water is one of the ecological services of the park most appreciated by the resident populations.

improved woodstoves, improvement of soil-use techniques, etc.
- More recently, reforestation on a larger sustainable scale for subsistence and economic purposes.
- Environmental education: production of *Kacheche* magazine, radio broadcasts, films, youth supervision, ecology campaigns, courses in environment public awareness, shows, forums, etc.
- Support for the communities: support for the associations through rotating credit, support in the form of ploughing tools and equipment, micro-projects, bee keeping, etc.
- Support to the ICCN for community conservation in the field: support to patrols along the park boundaries to prevent conflicts, meetings with communities for the follow-up of illegal use of park resources, training in management and prevention of conflicts.
- Protection of the integrity of Virunga National Park: participatory demarcation of the boundaries (see Chapter 25), placing of signs and boundary markers with the use of GPS and satellite maps, green belts, peaceful recuperation of invaded land, etc.
- Improving skills for the communities: training in nursery management, organisational management, improving woodstoves, etc.
- Improving skills for the ICCN: training in the use of GPS, of computers, etc.
- Women and environment: support for gardening, small animal husbandry, wood saving techniques, micro-finances, etc.

Women are key partners for community conserva-

Involving local communities in the protection of Virunga National Park

17.8 Evolution of partner associations of the ICCN around Virunga National Park

Year	Number of associations becoming partners	Number of members
1990	1	40
1991	1	25
1992	5	185
1993	2	37
1994	3	66
1995	11	282
1996	13	620
1997	7	150
1998	8	273
1999	9	277
2000	14	416
2001	2	28
2002	1	17
2003	5	110
2004	4	113
2005	1	41
Total	**87**	**2680**

17.9 Distribution of partner associations

Category of association	Number of members
Women's associations	34
Girls' associations	49
Youth associations (mixed)	4
Total	**87**

17.10 Distribution of associations by sector

Sector	Number of members
Rutshuru Hunting Preserve	21
Southern Sector	59
Northern Sector	7
Total	**87**

tion in Virunga National Park. They are more forthcoming than men, particularly given the conflict that has characterised this region of DRC these past years. Although not directly responsible for the pressures on Virunga National Park, a study (Balole et al., 2001) shows that most demands for firewood and other ligneous resources come from women, who use men and children to extract the resources. In some cases, women also actually participate in the transport chain for poached products. But they are generally more risk averse, and so are interested in the alternative solutions proposed by community conservation. Therefore, women are both a cause of Virunga National Park's problems, and a stakeholder in finding its solutions.

Due to the series of conflicts in eastern DRC, some sectors of the park, in particular the Northern Sector and the eastern flank of the Central Sector, have been difficult to reach. This explains the low number of NGO partners and limited program activity in these zones.

17.5 Assessment of community conservation in Virunga National Park

We will not assess in detail the co-management of the fishing resources of Lake Edward, as this is addressed in Chapters 10 and 13. While the model itself is not questioned, the fact that the system was not adapted to meet the political and socio-economic changes of the past 20 years is examined. At present, the fisheries are not suited to co-operative management as the rules are not being respected by the increasing number of external players (army, other state bodies) using the site. The ICCN's inability to enforce the rules underlines the limits of co-operative management in this context.

The fisheries aside, the success of the move from a period of absolute protection of the park to a policy of involving local people in conservation varies from one area to another. It is a dynamic process because the change is sometimes difficult and requires time, materials, finance and manpower. But it is clear that to succeed, community conservation needs different strategies for different cultures. This would encourage local communities comply with and participate in the conservation and management of natural resources. The past 20 years of community conservation at Virunga National Park have not solved all the problems that have arisen. But a recent evaluation of the community conservation project (Kalpers and Braun, 2005) showed that a large number of people around the park have maintained an astonishingly positive attitude, despite the very difficult context of these past years. This is largely thanks to the community conservation activities and public awareness.

The different indicators of success include:
Qualitative indicators:
– Permanent collaboration between the ICCN and local communities,
– Image of the park (and the ICCN) improved among the communities,
– Respect of park boundaries,
– Improving the skills of the associations and the ICCN in terms of community conservation methods (for example, the joint training of both rangers and local conservation leaders who work together at all the stations of Virunga National Park),
– Existence of committees dedicated to the search for solutions to problems related to the park,
– Awareness activities targeting the political, administrative and traditional authorities concerned with the park's problems.

Quantitative indicators
– 87 associations working to meet ICCN nature conservation objectives through the WWF- PEVi.
– Tens of thousands of people have voluntarily agreed to evacuate the park after illegally occupying it.
– Participatory demarcation to protect the integrity of the park. More than 460 signs marking park boundaries put in place in agreement with communities. This follows a long period of ignorance and refusal to recognise these boundaries by the same local communities.
– Community infrastructures supported by partners of the ICCN.

- Community members have received training in different fields.
- Response of local populations to the problem of firewood and timber through reforestation: establishing nurseries for seedling production, and community and private plantations.

17.6 Lessons and challenges

It is important to capitalise on the lessons learned about community conservation at Virunga National Park. These are:
- The need to carry out preliminary studies aimed at a better understanding of the socio-economic context. Studies must also examine the attitude of neighbouring populations to the protected area and the institution in charge of it.
- The participation of the population at all levels of intervention is a guarantee of success. During the boundary demarcation process, in those areas where the principle of demarcation was understood, the signs and boundary markers were respected. This was not the case on other sites where the population was not properly consulted.
- Community conservation has changed the negative perception that neighbouring populations had of the park. Thanks to the changes in local perceptions, these populations have started supporting park surveillance activities by informing the park staff of the location of poachers.
- It is vital the ICCN plays the leading role so the communities see the benefit to the national park and of ICCN. Communities will feel abandoned by ICCN if they perceive that external actors (a donor or an international conservation NGO) are in charge.
- It is not yet possible to establish a positive correlation between community conservation and a reduction of pressure on the park's resources. Ironically, a study by Musombwa et al. (2005) showed that the great majority of charcoal used in Goma came from the park's Southern Sector which had received the greatest investment in community conservation. However, one has to take into account the peculiarity of this region where demographic pressure (large, concentrated populations in very small areas) and the geographic configuration (presence of active volcanoes) can mean there are no alternative forests from which to get wood. External factors such as the army's role in illegal trading complicate the picture. It is also important to consider that we cannot know what the current situation would be if the community project had not intensively encouraged reforesting.
- The necessity of harmonising community activities with the conservation needs of Virunga National Park. The tendency is to respond to the needs that are identified as a priority by the communities, but that risks investment in activities that do not have a direct impact on the park's conservation. It is therefore necessary to pay particular attention to appropriate, focused activities.

The ICCN's current control over the park is the combination of two factors. On the one hand, it is due to partners' direct support for the restoration of its authority, as a state institution charged with management of national parks. On the other hand, it is due to the ICCN's support of neighbouring populations. However, community conservation has some way to go before it can play a genuinely important role in Virunga National Park. The challenges include:
- The establishment of a long-term national strategic plan adopting community conservation as a management tool for protected areas within the whole of DRC. This would formally define the relationship between the ICCN's centralised management approach and the decentralised management approach for community conservation. Wardens and rangers will need further training in community conservation techniques.
- The establishment of a clear legal framework for the management of natural resources and the sharing of their benefits with the park's neighbouring populations (see the Forest Code of DRC).
- The use of community conservation in the struggle against poverty. Sustainable development of communities and sustainable conservation of the park's resources go hand in hand. This should be done in conjunction with other state services that have mandates for the population's economic development.

17.11
A training workshop held by a local NGO on the principles of community management of natural resources.

17.12
One of the meetings bringing together village chiefs, villagers, the ICCN and project partners to agree on the management of the buffer zone in the Rutshuru territory.

17.13
Support for wood production and agro-forestry is one of the most frequent demands on the ICCN.

17.14–15
Great efforts are taken to promote awareness among children, the managers of tomorrow.

17.7 Perspectives for community conservation in Virunga National Park

The introduction of community conservation has been particularly challenging in Virunga National Park. For the past 15 years the park has been in a crisis situation, due to repeated armed conflicts and the subsequent difficulties in protecting the park's natural resources. There has been no public funding for the work, which has been almost entirely financed by WWF. These extra challenges are important to consider when comparing the success of community conservation in Virunga National Park compared to other countries where there has been great progress in community conservation.

We need a national policy of nature conservation in which the community conservation is clearly marked as a major strategy. This national policy must lay out each participant's responsibilities: government, the ICCN, donors, conservation partners, the private sector and local communities. Furthermore, the policy must take on board the fact that international NGOs will not continue their community conservation support indefinitely. Eventually the ICCN will need the budget to take it on. The ICCN already works with its partners on this understanding.

The good news is that a national strategy for community conservation has just been finalised and is largely inspired by lessons learnt from Virunga National Park, as well as from Garamba National Park and elsewhere. The terms of reference for Community Wardens are also in the process of being properly drawn up to make this post a permanent fixture of the ICCN. The necessary addition of a Community Conservation Section within the ICCN Headquarters in Kinshasa is also being considered.

The restoration of Virunga National Park remains the immediate and absolute priority. Co-management of its internal natural resources, as is happening in Bwindi, is not possible at present. Therefore, for the immediate future, community conservation activities must concentrate around the edges of the park. Buffer zones, which have unfortunately become very rare, should be used as practice areas for co-management between the ICCN authorities and the communities. The development of such partnerships would create greater trust and understanding, with opportunities for both to learn from each other's experience. Once these first steps have been completed, co-management of some of the resources inside the park should be considered.

Ephrem Balole & Samuel Boendi

17.8 Concluding remarks

Centralised management remains a valid organisational model for the running of Virunga National Park from a legal point of view. The ICCN is a national institution responsible for enforcing the laws made to protect the park's natural resources. However, at the same time the reality of the day-to-day running of the park and the role it plays in the lives of local people make management via a community conservation model or a neo-liberal approach, equally vital.

The neo-liberal element is important for ensuring that conservation efforts have a material benefit for local communities, and that they are therefore lasting. In 1994 a scheme giving communities a share of gorilla tourism revenues was successfully introduced. Sadly, it only functioned very briefly because the Rwandan refugee crisis occurred the same year. As part of the effort to regenerate tourism in the region, a transparent system for sharing revenues with local people should be reinstated. In the same way, a neo-liberal approach should be applied to ensure the sustainable management of the Lake Edward fisheries, specifying the responsibilities of each of the parties.

Community conservation promotes the participation of surrounding populations in the management of some of the resources of protected areas. The experience gained in community conservation in Virunga National Park now needs to be drawn into a national strategy for DRC. This strategy should widen the definition of conservation to mean 'development that has a maximum impact on both the poverty of populations living near protected areas and on the protection of natural resources'.

The dual operation of the centralised and decentralised models requires a respect for the principle of subsidiarity which states that 'in matters of decentralisation, responsibility must be given to the hierarchic level best able to deal with it'. This entails a division between the Government, the ICCN, local communities and the private sector, each with responsibility for the role they are most capable of fulfilling, no more and no less. For this we need a new national nature conservation policy.

17.16
Each plant species in this garden originally came from the park and each has a medicinal value.

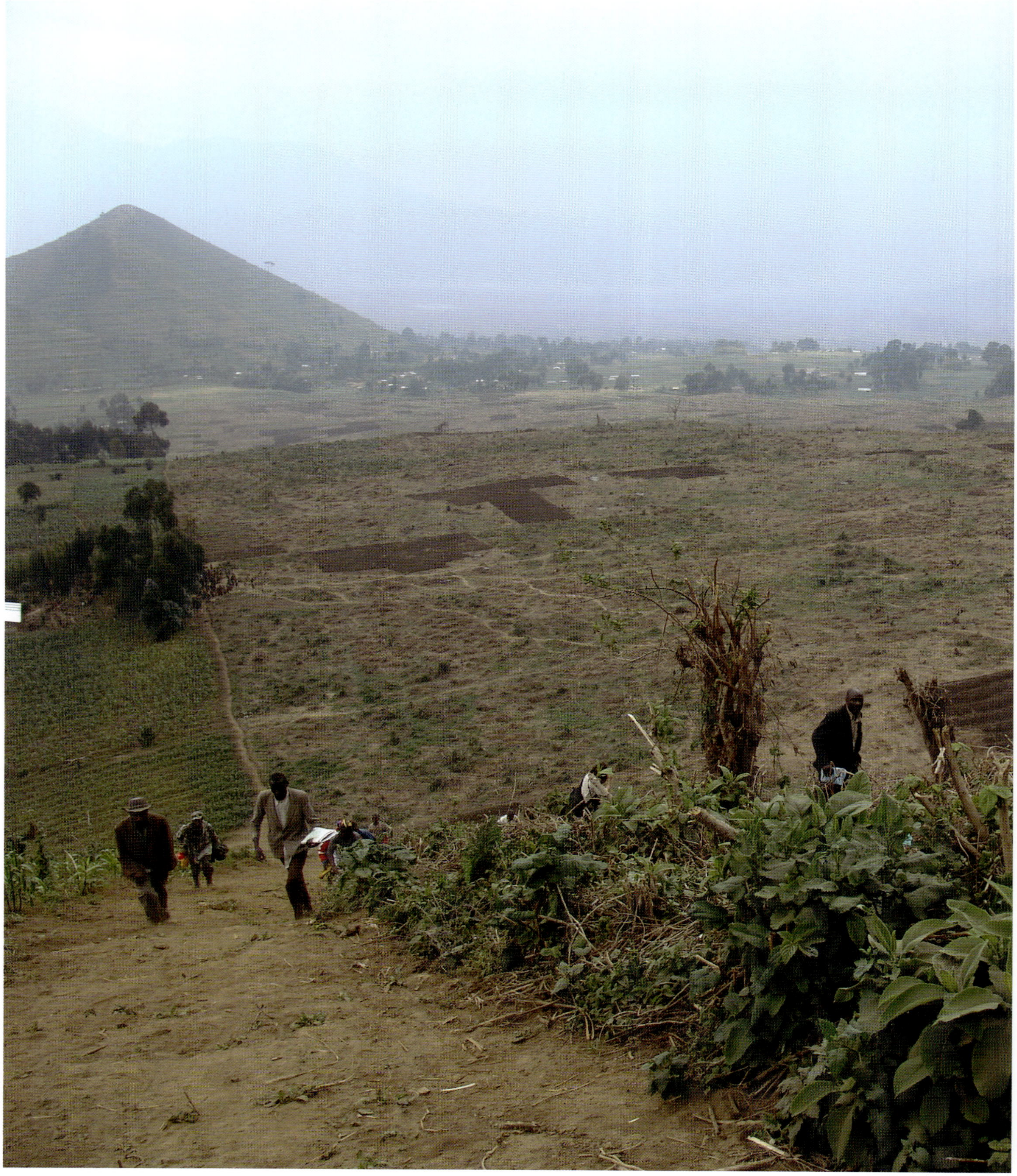

18 Managing a National Park in crisis

Robert Muir & Emmanuel de Merode

The nature of resource depletion in sub-Saharan protected areas remains a poorly understood process. Yet it is becoming increasingly clear that a significant proportion of the damage is not gradual but the result of sudden, catastrophic and unexpected events (Martin and Szuter, 1999; Shambaugh et al., 2001). For more than a decade, and perhaps more than in any other protected area in Africa, the wardens of Virunga National Park have been forced into what can only be described as crisis management. A series of conflicts have inflicted untold damage on the park's natural resources. In this chapter, we examine three recent crises for Virunga National Park. We recount the actions taken and draw on the experiences in the field to develop some recommendations for the future.

18.1 A background to the recent crises in Virunga National Park

The origins of recent crises are mostly political and largely explained by the park's position at the heart of the Great Lakes, a deeply unstable region caught up in socio-political turmoil. The recent history of the crises is explained in Chapter 4 by Kalpers and Mushenzi, and again in Chapter 16 by Balole et al. The politics of access to natural resources has played a determining role in starting and sustaining the conflict, thereby rendering the resources of Virunga National Park all the more vulnerable (United Nations, 2001).

Crises are usually unexpected events associated with a threat to stability (Davies and Walters, 1998). Since the Rwandan genocide in 1994, a continuous succession of crises has contributed to the death of what is now estimated to be over 100 Virunga Park rangers. During 2004 alone there were thirteen armed attacks on patrol posts and park stations, leaving three rangers dead or disabled, one patrol post completely destroyed and the remaining rangers in the affected areas destitute, without food or accommodation. These events had a marked impact on the morale of all the park rangers, severely limiting their ability to carry out their duties. The problem was compounded by the *Institut Congolais pour la Conservation de la Nature*'s (ICCN) inability to respond effectively to this crisis situation, even with emergency humanitarian relief provided by international partners.

A crisis need not in itself have damaging consequences; however if associated with uncertainty, the resources of the responsible organisation may be overstretched and consequently it will be unable to manage the situation effectively (Stubbard, 1987). Furthermore, response time is critical; failure to take effective action quickly can incur rapidly increasing management costs over time (Greening and Johnson, 1996). As a general rule, a lack of solid contingency planning, combined with financial, institutional and political weakness, render an institution incapable of responding effectively to a crisis.

In this chapter we examine the response to three major crises between 2004 and 2005. The first was a success, thanks to careful but rapid planning and execution. The other two were failures. Thus, to begin with we assess the events surrounding the systematic destruction of gorilla habitat by Rwandan farmers in June 2004. We then examine the anthrax crisis for the hippo population of Lake Edward in late 2004. Finally, we look at the direct threat to park rangers and their families at Rwindi following renewed clashes in Kanyabayonga in December 2005.

18.1
The crisis in the Mikeno Sector in May-June 2004 had a dramatic impact on the park. The right-hand section of this photograph shows the complete deforestation of the national park area. This same area contained dense forest only a few weeks before. One can clearly see the park boundary, with established cultivation on the left side.

Event

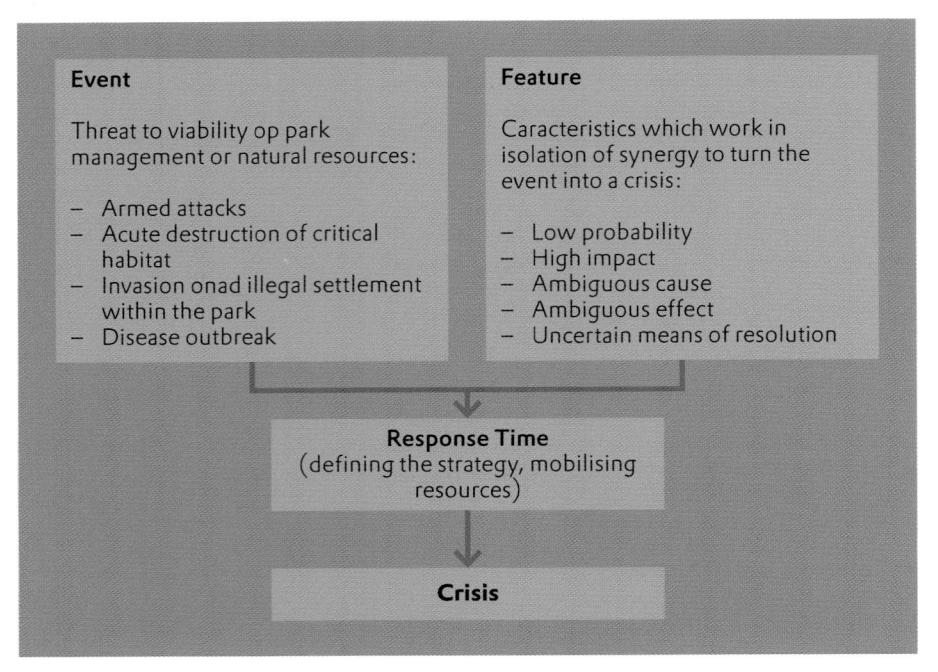

Threat to viability op park management or natural resources:

- Armed attacks
- Acute destruction of critical habitat
- Invasion onad illegal settlement within the park
- Disease outbreak

Feature

Caracteristics which work in isolation of synergy to turn the event into a crisis:

- Low probability
- High impact
- Ambiguous cause
- Ambiguous effect
- Uncertain means of resolution

Response Time
(defining the strategy, mobilising resources)

Crisis

18.2 Virunga: a history of crisis management

Within Virunga National Park a crisis is any event that threatens the viability or integrity of the park's natural resources or management structures, and which exceeds the ICCN's day-to-day abilities to respond appropriately to resolve the problem. The situation in Virunga National Park is complicated by its shared borders with Rwanda and Uganda. The civil war in Rwanda resulted in the displacement of people throughout this border area, and pressure from tribal groups has further complicated the situation. Superimposed upon this complex situation are wider issues relating to the battle for politics and influence in an unstable region with significant mineral wealth.

18.2.1 Mikeno Crisis

On June 10, 2004 park rangers sent a report to the ICCN stating that 'thousands' of Rwandan farmers had entered the Gorilla Sector of Virunga National Park and had started cutting down the forest. The ICCN called upon the international conservation community to intervene, and aerial reconnaissance was immediately carried out by the European Union (EU) and the Frankfurt Zoological Society (FZS) who were able to confirm that the destruction was taking place at an estimated rate of up to 2km^2 a day. This rapid deforestation of the Mikeno sub-sector was alleged to have been authorised by Rwandan military to reduce their vulnerability to surprise attacks along the border. While a need to clear a band 200 metres wide along the international border may have been understandably necessary, the United Nations Mission to Congo's (MONUC) International Team of Experts were quick to point out that the destruction of 'critical forest' within the park extended up to 12km from the international border.

Investigations carried out on the ground by ICCN and partner organisations found that up to 6,000 Rwandans were making a daily trip across the border, paying one US dollar to the Rwandan military and local chiefs for the right to clear one hectare of land. Within two months a 15km^2 area of forest had been either cut down or severely degraded. Large numbers of cattle were also being put to pasture on the cleared area, apparently with the intention of detonating landmines and booby traps. However, further investigations revealed that, once cleared, the land was being sold to Rwandan citizens for up to 1,000 US dollars per hectare, some of whom travelled from as far as Ruhengeri and Kigali.

The information collected on the ground and through aerial surveys, as well as one SPOT satellite image taken on June 6, was used to inform the international conservation community, the European Commission, UNESCO, USAID, diplomats and foreign officials who applied pressure on the Rwandan Government and local and regional authorities to bring a halt to the incursion and destruction of the forest. On June 27, two days before the MONUC Group of Experts were due to conduct an official onsite investigation, an order was given by the Rwandan military for all farmers to immediately evacuate the area.

With the area now clear, the ICCN requested that the international conservation community help fund the construction of a 20km dry stone wall, one metre high and one metre wide, along the denuded area to help re-establish the park limits. The wall would also prevent further movement of domestic livestock into the park, and provide a strong visual response to the recent incursion. Funds from the United Nations Environment Programme, the International Gorilla Conservation Programme and the World Wide Fund for Nature were pledged to complement those funds already provided by FZS and the EU, and work on the 'eco' wall started on July 6, 2004 and was completed within six months.

This rapid intervention by NGOs and other international bodies, to agree to both fund and start building the wall within two weeks, has had some very positive effects on the ground. Above all, it demonstrated to the local population the determination and commitment of those who have pledged to protect this World Heritage Site, and their ability to respond quickly and to take appropriate action in an emergency.

18.2.2 Anthrax Crisis

In October 2004, it was reported that 126 hippos in Uganda had died from a mysterious disease within one month, and that two people known to have fed on the meat had also died from what looked like a viral infection. The hippo deaths were reported to have occurred in Katako, on the northern shores of Lake Edward, the Kyamdura River, Kashaka, Kasenyi and the Kazinga Channel between Lake Edward and Lake George along the Congolese border. The Uganda Wildlife Authority mobilised a team made up of their own members and experts from Makerere University's Faculty of Veterinary Medicine and the Ministry of Agriculture, Animal Industry and Fisheries to try and identify the disease responsible. The team initially thought it could be rinderpest or anthrax, or possibly poisoning by heavy metals or soil-borne pathogens, but these were ruled out.

The Ugandan hippo population lives side by side with the Virunga hippo population, but very little is known about its movements across the border. Nevertheless, both sets of hippos on Lake Edward form a

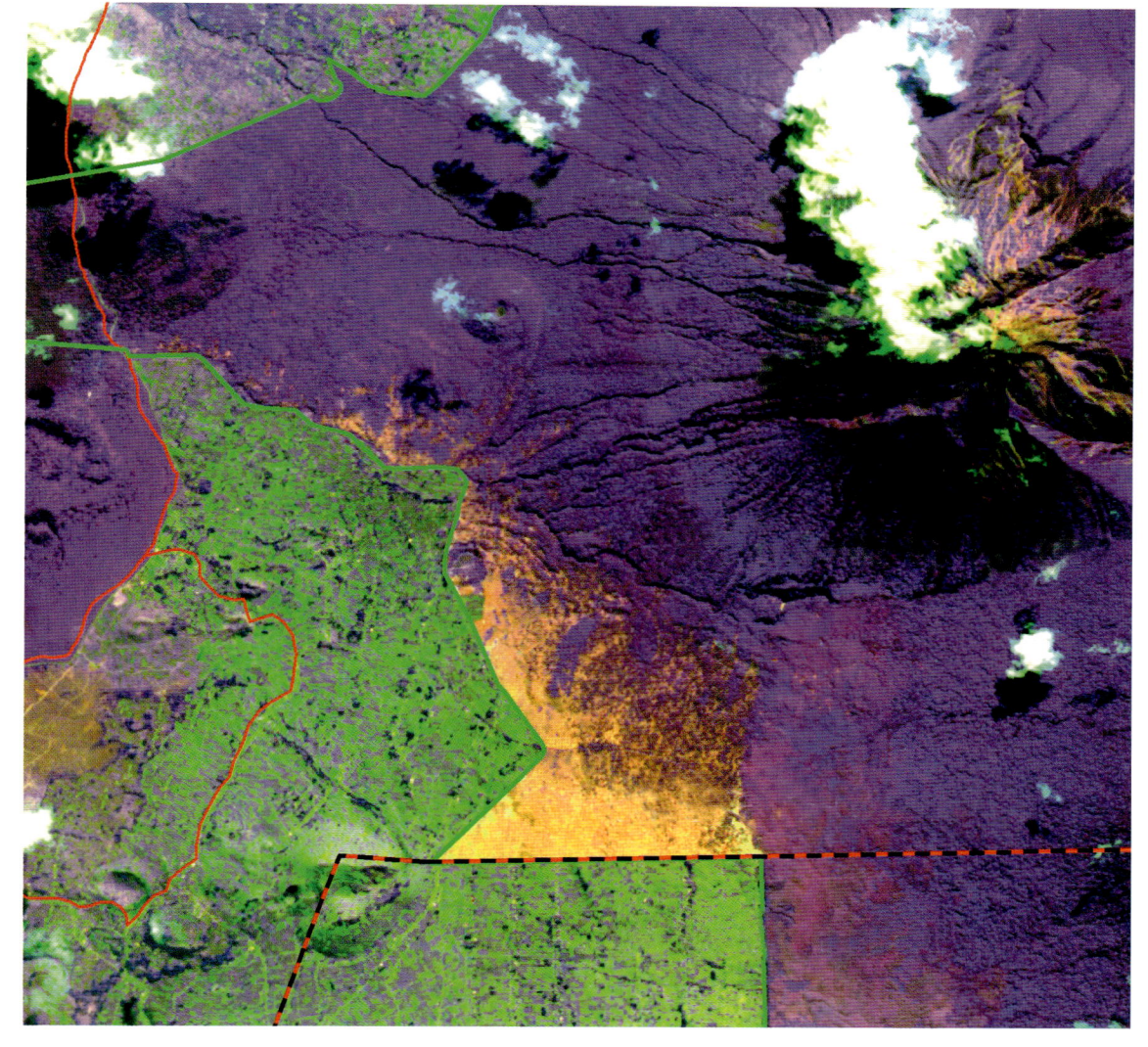

18.2
Goma town centre, two days after the Nyiragongo eruption in January 2002. This eruption provoked an intense crisis sparking the displacement of over 200,000 people. While many of these returned to Goma in the days that followed, the impact of the crisis had a long-term effect; the rebuilding of houses required substantial amounts of wood which were largely illegally harvested within the national park.

18.3
One of the greatest crises in recent years was in May-June 2004, when 1,500 hectares were destroyed by people from neighbouring Rwanda. This satellite image taken in July 2004 shows the Mikeno Volcano on the top right hand side of the image, with significant cloud cover. The area below the dotted line is Rwanda. The forested land is shown in purple, agricultural land in green. The deforested area following the 2004 land invasions are shown in yellow.

Managing a National Park in crisis   239

single population separated only by an international border which, naturally, has no demographic significance. Two carcasses were reported in Congo, both near Kasindi Port, close to the Ugandan border. But no diagnostic tests were carried out, and no measures, not even basic ones, were taken to prevent further spread of the possible disease; for example, the carcasses were not buried. Qualified ICCN personnel were not able to visit the site at the time because of a lack of transport, and as a consequence, there was no contact with the relevant local authorities or residents of the area.

It would appear that the epidemic never reached critical mortality levels in Virunga National Park, despite the fact that records show that over 30 per cent of the hippo population perished between November 2003 and August 2005. While anthrax could have been the cause of the hippo fatalities, the relatively small number of Lake Edward hippos probably played a major part in limiting the damage. Years of hippo poaching meant the population had plummeted to less than five per cent of its size in the early 1970s. Had the outbreak occurred at that time, when there were more than 30,000 hippos, the epidemic could have spread very quickly and caused significant damage.

Nevertheless, important lessons can be drawn from this incident. The ability of the ICCN and partners to detect a disease outbreak and to act appropriately proved almost non-existent. The response and the resources required to address a potential crisis of this nature were altogether inadequate. Had the epidemic not been reported and dealt with in Uganda, it may have been significantly worse.

18.2.3 Kanyabayonga Crisis

Following a number of weeks of increased tensions in the eastern province of North Kivu, military reinforcements from Kinshasa engaged troops from the *Armée Nationale Congolaise* – the military wing of *Rassemblement Congolais pour la Démocratie*, a former Congolese rebel movement allied with the Rwanda government.

Fighting officially broke out on December 15, 2004 forcing some 35,000 people to flee their homes. However, many others had fled some three to four days earlier. Some headed north towards Kayna and Kirumba, while others headed south towards Mabenga and Rutshuru. Among them were the 147 ICCN rangers and 519 women and children based at the park's headquarters in Rwindi, less than 10km away from Kanyabayonga. Due to rapidly rising insecurity, they left on foot on December 12 and headed for Kibirizi, Vitshumbi and Kiwanja.

Before leaving, the park rangers discarded their uniforms and equipment for fear of being mistaken for military personnel and targeted. Once they reached a town or village where they could take refuge, the rangers and their families struggled to find food. They relied heavily on the good will of the local communities who were also suffering as a result of the renewed fighting and struggling to survive.

By December 21, MONUC managed to negotiate a ceasefire and established a 10km buffer zone between Kanyabayonga and Lubero, but it was another 10 days before the rangers felt it was safe enough to start returning to Rwindi, some two and a half weeks after they had left.

On their return they discovered that everything within the station had either been destroyed or looted. All their personal effects had been stolen, along with all the linen, mosquito nets, clothes, medicines and food and kitchen utensils. The effect was that the station was entirely devoid of anything useful in a post-crisis situation. A compounding factor was that the ICCN and its partners were unable to provide any kind of humanitarian, logistical or financial support either during the crisis or once the rangers had returned to Rwindi. Humanitarian aid was eventually provided more than two months later through the *Commission des Urgences pour les Deplacés de Guerre* on March 1, 2005.

18.3 Crisis Management

Crisis management is a process of decision making, the purpose of which is to prepare an organisation to expect the unexpected. Crises tend not to occur as a result of a single factor, but rather as the result of a combination of factors which build on each other (Johnston and Stepanovich, 2001). The development of a comprehensive crisis management plan requires an overall vision and general agreement on necessary action. This is made difficult in Virunga National Park due to the dynamic mix of tribes and political factions within the region. These political complexities have to be accommodated if effective crisis planning and management are to take place.

A number of frameworks are available for effective crisis management. Key actions in crisis management programs tend to include strategic actions, technical and structural actions, evaluative and diagnostic actions, communication actions, and psychological and cultural actions (Pearson and Mitroff, 1993). The crisis management model highlights proactive rather than reactive responses to a crisis, including avoiding the crisis, preparing to manage the crisis, recognising the crisis, containing the crisis, resolving the crisis, and profiting from the crisis (Augustine, 1995). Other models include detection, preparation, damage containment, recovery and learning (Pearson et al., 1997). Drawing on these sources, we have created a condensed set of park management guidelines for crisis management. These guidelines fall into three categories: planning, actions and lessons.

18.4 Planning

18.4.1 Contingency Planning

One of the most effective ways to counteract a crisis is through contingency planning. In our model, planning includes both prevention and preparation for crises. Crisis management is brought about through strategic operational planning. It includes prioritising potential risks according to their likelihood and possible outcomes. The result then becomes the basis for contingency planning. The ICCN is responsible for contingency planning throughout Virunga National Park.

18.4.2 The Crisis Planning Team

A crisis planning team should include those able to take a strategic view as well as experts with hands-on operational experience. Effective crisis planning requires broad-based participation and support. Professionals

18.4
The rapid construction of a dry-stone wall by the communities around Mikeno helped to discourage land invasions. The value of the wall is that the park boundary is now clearly marked.

with an expertise in risk management and planning should be included within the team, along with park wardens and representatives from other disciplines or organisations with which the ICCN would have significant interactions during a crisis. Where plans involve the shutdown of some operations or even park stations, such as during the Kanyabayonga crisis, the repercussions need to be considered and prepared for as part of the planning process, with relevant personnel closely involved. Interdependent teams that come together in an emergency, but do not work together continuously over a period of time, do not tend to perform well in a crisis. They lack the coordination skills necessary to perform effectively and decision-making tends to become emotion-based, even under normal circumstances, exacerbating the crisis. Therefore training the crisis management team is an essential element of the crisis management process.

18.4.3 Threat Identification, Risk Assessment, and Planning Prioritisation

During this phase the ICCN should identify and assess threats so that it can prioritise its crisis planning efforts. The first task is to draw up a detailed inventory listing the threats faced by Virunga National Park. Their likelihood should then be assessed and agreed as well as their potential impact on park operations or the park's natural resources. These then need to be prioritised. Top priority for planning should then be given to the most critical areas.

18.4.4 Development and Review of Plans

The ICCN and partners should regularly (annually) review and update contingency plans, developing new plans for emerging threats. The events that will activate a plan also need to be identified and, where necessary, additional training carried out by the relevant partners.

18.4.5 Simulation Training of Plan Execution

Once the contingency plans are finalised, training will be necessary to ensure its smooth and successful implementation when a crisis occurs. All parties involved will need to be trained on all of the new procedures or practices. Where possible this should include walk-throughs, tabletop exercises and functional simulations. For the most critical areas, full rehearsals involving a combination of tabletop and functional exercises should be carried out to test the plans and the ICCN's ability to execute and monitor them. This testing process will also reveal any flawed assumptions in the plan.

18.4.6 Crisis Communication

Training and testing will go a long way to ensuring successful implementation of the plan, but robust communication is also key in a crisis situation. The timely notification of key personnel and emergency response teams through a tried and tested communications network is a vital element of crisis management. Where appropriate this should also include the dissemination of information to all staff and the local population. The ICCN should let those affected by the crisis know that the contingency plan has been activated and inform them if any immediate action is necessary.

18.4.7 Threat Monitoring

An important role for the ICCN management team is

18.5
The crisis following land invasions at Karuruma in the Northern Sector, by Hima cattle breeders from Uganda, was resolved peacefully with their departure in April 2006, following a series of negotiations with the ICCN. They left voluntarily after requesting and obtaining logistical support from the ICCN and its partners.

regularly surveying internal and external signals to detect early warning signs of any potential crises. In many cases advance warning may not be possible, and the implications of a threat may not be clear until well into a crisis. However the contingency plan should be implemented as soon as a critical event is identified. It is important to have a robust and reliable command and control system to ensure that information flows smoothly to the relevant authority for the speedy analysis and identification of any threat.

18.5 ACTIONS

The Chief Warden and other senior management staff have a vital role to play and must therefore be highly visible during a crisis. Even if the Chief Warden lacks the expertise to deal with a problem or crisis, he or she instills a sense of order and direction. Most importantly, the management will become the focal point for decisions and the coordination of effort. An interesting case in point occurred during the Kibumba crisis, where the Chief Park Warden, living in Rwanda, was isolated from the crisis on the ground by the unexpected closure of the international border between Rwanda and the Democratic Republic of Congo. The confusion between the government bodies, NGOs and the media, and the lack of direction as the crisis began to unfold caused by unsubstantiated reports and speculation from multiple sources, was exacerbated by the lack of leadership. The situation was fraught with the potential for the crisis to deteriorate further. The two principle lessons to be drawn from this crisis are the need for clear leadership and centralised communication.

18.5.1 LEADERSHIP

For successful crisis management there must be one identified individual in charge. If the Chief Warden cannot be present then a pre-identified named individual must be empowered to deputise on his behalf until either he or the crisis management team is in place. During a crisis, the leader can expect to be hit with a tidal wave of information (Stubbart, 1987) and must quickly reach decisions based on timely and accurate information. The crisis management team must support the leader in this process. With training, contingency plans and an established information network, the Chief Warden will have the necessary tools to assist him in his decision-making.

18.5.2 COMMUNICATIONS

A single spokesperson for the ICCN must be identified. During contingency planning communication channels need to be established between partner NGOs, embassies, the UN and other relevant international organisations. This should also include the media and local, national and international press agencies. If the crisis situation has the potential for armed conflict then a communications channel will also need to be established between the ICCN and MONUC, as well as the United Nations Office for Humanitarian Affairs (OCHA). For example, during the Kibumba Crisis the Director of OCHA proved an invaluable link with the Governor of the region who was in a position to give the necessary security assurances to the teams in the field. The park spokesperson should be responsible for all media liaison including the analysis and dissemination of open source material, in particular keeping ICCN staff and partner organisations informed of relevant media action and information.

18.6 Lessons

The learning stage is an opportunity to identify and evaluate what can be gained from the crisis (Pearson et al., 1997). What went right? How can it be reinforced in future planning? What went wrong? How can this be remedied? What planning assumptions were valid, and which ones failed? The learning stage is about profiting from the crisis. The assessment should be formalised, with park management and the crisis management team charged with documenting the findings and improving the crisis management process. A formal debriefing structure should be implemented to maximise productivity following a crisis event, and to avoid finger-pointing and assignations of blame.

Lessons learned from the Kibumba crisis demonstrate the importance of lobbying governments and key individuals, while the ability of the NGO and other partner organisations to react quickly in support of the ICCN to resolve the crisis proved critical. However, these were reactive measures that occurred in the absence of clear preparation or leadership. Crisis planning is about being proactive. It is also relatively inexpensive. Contingency plans can factor in assumptions concerning cost and the availability of financial resources. The ICCN's response to a crisis will, of course, be dependent on the means available to it to carry out certain mitigating actions. But this does not decrease the worth of good contingency planning. The actions described above apply to all crisis situations and cost little to achieve, as long as the planning has been completed before the crisis occurs. Through proper crisis management, therefore, the ICCN should be better prepared and better able to respond to crises as they occur, even during this volatile and uncertain period of national reconstruction.

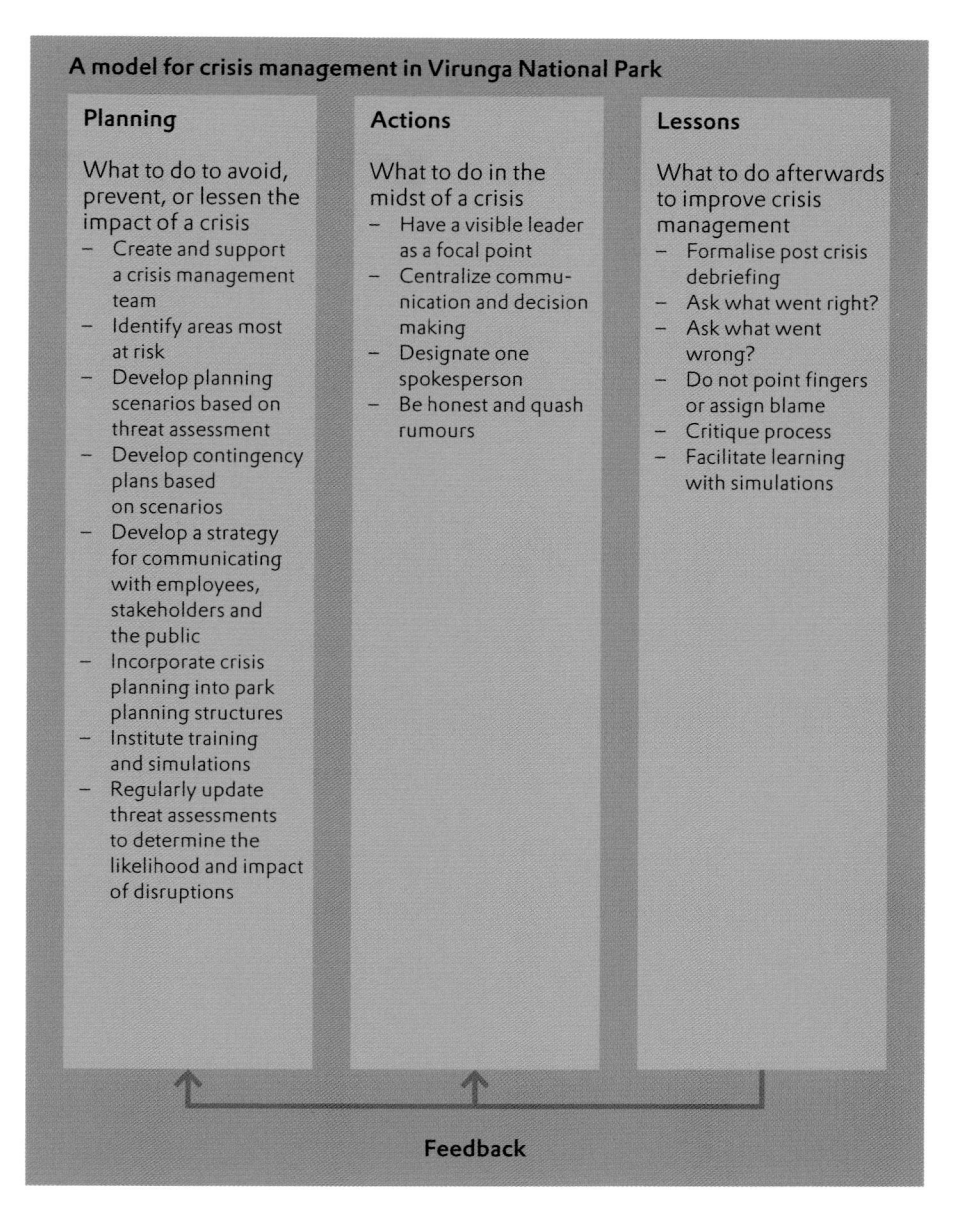

A model for crisis management in Virunga National Park

Planning	Actions	Lessons
What to do to avoid, prevent, or lessen the impact of a crisis	What to do in the midst of a crisis	What to do afterwards to improve crisis management
– Create and support a crisis management team – Identify areas most at risk – Develop planning scenarios based on threat assessment – Develop contingency plans based on scenarios – Develop a strategy for communicating with employees, stakeholders and the public – Incorporate crisis planning into park planning structures – Institute training and simulations – Regularly update threat assessments to determine the likelihood and impact of disruptions	– Have a visible leader as a focal point – Centralize communication and decision making – Designate one spokesperson – Be honest and quash rumours	– Formalise post crisis debriefing – Ask what went right? – Ask what went wrong? – Do not point fingers or assign blame – Critique process – Facilitate learning with simulations

Feedback

Jean Pierre d'Huart, Annette Lanjouw & Norbert Mushenzi

19 Recent evolution of the threats to Virunga National Park and a synthesis of lessons to be learned after 80 years

Jean Pierre d'Huart, Annette Lanjouw & Norbert Mushenzi

The preceding chapters have shown that the park's recent history is rich in events, both good and bad. They have also shown how the *Institut Congolais pour la Conservation de la Nature* (ICCN) and its partners have adapted their approaches, sometimes with great success and sometimes with limited results. This chapter brings together the different threats Virunga National Park must still confront, and examines the sum of lessons learned over the past 80 years. It also gives concrete leads and clear priorities for the park's regeneration.

19.1 The recent evolution of threats

19.1.1 Introduction

The park was created out of land acquired legally. By purchasing the rights after negotiations and inquiries about land vacancies and remuneration, by acquisition of abandoned land, and through the exchange of indigenous rights for compensation via socio-economic tools such as the *Coopérative des Pêches Indigènes du Lac Edouard* (COPILE, Indigenous Fishing Co-operative on Lake Edward).

In 1990, the state structure of the Democratic Republic of Congo (DRC) slowly began to collapse and the park authority was progressively discredited. Serious threats to the Virunga National Park began with interethnic conflicts in the Masisi area. Two military operations against the Maï-Maï, Mbata (slap) and Kimya (peace), shook the park. Not only were these battles fought inside the park, but many animals were slaughtered and infrastructure destroyed. In the Rwenzori Mountains it was Ugandan rebels who spread terror. In 1994, the park was hit by the arrival of swathes of Rwandan refugees who gathered in five camps around the Southern Sector of Virunga National Park, at Mugunga, Kibumba, Katale, Kako and Kayindo. Then came the rebellions, further aggravating the park's problems. In 1996 the 'liberation movement' under the *Alliance des Forces Démocratiques pour la Libération du Congo-Zaïre* (AFDL), and then in 1998 the 'rectification movement' under the *Rassemblement Congolais pour la Democratie* (RCD).

Now some community chiefs, citing traditional land rights, have begun staking a right to park territory. Their influence has led to large numbers of local people taking up the cause, and the issue has now become an ongoing source of conflict between park authorities and neighbouring populations.

Thanks to the actions of ICCN's partners, some of these threats are beginning to diminish.

19.1.2 The threats and their causes

The persistence of the political and socio-economic crisis in DRC since 1990 has damaged the Congolese population's social structure and resulted in severe poverty. The people living around the park have been hit particularly hard. The problems for Virunga National Park are further exacerbated by the jostling among various authorities for power and access to the wealth of natural resources while ICCN has become weak. The result has been a plummeting respect for the law. The combined impact of the ethnic wars, the arrival of the Rwandan refugees and the rebellions was a loss of control over the park by its managers

19.1
The capture of live animals for the pet trade is not as rife as poaching for meat, but the threat is real for species as sensitive as the chimpanzee. The family of this baby chimpanzee, found in Goma market in August 2006, was very probably slaughtered during its capture.

19.2
This young Mountain Gorilla was captured in a metal snare. He was freed, but the wires remain encrusted in the flesh of his left hand. Snares present a great threat to many species since any animal can be caught in one.

coupled with increasing pressure from surrounding residents.

The following are the threats faced today, whether they are direct or indirect:

Invasion of the park for agriculture, clearing and deforestation

Poverty in rural areas is a major preoccupation. Unemployment related to the closing of several mining and agricultural units, and the constant quest for land for production of raw materials such as coffee, papain (a digestive enzyme, product of the papaya tree) or wood, is pushing ever-increasing local communities into new fertile territory. The regional demography is such that arable land is insufficient and soil productivity is decreasing because of bad cultivation practices. Some local villagers believe the park, created out of land ceded by their ancestors, could be their last bastion of hope.

The illegal fishing villages and the increase in poaching

There has been a rush of people to the shores of Lake Edward due to the incorrectly-held belief that it holds unlimited fish stocks. Thousands of illegal fishermen have settled around the lake to practise artisanal fishing. They are encouraged by the shameful lack of surveillance by ICCN patrols. The situation is aggravated by a lack of strategic management of the official fisheries by tribal chiefs. Unsupervised military personnel and the presence of armed bands create further problems.

Enlargement of former villages and creation of new ones in the park

Internally displaced people and refugees are the main source of migration into the park and its surrounding area. Nevertheless, an ignorance of park laws, illegal land subdivisions by politicians, trafficking influence and the avarice of some village chiefs are the root cause of the population explosion in the villages around the park. For example Vitshumbi, Kyavinyonge, Nyakakoma and Tongo are all villages that have expanded massively. Lubilya (Kasindi frontier) is a new village, developing rapidly with intense commercial activity.

Violations of park boundaries

Besides encroachments due to demographic pressure and the need for resources, park boundary violations are also the result of bad feeling and provocation from tribal chiefs or politicians with various motives. Of equal importance is the presence of rebels and other armed bands and their lack of understanding or respect for the boundaries.

Charcoal making

A lack of forest management and reforestation strategies, and poor financial management of forestry beyond the park are significant threats to the park. This, coupled with demographic growth, means there is a critical lack of vital firewood for communities. The increasing urbanisation of local populations has particularly boosted the demand for charcoal. A recent study (see Chapter 14) estimated that 130 tons of *makala* (charcoal) are delivered to Goma each day.

Grazing

Some park land has been used as pasture for livestock. The introduction of cows into the park is related, not only to the personal interests of some politicians, but also to the poor management of pasture outside the park. In 2000, the Karuruma sector (Semliki Valley) was overrun by Hima cattle breeders from Uganda and more than 3,000 cows. In 2003, Kilolirwe, on Nyiragongo's western flank, was inhabited by thousands of displaced Congolese and their cattle from Rwanda.

Military camps

Military camps and posts started to appear within the park as a result of rebel activity. They pose a significant threat as the influx of undisciplined armed men has led to insecurity, deforestation and illegal hunting of large mammals. This has been the case in the Nyaleke and Mugunga sectors and in the Rwindi Plain.

Presence of outlaws

Bands of outlaws have lived in the park since the beginning of the interethnic wars in the Masisi area and the start of rebel activity. The only available source of protein for these armed groups is hunted game, while trade in bush meat and charcoal is their main source of income. It is too dangerous for the poorly armed rangers to patrol the areas they occupy.

19.1.3   The dynamics of the threats

Since 1990 the threats to the park have multiplied. Political crises have rocked the whole country and the weakness of North Kivu state authorities has also had a very negative impact on the management of Virunga National Park.

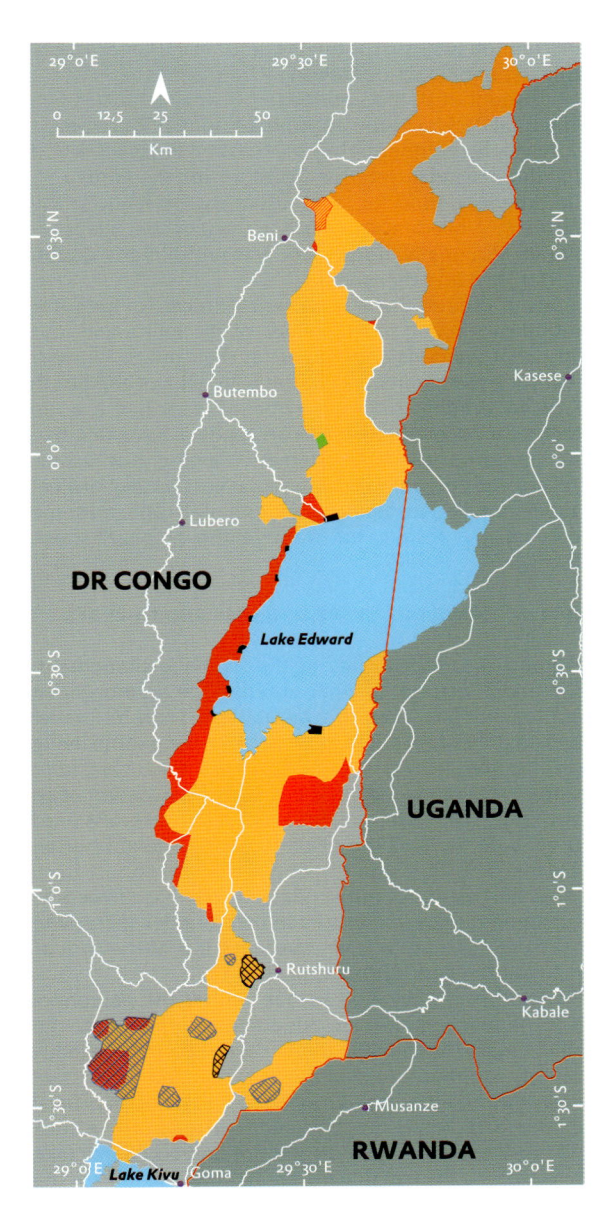

19.3
Map of the threats to
Virunga National Park in 2001,
as understood by the rangers.

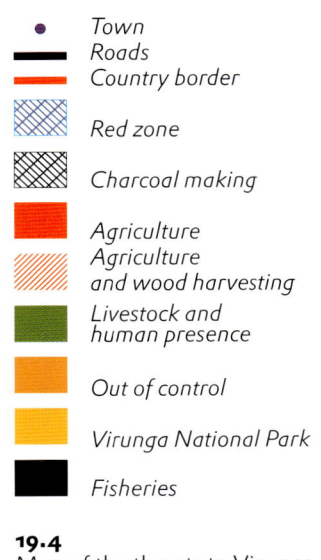

- • Town
- — Roads
- — Country border
- ▨ Red zone
- ▩ Charcoal making
- ■ Agriculture
- ▨ Agriculture and wood harvesting
- ■ Livestock and human presence
- ■ Out of control
- ■ Virunga National Park
- ■ Fisheries

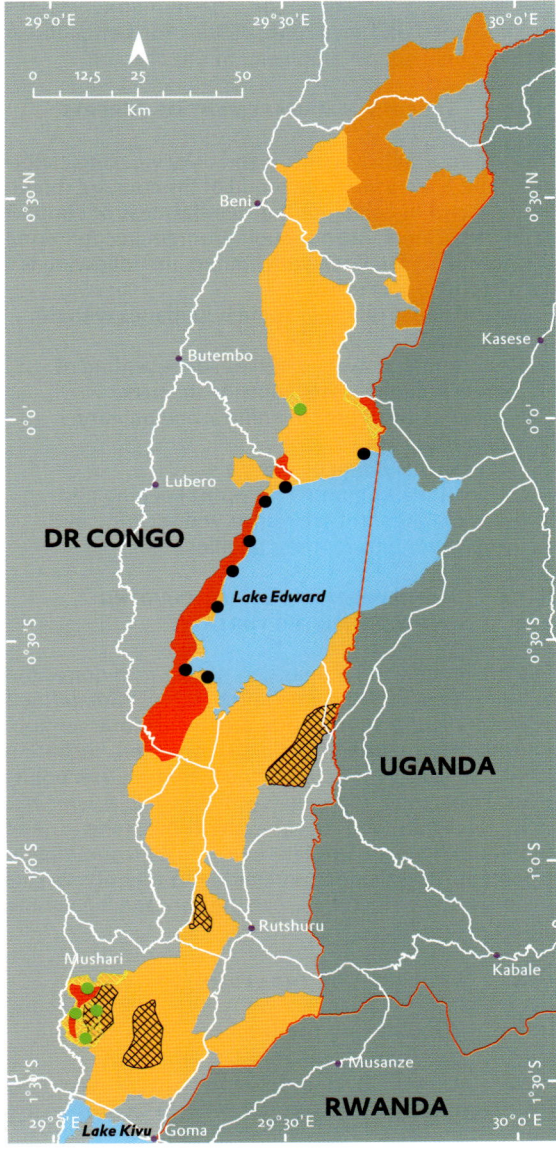

19.4
Map of the threats to Virunga
National Park in 2006,
as understood by the rangers.

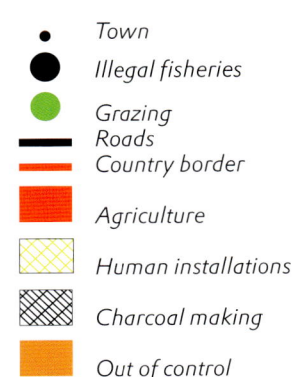

- • Town
- ● Illegal fisheries
- ● Grazing
- — Roads
- — Country border
- ■ Agriculture
- ▨ Human installations
- ▩ Charcoal making
- ■ Out of control
- ■ Virunga National Park

1990–1995: Military operations carried out against the Maï-Maï and the Hunde militia groups at Masisi and Kanyabayonga resulted in violent skirmishes in the park. After the military withdrew, the Maï-Maï settled in Rwindi and at the entrance to Vitshumbi. The only source of food was meat and fish from the park. Trade in elephant and hippopotamus ivory was also organised. There was wide-scale looting, decimating park infrastructure.

1994: The Rwandan civil war brought around one million refugees into the area in and around Virunga National Park. They were eventually gathered in several camps on the border of the park's Southern Sector, in violation of international norms. The concentration of such a large number of desperate people resulted in enormous deforestation, poaching and destruction of wooded areas.

1996: The so-called 'liberation movement' by the AFDL triggered the displacement of internal populations. The resultant installation of military camps and posts in several areas of the park was linked to the theft of equipment and vehicles, the proliferation of firearms, the disarming of the rangers and the weakening of ICCN. For a long time patrols had become impossible but the situation only worsened as military leaders posed as tribal chiefs and took complete control.

1998: The so-called 'rectification movement' begun by the RCD, aggravated the situation further. In the area around Beni alone, around 15 per cent of the park's area was seized for agriculture. The numbers of hippopotamus and elephants continued to decline (Chapter 8).

19.1.4 PRIORITY LESSONS

Bowed beneath the weight of these many threats, Virunga National Park is in a very precarious position. An energetic revival is needed, and quickly. The DRC Government and the international community must take responsibility for saving this World Heritage Site. Park managers are ineffective against the many different groups who cannot or will not observe the laws protecting the park. The weaknesses and failures of

Recent evolution of the threats to Virunga National Park and a synthesis of lessons to be learned after 80 years

247

park management since 1990, the numerous lessons to be learned and recommendations for better threat management appear at the end of this chapter.

19.1.5 Success in management of threats

Despite enormously difficult and risky working conditions, there is reason to be proud of the courage shown by ICCN and to celebrate the steadfastness of support from its partners. Even though progress in countering threats has, so far, been very limited, the following is a list of definite successes:
- hysical demarcation of some boundaries with the support of local people,
- Restoration of invaded spaces and resettling of people in suitable areas outside the park,
- Raising of awareness in political and military authorities,
- Active involvement of tribunals,
- International community support keeping hundreds of rangers at their posts and securing basic living conditions for their families.

19.1.6 The importance of anticipating and monitoring threats

The prevention and monitoring of threats requires the establishment of a management protocol with defined indicators and a strategy for preventive intervention. All interested parties should be consulted, including tribal chiefs, local people, partners, military and politicians. Long-term activities (such as public awareness, communication and development programs) need additional finance. The law needs to be competently enforced. A permanent committee to identify threats and monitor conflicts between the park and the people must be formed.

19.2 Synthesis of the lessons

19.2.1 Analysis of successes

It was a groundbreaking decision that was taken 80 years ago to form a national park to conserve the biodiversity of the Kivu region. Albert National Park was created with the objectives of protecting fauna and flora, allowing scientific research and bringing in tourism compatible with conservation. Human influence on the park was to be controlled, and there was only marginal consideration for the needs of a growing human population outside. Human activities in the park were restricted to those related to protection and research. This policy made the park both a museum and a living laboratory, and the scientific discoveries were numerous. The successful management of that time provided positive lessons that were followed for some decades afterwards. Through development and management of tourism, the natural riches of DRC became widely known across the world and the park attracted thousands of tourists. The fishing in Lake Edward provided employment, important protein for local people and helped local communities understand the value of conservation. The success of the park was a determining factor in the success of the region. More recently, co-ordination with neighbouring protected areas, trans-frontier management and recognition of the importance of shared resources have shown, albeit in a limited way, that Virunga National Park plays a key role in regional co-operation.

Throughout the periods of conflict, Virunga National Park's resources proved essential for people's survival, even though they were taken illegally; wood for the Rwandan refugees, pasture for the Hima cattle breeders, food for local residents and land for the farmers. While far from perfect, the level of collaboration in Virunga National Park was without any doubt, better than on other sites.

Even though the war weakened ICCN, its long tradition of training in the values of discipline and loyalty kept wardens and rangers at their posts. As most wardens were trained at Garoua, Mweka or the South African Wildlife College, it meant ICCN had a good technical base to rely on. The rangers have not had the benefit of regular, systematic training, but technical sessions organized through a UNESCO/UNF project and paramilitary training sessions have produced some very positive results (for example, the 'elephant shock team').

19.2.2 Analysis of weaknesses and failures

Among the deficiencies in the park's management, the most glaring is undoubtedly the authoritarian approach it has taken to conservation. ICCN has never had an official mandate to work with the people living on the periphery of the park. Therefore, apart from some NGO-managed projects, there has never been any large-scale attempt to raise public awareness or to contribute towards the development of neighbouring people. The knock on effect is a lack of support from local and national authorities, which has grown as surrounding populations have come under increasing pressure. This point is vital because it is at the root of many of the park's present problems.

Another observation is that neither the personnel nor the management are adequately prepared for the challenges from outside the park (environmental education, community development, reaction to massive invasions and looting). On the strict basis of its official mandate, ICCN has concentrated its efforts on the interior of the park only. ICCN did not and could not take into account the fact that the greater the struggle for the people living outside the park, the greater the threat to the park. The park represents a means of survival for people. In addition, some local authorities have used the park's rich resources as a tool in their power struggle. ICCN can no longer afford to ignore this dimension of the threat to the park.

Apart from the fish from Lake Edward, most people in the area have never really had any tangible benefit or significant return from Virunga National Park. Employment and revenue from tourism and visitors' permits have never formed a significant element of the local economy, despite the fact that the park has the highest potential returns in the whole country. Local people, therefore, see little advantage of the parks existence except as a source of natural resources. If a better system for sharing the benefits were created, with greater overall involvement of local people, the general attitude towards the park would be more positive, as demonstrated elsewhere in Africa.

A policy of absolute protection is the most natural approach possible, but it operates best over very large areas. At Virunga National Park, this policy has brought about some unpredicted effects. For example forest has returned to the savannah areas. In addition, natural processes have been affected by the isolationist approach. For example due to the loss of connecting corridors in the west, as a result of population explosion and an increase in agricultural land, some ecological processes such as genetic exchanges of forest species of the Semliki Valley have evolved in isolation. The park's connection with contiguous areas in Uganda and Rwanda fortunately prevented an evolution in complete isolation.

The legislation and rules, developed long ago with a protectionist policing perspective, are no longer valid today. ICCN needs an updated role and legal responsibilities if it is to adapt to present day challenges such as better managing buffer zones, controlling the trade in bush meat outside the park, acknowledging the authority of judicial police officers and so on. Updated legislation is absolutely necessary to bring about the institutional improvements needed.

Despite the emphasis on ICCN's important scientific role, there is no central research policy and no effort to capitalize on the results of past research (for example, a policy on burning grasslands based on the results of zoological and botanical research is oriented towards a local optimization of the large mammal fauna).

After 80 years, there is still no management plan for park improvement, although this is a condition imposed on all World Heritage Sites.

19.2.3 INSTITUTIONAL ASPECTS AND DECENTRALIZATION

Many lessons can be learned from the strong, centralized decision-making in place up to independence in 1960. This permitted rapid but authoritarian development. The IPNCB *(Institut des Parcs Nationaux du Congo Belge)* benefited from strong political support and royal endorsement. Staff members were mostly Belgian, with only a few Congolese wardens (including the late Anicet Mburanumwe) trained before independence. Virunga National Park would benefit from a more efficient management now if the Institute had invested more in the local staff then and if the Belgians, and ICCN after them, had accorded more importance to training.

ICCN was created at independence as the INPN, but restructuring from the Headquarters in Kinshasa took time. In the meantime the poorly prepared park suffered in the 1960s rebellion. After a period of relative stagnation between 1960 and 1970, resulting from very limited external support and the weak capacity of ICCN, the then President Mobutu called upon the Belgian Co-operation and the European Union, resulting in significant support (See Chapter 3). The management structure of Virunga National Park was firmly re-established, the paramilitary structure was improved, and a large number of patrol posts were created to provide control points for the rangers. The development of tourism and scientific research made considerable progress. New wardens were sent to the *Ecole de Faune,* in Garoua, for training.

Between 1985 and 1995, Virunga National Park entered a progressively difficult period due to the socio-economic problems of the country. There was a drastic reduction in funding and salary payments ended. Gorilla tourism became the only substantial source of revenue. After an institutional review requested by the World Bank in 1991, ICCN was decentralised with two Regional Headquarters, one in North Kivu and one in South Kivu. The Provincial Headquarters have an important role to play, but their responsibilities need significant clarification as the current situation can be frustrating for wardens.

During the 1998–2003 conflict, the park was split into two down political lines after falling into the hands of the fractured rebel movements: RCD-Goma and RCD-ML, with each taking over a section of park management (Chapter 4). In spite of the support of NGOs and UNESCO, this troubled period witnessed the net weakening of the park; numerous political and military interventions in park management; appointment of staff without technical reference; loss of control of the Headquarters in Kinshasa; divisions among the staff. The rangers, disarmed and insecure, pursued their work as far as they could, given their limited means. A near total absence of transparency in decision-making and revenues at ICCN Co-ordination in Goma made it much more difficult to target support from partners.

Despite the courage and loyalty of staff during these crisis years, the general disorganisation and lack of funding led some rangers and other staff to start making money from the park's natural resources. The anarchy in the region, favoured by some politicians and military personnel, served to encourage poaching and trade in bush meat and charcoal. The disorganisation of ICCN and lack of political support led to a culture of impunity that still prevails today. Even some ICCN leaders have abused their power and access to trade the park's natural resources for political support or money. The continued confidence of staff in ICCN rests in part on the decisions that will be taken concerning these embezzlements.

19.5
Alongside agricultural invasions and charcoal production, poaching is a serious threat to Virunga National Park. Poaching is mainly carried out by men in uniform, either rebels or the regular army, as shown here when the FARDC killed five elephants in the Rutshuru Hunting Domain on May 13, 2005.

19.6
This young okapi was confiscated by military from smugglers coming from the forest in Ituri and returned to ICCN in Goma in June 2006. Okapis are unfortunately an extremely sensitive species and this individual died a few days later.

19.2.4 Role of external partners

Between 1960 and 1970, the park had very little support from external partners. The only aid came from the Food and Agriculture Organisation of the UN (FAO), paying for infrastructure development. Then, significant support from the Belgian Co-operation came in two tranches; the first from 1971 to 1975 and the second from 1978 to 1982. This allowed the establishment of new infrastructure and management plans. A great many objectives were achieved, but there were still only a very limited number of Congolese staff involved in responsible positions. Other support made possible other improvements to equipment and infrastructure, including funding from the *Fondation pour Favoriser les Recherches Scientifiques en Afrique* (airplane, photographic material, reference works), the Frankfurt Zoological Society (FZS) (laboratory at Lulimbi), and the World Wide Fund for Nature (WWF) (vehicles, ferry on the Rutshuru River).

The 1980s heralded some important projects; these included tourism and conservation around the mountain gorillas in the Mikeno sub-Sector (International Gorilla Conservation Programme, IGCP), chimpanzees at Tongo (FZS), *Kacheche* environmental education (WWF), and Virunga Programme of the European Union. A more integrated relationship developed between ICCN and its partners, thanks to more clearly defined co-operation contracts. While support for the development of the park was considerable, support for the functioning of the institution and for training was comparatively small. Unfortunately, the disintegration of the country at the end of this period quickly reduced the effectiveness of ICCN.

From the beginning of the rebellions in 1996, aid from partners became critical. In the absence of any support from central government, this aid was the only means of survival that ICCN had left. For four years, these partners provided supplements to staff and covered operating expenses. Working together, in 2000, they obtained a UNESCO/United Nations Foundation project, becoming the only source of funding for staff and their families. The war, therefore, brought ICCN's partners together where previously they had generally worked separately. ICCN established the *Comités de Coordination du Site* (CoCoSi) as a platform well adapted to the shifting circumstances of this period (Chapter 20). However, harmony between its members was far from perfect achieved; the approach of some partners lacked transparency, weakened the cohesion and fuelled tension between the Co-ordination of ICCN in Goma and the Headquarters in Kinshasa. This period demonstrated once again the pressing need to strengthen and unify ICCN to take the leading role. It also highlights the need to support the donors in the development of a comprehensive strategic plan and widely distributed management plans.

19.2.5 Necessary manpower and protection strategy for Virunga National Park

During the period that preceded independence, the rangers had lances instead of firearms. Respect for the park was down to the collaboration of tribal chiefs, and their allegiance to the King and to the various benefits that the administration was able to give them. Pressure from surrounding populations was weak, the animal biomass high, and police authority strong.

After independence, ICCN was able to maintain the park adequately for it to be made a World Heritage Site and that brought about strong tourist development. But in the 1980s the local population had expanded to such a degree, and deforestation and poaching increased, that the park became increasingly difficult to manage. Nevertheless, the boundaries were generally respected and ICCN managed to keep a significant number of rangers at dozens of patrol posts.

The subsequent decrease in funding and the ever-increasing pressure on the land and resources from neighbouring communities (accentuated by the country's political divisions) created further pressures, but ICCN did not evolve with the changing situation. War and insecurity aggravated the problems, and the disarmed rangers gathered in the stations rather than take the risk of patrolling the park. The security situation prevented the rearming of the rangers and the military imposed their authority on the park by force. Though they were confronted by many pressures, the staff had no training in any other protection strategies. At the end of the period of conflict training in law enforcement was conducted at Ishango and a community conservation strategy was developed with ICCN in 2002.

The staff of Virunga National Park was unable to manage the political division of the park, the migration of a large number of people, the increasing settlements around the fisheries and the imbalanced support for the park. For example, the Southern Sector, which harbours the Mountain Gorillas, received much more support and backing than the Central or Northern Sectors. The rangers of these two latter sectors, demoralized and disarmed, could not prevent people from settling in the park. The park became more and more fragmented, and the connecting corridors, necessary for the survival of large mammals and the dynamics of the ecosystems, were greatly threatened. It is thought that some species survived by migrating into neighbouring protected areas in Uganda and Rwanda.

It is urgent to 'reunify' the park with the support of the authorities and the backing of the neighbouring protected areas, in preparation for the development of solid strategic plans.

19.2.6 Suggestions for integrating the lessons learned into an optimal mode of management for Virunga National Park

It is clear ICCN needs restructuring to include a strong Provincial Headquarters with a clear-cut role. A review, begun in June 2006, must also determine the optimal manpower (quality, quantity, deployment) and a serious and coherent strategy for the management of human resources. In addition there needs to be a new management plan, laying out renewed methods of protection and monitoring in the park, and thematic plans for now crucial areas such as community conservation, environmental education and management of tourism. These plans should make it clear which issues most need partner support. More widely, the legislation behind ICCN must be reviewed and brought up to date. ICCN needs a legal mandate legitimising work with local communities.

Management priorities need to be established if the long and increasing list of threats is to be dealt with. They include:
- A major project for restructuring and optimising the control and management of the fisheries,
- A clear definition of ICCN's responsibility in the development and management of tourism, in the transparent management of the revenues generated, and in the transparent redistribution of the benefits,
- Improving trans-frontier co-operation according to an established plan,
- Co-ordination with regional development actors to identify areas where they can work together. In particular, the park should expect a specific pressure on its wood resources. Working with exterior partners on mass wood production must be seen as a useful investment in terms of conservation of the park and as a contribution to the well-being of local people;
- Bringing together scientific research, the reconstitution of a reference library, the identification of future research priorities and publicity around the need for research and scientific co-operation;
- A clear-cut strategy identifying the best methods of park protection, relative to the deployment of its stations and patrol posts.

Guy Debonnet, Sivha Mbake & Véronique Tshimbalanga

20 Achieving better co-ordinated international support for the protection of a World Heritage Site

Guy Debonnet, Sivha Mbake & Véronique Tshimbalanga

Virunga National Park is not only a national treasure but a site of global importance.

Successive attacks on the park, first following the deterioration of the socio-political situation in the Democratic Republic of Congo (DRC) and later during the years of conflict, have been well documented in preceding chapters. Though the park sits in the middle of a region that was plunged into an unprecedented humanitarian crisis, its supporters did not give up. They persisted and persevered and established systems to maintain protection of the park. In the process, a unique partnership was formed between the *Institut Congolais pour la Conservation de la Nature* (ICCN) and its international partners. The aim of this chapter is to explain the origin of this partnership, to analyse its strengths and weaknesses, and to put forward ways in which collaboration can be improved.

20.1 Virunga National Park: a Congolese heritage, a global heritage

Virunga National Park is first and foremost Congolese. However, in 1979, its importance and value to the world at large was formally recognised.

The process began in 1972 when the 17th Session of the General Conference of UNESCO adopted the Convention concerning the Protection of the World Cultural and Natural Heritage. The Convention stated that some natural or cultural sites are of such exceptional value, their preservation is in the interests of all of mankind. The Convention states that, in view of the scale and severity of the new dangers that threaten our world heritage, it is the responsibility of the entire international community to participate in the protection of cultural and natural heritage sites that have this exceptional universal value. Thus, not only does the Convention recognise the global importance of some natural and cultural sites, but it also affirms that their protection is a duty shared by all the states that join the Convention.

The Democratic Republic of Congo (DRC) was among the first states to ratify the World Heritage Convention in 1974. The Convention led to the establishment of a list of cultural and natural sites: the World Heritage List. In 1978 the first sites were inscribed on the list. Virunga National Park was added the following year, in 1979. It was the first World Heritage Site in DRC.

The global value of the park explains, in large part, the support it has received since the 1980s from non-governmental organisations (NGOs) with a particular interest in conservation, such as the World Wide Fund for Nature (WWF) or the Frankfurt Zoological Society (FZS) and numerous others have followed. Compared to the considerable budgets available from bilateral donors, the sums mobilised by conservation NGOs are not generally as substantial, but their support is more stable and usually more flexible. In addition, compared to cooperation projects which often run too briefly to lead to sustainable protection activities, the conservation NGOs engage in the long term and they tend to continue throughout periods of difficulty. This latter point has been key in the protection of Virunga National Park over the last ten years. Perversely, the essential nature of their support can be seen from looking at what happened on an occasion when they did pull out. During the early 1990s the park lost all forms of funding: State subsidies disappeared (see Chapters 3 and 4), then the donors pulled the plug. Shortly afterwards, tourist revenues also plummeted. This triple blow left the ICCN and its partners with the gigantic task of protecting this World Heritage Site singlehandedly, without support.

20.1
The Rwenzori Mountains are a world heritage shared between DRC and Uganda. They fall into DRC's Virunga National Park and Uganda's Rwenzori Mountains National Park.

Generally speaking, the partners' interventions have had the effect of boosting national efforts to conserve the natural resources of Virunga National Park. However, for political reasons, this intervention can be limited. There are some areas in which they cannot get involved and still safeguard their neutrality, for example, the arming of rangers. Partners have found it extremely difficult to convince donors to provide support for urgent interventions within a war context.

20.2 Virunga National Park, a world heritage in danger

We will not return to the crisis of the early 1990s described in Chapter 4, nor to the negative impact of it on the people living in the Virunga National Park area. However, the cumulative effect of these events was that in October 1993 UNESCO sent an evaluation mission to the park which agreed that it should be put on the World Heritage Sites' Danger List. This was formally carried out at the 18th Session of the General Conference of UNESCO in 1994.

Despite this, the dangers multiplied in 1996 when the conflict of the Great Lakes sparked a war that spread to the entire country, fed by internal conflicts and external interventions. The Government of Kinshasa quickly lost control of two thirds of the country, which were taken over and 'administered' by different rebel groups supported by Rwanda and Uganda. Based in Kinshasa, the *Institut Congolais pour la Conservation de la Nature* found itself unable to do anything about the loss of biodiversity in the protected areas in the east of the country.

20.3 A concerted international effort to save a world heritage in danger

In the aftermath of this disaster, the UNESCO World Heritage Centre organised in 1999 a meeting on neutral territory in Naivasha in Kenya. Attendees included the leadership of the ICCN from Kinshasa, the senior wardens of the four World Heritage Sites in the areas controlled by rebel groups (Virunga, Kahuzi-Biega and Garamba National Parks and Okapi Faunal Reserve), the conservation NGOs and the German Cooperation, which had continued to support these four sites.

The meeting drew a sombre conclusion. In most cases the rangers had been disarmed. The militias and rebel groups operated with impunity, exploiting the natural resources and preventing ICCN personnel from doing their work. The ruinous impact of the war on local people meant that they had had to turn to the resources of these sites for their survival. Numerous groups of displaced people had settled inside or around the edges of the protected areas in an effort to find refuge away from the hostilities. The now unsalaried park personnel were entirely dependant on donations from partners.

The result of the Naivasha meeting was a decision to alert the international community and, notably, to use the World Heritage Convention to make the states engaged in conflict in DRC recognise their responsibilities in virtue of article 6, paragraph 3 of the Convention,[1] and to mobilise international support for the protection of the sites, in conformity with article 6, paragraph 1.2.[2]

The Naivasha meeting also led to the successful formulation of a plan for urgent intervention to preserve the integrity of the five World Heritage Sites. The plan advocated a two-tiered approach, providing direct support to the ICCN teams in the field at the same time as initiating 'conservation diplomacy' to make all parties involved in the conflict aware of the importance of preserving the World Heritage Sites.

The Chief Executive Director of the ICCN, an ICCN senior warden and a representative of the partners formed a team empowered to present this plan at the next meeting of the Bureau of the World Heritage Committee. The conservation NGOs were key to the implementation of the plan, given that they had the structural ability to act within very difficult conditions.

As a direct result of the meeting the United Nations Foundation (UNF) – a private foundation managing the donation to the UN of one billion US dollars by Ted Turner – expressed a strong interest in financing this audacious and risky program. A second meeting was then organised in Nairobi, to decide how the program should be carried out.

It was effective collaboration that made this operation so successful. The program was developed by the authorities of the ICCN; the wardens who were confronted on a daily basis with the difficulties of site protection in a conflict; the conservation NGOs armed with long experience of working in DRC and with long-term commitments to these sites; and UNESCO, supported by the political and moral authority conferred on it by the Convention.

However, it was not easy. The ICCN was not spared the impact of the general collapse of DRC's state structures. Therefore, progress was hampered by poor communication between ICCN Headquarters and the wardens, as well as the NGOs in the field. The weakened ICCN was unable to play its coordination role. Also, the NGOs worked in relative isolation from each other, even within the very same site. The creation of good coordination for this initiative financed by the UN Foundation was, therefore, crucial.

20.4 The different levels of coordination

Three organisational levels were established. They played a critical role in the management of this programme.
- At the international level, the World Heritage Centre of UNESCO was the UN body responsible for supervision and general coordination of the project, as well as mobilisation of the international community of the Convention for 'conservation diplomacy' activities.
- At the intermediate level, a Core Group was made up of the principle stakeholders. Its secretariat, based outside DRC, was assigned the task of ensuring communication on the state of conservation of the sites and coordination of planning activities.
- At the site level, a *Comité de Co-ordination de Site* (CoCoSi) was tasked with preparing work plans for activities executed at each site and ensuring coordination between the ICCN and all the partners on the ground.

Within the framework of the project, the CoCoSi was seen as an ideal platform from which to concentrate on the particular needs of each site, to co-ordinate the activities in the field and to make strategic conservation decisions. It is at the level of the CoCoSi that specific work plans for the site were established. These plans comprised of objectives, activities, results, budgets and the commitment of the partners.

20.5 ADVANCES DUE TO THE PRINCIPLE OF THE CoCoSi

Besides its original coordination function, over time the CoCoSi gained additional powers, some of which led to remarkably successful cooperation between the ICCN and its partners in the resolution of specific problems.

An example is the Common Pool. The park's senior staffing costs were not covered by UNESCO finance. At the time the UNF-UNESCO programme was launched, the International Gorilla Conservation Programme (IGCP) and the Dian Fossey Gorilla Fund-Europe (DFGF-Europe) were the only ones paying the rangers and other staff, and then only in their respective sectors of intervention within the park. Some sectors had no partners affiliated to them. It was recognised that guaranteed and regular financial support was needed for all staff. As a result, four partners (IGCP, the Wildlife Conservation Society –WCS–, WWF and ZSL) who were all working on projects within the park grouped together to create the Common Pool, a fund fed by their own contributions. Initially intended to cover costs of senior ICCN staff, this fund quickly grew to become a means of organising and funding operating costs for each station. After two years, the four partners were able to use the success of the Pool to support a proposal to the MacArthur Foundation for greater financial support. The Foundation agreed to cover a proportion of the operating costs of the park. This example shows how a system adopted through collaboration between partners acted as a catalyst for obtaining significant support of a new donor for the whole site.

The development of a logical framework to plan interventions of the program, financed by the European Union, sprang from the same principle. Several partners, including some who, like the WCS, did not have management responsibility for these funds, were consulted and conferred responsibilities for the implementation of the program.

These successes are, without a doubt, due to the successful way the CoCoSi brings together conservators and key partners of the ICCN at Virunga National Park. The CoCoSi constitutes the only formal forum in the park for securing supplementary funds for strategic use. This may be because the CoCoSi lends 'validation' to proposed programs, giving them a heightened credibility in the eyes of potential donors.

20.6 ASSETS ACQUIRED BY THE UNF-UNESCO PROGRAM FOR THE PROTECTION OF THE WORLD HERITAGE SITES

The UNF-UNESCO program produced tangible results for Virunga National Park, not only in conservation terms, but also in strengthening the ICCN's partnership with conservation NGOs.

Thanks to this program, the morale and motivation of the rangers was boosted. They were encouraged to keep going throughout four years of conflict, from 2001 to 2005, thanks to regular payments, rations and medicines. Certainly, the ICCN's ability to wipe out illegality in Virunga National Park was limited by the lack of state funding and the lack of cooperation from armed groups who enjoyed strong local political support. However, the continued presence of the ICCN in the field sent out a very strong signal to local people and local authorities; the park had not been abandoned and the ICCN was doing its utmost to maintain respect for the law.

High-level interventions from the UNESCO Directorate, the DRC Presidency and many others also reduced the number of park invasions and the abuses of its wildlife by military and some local authorities.

'Conservation diplomacy' also had some impact. UNESCO sent several missions into the field to make contact with the parties engaged in conflict and, in particular, the military and political authorities on the ground. The missions' tasks were to mobilise support for the park's conservation and resolve pressing problems threatening the park's integrity. The success of these activities depended on the long-term follow-up by conservation NGOs and the ICCN. One example of a 'mission successful' was the evacuation of most of the people who settled illegally in the park. They were largely local people who had seen an opportunity to profit from the anarchy by enlarging their farmland with portions of the park, but there were also displaced people from Masisi, and refugees from Rwanda and Uganda. This problem is still not completely resolved, but the difference now is that the authorities have the will to address it and to restore the integrity of the park. Joint projects to remark the park's boundaries, involving local authorities and representatives have resulted in further evacuations of the land. Overall, it is possible this program, executed by the WWF and financed by UNESCO and other donors, could lead to both a restoration of the integrity of the park and the clarification and modernisation of its boundaries; such improvements could help to reduce conflicts with neighbouring populations (see Chapters 24 and 25).

Renewed cooperation between the ICCN and its partners, and between the partners themselves, is another important benefit. The knock-on achievements of the CoCoSi, discussed above, are only part of the success story. Another bonus has been the increased interest of donors who have regained confidence in the work of

20.2
In September 2004, an important UNESCO meeting took place in Paris to discuss four years of collaboration to save the five world heritage sites in danger in the DRC. This was the occasion for the Congolese Government to reaffirm its commitment to re-establish the security of these sites, to give the ICCN the necessary logistical and financial means and to give firm instructions to stop all poaching by the armed forces.

20.3
Leopards, though rare, still inhabit all sectors of the park. They suffer both from direct persecution and the depletion of their prey.

the ICCN and its partners. The extra finance this brings to the park is fundamental to its reconstruction. After the push to protection in crisis, the push to reconstruction in times of peace requires a large financial effort which cannot be sourced, initially at least, from central government or tourist revenue.

One last way in which the CoCoSi could still improve its impact would be to support fundraising for activities which are unfunded at present but nonetheless very important. The CoCoSi should provide help in writing proposals to be submitted at local and international levels, an activity that requires time, energy and expertise.

20.7 Towards consolidation and a revaluation of assets: Obstacles to surmount

20.7.1 The continuation of the coordination developed under the UNESCO program

The CoCoSi is now well anchored into the work of the ICCN and partners. There are, on average, three or four CoCoSi meetings each year, held in Goma or Beni. All partners participate in the meetings and make important decisions that inform their work.

The Core Group, the national coordination platform, was enlarged in 2003. It was the launch pad for the formation of the *Coalition pour la Conservation au Congo* (CoCoCongo). The CoCoCongo brings together the ICCN (Headquarters and heads of its individual sites), the ICCN's technical partners, main donors and representatives of the ministries and other institutions that have more or less direct ties with the work of the ICCN. This platform is presided over by the ICCN and animated by various thematic working committees. The Core Group meets annually in Kinshasa, for year-end reviews, decision-making on broad conservation strategies for the DRC, validation of programs, and a vital exchange of information and experience between members.

20.7.2 Improvements needed

Despite these successes, improvements in coordination systems are still necessary.

Overall planning and coherence remains far from perfect. The development of joint plans is not very effective due to a lack of team working and limited planning ability. Even though most of the conservation NGOs working in Virunga National Park work together quite well (and a net improvement has been observed, even extending to the elaboration and execution of joint proposals), some partners still only participate in the CoCoSi in theory. Poor communication means the work of various bodies remains insufficiently co-ordinated.

There is also a lack of transparency, for a variety of reasons. The partners have limited budgets though presently they have to finance most of the park's activities. Budgetary planning can create confusion as the funds available do not always match budgetary needs. The partners themselves are sometimes unwilling to share information with each other. Finally, some partners do not respect commitments made within the CoCoSi with respect to other partners and the ICCN.

The ICCN's own lack of financial transparency is a particular problem. It is almost impossible for partners to obtain information from the ICCN on the revenue it receives from park activities, such as tourism, toll fees, fines, rents from different buildings and so on. This creates a sense of double standards: many partners believe that in return for transparency on the funds they are investing in the park, there should be a corresponding level of openness from the ICCN.

The ICCN has yet to produce a strategy document listing its priorities, due to a lack of technical ability among other things. The risk is that various partners set their own priorities, which are not necessarily the right ones for the park overall, creating confusion and frustration for all concerned and an overall disrespect for any system of prioritisation.

Finally, significant progress has been made on common fundraising initiatives. Examples of note include the Common Pool, the European Union program, the WCS-WWF program on the resolution of conflicts, the CARPE program (includes the WWF, WCS, SNV, African Wildlife Foundation (AWF) and the Africa Conservation Fund (ACF) and the new phase of the UNESCO program. However, a strategic plan, a restructuring plan or a business plan that could serve as a basis for the mobilisation of resources have yet to be finalised. The development of these plans should be considered a priority for the next three to five years.

20.8 CONCLUSION AND PERSPECTIVES FOR THE FUTURE

Thanks to the World Heritage Site program in DRC and to the other conservation activities conducted by NGO partners of the ICCN, thanks also to the courage and the devotion of the ICCN personnel in the field and that of their partners, irreversible degradation has been avoided in DRC's World Heritage Sites and in particular in Virunga National Park. Improved coordination and cooperation between the different parties and the proactive use of the World Heritage Convention were key to this success.

Nevertheless, the battle is far from won. Virunga National Park remains one of the most threatened of the World Heritage Sites. Political instability and insecurity still threaten the lasting protection of the park. However, it is important to emphasise that the conditions now appear to be in place for a change from damage limitation to restoration of the park. There are new donors supporting the park and there is the political will to resolve some of the problems.

It is also encouraging to observe that the ICCN has capitalised on the experience acquired in Virunga National Park and in other World Heritage Sites. The CoCoSi is now the coordination mechanism for all protected areas managed by the ICCN. In addition the ICCN, inspired by the Core Group, has established the CoCoCongo to ensure better coordination of the activities of all ICCN partners. The CoCoCongo has created an important lobbying group for the protection of biodiversity in DRC, a group that should be able to influence decisions about the development of the country to include the preservation of its protected areas.

The priority now is the development of a coherent, strong strategy and reorganisation plan for Virunga National Park which will provide the direction and priorities needed to make the most of the work and support of ICCN's partners.

1 'Each one of the States party to the present Convention engages to not deliberately take any measure susceptible of damaging directly or indirectly the cultural and natural heritage taken up in articles 1 and 2 which is situated on the territory of other States party to this Convention.'

2 'In representing fully the sovereignty of the States on the territory of which is situated the cultural and natural heritage taken up in articles 1 and 2, and without prejudice of the real rights foreseen by the national legislation, the States party to the Convention recognise that it constitutes a universal heritage for the protection of which the entire international community has the obligation to cooperate.'

21.1
Park rangers are the pillars that support the Institut Congolais pour la Conservation de la Nature; *strengthening their capacities is a necessary step towards an overhaul of the institution.*

21 The need for an institutional reorganisation of the ICCN at the end of the conflict period

Jean-Pierre d'Huart & José Kalpers

In spite of being the oldest, most celebrated and most diverse of the national parks within the Democratic Republic of Congo, Virunga National Park depends, as does the entire network of parks and reserves, on the quality of the management institution, on the clarity of its mandate and on the means at its disposal. The parks of the Congo have survived, with varying degrees of fortune, the many shocks that have marked the country's history. The wisdom, realism and flexibility that the managing authority has shown in adapting to new situations are among the qualities necessary for the survival and development of the parks, in particular during difficult times. From the early 1990s, the ICCN, just like most of the state-owned institutions of Congo-Zaire, has had grave functional and structural problems. The troubles that have deeply shaken the Democratic Republic of Congo over the past decade have had a serious impact on the people, the wildlife and the vegetation of Virunga National Park and other protected areas; but those problems have also emphasised and accelerated the weakening of the organisational structures of the ICCN. Reconstruction of the state offers the ICCN the opportunity to embrace a new institutional model whose structure would take into account past weaknesses and whose efficiency would match the natural heritage it is responsible for.

21.1 Brief history

From the creation in 1929 of the first institution for the management of protected areas in the Congo – the *Commission Administrative du Parc National Albert* (PNA) –, the objective of the park and the nature of the *Institut des Parcs Nationaux du Congo Belge* (IPNCB, the Institute of National Parks of the Belgian Congo) were very clearly defined. The decree that established Albert National Park gave it a 'strictly scientific objective', and the management responsibilities of the IPNCB were expressed by King Albert I in these terms: 'here you also have a monument to conserve, a monument that nature has constructed over the course of thousands of years and which is offered as it was formed, to this day, and from the first ages of the history of the world.' *(Ici vous avez aussi un monument à conserver, un monument que la nature a construit au cours des millénaires et qui est donné tel qu'il s'est formé, jusqu'à nos jours, depuis les premiers temps de l'âge du monde.)* A few years later, in 1933, the Duke of Brabant (future King Léopold III) specified that the national parks must be long-term observatories of natural evolution, as well as places of contemplation and rejuvenation; he assigned a triple responsibility to the Institute (IPNCB, 1937):
- Total conservation 'with the most attentive vigilance';
- Scientific exploration with a view to the progress of knowledge;
- Opening of some parts of the parks to controlled tourism.

These cornerstones, upon which were founded all the research and development activities undertaken by the IPNCB before the Congolese independence, mainly originate from an outstanding visionary scientist, Victor Van Straelen. In 1937, he wrote an action plan for the IPNCB that was both static (the maintenance of the absolute integrity of the protected areas) and dynamic (the inventory and description of species, the observation of ecological phenomena and biological cycles). The mission of the parks was summarised as follows: 'The total reserves will constitute islands of primitive life, limited by regions subjugated by man. Nature in its infinite complexity can be studied without any influence from artificial factors. (…) The surveillance of a total Nature Reserve will prove to be a vast undertaking of experimental ecology.'

The recommendations where then (IPNCB, 1942):
- to create parks of a large enough area to include complete biotic units and to allow the unfolding of natural cycles,
- to make the boundaries of these parks coincide with, wherever possible, natural barriers,
- to establish the parks where human density and economic value are 'as low as possible, so that the

21.2
On-the-job training is an essential element for maintaining the quality of work in the ICCN and must be integrated in the restructuring programme of the Institute.

creation of a National Park will have the minimum impact on the indigenous populations and the development of colonisation'.

The conceptual considerations of Van Straelen did naturally shape the structure of the institution in charge of their implementation, as well as the legislative arsenal which would become their reference (and, as far as the boundaries of the Virunga National Park are concerned, still is). The *Institut des Parcs Nationaux du Congo Belge,* with its objective to 'remove from human interference a natural area as vast as possible and to study the evolution of that area when left to itself', developed into a science-based institution with a strong component of *in situ* protection. While mainly financed by the Ministry of the Colonies, the IPNCB was endowed with a legal personality that gave it considerable independence to obtain complementary financing for its activities and publications.

The administrative body of the IPNCB was the National Parks Commission, composed of a President, a Secretary, and 24 members designated by the King. The great majority of these people were originally from scientific institutions and, to reflect the international character of conservation, a third of them came from foreign institutions. The executive body of the institute was the Executive Committee *(Comité de Direction)* which met each month; the National Parks Commission delegated its President, its Secretary and six members including the representative of the Ministry of the Colonies. The IPNCB also included two local Advisory Councils *(Comités Consultatifs)* one based in Kivu and one in the United States. Whereas most of the missions and activities took place in Albert National Park, the Institute kept this centralised structure until independence (IPNCB, 1942).

The first five-year report *(Premier Rapport Quinquennal)* of the IPNCB revealed that the administration of the parks from Brussels presented numerous problems related to distance and slow communication between Belgium and Albert National Park; difficulties in the supply and transport of equipment; difficulties with staff training; and problems with roles and responsibilities. Albert National Park was originally divided into two sectors: the Southern Sector, where the Chief Warden *(Conservateur Principal)* officiated from Rutshuru/Rumangabo, and the Northern Sector, with an assistant warden *(Conservateur Adjoint)* based at Mutsora. At first, military officers held these positions, then Belgian scientists. In December 1936, two other Europeans completed the managerial staff, a Building Manager and an Officer in Charge of Visits, based at Rwindi Camp. By the end of 1939, the African surveillance staff numbered 28 rangers in the Northern Sector, and 21 rangers and 32 assistants in the Southern Sector.

In spite of its relative distance from metropolitan areas, the Albert National Park staff succeeded spectacularly in implementing the priority programme, which was to: officially demarcate the park; establish basic infrastructures; ensure protection; and bring its support to several major exploratory missions. Starting in 1934, public interest for the Rwindi–Rutshuru Plain stimulated an increasing number of visitors to the region (nearly 900 in 1938).

The Second World War significantly hindered the development of Albert National Park but, between 1945 and 1960, the scientific vocation of the institution remained its focus and activities began again with renewed vigour. Between 1930 and 1960, no less than 15 exploration and study missions were sent to the national parks, 10 of them to Albert National Park alone. Statistics of the time show that the *Institut des Parcs Nationaux du Congo Belge* published 298 books and scientific articles representing 493 studies (of which 169 were for Albert National Park alone), totalling 32,000 pages in which 4,269 new species were described. The photographic library of the Institute contains 60,000 documents (Anon., 1964). Likewise, at the same time, numerous other studies, maps and articles about the parks were published by other scientific institutions.

Between 1960 and 1967 – the year of publication of the *Ordonnance-loi,* legislation creating the *Institut des Parcs Nationaux du Congo* – Albert National Park and the other parks suffered pressure from poaching and encroachment that the rangers only contained with much difficulty. The early 1960s witnessed a considerable decline in the security situation. With Dr. Jacques Verschuren, *Chargé de Mission* and the only European remaining in Albert National Park between 1960 and 1961, the first Congolese wardens (Mburanumwe, Munyaga, Kajuga and Bakinahe) attempted to organise the staff. They behaved heroically in the defence of the park and several rangers lost their lives during the clashes. The park was supported by providential but irregular donations from the Zoological Society of Frankfurt, the WWF, the *Fondation pour favoriser l'étude des parcs nationaux du Congo* (Foundation for the Promotion of the Study of the National Parks in Congo) and a few patrons. Even though the central government paid the staff salaries, there was no longer a central structure capable of directing the institution and, in the end, the destruction of the park was feared (Verschuren, 2001).

In 1969, President Mobutu called for support from the Belgian aid agency to revive the national parks. Dr. Verschuren, in charge of this bilateral co-operation project, was appointed director of the new *Institut pour la Conservation de la Nature au Congo* (Institute for Nature Conservation in Congo) which became, successively, the *Institut National pour la Conservation de la Nature* in 1972 and the *Institut Zaïrois pour la Conservation de la Nature* (IZCN) in 1975. Benefiting from vigorous political, technical and financial support, the Institute for Nature Conservation in Congo soon became a strong institution. In 1973, the administrative, financial management services and environment concerns were concentrated in Kinshasa while the technical and scientific management were based in Goma. In 1974, a veterinarian and management department was added, only to be abandoned shortly afterwards. In 1978, the IZCN was removed from the supervision of the Office of the President; a new piece of legislation, an *Ordonnance-loi,* created a Board of Directors and a Management Committee, and placed the IZCN under the responsibility of two ministries, Environment and Portfolio (INCN/IZCN, 1973–1977).

Throughout the 1980s, with the development of many new national parks and the availability of regular funding, the personnel of the *Institut Zaïrois pour la Conservation de la Nature* increased and a growing number of technicians were trained at the *Ecole de Faune de Garoua* in Cameroon.

21.2 THE INSTITUTIONAL REVIEW OF 1990–1991

With the aim to provide financial support to IZCN, the World Bank ordered an international review of its structure and management methods. For several months, a number of consultants criss-crossed the country and analysed the functioning of the Institute. Among the conclusions of this study (McPherson, 1991), we can mention some of the points most relevant to the management of protected areas:
- the Institute must increase its efforts to promote conservation in the long term, particularly in matters of environmental education, and increase collaboration with the populations in peripheral areas;
- greater strategic attention must be paid to the programmes, personnel and budgets allotted to research and tourism;
- the organisational relations between Wardens, Liaison Offices, Regional Co-ordination, Technical Direction and the *Président Délégué Général* must be clarified;
- the organisation and management of the Institute must be less centralised in order to increase its efficiency, with a recommendation to create four regional Directorates (Goma, Lubumbashi, Kisangani, Kinshasa);
- the structure must include a new department in charge of Projects and Planning;
- the status of personnel must be reviewed, and a collective agreement worked out;
- an in-depth reorganisation of the personnel must be considered; and programmes of the Technical Directorate, the Research Directorate, and the Directorate of Domains and Reserves must be organised into a single technical and scientific Directorate;
- the functions of the Regional Co-ordinator and of the Liaison Offices must be clarified and streamlined; the Regional Directors must have full authority over the administrative and technical operations of the protected areas under their jurisdictions;
- the system of evaluation of wardens must be refined; the level of competence and the number of rangers must be appraised on the basis of precise criteria (the number of rangers must be decreased from 429 to 200 in Virunga National Park);
- a percentage of the revenues must be retained and used directly for park operations and/or for the surrounding communities;
- the optimal level of overall financing in the protected areas must reach USD 778/ranger/year for salaries and associated costs, and USD 30/year/km^2 for operational costs; the targets of estimated annual financing would then be USD 2,300,000 for the protected areas, plus USD 360,000 for the Directorate General, plus USD 100,000 for each of the three Regional Directorates: in total, around USD 3 million per year;
- stronger co-ordination and planning with partners must be ensured through a Consultative Committee.

The study concluded that the Institute did not have adequate funds for delivering its mission and that it needed support from donors. In exchange, the Institute would have to improve its accounting and budgeting processes, and allocate adequate funds to cover running expenses in the protected areas.

Following the unrest that began in September 1991, several donors, including the World Bank, suspended their assistance to conservation and the study's recommendations were never fully implemented. Discussions with several partner organisations which were still supporting IZCN led to the establishment of two Regional Directorates in 1992, in North Kivu and South Kivu. A system of revenue redistribution helped the park stations operate at a minimum level. In 1997 the Regional Directorates became Provincial Directorates.

The recurrent instability that then affected the country had a deep impact on the evolution of the IZCN and the development of the parks. In addition to several important partners pulling out, the government was no longer able to manage the ongoing budget of the Institute, and salaries were only paid occasionally. The Institute became increasingly reliant on external aid.

A few informative documents provide interesting figures regarding the personnel and the evolution of the workforce:
- 1973 total INCN: 1065, DG (Directorate General) Kinshasa: 20 (INCN, 1973)
- 1990 total IZCN: 1334, DG Kinshasa: 107 (McPherson, 1991)
- 1995 total IZCN: 1909, DG Kinshasa: 69 (ICCN, 2003)
- 2006 total ICCN: 1917, DG Kinshasa: 72 (ICCN, 2006)

It is noteworthy that between 1991 and 1995, the number of staff at the Directorate General was reduced by around 35%, while, at the same time, the field personnel increased by around 45%. In Virunga National Park, the workforce numbered 542 in 1986, 429+ in 1991, 460+ in 2003, and 689 in 2006 (d'Huart, 1987 and 2003; McPherson, 1991; ICCN, 2006). There is some

21.3
ICCN staff receive a training in the handling of GPS, used during monitoring and law enforcement patrols.

21.4
Keeping a conservation spirit and an institutional memory helps maintaining the workforce committed in case of crisis. Director Déo Kajuga is interested in ethnobotany and has worked at the ICCN for over 30 years.

confusion as to the real list of personnel because some names have turned out to be fictitious and the status of others is uncertain. In 2006, personnel administration in Virunga National Park was still unclear because 150 staff members employed during the war were not officially confirmed by the Directorate General. Since 1998, the PARCID project (*Projet d'Appui et de Renforcement des Capacités Institutionnelles de la Direction Générale,* Project of support and reinforcement of the institutional capacity of the Directorate General) of the GTZ, German Technical Co-operation) has assisted the ICCN in its efforts to rationalise personnel. Of more than 100 staff, some 60 have officially retired with a final bonus paid. The project has also contributed to the restructuring and reorganisation of the Directorate General (GTZ, 2000).

21.3 THE IMPACT OF THE WARS ON THE ORGANISATION OF VIRUNGA NATIONAL PARK

Between November 1996 and 2002, the war and the rebellions caused a rupture between Kinshasa and the east of the country, and therefore between the Directorate General and Virunga National Park, first temporarily, then with a longer lasting effect. In 2000, the rebel movement of the *Rassemblement Congolais pour la Démocratie – Goma* (RCD-Goma), took over the control of a section of eastern DRC, and created a *Coordination ICCN* at Goma in order to replace the Directorate General and to supervise the management of Kahuzi-Biega National Park, as well as a large part of Virunga National Park. On the other hand, the RCD-ML *(Mouvement de Libération)* controlled a territory that included the northern part of Virunga National Park and gave its responsibility to its own Minister of Agriculture. The park and its personnel were then, from that point on, divided between two different rebel movements, one supported by Rwanda, the other by Uganda. The Provincial Directorates of Goma and Bukavu were nevertheless maintained.

From the beginning of the unrest, looting, requisitioning and destruction contributed to heavy loss of park equipment and vehicles. Disarmed rangers left their patrol posts and regrouped in the park stations. Between 1996 and 2005, more than a hundred rangers died to protect the park. Without clear instructions, without equipment and without pay, some deserted. Supported by several NGOs (IGCP, FZS, WCS, WWF), the Southern Sector of Virunga National Park continued to function despite the immense difficulties brought about by the situation. Apart from occasional support from the WWF *Projet Environnemental autour des Virunga* (PEVi -Kacheche) and the ZSL, the Central and Northern Sectors were subject to intense poaching, local intrusions and an anarchic exploitation of natural resources.

The ICCN staff had an allegiance dilemma; while wishing to keep contact with the Directorate General and the network of other protected areas, they had to submit to the ruling political authority of the time. The fracture with Kinshasa triggered some delicate situations, since the ICCN Coordination under RCD-Goma made decisions about personnel without consulting the Directorate General where all the employee databases were held. This situation caused great confusion, as much at the stations and Provincial Directorates level as for the partners working to support the ICCN.

Fortunately, during the second part of the war years (2001–2004), the ICCN received some essential assistance from the UNF/UNESCO project in support of World Heritage Sites. The partners of the Institute worked in close collaboration with the ICCN through the *Comité de Coordination du Site* (CoCoSi). The rangers' allowances were directly paid by UNESCO while the wardens and other senior staff were covered by other conservation partners. Despite the appalling security and the political meltdown, a minimum level of support was assured throughout the war and contributed to the survival of Virunga National Park.

21.4 TOWARDS A RESTRUCTURING OF THE ICCN: THE STAKES AND PERSPECTIVES

Although the ICCN has suffered greatly, in part from a slow institutional decline that began in the 1980s, and in part because of the repeated political and military crises of the past decade, the Institute still retains some potential for a promising new beginning.

Above all, the ICCN has a long history since its creation as an institution in charge of protected areas at the national level. This tradition has undoubtedly allowed the development of a genuine culture of conservation, coupled with a collective institutional memory, as valid at the central level as in the field. This *esprit de corps* has also enabled the emergence of a real sense of pride at the very heart of the institution, as well as at the national level (Inogwabini *et al.,* 2005).

At its establishment, the ICCN was conceived as a hierarchical structure, with relatively clear organisation charts and fairly solid job descriptions. One can also find a very sharp sense of discipline, present at all levels and in particular in the parks, which adopted a quasi-military management style. The communication networks and procedures, whether by mail or by radio, are very developed and, in principle, allow intense exchanges between the stations and the Directorate General. This structure and this discipline, even though they have been worn over time, still provide a solid and efficient framework to the entire institution.

As a public enterprise, the ICCN has a fundamental advantage; the Institute was indeed among the first conservation agencies in Africa to be created as a parastatal with a high level of autonomy. Elsewhere in Africa, especially in the French-speaking countries of Central and West Africa, the conservation mandate was usually carried out by ministerial departments. Today, a comparison between the DRC and many countries that have enjoyed a much higher level of peace still puts the ICCN on top in terms of institutional setup. In French-speaking Africa, the ICCN is often taken as a model of an autonomous, hence more efficient, parastatal agency, as is also the case in most of English-speaking Africa which has greater experience in this field.

Human resources represent another essential asset for the ICCN: again, in comparison with other countries in the region where similar institutions can only afford to mobilise two or three rangers per park, the ICCN deploys several dozen rangers in the field, and in the case of Virunga National Park, several hundred. Of course, quantity doesn't necessarily mean quality, but these figures nevertheless show the importance the Congolese authorities give to nature conservation. The people that make up the ICCN are its most impor-

21.5
Virunga National Park (Democratic Republic of Congo), Volcanoes National Park (Rwanda) and Mgahinga Gorilla National Park (Uganda) all share the extinct volcano Sabinyo.

tant asset, even if the number of its employees must be rationalised and their technical capacities developed.

The ICCN must nevertheless address a series of constraints and weaknesses that represent as many obstacles for its institutional development.

The political climate of the country is still relatively unstable, which of course creates some uncertainty amongst the national institutions in general, and the public enterprises in particular. Their management regularly comes under scrutiny and reorganisation plans are often discussed, which only causes a growing sense of insecurity within the ICCN as an institution.

This transition period, which has now gone on for several years, has also contributed to weaken the institutions in charge of environment or conservation. These issues naturally appear far down the list of government priorities, at a time when political parties are more preoccupied with power-sharing, starting up the economy and normalising the public sector. It is feared that, during the long period of economic recovery that is taking place in the DRC, incentive measures will be granted to highly productive economic sectors, such as mining or forestry, and that conservation would take a back seat against other types of land use.

One of the main recommendations of the institutional study by the *Institut Zairois pour la Conservation de la Nature* (IZCN) in 1991 was to decentralise some of the functions of the Institute. As mentioned earlier, an initial decentralisation actually began in 1992 with the creation of two Regional Directorates, which became later Provincial Directorates. In practice, however, the ICCN remains highly centralised, and decision-making is rarely, if ever, delegated to people in the field such as the heads of parks or the heads of stations. This situation brings about communication issues and confusion, while hindering the decision-making process.

There is further confusion between the political and the technical arms of the institution, a situation probably related to the current instability. Drafting and designing policies for the conservation and the management of protected areas are, in principle, the responsibility of the supervising ministries, while their implementation in the field is normally the duty of an institution such as ICCN. In practice however, the ICCN has to fill in the gaps in policies or guidelines, and this situation presents a danger to the extent that it behaves as an auto-regulating system.

A new institutional review of the conservation sector, funded by the European Union and the World Bank, is in progress as these lines are being written. This review will produce an analysis and a diagnosis of the ICCN and a proposal for a reinforcement programme that is anticipated by the ICCN and all its partners. Without prejudging the results of this review, we propose hereafter some reflections for the future.

21.5 Indicative guidelines for a restructuring

There is no lack of countries where conservation agencies have set up restructuring programmes (Kenya, Uganda, Rwanda, Benin, etc.) (ORTPN, 2004; UWA, 2002). These examples provide a good insight on lessons and principles that could guide us in a potential institutional reform of the ICCN.

To start with, there must be a clearly expressed, sincere political will at the government level. This must be the result of a genuine appropriation of the principle of restructuring, not just the result of pressures imposed by partners or donors. This attitude is probably the most important condition to meet as it enables the mobilisation of political support for conservation, a key ingredient for the survival of the national parks.

Restructuring a public enterprise like the ICCN must be an inclusive process, taking into account the greatest possible number of stakeholders. All parties must be consulted in the debates and discussions bearing on aspects as fundamental as methodology, time-frames, steering structures or political and technical options. This principle makes for a slower process but, in return, it gives a greater chance of success in the long term while sticking to the principles of good governance. It would be a priority to consult, for example, government agencies directly or indirectly involved in conservation, research or tourism; provincial authorities, local and traditional authorities, interest groups and associations present at the periphery of the parks and reserves; national and international NGOs; donors and aid agencies; the national and international private sector.

The mandate of the new institution must be clarified, and this requires a complete revision of the set of laws pertaining to biodiversity conservation and protected areas. The current laws present numerous loopholes and are a cause of confusion in areas such as environmental management, biodiversity management, hunting, tourism inside the protected areas, and so on. The loopholes find their origin in the incremental evolution of the legal texts, which were amended by repeated small touches over the years, progressively creating grey areas and room for interpretation. Ideally, a new institution should benefit from a new and stronger legal framework that would clarify the discrepancies and the incoherence currently experienced.

Closely linked to this legal revision, the institutional setup must also be properly addressed. It would indeed be irrelevant to put a new institution in place if its mandate is not clearly defined in relation with sister agencies. It is necessary, in particular, to clarify aspects such as tourism management in the protected areas, where misunderstandings regularly arise between the protected area agency and the national tourism board.

As a public institution, the ICCN must conform to and execute national policies regulating the public service. National directives take precedence in this field; they are sometimes further comforted by international institutions such as the World Bank or the International Monetary Fund. It is possible that the present period of transition will lead to a broad reform of the state administrations, which the ICCN will have to take into account. Therefore, it is best to avoid rushing into a restructuring that may be questioned later in light of an overall reform.

Even though policy making is the prerogative of the relevant ministries, the ICCN must play a leading role by providing technical expertise and assuming a proactive, rather than a passive, stance. The Institute must also draw up sets of guidelines and strategies covering thematic aspects of management such as law enforcement, community conservation, tourism management in protected areas, veterinary interventions, research and monitoring, problem animal control, and so on.

From the first appraisal and initial discussions on the restructuring of the Institute, an in-depth analysis of the possibilities of decentralising the management structures must start. There are numerous options ranging from total centralisation (the situation that prevailed for many years) to complete decentralisation where the Directorate General would co-ordinate activities and define guidelines and strategies, while the different protected areas would have a large degree of autonomy at the field level. There is no blueprint solution in this very complex area, and many opinions have to be taken into consideration while conducting a detailed analysis of the pros and cons of all the institutional setups possible (Bensted-Smith & Cobb, 1995).

The theme of human resources requires much attention; first in assessing the actual needs in personnel, then in the recruitment and, finally, in the development of capacities. Broad guidelines are needed to take into account the following elements:
- Analyse the existing human resources, using an external company
- Develop new organisation charts and terms of reference for the different positions
- Ensure transparent recruitment, using an independent company
- Recruit senior management staff first, then recruit junior staff members and the rest of the personnel
- Following recruitment, analyse training needs
- Set up an immediate training programme for the most urgent needs identified
- Design and implement a five-year training strategy.

During the restructuring process, and when the senior management staff has been recruited, it will be necessary to strongly focus on the definition of a new vision for the institution. Senior management, as well as key stakeholders outside the Institute, will be mobilised to develop a strategic plan (between five and ten years) detailing the objectives, expected results and activities for each department or section of the Institute. A range of indicators will be carefully identified, partly to verify whether the objectives are being achieved, and partly to conduct a continuous evaluation of personnel performance. The strategic plan will drive the Institution's planning, since it will enable the definition of annual operation plans, quarterly and monthly plans, and so on. Each protected area will also require its own general management plan that should be in coherence with the strategic and operational plans of the Institute.

Regarding the structure of the new institution, and in addition to the traditional departments already in place within the ICCN, it will be necessary to create a series of new departments:
- Planning: design and follow up of strategic and operational plans, co-ordination of partners, liaison and communication with international conventions and programmes;
- Community-based conservation: all aspects related to revenue-sharing, integrated conservation and development, environmental education and awareness, conflict management;
- Research and monitoring: definition of monitoring guidelines and research priorities, centralisation and compilation of data, geographic information systems. This department must work hand-in-hand with the planning department in order to maintain the feedback loop.

Annette Lanjouw, Anecto Kayitare & Andrew Plumptre

22 Transboundary Natural Resource Management in the Virunga Region of the Albertine Rift

Annette Lanjouw, Anecto Kayitare & Andrew Plumptre

Virunga National Park is contiguous with no less than five national parks in other countries – four in Uganda, and one in Rwanda. These, in turn, are contiguous with other protected areas – national parks, wildlife reserves and forest reserves. Notwithstanding their close social and economic ties, together with a long shared history, these three countries also share a large landscape of protected areas. Therefore, the implementation of an effective transboundary strategy is essential to the long-term conservation of Virunga National Park. This chapter retraces the development and objectives of a transboundary strategy and then examines the current situation. Finally, the authors, all of whom are, or have been, directly involved in the implementation of the transboundary program in the field, review the successes and failures of the strategy.

22.1 Objectives and background to transboundary collaboration in the Virunga Region

Transboundary Natural Resource Management is 'any process of cooperation across boundaries that facilitates or improves the management of natural resources, to the benefit of parties in the area concerned' (Biodiversity Support Programme, 1999). Virunga National Park forms the backbone to a contiguous assemblage of protected areas stretching from the northern shores of Lake Kivu to the southern shores of Lake Albert. These protected areas include the Parc des Volcans, Mgahinga Gorilla, Bwindi Impenetrable, Queen Elizabeth, Rwenzori Mountains, and Semuliki National Parks, among the oldest in Africa. They cover a significant proportion of the Albertine Rift across three international boundaries.

However, it was not until the early 1990s that mechanisms were developed for a transnational approach to the ecosystem as a whole. Up to this point traditional systems of national conservation and management were all that existed, focusing only on the portion of the biodiversity under the sovereign rule of the respective country. This ignored the fact that ecosystems have natural processes and needs which do not respect national borders: They depend on effective management in all three countries. When the transboundary approach was launched it included the three national protected area authorities: Rwanda's *Office Rwandais du Tourisme et des Parcs Nationaux* (ORTPN), DRC's *Institut Congolais pour la Conservation de la Nature* (ICCN) and Uganda's Wildlife Authority (UWA). The approach initially covered the Virunga Volcanoes and Bwindi Impenetrable National Park region within the Greater Virunga Landscape. From 2003 transboundary collaboration began for the remaining areas contiguous to Virunga National Park.

The approach centred on field-level collaboration, with links extended to higher political levels, only once this collaboration was effective and produced clear results. Often transboundary collaborations focus on top-down formal agreements between governments to work together. But the bottom-up approach taken in this case has meant that effective collaboration on the ground can precede the formalised adoption of the process by the government of each country.

This is not to say that high-level involvement in the formation of the transboundary process has not occurred. In a meeting in 2001 bringing together the Executive Directors of the protected area authorities the ORTPN, the ICCN and the UWA, the management objectives for transboundary collaboration were identified. These were:
– To guarantee the long-term conservation of wild-

22.1
Bwindi Impenetrable Forest National Park is one of the parks that make up the Greater Virunga Landscape.

22.2
Rangers from all three countries that share the Virunga Massif receive training in transboundary gorilla monitoring.

life resources in the parks that form the Virunga ecosystems,
- To ensure that expertise and experience between the ORTPN, UWA and ICCN are shared and pooled on a good neighbourly basis,
- To increase the local and international profile of this important conservation area, thereby greatly enhancing its potential as a tourist destination,
- To realise the full economic potential of the parks and surrounding areas, bringing economic benefits to Rwanda, Uganda and DRC especially to the local communities adjacent to the parks,
- To ensure that joint promotional campaigns stimulating a three-way flow of tourists are developed, thereby increasing the tourism potential for Rwanda, Uganda and DRC, including steps to facilitate the freedom of movement within the Virunga Transfrontier Forest,
- To comply with the requirements of international law regarding the protection of the environment,
- To integrate, as far as possible, the managerial, reservation, research, marketing and other systems of the ORTPN, UWA and ICCN in respect of the parks.

The overall objectives for transboundary collaboration in the Central Albertine Rift, however, are much broader and have brought together conservation with development objectives. In 2004 the overall objectives were:
- Cooperative conservation of biodiversity and other natural and cultural values across boundaries,
- Promotion of landscape-level ecosystems through protected area planning and management,
- Advocating integrated bioregional land-use planning and management to reduce threats to protected areas,
- Establishment of a common vision for transboundary collaboration,
- Building trust, understanding and cooperation among wildlife authorities, nongovernmental organisations, communities, users and other stakeholders to achieve sustainable conservation and thereby contribute to peace,
- Sharing of regional resources, management skills, experience and good practice to ensure efficiency and effectiveness in managing biodiversity and cultural resources,
- Enhancing conservation benefits and promoting, at a transnational level, awareness and sharing of these benefits and conservation values among stakeholders,
- Strengthening cooperation in research, monitoring and information management programs,
- Ensuring that conservation of biodiversity in the region contributes to the reduction of poverty.

The donor community provided significant resources enabling transnational collaboration to take place. Contiguous forest reserves in Uganda were also identified as one of six key areas for conservation as part of a strategic planning process for the Albertine Rift region, bringing together a wider set of NGOs with the protected area authorities.

22.2 Justification for a transboundary conservation approach

22.2.1 An ecosystem divided by international boundaries

The Greater Virunga Landscape is covered by a range of forest, savanna, mountain and aquatic habitats spanning the borders of eastern DRC, north-western Rwanda and western Uganda. The international borders separate the contiguous national parks, including the Virunga National Park (DRC), Volcanoes National Park (Rwanda), Rwenzori Mountains, Queen Elizabeth, Bwindi Impenetrable and Mgahinga Gorilla National Park (Uganda).

The major threats to the main contiguous protected areas have been the destruction of vegetation and habitat for agriculture to meet the food needs of the growing human population, and large-scale poaching. In Rwanda and DRC these threats are in part the result of the displacement of people as a consequence of civil unrest, as well as economic breakdown, forcing people to rely on the land for food and shelter. People are living an increasingly impoverished lifestyle in much of the region and rely on the forest for firewood and construction wood, resources for food, and water.

22.2.2 Population pressure

One of the prime threats to the protection of these forest blocks is encroachment by local populations. The numbers of people living in the areas surrounding these forest blocks are probably among the highest in Africa. The average population density for Africa in 1989 was 21.2 people per square kilometre (World Resource Institute, 1990) whereas the rural population per capita densities around the habitat of the Mountain Gorilla average 420 people per square kilometre (Lanjouw et al., 2001).

As over 90 per cent of the population practice subsistence farming, some access the protected area to supplement their livelihoods. The national parks do not have buffer zones between the local communities and the park resources.

22.2.3 Regional instability

In addition to the already highlighted threats, transboundary natural resource management has been further challenged by recent political events. Uganda and more recently Rwanda are emerging from extended periods of internal conflict. DRC is struggling with what has been described as the world's most deadly conflict since World War II in which an estimated 5.4 million citizens have died since 1996 (IRC 2008). These conflicts have resulted in massive population movements, especially after the genocide in Rwanda in 1994 in which over two million people fled into DRC, Burundi, Tanzania and Uganda (Joint Evaluation of Emergency Assistance to Rwanda, 1996). The refugees spent more than two years in camps with large numbers based at the foot of the Virunga Volcanoes of DRC and in the town of Goma.

Due to the series of conflicts in the Virunga Volcanoes region, the primary focus has been on supporting the protected area authorities to continue conservation activities. Faced with a multitude of problems, conservationists were forced to limit themselves to a 'reactive' attitude, able only to follow events as they developed and intervening only where security conditions allowed and finances permitted (Thorsell, 1991). This obligatory focus on 'reactive' intervention has led conservationists in Virunga to develop a very flexible regional framework, allowing it to respond to the dynamic political and economic situation and the changing priorities on the ground.

22.2.4 Rationale

The Greater Virunga Landscape is home to an exceptional diversity of species, many of which are endemic to the Albertine Rift, and which attract a great deal of international and national attention. Mountain Gorillas in particular are found in only two blocks of forest. None exist in captivity. There are approximately 380 gorillas in the Virunga area, and 320 in the Bwindi area (Kalpers et al., 2003; McNeilage et al., 1998). The density of the human population living around these two forest blocks limits the habitat of the gorillas. Significant habitat expansion is not a reality. The survival of the gorillas, and their habitat, therefore depends on the maintenance of the integrity of the remaining forest. As a flagship species for the area, ranging across the three borders within the relatively small blocks of forest, the gorillas must be protected equally effectively from all three sides. Similar approaches must be applied and the authorities must join together to ensure the protection is effective.

The forest plays an important water-catchment function, as well as ensuring soil stability in the region. The forests of the Greater Virunga Landscape are therefore not only important as the natural habitat for a large diversity of wildlife, including a number of endangered species, but they are also important for maintaining the ecological processes necessary for the agricultural livelihoods of the people in this region. As the forests are shared by three countries, it is necessary for the three governments to work together to ensure that management and conservation is effective.

22.2.5 Critical Threats

The sources of stress or threats that were identified during the exercise were ranked as very high, high, medium and low.

22.3 Key mechanisms developed

22.3.1 Phased approach to regional collaboration for mountain gorilla sites

The collaborative transboundary program was first launched for the Virunga Volcanoes area together with

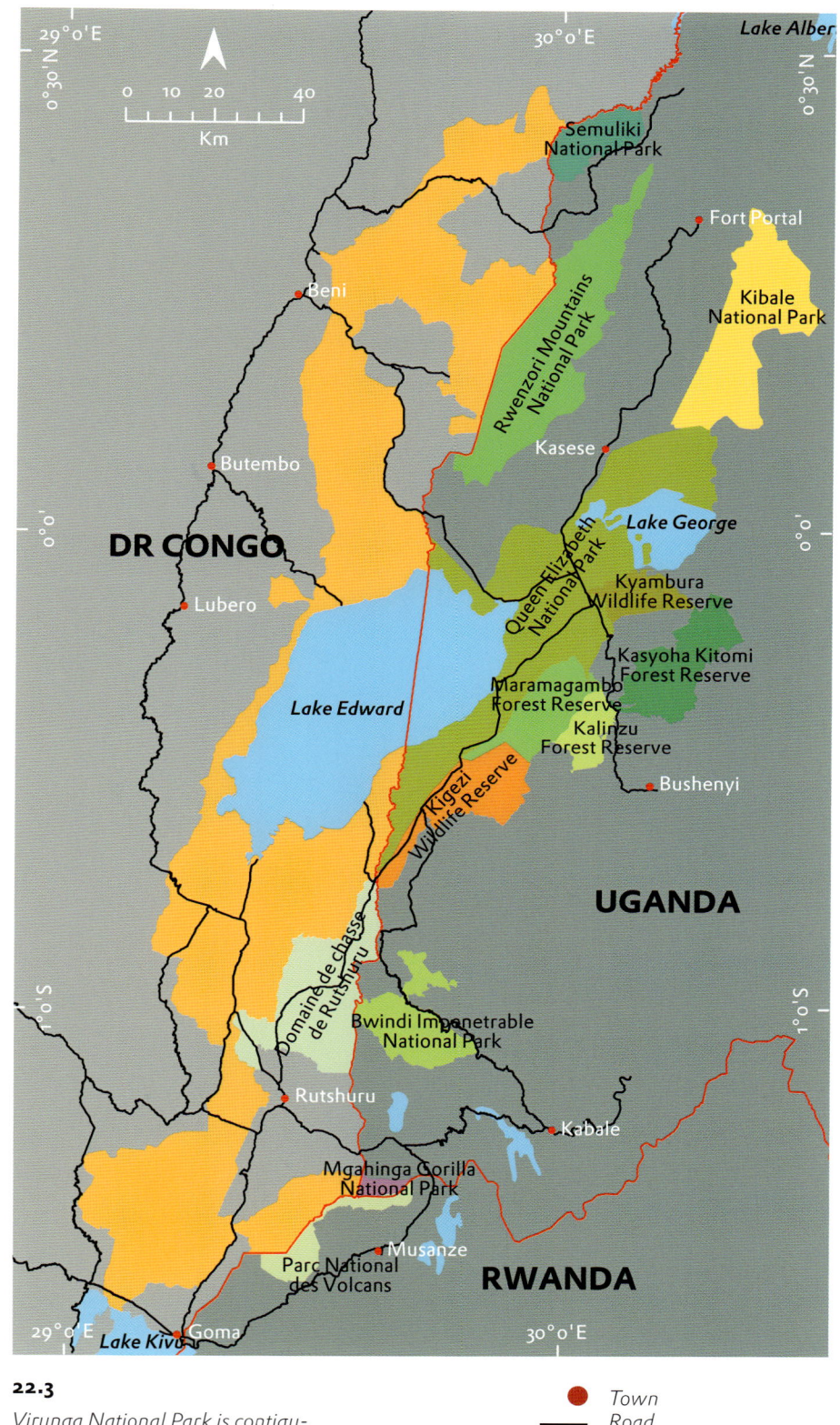

22.3
Virunga National Park is contiguous with several parks in Uganda and Rwanda. As such, it forms the backbone to around 15 transboundary protected areas, generally referred to as 'the Greater Virunga Landscape'.

● Town
— Road
— International boundary
▇ Virunga National Park

22.4 Summary Table of Conservation Targets and Goals in the Virunga Heartland

Targets	Goals
Vegetation succession on lava fields	• Protect the integrity of volcanic succession • Reduce and halt the utilisation of threatened species • Understand the dynamics of succession on the volcanic substrate
Savanna ecosystem	• Halt the illegal occupation of park lands • Remove exotic species and restore the ecosystem • Harmonise legislation at regional level to facilitate collaboration
Medium and high altitude forest zones (including bamboo zone and corridors)	• Maintain and extend the integrity of existing forest gradient • Restore degraded forest areas where feasible • Enhance the capacity of existing institutions to protect and manage • Sustainable use of bamboo and provision of alternatives where appropriate • Establish regulatory mechanisms for sustainable use and protection • Promote on-farm substitution of forest/bamboo species • Restore corridors and where necessary enlarge them • Establish corridors where they do not exist
Reptiles, amphibians and invertebrates	• Improve the understanding of the distribution and population sizes of reptiles and amphibians in the region • Improve protection of valuable areas (including wetlands) for reptiles and amphibians
Avi-fauna	• Increase understanding of migratory sites for birds • Increase protection of high-altitude swamps and wetlands • Maintain migratory stopping points • Elevate status of Bwindi Impenetrable NP and Mgahinga Gorilla NP to Important Bird Area
Large forest mammals such as the forest elephants, bush pigs, Giant Forest Hog, duikers and buffaloes	• Improve understanding of population size, composition and distribution • Understand their ecological role in maintaining forest structure and other roles; functions and diversity
Gorillas and Chimpanzees	• Understand forest carrying capacity and primate population dynamics • Increase protection through law enforcement, education and sensitisation • Increase protected status of Sarambwe contiguous to Bwindi and Echuya FR (against the border with Rwanda) • Maintain and establish corridors • Enhance transboundary collaboration of their shared habitat
Hydrological functions (Springs and water-courses)	• Preserve and protect all existing springs and water courses within park boundaries • Identify, restore and protect springs and water courses outside parks and where feasible maintain/improve regimes • Protect the watershed from erosion
Wetlands and Lakes (including fish and fisheries)	• Maintain flow regimes within and between wetlands to maintain hydrological connectivity • Increase the extent of riparian and lacustrine vegetation • Maintain and improve lake and river water quality and species diversity within the wetlands and lakes • Enhance regulation of fisheries
Soil outside protected areas	• Control against erosion • Improve soil productivity • Improve agricultural and land tenure policy

Bwindi-Sarambwe at the initiative of the three protected area authorities in Rwanda, DRC and Uganda, with international support. Propelled by the necessity to protect the habitat range for the mountain gorillas, the governments of the three countries gave the International Gorilla Conservation Programme (IGCP)* the mandate to develop a transnational framework and mechanisms for collaboration. It was independently agreed by the responsible ministries in each country that the three protected area authorities (ORTPN, ICCN and UWA) would participate as national representatives and form an integral part of the team. To fulfil its mandate, the IGCP has developed a regional strategy with a phased approach.

22.3.1.1 Phase 1: Field-based coordination and collaboration

In the preliminary phase, the focus is on harmonisation and coordination of management approaches, and development of field-based, informal mechanisms for collaboration. The protected area authorities from the four parks work as a team to manage the forest blocks as a shared unit and thereby strengthen conservation impact. This phase emphasises regular communication between wardens and management staff of the parks, sharing of information on the situation in the four parks, and joint planning and implementation of activities.

22.5 Critical and major threats	Ranking
Agricultural conversion (and resulting habitat loss)	Very High
Overfishing using poor methods	High
Nutrient overload and erosion (agricultural practices)	High
Cattle/livestock grazing	Medium
Charcoal production/ fuel wood collection	Medium
Poaching for both food and trophies	Medium
Harvesting of timber, bamboo and Non Timber Forest Resources	Medium
Insecurity/conflict	Medium
Pollution/waste management	Medium
Clay & sand mining/brick production	Medium
Resettlement & displacement of people/urbanisation	Medium
Fire	Medium
Disease	Low
Need for water by population	Low
Roads	Low

22.3.1.2 Phase 2: Formalisation of regional collaboration

The existence and use of the harmonised approaches in the three countries will facilitate the second phase of the strategy—formalisation of the transborder collaboration and regional policies. The second phase, however, is dependent on a minimum level of political harmony among the official governments of the three countries, and this has been a major constraint in the region for the past ten years. In the Virunga Volcanoes region transboundary conservation has been developed through field-based collaboration, rather than official agreements. Experience shows that formally designated protected areas are frequently far from effectively protected on the ground. Ideally, the two should complement each other, and this is the goal of the phased approach. Formalisation of the field-based coordination and collaboration is necessary in order to ensure that the principles are institutionalised and not dependent on individuals who know and trust one another. In order to provide both the structure and principles for sustained collaboration over time, and through changing political and economic circumstances, the processes and activities involved in regional collaboration must be included in strategic and operational planning, and time and resources must be allocated to these activities.

22.3.1.3 Phase 3: Formal designation of a transboundary protected area

A final phase could involve the signing of a formal agreement among the three governments, establishing a transborder protected area (TBPA). The agreement would outline the legislative background of the TBPA, define its purpose, describe the parties and the endorsing partners, and define the protected area, its structures (a joint commission or other mechanism) and modes of operation. Such an agreement would strengthen and provide political support to the institutions involved in regional collaboration, and facilitate the evolution and adaptation of collaborative structures and approaches over time.

By focusing on the protected area authorities, strengthening their ability to effectively manage the parks and demonstrating the potential economic as well as ecological value of the forest, the importance attributed to environmental issues has slowly increased over the years. As a consequence of the emphasis on informal, field-based mechanisms for collaboration, the political tensions and conflicts in the region have not impeded transnational collaboration throughout the past ten years, and this has strengthened the impact of environmental activities.

With the conservation authorities of each area, a structured process of collaboration was adopted with the primary aim of ensuring the management of protected areas and conservation also contributed to development objectives in the region. This structured process is based on:
- Communication and joint planning mechanisms,
- Joint activities and capacity-building,
- Involving local authorities and local people in conservation,
- Research and monitoring for improved management,
- Tourism and promoting a cross-boundary destination with shared attractions,
- Involving protected areas in the improvement of regional security.

These processes of collaboration are being implemented by the three protected area authorities in the four national parks currently involved in the Transboundary Framework.

22.3.2 Regional collaboration within the Greater Virunga Landscape

In 2003 the Wildlife Conservation Society (WCS) provided technical and financial support to begin a similar process of collaboration between the ICCN and the UWA to extend this cross-border work to neighbouring protected areas along the international border between DRC and Uganda (Queen Elizabeth, Rwenzori Mountains, Semuliki, and Kibale National parks, Kasyoha-Kitomi and Kalinzu Forest Reserves, Kyambura and Kigezi Wildlife Reserves) within the Greater Virunga Landscape. These additional areas form about 80 per cent of the Greater Virunga Landscape and have historically been overlooked with the emphasis on mountain gorilla conservation. However, these areas have huge importance in their own right, including some of Africa's most important protected areas for biodiversity conservation. Together with the Volcanoes-Bwindi region they have the potential to become a major tourist destination in Africa.

Many of the transboundary issues that this area faces differ from those found further south. Elephant poaching in DRC is currently a major problem and ivory is being taken out through Uganda to destinations in the Middle East and possibly the Far East. The impact of armed groups and the military in this region is also much more of an issue than in the mountain gorilla region, with military and armed groups targeting ICCN rangers who are trying to stop the poaching. Illegal settlement in Virunga Park by Ugandans is another issue that has been tackled. Soon after the transboundary collaboration started, the ICCN was able to remove 20,000 illegally settled Ugandans from the Lubilya region (north of Lake Edward) with the help of the UWA and local civic leaders in Uganda. Prior to the collaboration, the ICCN had not been able to evict these farmers.

It is clear that the transboundary collaboration process is expensive, costing over 3 million US dollars in 2004. There is a need to assess whether the amounts invested are really leading to more effective management of the Greater Virunga Landscape, and whether it might not be better to invest in each country separately or whether there is an intermediate level of funding that would work better. These assessments are important to avoid disinterest and disengagement by national stakeholders. As such, a Core Secretariat is developing a conservation plan that will specifically address issues that concern the management of this landscape.

The Core Secretariat for transboundary collaboration has been created by UWA, ORTPN and ICCN, including the Executive Director and a technical support person from each of the three countries, with WCS, the Albertine Rift Conservation Society, and with IGCP as the facilitator. The mandate of the Core Secretariat is to finalise and institutionalise the Greater Virunga Regional Management Plan and implementation strategies and to facilitate the collaborative processes with local stakeholders, protected area field staff, partners, authorities and governments. The process of planning in the Greater Virunga Landscape is based on analysis of conservation targets, threats and strategies in the Albertine Rift region.

22.3.3 Mechanisms and priorities established for transboundary collaboration

22.3.3.1 Regional-level communication, planning and cooperation in protected area activities

One of the key tools for collaboration is communication. For this reason, the transboundary program emphasises open communication between the three protected area authorities. Specific mechanisms have been developed to enable a regular exchange of information and joint planning:

- **Regional meetings:** These quarterly meetings bring together key protected area personnel of the three countries as well as technical partners. The regional meetings look at technical issues, allowing the exchange of information relevant to the conservation of these ecosystems and stimulate discussions on specific issues. In order to maintain neutrality, each country hosts a meeting on a rotational basis.
- **Wardens' coordination meetings:** One of the many products of the regional meetings was the development of more regular meetings between the wardens of the four protected areas to tackle the technical issues in a more focused manner. Several joint patrols and cross visits have been implemented as a result of these meetings.
- **Joint patrols:** Protected area authorities (PAA) regularly conduct joint surveillance and anti-poaching patrols. During the patrols, park staffs come together to patrol border areas to share information and logistics. They work as a team, and the patrols tend to be successful because they increase the patrolling of border areas and the exchange of relevant information.
- **Cross visits:** Due to the differing political and economic situations of the three countries, wildlife officers are funded to visit sites in the neighbouring countries to gain a better understanding of the different challenges for protected area conservation.
- **Gorilla census:** Censuses of the gorilla populations in Bwindi and the Virunga have involved staff from the park authorities in the three countries and conservation organisations. Not only did this result in training of the park staff of all three countries but it also strengthened the regional links between them.

22.3.3.2 Regional-level ecological and law enforcement monitoring

The foundation for effective management and conservation of the forest is a strong understanding of the threats to the forest and the needs of key species within that forest. These threats change over time. Thus, a three-pronged information gathering and monitoring program was developed to inform park management. This Regional Information System (RIS) was established to be implemented and used by the three protected area authorities responsible for the management of the Virunga-Bwindi forests.

The first component of the RIS is ranger-based monitoring (RBM) inside the forest, started in 1996. The objective is regular monitoring of the forest by park rangers in order to understand the illegal human use of the habitat (law enforcement monitoring – including poaching, woodcutting, etc.), ecological processes in the forest and distribution and habitat-use of specific key species. The monitoring feeds directly into the day-to-day management of the park and enables surveillance and specific interventions to be based on solid data. This can include where to send patrols based on poacher activity, availability of seasonal resources and presence of snares. It can also include the movements of key species, such as the gorillas and their use of the habitat. The RBM has produced effective field maps for the park staff and patrolling rangers, using topographic features and toponyms. At present, the data are being analyzed in each park, as well as at the headquarters of the protected area authorities. A centralised, regional database is also being developed so that the data will be available for the entire ecosystem, thus allowing park staff to manage the area as one ecological unit.

More recently, the RBM has been expanded to cover the whole Virunga National Park. This means that data can be collected throughout the park and the threats and issues better understood on a park-wide scale. The

22.6
Transboundary protected area management requires a large number of meetings between ICCN, UWA and ORTPN—the institutions responsible for protected area management in the DRC, Uganda and Rwanda respectively. Here, we find a joint planning meeting for the monitoring of key species in the Greater Virunga Landscape.

process of data collection is now managed by ICCN with a centralised database in Goma. As it is currently difficult to access large parts of the park, information provided by the rangers on the ground is invaluable for managers and their NGO partners. The process of RBM needed to be put within a framework for the park and in 2005 ICCN called a meeting, facilitated by WCS, to develop this, with NGO partners, leading to a monitoring plan for Virunga Park (see Chapter 23 on monitoring).

The second component of RIS is socio-economic monitoring outside the forest. This includes the collection of data on numbers of people living near the forest, their distance from the forest, their livelihood activities and their needs and dependence on resources from the forest. These data are analyzed in conjunction with the illegal activities in the forest, to gain a better understanding of the overall threats to the forest and its wildlife and which factors are driving these threats.

The third component of the RIS is remote sensing of the gorilla habitat to detect changes in vegetation cover and land-use over time. This is being conducted in conjunction with the European Space Agency and the UNESCO World Heritage Centre. Through satellite imagery, the changes in forest cover and impact of the human population can be measured and understood in conjunction with the socio-economic and ecological data.

22.3.3.3 Regional-level development of skills and ability

Training is a major component of any institutional-strengthening program and has been a primary focus of IGCP and WCS. By applying similar approaches in the three countries and maximising the opportunities for exchange of skills between the countries, transboundary conservation has been strengthened.

22.3.3.4 Regional-level cooperation in collaborative activities

The protected area authorities have regularly conducted joint surveillance and anti-poaching patrols. In the joint patrols, the staffs of contiguous parks come together to patrol the border areas, share information, logistics and work as a team. These border areas are often very vulnerable and operations in the Volcanoes-Bwindi region have recently involved the military of all three countries, thus bringing together not only park staff, but also military personnel.

The 1998 census of the gorillas at Bwindi Impenetrable National Park and the 2003 census of gorillas in the Virunga massif involved staff from the UWA, ORTPN and ICCN. Again, the objective was training of park staff in all three countries as well as strengthening the regional links among them.

22.3.3.5 Analysis and presentation of achievements at international conferences

In 1997 the protected area authorities from the DRC, Rwanda and Uganda jointly presented a paper at the International Union for Conservation of Nature (IUCN) International Conference on Transboundary Protected Areas as a Vehicle for International Cooperation (Cape Town, South Africa). The conference, therefore, not only stimulated the governments of Rwanda, DRC and Uganda to continue efforts at strengthening collaboration on the ground, but also showed the strength of collaboration already in effect.

The IUCN meeting was followed in 1998 by an International Symposium on Parks for Peace, in Bormio, Italy. At this meeting, the participants generated material for the development of draft Guidelines for Transboundary Cooperation in Protected Areas, and a draft Code for Transboundary Protected Areas in Times of Peace and Armed Conflict. The IUCN World Commission on Protected Areas (WCPA), the Programme on Protected Areas and the IUCN Commission on Environmental Law are currently finalising these tools to be published as Transboundary Protected Areas for Peace and Cooperation under WCPA's Best Practice Protected Area Guidelines Series.

Finally, in 2003, the case study of Central Albertine Rift transfrontier collaboration was presented at the IUCN World Parks Congress in Durban, South Africa.

22.3.4 Economic mechanisms

North-western Rwanda, eastern DRC and south-western Uganda have a large proportion of the population living below the poverty line, with insufficient land to meet their most basic needs (Waller, 1996). Very few alternatives exist to subsistence agriculture on steep slopes and plots that are too small to feed the average family. Numerous efforts have been made, and consultant hours spent, searching for alternatives for the local people in this region. Tourism, and more specifically, nature-tourism, offers one of the few viable options. Although a fragile industry, easily affected by political, economic and social changes, tourism nonetheless has real economic potential for the region.

The risk of tourism, however, to both the mountain gorillas and their habitat is also considerable. The potential transmission of diseases from humans to gorillas, thus possibly infecting the entire gorilla population, poses one of the greatest threats to gorilla conservation (Homsy, 1999). Transmission of diseases is not only a potential risk between tourists or researchers and gorillas, however. It is equally, if not more likely, between gorillas and poachers, local farmers, harvesters of natural resources, park staff, military and rebels. Efforts to sensitise some of these groups are underway by IGCP and partner organisations, including the Institute of Tropical Forest Conservation (ITFC).

Common rules have been established and applied in all three countries, to manage and control tourism. These rules focus on reducing the risks of disease transmission, overexploitation of the gorillas for tourism and reducing the stress to the gorillas. Having the same rules at each tourism site strengthens collaboration and reduces competition among the three countries, which jointly developed the tourism rules and regulations. Common approaches are also being applied with respect to interpretation and development of joint messages for conservation, handling procedures, and training for tourism staff. These activities are forming a fledgling regional tourism approach, which will be strengthened once peace and political stability return to the region.

To spread the economic benefits of tourism to the communities around the parks in Uganda, IGCP and its partners have worked toward developing tourism-linked enterprises for the local communities. Similar enterprises are now being developed in Rwanda, again to ensure the flow of economic benefits to the communities and to strengthen the links between the local people and the parks in both countries.

22.3.5 Social aspects

Activities developed at a *national* level to integrate local communities into the management of protected areas, and to spread the benefits of conservation (both economic and ecological) to human populations, have strengthened the potential for learning, collaboration and cooperation *across international borders*. This type of contact can help fortify the cultural ties among communities that have been restricted by country borders or alienated by political conflict. In so doing, they can support social and political stability in the region. In the context of the Virunga, the years of war and conflict have destroyed the cohesion of the community and exacerbated tensions between local peoples and authorities as well as neighbours. The work to involve communities formally with protected area management, and share benefits with them, as well as to facilitate cross-border linkages, will contribute to improving these relations.

22.4 Successes and failures: Achievements and impacts

The Greater Virunga Landscape region is still in the midst of acute political turmoil and conflict. This is causing severe pressures on the environment through the unsustainable exploitation of natural resources and breakdown of social and economic structures which had been protecting the environment. This has been a major challenge for the implementation of transboundary collaboration.

22.4.1 Politicisation of transboundary work

One of the greatest challenges has been the political nature of the difficulties in the region. Given that much tension exists between the three countries, any work bringing together representatives of the countries has led to the politicisation of their discussions and activities. It has been extremely difficult for them to move forward on what are obvious and seemingly uncontroversial ideas, due entirely to the suspicion that political motivations may be behind these initiatives.

22.4.2 High staff turnover

Due to the war and the political challenges in the DRC, the ICCN has had a high staff turnover. In addition, there has been a significant institutional review and restructuring of the protected area authority in Rwanda which has also led to numerous staff changes. Despite the many justifications for these changes, they have caused problems with continuity, institutional memory and losses in the investment in relationships, essential for effective transboundary protected area management.

22.4.3 Staff fatalities

The conflicts in the region have led to many fatalities, including over 100 rangers of the protected areas in the Virunga region. Staffs from all three countries have been killed during the conflicts and this has led to an enormous loss of capacity, experience and motivation. It is a continuous challenge for the park staffs of the DRC, Rwanda and Uganda to survive and effectively protect the wildlife and natural resources in this region.

22.4.4 Different approaches and incompatibility of tools

Different experiences and objectives have led the different partners of the protected area authorities in Rwanda, DRC and Uganda to develop different tools for conservation. This is most notable in the monitoring and collection of data on ecological, socioeconomic and law-enforcement parameters. Although each of the tools has its strengths, the protected area authorities are often confused and overloaded with demands for data. They are encouraged to use numerous different methods and technologies and the result is excessive demand on field staff. An example has been the simple ranger-based monitoring approach. This can be done using GPS and computers but it can still be effective with just a pencil and paper. In addition, they could use the cyber-tracker system, the Palm-pilot hand-held data recorder, the MIST system developed by GTZ in Uganda and various other tools to complete the same task. Data is regularly requested on all the following areas: ecological, law-enforcement, socio-economic, large mammal, disease and health and data required for ground truthing spatial imagery. The lesson is that to be effective, data collection and monitoring have to remain reasonable and realistic.

Despite these challenges, considerable progress has been made by the transboundary conservation program. Progress includes:
- Bilateral and trilateral meetings between the protected area authorities of the three countries and the four parks,
- Communication network and system for regular information exchange between the three countries (regional meetings, warden's coordination meetings, etc.),
- Regular joint patrols between field-based park staffs of DRC, Uganda and Rwanda,
- Improved understanding of the habitat, priorities for conservation and threats, through collection, analysis and exchange of information at a regional level (Regional Information System). This includes data on human use of the forest, key species in the forest, socio-economic indicators outside the forest, and habitat change through remote sensing data,
- Skills-building of protected area authority staff,
- Improved communication and relations between the three countries,
- Increase in flow of benefits to local communities.

22.4.5 Lessons learned

Looking at the transboundary work implemented in the Greater Virunga Landscape region, and placing it in the political and social context, a number of lessons can be identified. However, most of the transboundary natural resource management lessons cannot be examined in isolation from the context of conflict in the region. Therefore, a great deal of overlap exists between lessons learned on the potential and importance of focusing on conservation during conflict, (Cairns, 1997; Lanjouw, 2003) and the potential and experience of transboundary natural resource management (Kalpers and Lanjouw, 1997; Muruthi, 2000).

22.4.6 The transboundary natural resource management continuum

Effective conservation involves the reduction of threats to natural resources, ecosystems or species. When those threats come from more than one side of a border, it is necessary to focus on threat reduction at a transnational level. Given the sovereignty of nations, this will require formalised coordination and, where possible, formalised collaboration of conservation activities. The stronger the will and ability to work together, the more effective the conservation will become.

If we look at transboundary cooperation as a continuum: At one end, efforts can simply be made to make conservation approaches in each country harmonious, or non-conflicting. At the other end is full transfrontier management of one shared ecosystem, or a formal Transboundary Protected Area. When strategies move along the continuum towards increased collaboration, the habitat can be managed more effectively. However, the ability to conserve effectively can be hampered if there are too many different people, institutions and sectors involved, For this reason it is not always possible, or even desirable, to establish full transnational management of an area as one shared unit. Such high-level political involvement can delay, or even impede effective collaboration on the ground. Therefore, effective transboundary natural resource management could be defined as the point on this continuum where the level of collaboration results in an optimum level of positive conservation outcomes, relative to costs. It needs to be flexible over time and evolve based on needs and opportunities. The development of human and institutional capacity is a critical factor. Collaboration across borders only happens among people, either as individuals or as members of institutions. To collaborate effectively, a basic level of trust and understanding is required. In addition, the institutions need to be strong enough to coordinate their activities with others. A strong institution has good capacity and a clear understanding of the issues involved. Once mechanisms for effective coordination have been developed, and institutionalised, collaboration becomes routine.

There is no lack of examples showing the uselessness of 'paper parks'. Formally designated parks or policies, with no basis on the ground, have little impact on conservation and natural resource management. Working out the complex mechanisms, both institutional and personal, that make collaboration work on the ground is the fundamental basis of effective conservation. Once established and implemented by all parties, formalisation of these mechanisms and relationships is often a much simpler process. IGCP and WCS took this 'bottom-up' approach when tasked with developing transboundary activity in the Virunga region because they knew that it is the people in the field who are usually the most motivated to find realistic and practical solutions because they generally want to do their jobs as effectively and efficiently as possible. Furthermore, by involving the many stakeholders on the ground, and ensuring their needs are met, transboundary processes can become sustainable. IGCP and WCS have emphasised the importance of working with the protected area authorities on the ground – the wardens and rangers. But it is also important to involve community representatives, local authorities, and conservation and development partners. Now the groundwork has been laid, the transboundary program is gradually being recognised by the protected area authorities' headquarters, representatives of the parent ministries and higher government levels in each country involved. The result has been overriding support and interest in the aims and achievements of the program, despite the fact that, at present, formalisation of the process is not yet possible.

Transboundary collaboration is a process rather than a goal. The formal designation of a transboundary park is not the trigger for collaboration – what is important is that the relevant people and organisations are already working together, communicating and coordinating activities, developing joint plans and implementing joint or coordinated activities. A framework for greater collaboration and a greater sense of shared aims is evolving, involving people from all three countries. One of the major outcomes of regularly bringing together wardens from different countries is the development of friendships between them. It is these friendships, and the shared desire to conserve the larger landscape, rather than any formal agreement, that have resulted in action thus far. These friendships will be the conduit for greater transboundary collaboration in the future.

* The International Gorilla Conservation Programme was established in 1991 as a regional program in the Virunga-Bwindi region, involving a coalition of three international conservation organisations: the African Wildlife Foundation (AWF), Fauna and Flora International (FFI) and the World Wide Fund for Nature (WWF).

Policy level
Executive level
Implementation level
Technical/advisor level

```
┌─────────────────────────────────────────────────────────────────┐
│  National Ministries                                            │
│  (Environment/Tourism/Foreign Affairs)                          │
│  MOU                                                            │
│  Formal classification of the Transboundary Protected area      │
└─────────────────────────────────────────────────────────────────┘
                              ↕
┌─────────────────────────────────────────────────────────────────┐
│  UWA/ICCN/ORTPN                                                 │
│  MOU                                                            │
│  Role: to establish and implement a policy for a collaborative  │
│  framework                                                      │
│  To develop a Transboundary Management Plan                     │
└─────────────────────────────────────────────────────────────────┘
           ↕                                          ↕
┌──────────────────────────────┐      ┌──────────────────────────────┐
│ Wardens' Joint Activities/   │      │ Mechanism for the promotion  │
│ Wardens' Committee           │ ←→   │ of Institutional Linkages    │
│                              │      │                              │
│ Roles: sectoral planning and │      │ Role: to establish and       │
│ execution of the regional    │      │ manage collaboration with    │
│ framework, Warden's Committee│      │ stakeholder institutions     │
└──────────────────────────────┘      └──────────────────────────────┘
                ↕                                    ↕
              ┌─────────────────────────────────────────┐
              │ Regional Transboundary forum for        │
              │ conservation (key stakeholders)         │
              │                                         │
              │ Core secretariat for coordinating a     │
              │ regional framework                      │
              │                                         │
              │ Role: compilation of the Regional       │
              │ Management Planning framework,          │
              │                                         │
              │ Monitoring and evaluation               │
              │                                         │
              │ Communication                           │
              └─────────────────────────────────────────┘
        ↙           ↕              ↕              ↘
┌──────────┐ ┌──────────────┐ ┌──────────────┐ ┌──────────────┐
│ IRS group│ │ Regional     │ │ Regional     │ │ Security     │
│          │ │ enterprises  │ │ tourism group│ │ Group        │
│ Role:    │ │ and community│ │              │ │              │
│ research │ │ conservation │ │ Role:        │ │ Role: define │
│ and      │ │ group        │ │ implementing │ │ and develop  │
│Monitoring│ │ Role:        │ │ a regional   │ │ Protected    │
│          │ │ conservation-│ │ approach to  │ │ Areas's roles│
│          │ │ related      │ │ tourism      │ │ in           │
│          │ │ enterprise in│ │ planning and │ │ strengthening│
│          │ │ the community│ │ development  │ │ regional     │
│          │ │ conservation │ │              │ │ security     │
│          │ │ awareness    │ │              │ │              │
└──────────┘ └──────────────┘ └──────────────┘ └──────────────┘
```

Andy Plumptre, Maryke Gray, Marc Languy, Léonard Mubalama, Déo Kujirakwinja & Déo Mbula

23 Towards a comprehensive research and monitoring program for Virunga National Park

Andy Plumptre, Maryke Gray, Marc Languy, Léonard Mubalama, Déo Kujirakwinja & Déo Mbula

Effective management requires rigorous monitoring. Management experts use what is called the project management cycle in which an intervention is used, it is monitored and lessons drawn from the results of monitoring are used to refine the intervention. Park management is similar to this kind of project management, but the monitoring is all the more important because of the many unknown variables that affect complex ecosystems. Park management is a process of trial and error, and the failures are as useful as the successes in improving management strategies. The ICCN, with its partners, recently developed its first plan to monitor and therefore improve the effectiveness of management of Virunga National Park. The plan brings together the various types of monitoring activities that have been tried and tested in the national park over many years. In this chapter, the authors look at the different monitoring techniques that are in use in the park and then introduce the new plan as a comprehensive monitoring system for the whole park.

Research and monitoring are quite often, and with good reason, intimately tied to protected area management programs. Each provides key information for the other. Thus, the main areas of research and monitoring are presented at the end of this chapter, with reference to relevant management activities over the next 10 to 20 years. A detailed historical overview of research in Virunga National Park was given in Chapter 5.

23.1 The history of monitoring in Virunga National Park

Despite Virunga's long history, a comprehensive system for monitoring the effectiveness of park management activities has only just been established. There are three main areas where monitoring has been used historically: 1) large mammal population dynamics, 2) vegetation change, 3) climate.

23.1.1 Monitoring large mammal populations

The monitoring of large mammals was reviewed in some detail by Languy in Chapter 8. The author primarily focused on the large savanna mammals and noted the decline of species such as elephants, buffaloes, topis and hippos in Virunga National Park. There has also been extensive monitoring of the mountain gorillas in the Mikeno sector of Virunga National Park (Gray *et al.*, in press; Scholley, 1991; Aveling and Harcourt, 1984; Weber and Vedder, 1983; Harcourt and Fossey, 1981). These results are documented in Chapter 12. Verschuren (1986) provides some data on the changes in bird populations, suggesting an increase in vulture and marabou stork populations, probably related to an increased number of large mammal carcasses due to rising levels of poaching.

23.1.2 Vegetation monitoring

As shown in Chapter 7, Verschuren (1986, 1993) documents the changes in park vegetation by comparing fixed point photos taken in 1934 (De Witte) and subsequently at regular intervals (Verschuren). These photos indicate a range of outcomes. In some areas, a marked increase in woody vegetation is probably the consequence of a corresponding decrease in large mammals. Likewise, there is an increase in grass height in areas formerly grazed by hippos. Other areas, where animal population numbers have not declined to the same extent, have remained largely stable. Chapter 7 provides some striking examples of time-series photographs. In the same chapter Van Geysel describes changes in vegetation in different geographical settings over time.

23.1.3 Climate monitoring

Rainfall and temperature have been monitored over many years at the meteorological station in Rwindi. Data are available from the late 1950s through to the mid-1980s (Verschuren, 1986) which suggests a slight increase in rainfall over time, as well as an increase in average and maximum temperatures. It is difficult to assess whether the changes in temperature are real

23.1
ICCN officers and a partner project undertaking a ground truthing exercise for satellite imagery. These images provide an opportunity to monitor long-term developments, especially illegal park invasions.

because the changes are very small and could be the result of errors in thermometer reading. Other data are available from other stations in Virunga National Park, but these need to be collated and analyzed.

23.2 Recent developments in the monitoring program in Virunga National Park

Several recent initiatives have improved the monitoring system of the park. Recent large mammal surveys (Mushenzi et al., 2003) build on earlier surveys to provide an accurate picture of the changes in population abundance over time. The International Gorilla Conservation Programme (IGCP) (a coalition between the World Wide Fund for Nature—WWF, the African Wildlife Foundation—AWF, and Flora and Fauna International—FFI) helped to develop a Ranger-Based Monitoring programme (RBM) that provides daily information on patrol ranger observations of key species and illegal activities in Mikeno Sector of the park, useful for the effective deployment of patrols. Other methods used include boundary monitoring and monitoring the extraction of natural resources from fixed points, together with the analyses of vegetation and land-use change using aerial photos and satellite images.

The advantage of using a range of different monitoring techniques is that it is sufficiently flexible that, should funding be constrained, the suspension of one form does not affect the continued implementation of other less costly forms. Nevertheless, the disadvantage of this is that these modular initiatives have not always been properly integrated. As a result of this lack of coordination, in 2005 the various partner organisations of Virunga National Park met to develop the first monitoring plan for Virunga, with the objective of providing a comprehensive system for all monitoring activities. This system is explained later in this chapter.

23.2.1 Development of Ranger-Based Monitoring (RBM) in the Virunga Volcanoes

In 1997, IGCP developed RBM as a key management tool for the three national park authorities of Uganda, Rwanda and the Democratic Republic of Congo (DRC) who share responsibility for the conservation of the mountain gorilla habitat which straddles the borders of all three countries. The aim was to provide 'a simple tool for monitoring and managing the forest ecosystem of Virunga and Bwindi'. The name RBM was established because of the strong focus on routine observations by patrolling rangers, who used this information to guide their day-to-day conservation activities.

The system covers a number of areas: the seasonal nature of illegal activities, changes in the intensity of illegal activities over time, the distribution of illegal activities across the protected area and the degree of surveillance offered by patrols. The system allows managers to determine the best deployment of surveillance efforts across the protected area. Data on key species are also collected, including detailed information on the gorilla groups habituated for tourism (recording the population dynamics, distribution and diet), thus providing an up-to-date assessment of the status of the gorilla populations over the whole ecosystem. RBM data is coupled with socio-economic data collected outside the park to gain greater insights into the need for resources of local communities and the impacts this has on the life of the park. This allows the authorities in Uganda, Rwanda and DRC to better understand the interplay between the needs and behaviour of the local population and the threats to the forest and its wildlife.

Despite these benefits, RBM is essentially an opportunistic system, based on ranger patrol observations, and it, therefore, has a number of inherent weaknesses. The data provided do not strictly adhere to formal scientific sampling frameworks. The primary objective of the patrolling rangers is law enforcement and this, not monitoring, dictates where and when patrols are deployed. If rangers only patrol in areas with illegal activities, the risk is the data they collect will be biased. Moreover, the rigor with which data are collected will vary depending on the other activities being pursued by the rangers, and less rigorous recording of data can lead to an underestimation of illegal activities. There is no allowance for variation in observer precision, and no provision for detection rate differences according to habitat.

Nevertheless, it is important to emphasise that effective protected area management requires an optimal relationship between available resources and survey accuracy. The RBM system is well adapted to the needs of park managers as it provides extremely timely information at minimal cost, and the data are systematically collected. It is easily integrated into an automated system of analysis, and then processed and distributed to the different national parks' management teams. The RBM is simple, requiring only readily available resources which ensures it can be maintained when funding is scarce or during periods of armed conflict.

23.2.2 Expansion of Ranger-Based Monitoring across all of Virunga National Park

The success of RBM in the Virunga Volcanoes region has led to its extension across the rest of the protected area. The preliminary project was originally established in the context of UNESCO's Law Enforcement Monitoring Program, but with only limited results. In early 2004 ICCN worked with the Wildlife Conservation Society (WCS), IGCP and WWF to standardise the data-gathering process, so that data from the Virunga Volcanoes on illegal activities and data from other sectors of the park were collected in a comparable fashion. The aim was for rangers to be able to collect the same range of data in any part of the park. Training in data collection and GPS use was carried out and all patrol posts received the equipment necessary to implement RBM in all park areas where the security situation allows.

The WCS has supported the RBM program outside the Mikeno Sector since 2004, and three years of data has already been entered into the monitoring database. ICCN now has an office in Goma with the necessary hardware and software and a staff member trained to develop and manage the database. Certain preliminary results have been analyzed for 2004 and are summarised in Figure 23.2. The distribution of data points indicate that only about half of the national park's area has actually been patrolled, the other areas were too insecure to patrol effectively.

Poaching has been a critical problem in Virunga

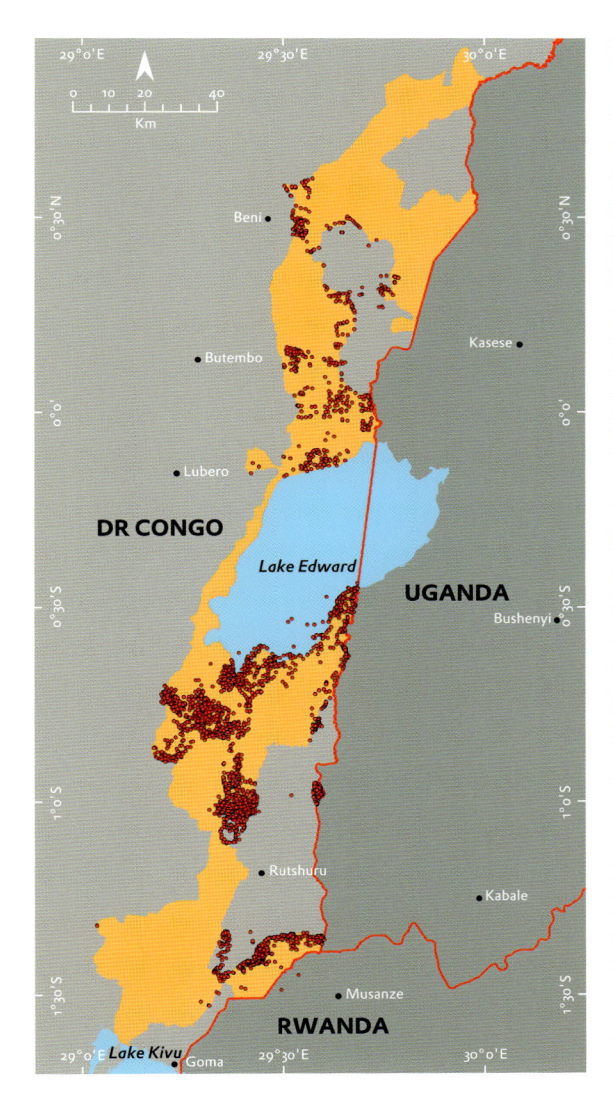

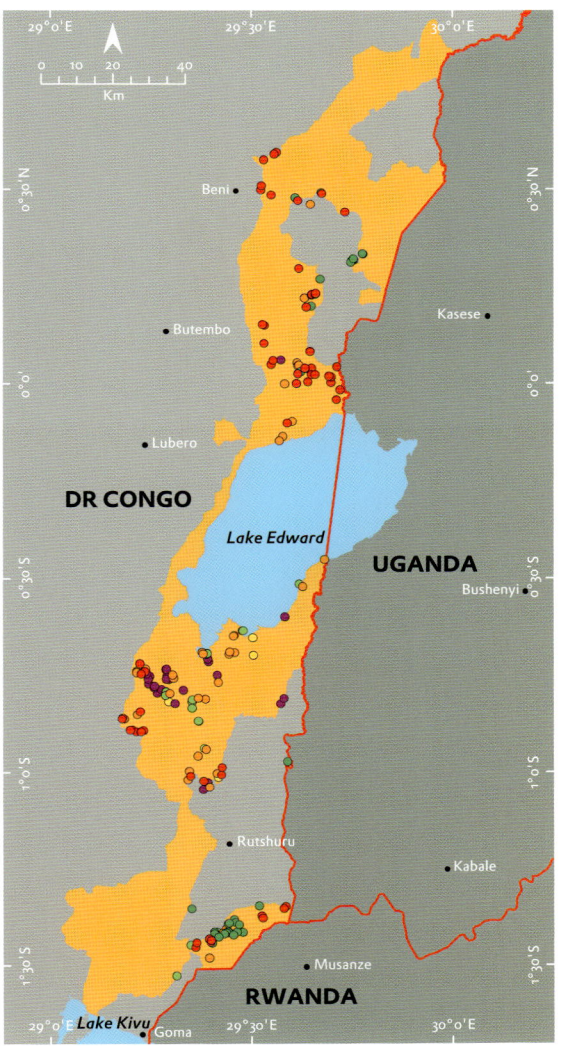

23.2
Ground covered by patrols in 2004. The areas surrounding active volcanoes and the Semliki forest, eastern part of the Central Sector and the west of the Rwenzori mountains, were barely visited because of the heavy presence of rebel military

- ● Town
- ― International Boundary
- ● Sites visited
- ▮ Virunga National Park

23.3
Signs of poaching reported by the rangers in 2004 in Virunga National Park. Gunshots, carcasses, camps and direct observations of poaching and snares.

- ● Town
- ― International boundary
- ● Metal snare
- ● Rope snare
- ● Poachers' camp
- ● Gunshots
- ● Military camp
- ● Hunters
- ● Armed contact
- ▮ Virunga National Park

National Park during recent years, especially in the Central Sector (Figure 23.3). Armed contacts between rangers and poachers have taken place to the south of Lake Edward and on the western shores of the lake.

Encroachment of the park for land for agriculture and house construction has been a serious problem. In 2004 settlers invaded large areas of the park (Figure 23.4) but, since then, several areas have been cleared by ICCN with support from its partners. Invaded areas not picked up through the RBM system (e.g. the western shores of Lake Edward, Kilolirwe) can still be detected using satellite imagery or aerial surveys.

Observations of key species were reported by rangers on patrol. Sightings of certain species are particularly interesting because they are either threatened or unusual animals to find in this particular landscape (Plumptre *et al.*, in press), such as lions and elephants (Figures 23.5 and 23.6). There are only small numbers of these species in the park and they require large expanses of land for their survival. Therefore, for viable populations to survive in Virunga National Park the entire landscape needs to be protected.

These maps show what can be produced when ranger data are accurately recorded. It is also possible to see the number of animals or illegal activities that are encountered for each patrol day, how this changes over time, and to assess the differences between the various sectors of the park (Figure 23.7).

The long-term objective of the RBM is to evaluate patterns of illegal activities and their distribution within the park. For the Virunga Volcanoes, a much larger database already makes it possible to make these analyses. The data reflects the significant human pressure in the area and indicates very clearly that illegal activities are frequent and consistent across the protected area. From 1998 there has been an increase in law enforcement activity, but unfortunately this has not significantly reduced the frequency of illegal activities in the areas of the park that are patrolled. Metal snares continue to be found throughout the Virunga Volcanoes Range, while other activities are more localised (e.g. bamboo cutting is more frequent in the Volcanoes National Park than in Virunga National Park).

RBM has proved extremely useful in providing critical information to managers during periods of conflict in the country. Due to their regular presence within the park, rangers are often able to access parts of the park that are too dangerous for wardens or members of the conservation groups. The rangers are, therefore, able to provide information which can be mapped giving an

23.4
Areas invaded by cultivation, identified by rangers on patrol. This does not include those areas that could not be visited by rangers for security reasons.

- ● Town
- ▬ International boundary
- ● Agriculture
- ● Human settlement
- ▮ Virunga National Park

23.5
Elephant observations, elephant dung and carcasses

- ● Town
- ▬ International boundary
- ● Elephants (observation)
- ● Elephants (carcasses)
- ● Elephants (dung)
- ▮ Virunga National Park

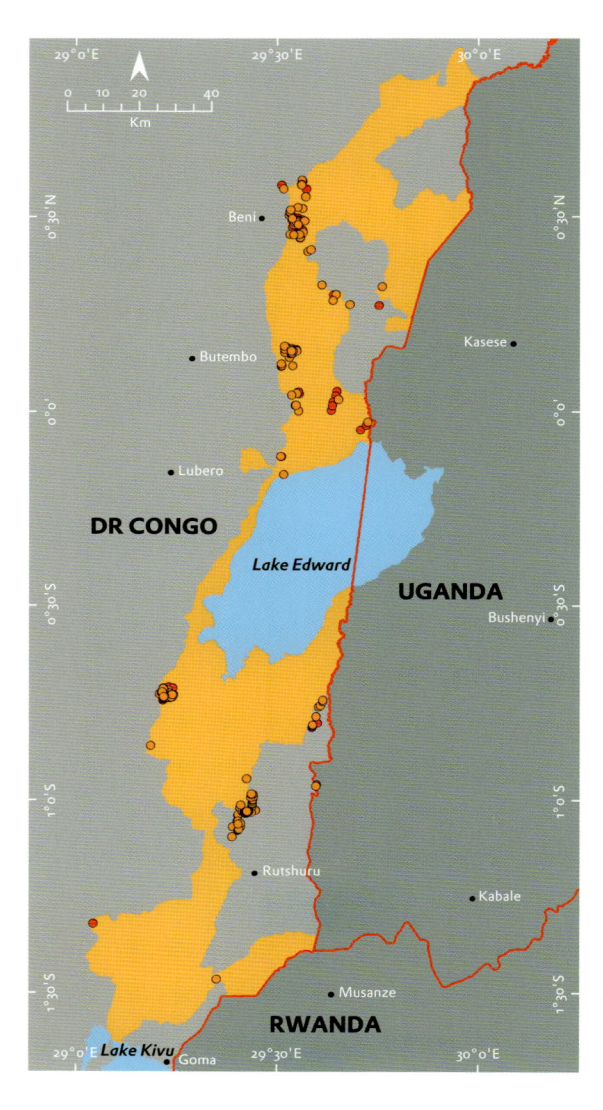

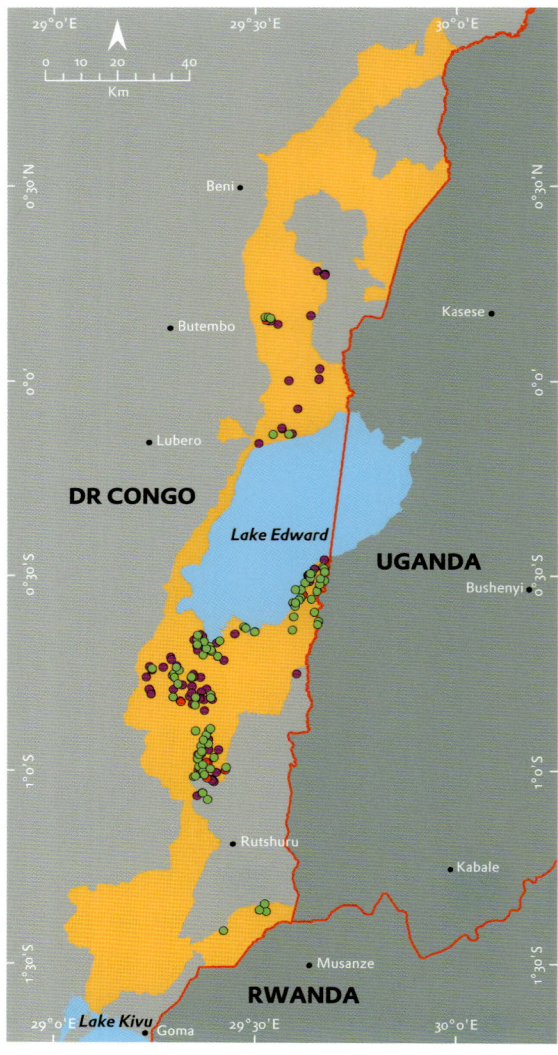

invaluable visual presentation of what is happening on the ground.

However as we have discussed above, RBM does not provide all the information needed for the effective management of the park. The inherent bias of the data needs to be taken into account during the analysis of results. Though the system is effective at keeping tabs on the levels of illegal activities and useful for detecting species of particular interest, for the sake of proper scientific rigor it is important to also use standard sampling techniques which provide accurate estimates and confidence intervals.

23.2.3 Monitoring the integrity of the park boundary by aerial reconnaissance

Park invasions are one of the most serious threats to the integrity of Virunga National Park. It is therefore essential that ICCN be in a position to assess the integrity of the park boundary (see Chapters 9, 24 and 25). There are over 1,100 kilometres of boundary around Virunga National Park, some of which is extremely remote and insecure, making it all but impossible for rangers to patrol on foot. Recently, a system has been devel-

oped to carry out regular monitoring of the boundary. Digital video or high-resolution still cameras are fixed to the undercarriage of aircraft so that they can take photographs of the underlying landscape. The cameras are linked to a GPS, which, in turn, is connected to an onboard computer. In this way, the system can upload the aerial photograph and the geographical coordinates of the centre of the photograph. By carefully calibrating the frequency of shots, it becomes possible to monitor over 400 kilometres of boundary in a single flight. On returning to base, the data can be downloaded onto a geographic information system (GIS) and analysed almost instantly to verify the degree of invasion, changes in land cover or the establishment of new settlements.

23.2.4 Landscape monitoring with high-resolution satellite imagery

The national park's considerable surface area also makes it difficult to cover the interior on foot. It is even more difficult to monitor the areas outside the park on the ground. As shown clearly in Chapters 9, 13 and 25, the use of satellite imagery provides information of immeasurable value for assessing land use change out-

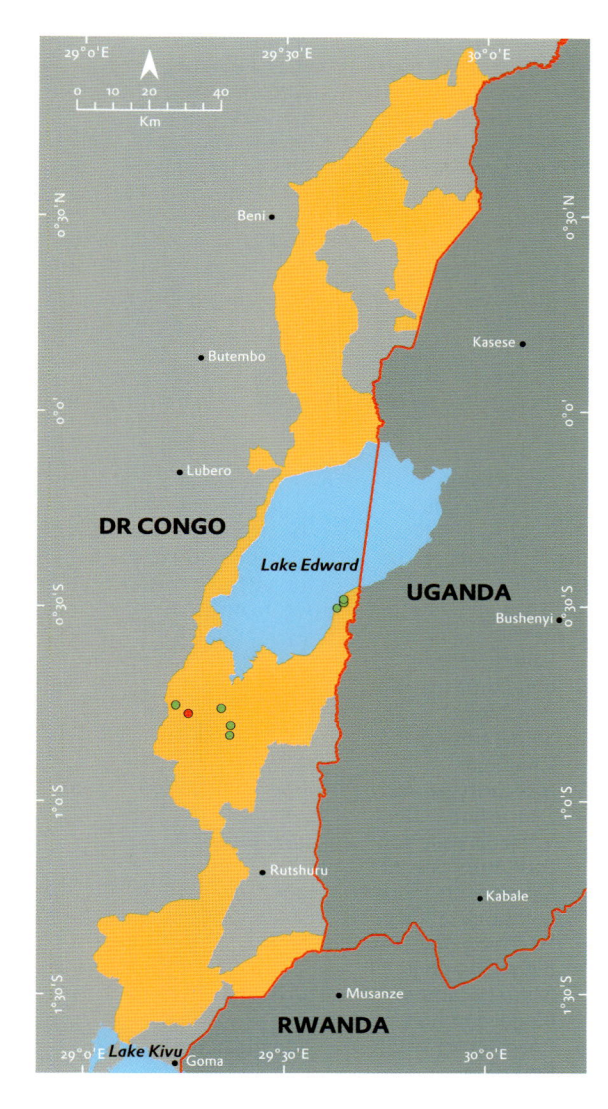

23.6
Lion observations and lion carcasses

● Town
— International boundary
○ Lions (observations)
● Lions (carcasses)
▬ Virunga National Park

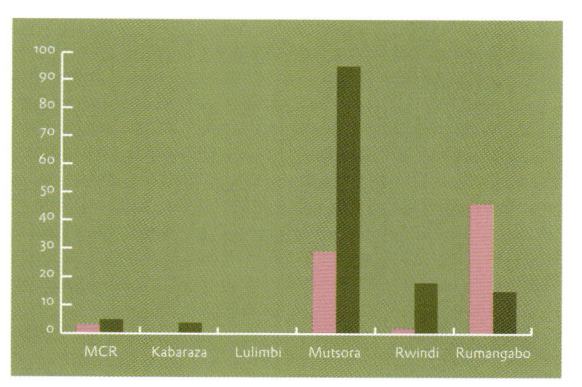

23.7
The number of snares recovered per day by patrols in 2004. This graph clearly shows that metal snares are much more frequently encountered in Mutsora. In Mikeno, rope snares are more common.

▬ Number of rope snares
▬ Number of metal snares

side the protected area. The use of satellite imagery has several advantages:
- It allows for a total coverage of vast areas, without logistical or security constraints,
- It provides data that is much more reliable and spatially accurate than aerial photographs,
- It allows for better change analyses over time (unlike other data collected on foot where systematic and random errors can be introduced making comparisons difficult),
- It is a means of objectively documenting a situation on the ground that cannot easily be contested by the various parties who may be involved in the changes that are observed (see Chapter 24).

The use of satellite imagery for surveying Virunga National Park is subject to constraints. Firstly, cloud cover in eastern DRC can render the regular recording of satellite imagery very difficult, especially in the mountainous areas. However, on the whole, satellite imagery in Virunga National Park has been successful and recent use of the technology has shown that it is possible to obtain a satellite scene without any cloud at least once a year.

A second constraint is the cost. Virunga has a long shape, which requires several scenes to cover. It is an advantage, however, that the park's orientation (north north east to south south west) corresponds with the flight path of the satellite, and so only six 60 kilometer by 60 kilometer (SPOT imagery) scenes are required for coverage of the whole park. At present, the total cost for a full mosaic of scenes covering the relevant area is 31,500 Euros or 40,000 US dollars. The same amount again is required if five meter resolution panchromatic (black and white) imaging is used. These can be merged with the ten-meter color images. While the cost is considerable, it has to be seen in relation to the 2,500,000 US dollars spent in Virunga National Park by donors each year, together with the 1,500,000 US dollars per year that tourism could yield (see Chapter 11). Given that the imagery only needs to be purchased once every five years, the cost is less than one per cent of the total cost of park management and could be considered a reasonable investment.

In 2004 and 2005 WWF acquired a full set of SPOT *(Satellite Pour l'Observation de la Terre)* images covering the entire protected area. This exercise demonstrated that a ten meter resolution was adequate to establish major land use changes across the landscape, such as the clear felling of the Mikeno forests in 2004 and forest destruction along the Semliki River. However, a five meter resolution is necessary to locate areas within the park that have been invaded, boundary areas under pressure, or the expansion or introduction of new fishing settlements along the western shores of Lake Edward (see Chapter 13). The complete mosaic of 18 scenes at five meter resolution taken at this time covers the entire protected area and also the area within 50 kilometres of the park boundary. As such, the mosaic includes the buffer zone around the park. We expect to acquire the same scenes in 2009–2010, which will help to set management priorities for the park.

23.2.5 TIMES SERIES PHOTOGRAPHS

Verschuren (1986, 1993) compiled a set of personal and other photographs taken in the early years of the park's existence, and at successive dates, to document habitat change within Virunga National Park. These photographs cover 30 or so sites and Chapter 7 details their exact location. Chapter 7 demonstrates the value of these photographs in depicting the habitat changes that have occurred. Most of those photos were repeated in 2005 and 2006 making it possible to detect changes over the years. It would be ideal if those same locations could be photographed at ten-year intervals.

23.2.6 Monitoring from fixed points

Early signs of deforestation or other illegal activities can be detected from fixed observation points, especially elevated positions. From the Rumangabo hill for example, charcoal burning can easily be detected. Once spotted, the ICCN can deploy patrols to the sites and take corrective measures. The mountainous relief of Virunga National Park makes it possible to locate a considerable number of similar vantage points, including those that could be used for surveillance over the Rwindi plains.

Observation posts can also be placed at strategic points along roads. These allow the monitoring of activities, such as the transport of charcoal or wood from the forest. WWF has already put in place a number of these observation posts which have made it possible to document the illegal exploitation of 98,200 cubic meters of wood (of which two thirds had already been converted to charcoal), over a period of 12 months in 2004.

23.2.7 Socio-economic surveys

The significant evolution in conservation policy and practice to involve close collaboration with local communities in recognition of the fact that national parks need to deliver reasonable benefits to them, means it has become necessary to monitor the socio-economic status of these populations together with their understanding and attitudes towards conservation. WCS worked with IGCP and CARE in 2002 to determine the basic socio-economic status of populations around the Virunga Volcanoes and Bwindi National Parks as well as Nyungwe. More recently WCS and WWF carried out similar work in the north west of Virunga National Park. These studies should be extended to other parts of the park and used to monitor people's behaviour and attitudes towards the park in relation to different interventions, with a view to encouraging greater participation by local communities.

23.3 The establishment of a monitoring plan for Virunga National Park

As shown above, several initiatives have been launched (or re-launched) over the past few years to establish formal monitoring in Virunga National Park. However each of the initiatives was implemented by an NGO with a vested interest in the project and the funding. It has become necessary to establish a general framework for park monitoring. Why is monitoring necessary? What needs to be monitored and why? How would the results be used? How would management adapt its strategy as a consequence?

In early 2005 ICCN and its NGO partners came together to develop a comprehensive monitoring plan for Virunga National Park. WCS proposed a method, based on their work in Uganda, designed to ensure that monitoring data was used by management staff to inform their decision-making. The logical process involved:

- Identifying targets (species, habitats and ecological processes) to protect within Virunga National Park,
- Identifying threats to these targets,
- Classifying threats based on the method proposed by Margoluis and Salafsky (2001) by defining the area affected by the threat, the intensity of that threat, and the urgency with which the threat needs to be addressed. (In the case of Virunga National Park, this classification was carried out separately for the park's different sectors because the threats differed in nature from one sector to another),
- Evaluating the strategies that ICCN and its partners currently use to address these threats,
- Identifying the parameters and monitoring indicators that are useful in monitoring the effectiveness of the strategy,
- Establishing the data collection techniques,
- Identifying the people responsible for data collection and analysis,
- Nominating the people who need to be informed of the results and when they should meet to establish the implications of the patterns identified.

In Virunga National Park, the plan identified 15 major threats to the park, affecting 30 target species considered key for the conservation strategy, together with 22 habitats and processes. In order of importance, these threats include:

- Land invasions
- Illegal construction
- Poaching
- Illegal fishing
- Charcoal
- Insecurity caused by the presence of armed militia
- Fuel wood collection
- Volcanic eruptions (a natural process)
- Bush fires caused by man
- Disease introduced by man
- Livestock in the park
- Poaching of infant gorillas
- Illegal logging
- Military camps involved in illegal activities
- Illegal bamboo harvesting
- Poaching of wildlife outside the national park
- Pollution
- Climate change
- Illegal water collection from within the national park

The monitoring parameters (the data that needs to be measured to establish if the degree of threat is changing) and the indicators (proxy measures that can be used if the threat itself cannot be recorded) have been identified for all of the threats listed above, except volcanic eruptions, climate change and pollution which participants decided needed to be examined through a research exercise and not a monitoring program. The impact of armed groups and military camps was combined into a single category. The difference between a parameter and an indicator is illustrated well for the case of poaching. The measurement of the incidences of poaching taking place (the monitoring parameter) is used to determine if the level of threat is changing. However, because generally the poached animals are removed and therefore all direct evidence of the threat is hidden, the number of carcasses, recovered snares and arrested poachers are used as indicators.

Research projects have also been developed to help managers better understand the underlying causes of the threats and to develop improved strategies to counter them. They include looking at the socio-economic factors that encourage poaching and illegal fishing; improving understanding of local people's use of the

23.8
Rangers being trained in the use of modern survey equipment.

23.9
There has been very little research into the invertebrates of Virunga.

Towards a comprehensive research and monitoring program for Virunga National Park

23.10 *ICCN rangers and their neighbouring colleagues undergoing training in monitoring techniques.*

park; the effectiveness of crop protection measures and the impact of illegal logging and charcoal production within the park.

The new monitoring plan runs for five years, from 2006 to 2010. It is highly dependent on RBM data. It also requires aerial survey data on large mammal populations and certain bird species, as well as encroached areas and boundary and fishing villages. Annual meetings are convened to examine the patterns of change in the degree of threat, and to establish whether necessary corrective measures are having the desired effect.

23.4 Towards a research program that is fully integrated into monitoring: the main axes

As discussed above, research and monitoring are intricately linked. In the context of Virunga National Park's monitoring program, the main priorities for research were identified by ICCN and its field partners. This was formalised during a meeting in June 2005, with the participation of the wardens and rangers in charge of monitoring from all of the main stations (Mutsora, Rwindi, Lulimbi, Rumangabo, *Domaine de Chasse* and Kabaraza), representatives from the Provincial Headquarters and representatives of the conservation partners of ICCN in Virunga National Park (WCS, WWF, IGCP, Dian Fossey Gorilla Fund International, Gorilla Organisation, Zoological Society of London, Frankfurt Zoological Society, Mountain Gorilla Veterinary Project).

In order to finalise the subjects for research, the participants were asked to name the various species or processes on which they felt more information would improve park management, with particular emphasis on the management plan. These subjects (see below) will be used to guide future data collection, in order to elucidate the main axes of research. The research plan is mainly expected to benefit management decision-making, but it can also be used in the promotion of ICCN and Virunga National Park and to increase their appeal to donors for the park's long-term sustainability. In this light, the role of the Protected Area Management Information System (SYGIAP), coordinated by the central headquarters of ICCN, is pivotal to the successful implementation of the new monitoring plan.

Other areas of research were considered vitally important and urgent, according to their location and the needs of the managers. Some of these research topics are specific, others more general, and they are grouped according to a number of themes as follows:

Crop damage
- The effectiveness of protective barriers,
- The impact of wildlife crop damage on household livelihoods,
- Initiatives that encourage local residents to tolerate wildlife and not to seek to destroy it.

Use of park resources
- How do local populations derive benefits from the park,
- Factors that encourage local populations to invade the park,
- What options are available to increase people's access to water, wood, and bamboo outside the park,
- The amount of illegal logging within the park and its impact,
- Changes in resource exploitation within the park.

Charcoal production
- Who are the participants and what is the impact (ecological and socio-economic) of charcoal production?
- Economic factors leading to charcoal production,
- Options for reducing or stopping charcoal production within the national park.

Poaching
- Economic factors contributing to poaching,
- Poaching intensity within the national park and impact on animal populations,
- Options and strategies for reducing or stopping poaching.

Illegal fishing
- Impact of over-fishing on local livelihoods,
- Attitude and behaviour change following the depletion of fish stocks,
- Options and strategies for protecting breeding sites,
- Economic viability of fish trading, and identification of the beneficiaries.

Human settlement
- Number of people settled within the national park, and their impact on the park's resources,
- Economic pull factors inciting people to enter the park,
- Push factors motivating people to leave the park,
- Improving park-people relations.

Military presence
- Impact of armed groups on wildlife and habitats within the park,
- Attitude of military personnel towards conservation,
- Options for reducing illegal activities carried out by military personnel.

Gorillas
- Extent of the baby gorilla trafficking,
- Strategies for putting a stop to the traffic,
- Carrying capacity of the Virunga and Sarambwe massifs for gorillas,
- Impact of gorillas on cultivation,
- Disease in gorillas, and preventative measures,
- Impact of bamboo cutting on gorilla movements.

Large mammals
- Importance of contiguous protected areas for large mammals,
- Population dynamics of key species (chimpanzees, elephants, lions, topis),
- Carrying capacity of the Mikeno Sector for elephants and buffaloes,
- Disease transmission from livestock to wildlife.

23.11 Research and monitoring topics on particular species

Species	Survey Method	Data gathering frequency
Chimpanzees	Forest surveys	3 years
Golden monkeys	Transects	
Migratory birds (aquatic)	Direct counts (4 sites)	Seasonal
Carnivores, bushbuck, duikers etc.	Camera traps	
Rwenzori duikers	Camera traps	
Habitats	Satellite imagery	10 years
	Aerial photos	
	Periodic static shots	
Bamboo studies (Mikeno, Tshiaberimu)	Field notes	6 months
Gorillas	Census	5 years
Hippopotamus	Aerial counts	3 years
Elephants	Aerial counts,	
	Forest transects	3 years

Other species
- Population dynamics of key species (crocodiles, shoebill stalks, vultures, migratory species),

Cattle breeding
- Number of cattle breeders within the park and reasons for their installation,
- Impact on the park,
- Possible options for their evacuation from the park.

Socio-economic and political
- Changing socio-economic status of protected area adjacent populations (including density and demographic aspects),
- Factors contributing to illegal activities,
- Improved crop yields.

Bush fires
- Impact of man-induced bush fires and their management,
- Options for fire-management.

Habitat
- Current distribution of habitats,
- Changes in habitat structure, especially savanna and bamboo,
- Impact of changing habitat on animal populations,
- Functional role of wildlife within the national park and impacts on habitat.

Stéphane Leyens, Carlos de Wasseige & Marc Languy

24 The contribution of geomatics to the management of Virunga National Park: Social and ethical issues and perspectives for the future

Stéphane Leyens, Carlos de Wasseige & Marc Languy

Geomatics is the combined use of the technologies of remote sensing via satellite imaging, Global Positioning System (GPS), and digital mapping via Geographic Information System (GIS). These have become essential tools for delimitation, demarcation, surveillance and, increasingly, the management of protected areas. In this chapter, the authors reflect on the challenges and advantages that the use of these technologies offer for the improved conservation of Virunga National Park. They also consider some of the ethical and social considerations in relation to the use of this technology.[1]

24.1 Introduction

The usefulness and even the necessity of spatial imagery to enhance the protection of Virunga National Park cannot be questioned. Protected areas across sub-Saharan Africa are increasingly under pressure from growing human populations, posing a complex challenge for their continued conservation. This is particularly true for Virunga National Park where conservationists have endured a very high population density and armed conflicts have devastated the Great Lakes region for 15 years. There are four separate issues. Firstly, Virunga National Park is located in one of the most populated regions in Africa and so the protected area is under severe human pressure. Secondly, the genocide in Rwanda in 1994 led to the establishment of immense refugee camps around the park, dramatically amplifying the level of human pressure on the natural resources of the park to this day. Thirdly, the park is located in a particularly unstable region where armed gangs and warlords loot natural resources at will. Lastly, a near total lack of respect of the law often renders the work of the park authorities on the ground almost impossible. Consequently, the number of illegal incidents in Virunga National Park is very high and the possibility of instilling respect for the legal boundaries of the park is extremely limited.

We will look at three examples of how geomatic tools have been used to aid the effective management of Virunga National Park, to demonstrate the increased role that these technologies can and should play in surmounting the above difficulties and protecting the park in the future.

24.2 The contribution of geomatic and mapping tools in monitoring Virunga National Park

24.2.1 The contribution of satellite images: The case of the deforestation of the Mikeno sector

The importance of remote sensing in the conservation of protected areas is no longer open to question. In the case of Virunga, remote sensing has been shown to be essential for a whole series of management activities and controls. Below we look at a number of scenarios where the use of satellite imagery has been particularly useful for the managers of Virunga National Park, before going onto look at the particular example of the deforestation of the Mikeno Sector in 2004.
- Firstly, in the delimitation of park boundaries; a satellite image provides formatted and objective information on the exact locations of landmarks that mark park boundaries. For example, if the

24.1
ICCN guards taking a reading with a GPS receiver.

Remote sensing

Literally, remote sensing refers to the techniques used to obtain information about a faraway object without establishing a physical contact with it.

Thus understood, the term could encompass simple vision or acoustics. However, the modern interpretation refers to the collection of techniques and processes used for the acquisition (the recording of the energy of an emitted or reflected electromagnetic ray) and the manipulation and treatment, manual or digitised, of satellite images or aerial photographs. The gap between aerial photographs and satellite images is narrowing. It is, in fact, sometimes difficult to tell the difference between an aerial photograph and a very high-resolution spatial satellite image. The visual interpretation of aerial photographs or satellite images, commonly referred to as photo-interpretation, also pertains to remote sensing.

Global Positioning System (GPS)

GPS is a system of position location by satellite put in place by the Secretary of Defence of the United States. The technique involves a set of 24 satellites orbiting at a distance of about 20,000 km from the earth's surface and handheld receivers that capture their signals. Wherever the user might be on earth, the GPS receiver can instantly show the co-ordinates, in longitude, latitude and altitude, of his position. The accuracy of the position may vary from some ten metres to a few millimetres depending on the following factors: the intrinsic quality of the receiver; the methodological choices made for filtering the information sent by the satellites for the calculation of the position and the registering of co-ordinates; and the possibilities of correcting the co-ordinates by external means (WAAS, Wide Area Augmentation Service, a GPS correction system). The GPS receiver is also a navigation tool that can plot the direction the user should take to reach a given destination, provided it has been encoded beforehand.

boundary is a river, it is easy to follow its course and to record its outline on the satellite image as a digital file. If one were to follow it in the field various obstacles might get in the way. This digital outline can then be used to make up a map to officially set down the park boundaries for all to see.
- Secondly, regular satellite imaging of the same area allows conservationists to observe any changes in land use over time. This kind of monitoring is particularly useful given the growing external pressures from humans living inside and outside the park. In the same way as aerial photographs at the end of the 1950s were used, the first satellite images, from the 1970s, have permitted a study of the evolution of Virunga National Park and long-term land use (see Chapter 9). As the number of earth-observation satellites continue to increase it is becoming technically possible to obtain high-resolution images nearly every three days. Assuming favourable atmospheric conditions, almost continuous monitoring is possible.
- Thirdly, satellite images allow park wardens to obtain information about otherwise inaccessible areas of the park without risk and at a reasonable cost. Due to the occupation of some areas of Virunga National Park by rebel groups, it is not safe for rangers to patrol everywhere. Satellite images allow park managers to maintain surveillance of those areas nonetheless.
- Lastly, the images provide an indisputable visual record of any major ecological degradation and a means to measure its extent. They can, therefore, provide evidence that can be used to alert park managers and to convince politicians of a need to act to protect ecosystems. Satellite images can therefore contribute significantly towards the resolution of crisis situations.

One example of the vital role satellite images have to play in the conservation of Virunga National Park is provided by the deforestation of the Mikeno sector between May and June 2004 (Chapter 18). The problem began when 6,000 people, mainly from neighbouring Rwanda, invaded the park. They progressively cut and burned the forest, so they could graze cattle and farm the land.

A report prepared by various non-governmental conservation organisations described the event as 'exceptional, disastrous and disturbing'. Even though the report was written based on the accounts of people on the ground, it could not escape accusations of subjectivity and they had no way of irrefutably measuring the extent of the damage.

But thanks to satellite imagery it was possible for park managers to obtain evidence that clearly showed the state of the land before and after the invasion. An image taken in January 2003 showed the Mikeno sector and the two distinct areas of ground occupation, the cultivation mosaics on the one hand, and the forest vegetation on the other. On June 7, 2004, another image was taken at the peak of the devastation, following programming of the Spot 5 satellite to produce a ten-metre pixel picture. It took a few days for the image to reach the Geomatic Unit of the *Université Catholique de Louvain,* but once there, within a few hours spatio-maps were constructed providing objective evidence of the extent of the deforestation. They revealed that in less than one month, 700 hectares of forest had been degraded or cleared. A second image from the same satellite acquired almost a month later on July 3, 2004, revealed that 1,500 hectares of former forest had been occupied by men and cattle.

While the crisis was resolved on the ground, the satellite images were vital in giving the ICCN *(Institut Congolais pour la Conservation de la Nature)* and the ORTPN *(Office Rwandais du Tourisme et des Parcs Nationaux)* the information they needed to confirm the extent of the damage and the seriousness of the situation and to garner the support they needed to bring it to an end.

Geographic Information System (GIS)

A Geographic Information System is often considered to be just software. The software is in fact just one of three elements of the System:
- *An efficient data processing environment comprising one or several computers equipped with geographical information and mapping programs,*
- *One or more people trained in the use of data processing tools,*
- *Data available in a format compatible with the data processing environment chosen.*

The data and the staff are thus part of the system to the same degree as the data processing environment, which includes the geographic information program itself. A Geographic Information System is conceived to manage, depict, and analyse geographical data. It is supported by a geographical database made up of multiple information layers. These may correspond, for instance, to the hydrographic network, the network of roads and trails, the contour lines, the location of towns, villages and specific infrastructures, vegetation parameters, etc. More specifically, for the management of a protected area, the GIS database should ideally contain information about the distribution of the fauna; natural or semi-natural habitats; the location of illegal activities; the itineraries of patrols made by the guards and so on. Maps produced from a GIS are the result of compilations from this geographical database. Two types of information are associated with a GIS:
- *Geographic data, which gives the spatial reference,*
- *Descriptive data, compiled from the information associated with the geometric data. Geometric data are represented by three different possible entities: the point, the line or the polygon.*

The choice of the type of entity used will depend on the function of the information to be represented. For example, a village could be represented by a point (geometric data). The data describing it (name, total population, the number of schools, and so on) will be collated in the table of attributes. A GIS does not serve just to edit mapping documents, but also to perform analyses and to work with spatially referenced data. Because of these qualities, the GIS is a genuine tool in assisting decision-making processes and in enabling monitoring of the evolution of many of the characteristics of the national park and the activities that take place there. The GIS also integrates data gathered by GPS, satellite images and all other digitised and geocoded documents (aerial photographs, scanned maps and so on).

The satellite images were included in reports brought before various national and international bodies (UNESCO and the governments of Belgium, the United States and Rwanda) to ensure that political pressure was brought to bear to maintain the evacuation so that the farmers would not return.

The quality of the images and the speed with which they were available certainly contributed to resolving the Mikeno crisis and to avoiding any further threat to the same site or nearby areas. It is not the first time satellite imaging has been used in an ecological disaster, but this example shows the objective way satellite imagery can be used to document a disaster as it happens. The objectivity and irrefutability of the images is particularly important in a climate of political instability where interested parties may try to manipulate or deny the facts. Several months after this crisis was resolved, a number of authority figures admitted that 'the conservation NGOs with their satellites' had prevented the furtherance of some political agendas that were not in the interests of the conservation of Virunga National Park.

In addition, this crisis reminds us that satellite imagery is not just an illustrative tool. The growing accessibility of satellite imagery on the internet makes it difficult for decision-makers to ignore the powerful evidence they provide, ensuring more vigorous action against incursions.

Other areas at the heart of Virunga National Park are, at present, under sustained monitoring by remote sensing. Whether it's the illegal fisheries on the shore of Lake Edward, the deforestation within the Beni/Mavivi area or the degradation of gorilla habitat on Mount Tshiaberimu, remote sensing can provide such irrefutable proof of wrongdoing that few people would dare to challenge it. Though an increasingly important tool for the conservationist, it is also useful for tracking human development and the wellbeing of local populations.

24.2.2 The contribution of Global Positioning System to the demarcation of the boundaries of Virunga National Park

Participatory demarcation of the boundaries of Virunga National Park was begun in 2002 by the ICCN and the World Wide Fund for Nature (WWF). The aim was to identify, on the ground, the legal boundaries as contained in the legal documents of the park, and to ensure that knowledge of these boundaries was acquired by all parties. (The word *demarcation* used here means the process of identifying the boundaries – as spelled out in an existing legal text – followed by physically marking them. One should not confuse this with *delimitation* which relates to the definition or the establishment of boundaries through writing up and enacting a legal text.)

The *participatory* dimension was important as it would be unhelpful for the ICCN alone to identify the boundaries of Virunga National Park: there needed to be a joint acceptance of the boundaries by all concerned parties if they were not to be constantly called into question.

The technical solutions offered by GPS are a valuable aid in the identification and location of the landmarks described in the legal documents or recognised during the demarcation. The technology allows information to be placed in a geographical information system (GIS) (see box). Each location is stored in the memory of the GPS receiver before being transferred to the computer and integrated into a GIS. The GPS allows the collection of geographic data and also works as a navigation tool, serving as a type of modern-day compass.

With high-resolution spatial satellite images (printed on the background of a geographic map or displayed on a computer screen) and a GPS receiver, reconnaissance and identification work in the field is greatly facilitated. The GPS enables an area to be clearly marked out or a

31 January 2003   *7 June 2004*   *3 July 2004*

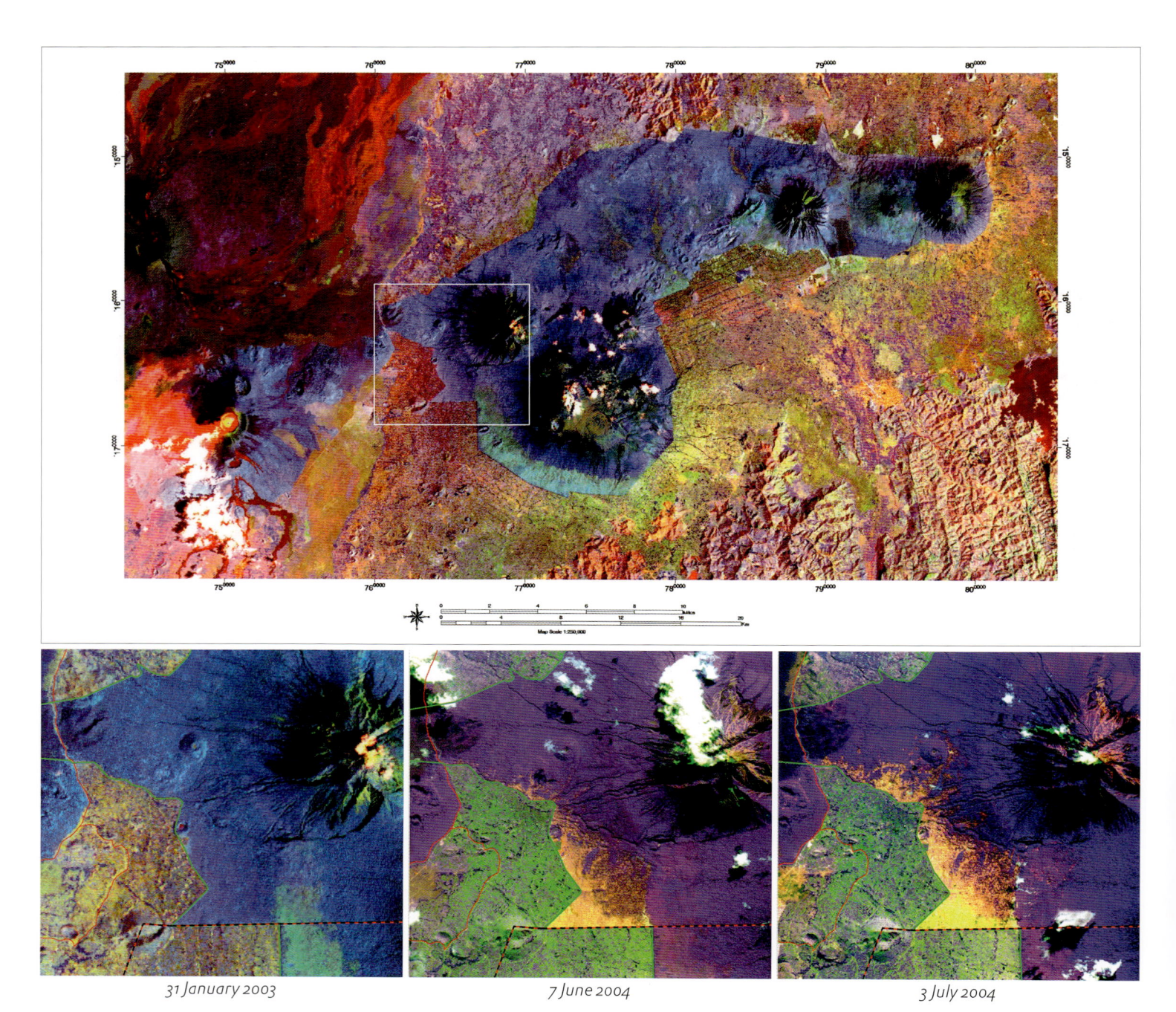

24.2
Satellite images of the extensive deforestation in the Mikeno sector during 2003 and 2004. In this false colour composition, mountain forest appears in purple and the cleared forest areas appear in yellow-orange.

landmark's exact location clearly identified with, at worst, a maximum imprecision of ten metres.

The points recorded on the GPS are connected to each other by linear segments. These segments can either follow the physical features of the territory, such as rivers, roads or the crests of hills or mountains, or it can connect two positions in a virtual, abstract way. For example, two hilltops can constitute landmarks through which the boundary must pass.

The GPS receiver works as a navigational tool in that a landmark that was previously registered and encoded can easily be re-found. For example, the co-ordinates of longitude, latitude and altitude of signposts placed during a demarcation mission are registered by GPS, with a precision of ten metres or better. This recorded information may then be used at a later date to check that each signpost is still in place, providing an efficient means of monitoring the effectiveness of border demarcation over time.

24.2.3 The significance of a Geographic Information System in the management of Virunga National Park: The case of relocation of populations.

Virunga National Park authorities are facing a growth in the illegal settlement of populations within the park. As a result of this, they have begun a dialogue with all concerned parties, including the populations themselves, to explore the possibilities of shifting these populations to areas outside the park.

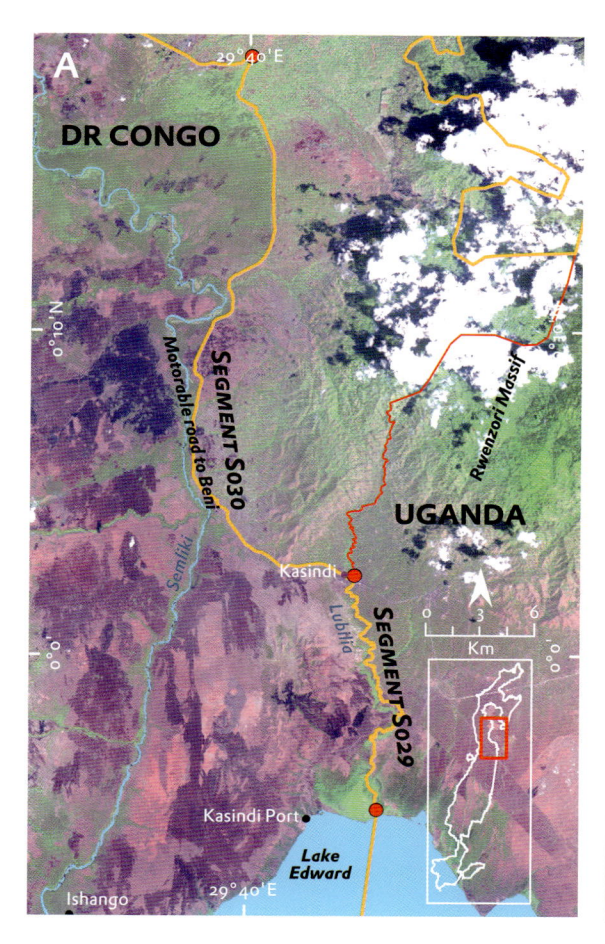

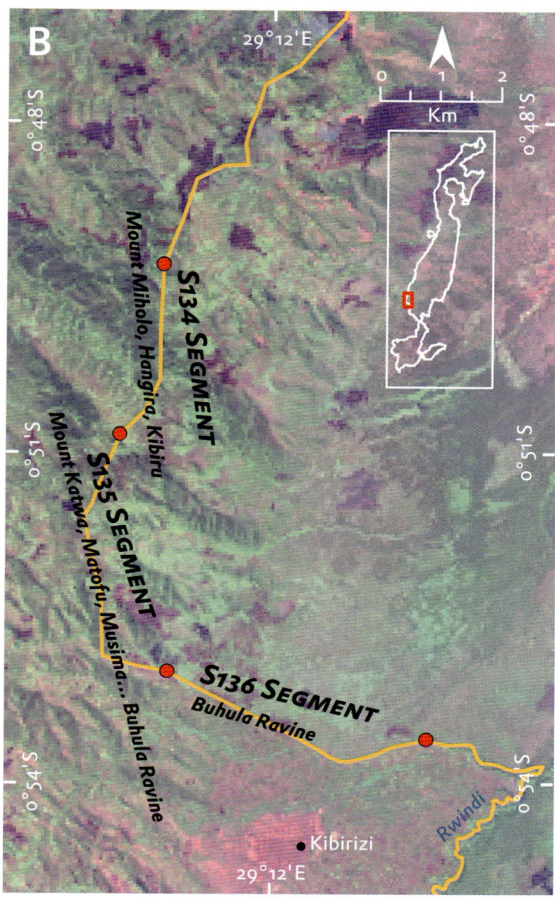

24.3
Maps showing the different techniques for delimiting park borders, using different landmarks joined by a continuous line:
A The line follows the landmarks of a river and a road
B The line is drawn between various hilltops.

- ● Patrol post
- ● Village
- ● Starting points and end points
- ━ International border
- ━ Virunga National Park boundary

The major challenge of this process is to identify resettlement sites for the populations who have decided to evacuate the park voluntarily. These sites also have to be accepted by other local parties, including park authorities, civil and tribal authorities and eventually the populations who would accommodate the new arrivals.

The usefulness of a GIS in the process of shifting populations lies essentially in its ability to:
– Produce analyses of various types of data,
– Foresee and evaluate a variety of factors affecting the relocation,
– Produce specific maps.

The GIS can be used to provide maps that give a visual aid that allows the affected parties a means of clearly weighing up the different options. Unlike a traditional map on paper, the map created from the GIS can be continuously updated and edited as new information comes to light. It is a user-friendly way to discover, analyse and document prospective relocation sites and a formidable tool in the preparation of field visits, an activity for which GIS cannot be substituted.

It is in this way, therefore, that the authorities identified a suitable resettlement site for the people illegally settled at the Kilolirwe site of Virunga National Park. Success is guaranteed only if the displaced population and those who will accommodate them do not find themselves worse off than before. Therefore, the proposed site needed careful evaluation and comparison with the needs of the populations involved. The spatial evaluation of the various characteristics of the habitats involved, and their integration into a GIS, permitted a choice of land according to well-defined criteria. Satellite images, aerial photographs, scanned maps and GPS points collected in the field are essential elements for making up the GIS.

After these field visits, the site of Bibwe was chosen as a possible resettlement site that offered a series of advantages responding to the expectations of the displaced population. Bibwe is located in the chiefdom of Bashali in the Masisi territory. The area is characterised by a succession of plateaus and hills with an altitude of nearly 2,000 metres. The climate varies noticeably according to altitude. The average temperature is 20°C at lower altitudes and 12°C in the high mountains. The region benefits from a regular rainfall that feeds an important network of small streams, and has soils with an excellent agricultural value. Hence, it has an environmental context favourable to the development of luxuriant vegetation and the practice of sustenance gardening (beans, maize, cassava (manioc), sorghum, peas, bananas, fruit trees, sweet potatoes, Irish potatoes, colocasia), and keeping livestock (cattle, sheep, goats, chickens, ducks, pigs). Moreover, since the region has a relatively low population density, the maximum capacity of the land has not yet been reached.

The sudden arrival of a great number of people in the region requires much thought about numerous aspects. The displacement of the population of Kilolirwe ideally will need to be accompanied by an improvement in their standard of living; it is a real opportunity to begin development activities and improve infrastructure in

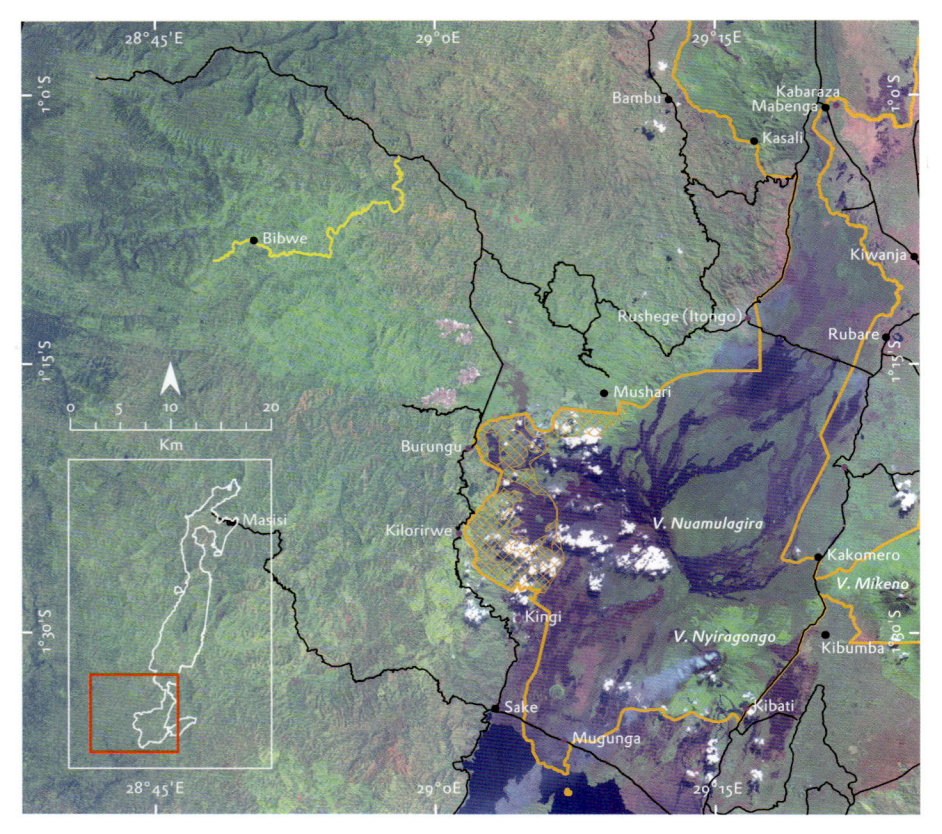

24.4
Showing Kilolirwe, the invaded area inside the park, and the location of the relocation site Bibwe.

- ● Village
- ● Patrol post
- ― Planned road (towards Bibwe)
- ― Road
- ― ViNP limit
- ▨ Invaded area

the region. The deficiencies can become proposals for funding projects, with the GIS used to evaluate exactly what work needs to be done and the extent of it.

The main developments required for the resettlement to the Bibwe region to be feasible are:
– Construction of housing for families who have decided to evacuate Virunga National Park,
– Food and non-food assistance to households during the first six months,
– Access of displaced households to health care,
– Access to drinking water,
– Rehabilitation and equipping of schools – an immediate need since these schools would be receiving the children of the displaced people,
– Reduction of poverty in the resettled households and support for the re-launching of revenue-generating activities,
– Rehabilitation of poor sections of roads and the construction of bridges with the aim of furthering access and integration of the region.

24.3 Geomatics and conservation: Some social and ethical issues

The case studies presented above, showing how geographic information techniques (remote sensing, GPS and GIS) contribute to the effective management of Virunga National Park leave little doubt as to their importance in nature conservation. However, the situation is more complex than it appears. It is well known that the application of a technique in a social field – and nature conservation is certainly a social field with its human history, institutions and impact on society – creates a new set of difficulties at the same time as resolving the original problems. In this section, we would like to offer a few thoughts on the social and ethical challenges that the use of geomatic tools present in relation to the management framework of Virunga National Park – and more generally, in the framework of nature conservation in developing countries.

24.3.1 Efficient conservation and equitable conservation

It has been recognised for several decades now that nature conservation can no longer ignore the broader context in which it is carried out. Barring the use of military force to preserve it, a nature area cannot be effectively protected unless conservation managers take into account the social reality in which it exists. Local and national socio-economic situations, cultural values, competent institutions, customary law and regional history are elements that a serious and efficient conservation program must integrate into its protocol.

In conservation agendas it is now common to accept that conservation and development are two sides of the same coin, including concepts such as 'participatory management' or 'community conservation'. These days the general idea is that conservation should work with and for the local populations. Although, in practice, all conservation players now accept this, the reality of the situation is that there are wide divergences as to how and with what aim. Some people are convinced that totally inclusive programs are the only way forward. Others dismiss these as utopian 'communitarist' strategies based on myths disconnected from reality (Anderson & Grove, 1987; Oates, 1999). Between these two extreme positions there is a large spectrum of more nuanced positions. However, a running theme is a general concern about justice. Whatever motivations the various players might have – be it 'humanist' or 'pragmatic' – all wish to ensure that conservation is equitable, ie: fair or just, for all. It can be said that awareness has formed that if conservation aims are to be achieved, if conservation is to be efficient, it must also be perceived as equitable by all those affected. Taking the issue of boundaries an example, it would be difficult to enforce a national park boundary without taking into account the needs of local people for access to wood, water or land. If the national park is their only source, the boundary will not hold as you have unfairly removed this resource. If you help them locate an alternative source, you have made the conservation fair and the boundary is more likely to hold.

24.3.2 What is justice? What is equitable management?

We define an action that is equitable or fair or just as an action (a strategy, a policy, a program) that gives to each his due (Perelman, 1990). This is a definition of (social)[2] justice on which everyone can agree. It is a fundamental definition since it states that justice is about ensuring *everyone* receives the benefits to which he/she is entitled. Therefore, justice is global as one must consider all the different parties concerned by an action. In this way park management can only be equitable or fair if it endeavours to guarantee to everyone concerned what he is owed. The complexity of achieving this level of fairness becomes clear when you consider that a conservation policy implies that some resources (land and the fauna and flora it supports) become inaccessible to some people in order for other resources (a natural heritage, a healthy atmosphere) to benefit all.

The shortcoming of this definition is that it says nothing about how to determine who is entitled to what. Theories of justice have developed around the question of whether it is possible to identify criteria to define this and, if so, how to proceed to that identification (Kymlicka, 2003). In order to gain a better idea of what equitable conservation management should look like we shall use a theory which defines justice as tripartite (Fraser, 2005): Justice requires an *equitable distribution* of resources; a *recognition of the identities* of those concerned and a *parity of participation*.

Breaking this definition down, we can see that *equitable distribution* derives directly from the principle of 'to each what he is entitled to'. But besides a pure distribution, this definition of justice requires a *recognition of the identities* of the different people involved so that we can distribute fairly to each. Without this the distribution could be mere blanket compensation without respecting that different people have different values. The following question then arises: How does one determine the values and the differences that should be recognised? We then reach the third pillar of this definition of justice: A democratic procedure is needed that endeavours to give a voice to all parties concerned by an action. Therefore a *parity of participation* in decision-making is needed to ensure that the distribution is fair. Equitable conservation management should incorporate all these elements.

24.3.3 The participants and their interests

Who are the parties concerned by the management of a protected area? And what are the issues that affect the value they place on the impacts of conservation?

There are three levels — local, national and international. At the local level, the populations living near or in a protected area are directly concerned by decisions about the management of the area. The 'revenue shortfall' resulting from restrictions related to nature conservation (sources of proteins, combustibles and building material), as well as revenue generated by some activities associated with the protected area (eco-tourism, employment in administration) or compensations in material goods (various infrastructures, schools, dispensaries) constitute the goods in need of redistribution. The impact that conservation policy has on the activities of these populations (hunting, fishing) and on their social representations (the symbolic tie with nature) can either endanger or, on the contrary, promote their identity (valorisation of a traditional knowledge, involvement in management), also affecting how much they value the goods to be distributed.

At the national level, it is the State, its institutions and its populations that are concerned. The redistribution of goods concerns the benefits the country enjoys as a result of conservation (eco-tourism, parallel investments by donors in infrastructure or education) as well as the losses that could result from conservation (forest, mineral or lake exploitation). National concerns around conservation centre around the image of a nation, with valorisation of the managing institutions and the national natural heritage on the one hand, and the loss of national sovereignty in favour of international management on the other.

Lastly, at the international level, mankind as a whole is a concerned party, both in a material way (redistribution of wealth such as biodiversity) and a symbolic way (recognition of a natural heritage common to the human species).

24.3.4 Geomatics as an equitable management tool in Virunga National Park

Moving from the illustrations given of the use of geomatic techniques in the management of Virunga National Park (point 24.2 above), we will emphasise some ways in which the use of remote sensing, GPS and GIS may be able to contribute to the development of more equitable or fair and, therefore consequently more efficient, conservation practices. The main advantages of geomatics are based around the objectivity of the data, the multiple layers of information that can be incorporated and considered and the access people might have to this information, allowing a greater participatory approach.

24.3.4.1 Objectivity of the data: Distal approach and proximal approach

In each of the three cases illustrated above, the high degree of objectivity and flexibility that geomatics techniques permitted was emphasised. The data collected and treated is precise and difficult to dispute. Contrary to field data, the satellite images used in the Mikeno crisis, for example, offered complete and objective information, which was easily quantifiable and facilitated rational decision-making, in turn imposing reason upon everyone. In addition, as is attested in the case of demarcation with GPS, the recourse to objective, precise, easy-to-share information is an important means of ensuring a participatory approach.

24.5
A sign placed in 2004 during participatory demarcation of the boundaries of Virunga National Park.

24.6
Concrete markers delimiting the national park. They were laid down over 40 years ago and remain very important physical markers of the boundary of the park.

Finally, the GIS, used in the defining of resettlement sites for displaced persons, meant that the many different considerations and values of the various parties involved could be considered and evaluated.

Here we return to the characteristics of equitable conservation. These require that information be accessible to all involved parties allowing wide participation; that there is the possibility to integrate multiple factors allowing the recognition of the identities, values and differences of all involved; and that it is possible to evaluate the different distribution options to ensure they are equitable or fair. The objective treatment of data is a necessary condition for proper management.

However, a word of warning here: Assuring the objectivity of the information is not so simple. First of all, objectivity is an ideal never attained. When the expert in geomatics establishes a map, the vectors that he will choose to represent objects; the key that he decides to use to distinguish the different elements; the colours that he uses are all subjective factors that he imposes on the *reality* that he describes. The motivation for the development of the map is another layer of subjectivity. Hence the importance of a participatory approach, in which the motivations of the different parties concerned can be 'made objective' or cancel each other out as they are drawn together in the mapping process.

It is also essential to keep in mind that the objectivity of the data does not predetermine the action that follows. Nor does it justify any particular action. Highly objective data has no *a priori* value greater than subjective data. Thus, a distal approach which can be considered more objective, such as remote sensing, cannot entirely substitute a proximal approach in the field, such as the use of GPS. In spite of its remarkable power of objectivity and quantification, remote sensing lacks the vital elements of information that only a proximal approach allows. These missing elements are mainly the causes and the reasons of the state of fact, and they often arise in a socio-economic domain; their knowledge is needed to tackle the problems of redistribution and identification.

24.3.4.2 THE SHARING OF INFORMATION: MATERIAL ACCESS AND COGNITIVE ACCESS

The case of the demarcation of boundaries with GPS is a good example of the participatory approach. A local arrangement starting with local landmarks from the official document of delimitation needs co-operation between the parties based on precise data that is understandable to everyone. In the same way, identifying relocation sites for the populations who had decided voluntarily to leave Virunga National Park is achieved with the participation of local, national and international players: With GIS laying out several options based on a multitude of factors and the different parties evaluating the different criteria in order to agree a solution. The tools of geomatics certainly provide a solid base on which to develop a participatory approach.

The crucial question concerns the level of access to the information for all concerned—'material' access and 'cognitive' access. *Material* access refers to the possibility of using geomatic techniques or of obtaining the data issued from these techniques. In the socio-economic and geopolitical context of Virunga National Park and, more widely, of the Democratic Republic of the Congo, it is reasonable to wonder whether the use of techniques that require a relatively weighty infrastructure and qualified personnel can be widely adopted. *Cognitive* access is about the possibility of understanding and using a given piece of information. Although a map edited by a GIS on the basis of a satellite image represents an interface that is relatively easy to read and interpret and the use of a GPS does not require expert knowledge, it is not clear that the entire population can participate in a procedure of decision-making enlightened by these techniques. We note, however, that different experiences give us cause to be optimistic, for example those of Turyatunga (2004) and Bassolé *et al.* (1999).

Another issue that has to be taken into account is that traditional African notions of space and property are very different from occidental perceptions. In fact, 'land property [in pre-colonial Africa] was not regulated by Latin-type laws relating to ownership. Instead the permanent availability in usufruct of the ground, or the right to use the ground which does not belong to anyone, is behind the settlement of human groups. African inhabitation has always been mobile, with continuous departures and arrivals. It is necessary to realise this to understand the absurd character of borders transposed from Europe onto Africa—geometric borders, artificial and sometimes imaginary' (Ki-Zerbo, 2004). In the light of this it is easier to see why some populations have difficulty regarding the respect of boundaries around a space that has become in some way private and the free use of which is forbidden to them. In addition, cartographic formalisation—a virtual line linking two landmarks—is significantly different from a more traditional open concept of space. For example, a cartographer would clearly separate two spaces with a single line without width, where the tribal chief would use naturally existing places, such as a clearing or a particular natural habitat, to mark out the same spaces. Consequently, we can ask ourselves if cognitive access to the information is sufficient among some of the African population for the data issued from the techniques of geomatics to be fully understood.

Finally, we have already discussed the risky geopolitical situation in Virunga National Park, which has rendered some areas inaccessible as a consequence. One of the interesting things about geomatics, and especially remote sensing, as we have already said, is that it provides a relatively easy way of gaining access to these difficult areas, by means that are sure, inexpensive and productive in terms of information collection. But perversely the very fact that surveillance could be carried out remotely, could lead to decision-makers blocking access to the field for proximal approaches by foot patrols of rangers, the importance of which we have also emphasised. Ironically, one geometric technique which increases access to information could cancel out the benefits of another decreasing the level of participation.

24.3.4.3 TOWARDS AN EQUITABLE USE OF GEOGRAPHIC INFORMATION TECHNIQUES

Despite the above analysis of geographic information, in which the advantages can be counterbalanced by the disadvantages, we set forth, in this last section, an overall hypothesis. That is, provided certain conditions are respected, the techniques of geographic information are levers for ensuring that the management of

Virunga National Park responds to the demands of social justice. In other words, geomatics provides levers than ensure much more efficient and equitable conservation policies. What are the conditions that need to be respected for this to be the case? What procedures would ensure a good use of geomatics?

Let us remember that the demands of social justice, following from our first hypothesis, mean that a distribution of goods (infrastructures, materials, education) in exchange for losses elsewhere (land, resources, symbolic places) is not enough. We also need to take into account the different values defended by the different people involved. However, equitable distribution via a recognition of different values cannot take place without the participation of all parties in the decision-making process. Participation, and therefore the proximal approach, is obligatory for justice to be achieved.

As a result, we believe that significant human and financial resources should be dedicated to developing the proximal approach in the field. The balance between GIS resources and information gathering resources on the ground could certainly be better. It is necessary to avoid shortcutting the proximal level, simply for convenience; it is important to exclusively rely on the distal approach only in real situations of risk that prevent field work.

In addition, greater participation can be assured if there are more inexpensive machines and software that are easy to handle and do not require great expertise. Internet access and a stable electricity supply are necessary if wider understanding and ability to use geomatic techniques is really to be achieved. Systems to ensure a scientific and appropriate use of these tools should be put in place. Effective participation will only take place if these systems incorporate both a technical apprenticeship, as well as a methodology that is easy to understand from a non-scientific cultural context in order to favour effective participation. In general, the just use of geomatic techniques requires a multidisciplinary approach in which the human sciences hold a significant place.

24.4 Conclusion

Faced with the anthropogenic threats that Virunga National Park is subjected to as a result of demographics and geopolitics, the remarkable opportunities offered by geomatics (GPS, GIS, remote sensing) represent a real hope for improving monitoring of the park. Not only do these technical tools provide the objectivity necessary for the development of a rational policy, but they also enable monitoring in sensitive areas where an approach on the ground involves risk. For these two reasons alone, these techniques are indispensable.

The use of remote sensing, central to the setting up of geomatic techniques, could result in a block on work directly in the field. However, GPS techniques that require physical access to the field are also crucial in understanding the social dimension of nature conservation. As long as certain conditions of installation and use are respected, we believe that geomatics is an essential tool in any nature conservation policy responding to the demands of social justice. It leads to a fair distribution of the goods at stake, determined by participation and an acknowledgement of the different groups affected by the conservation of Virunga National Park.

National, international and non-governmental organisations must take into account these factors when preparing their future management policies. In the middle term, there are two concrete courses of action. Firstly, an official vectorial file of the park boundaries should be compiled (see Chapter 25). This should be the geographical transcript of the legal document for which the landmarks, cited in the document and identified in the field, will all be properly defined by their co-ordinates (longitude, latitude and altitude). These co-ordinates will have been either collected in the field with the help of a GPS, or from a satellite image. The file will be a participatory work between experts and local representatives. It should be printable in map form for easier distribution. Secondly, the technical approach, that we advocate, with its participatory and multi-factor considerations, also allows the proper definition of buffer zones – transition zones between rural development and integral conservation. The buffer zone represents a crystallisation of the will to make social and developmental challenges compatible with efficient nature conservation and is part of a truly global ethical approach.

[1] This reflection was made thanks to the *Chaire Environnement* award from the *Université Catholique de Louvain*. The prize is sponsored by the Tractebel firm.
[2] One can distinguish social justice, which aims at the distribution of goods, rights and responsibilities, from civil and penal justice, which deals with wrongdoing, rehabilitation and punishment.

25.1
One of the old boundary signs dating from Belgian colonial times delimiting the national park. Many of them have been in place for decades, although more than 460 new signs were placed between 2003 and 2006 in collaboration with the local populations.

25 Towards the resolution of boundary conflicts and a modern definition of the limits of Virunga National Park

Marc Languy, Peter Banza & Zachary Maritim

Previous chapters have detailed the many difficulties the park has encountered, in particular the problems caused by invasions. By definition these intrusions are a violation of the park's legal boundaries. The re-establishment of park limits is thus a primary objective for the *Institut Congolais pour la Conservation de la Nature* (ICCN) and its partners. This will only succeed in the long-term if the local populations participate in the solution. This chapter examines the complex legal and physical contexts and offers some practical ways forward, via a process of participatory demarcation of limits and an updated definition of the park boundaries.

25.1 The challenge on Virunga's boundaries

Over recent years Virunga National Park has been experiencing an unprecedented number of intrusions. Between 2000 and 2004, over 160,000 people illegally entered the park to pursue agricultural activities and, to a lesser extent, to put their livestock out to pasture. This major problem has developed 80 years after the park was created and at the end of a long period of socio-economic crisis.

In order to counteract the rising level of intrusions, in 2002 the World Wide Fund for Nature (WWF) launched, through its Virunga Environmental Programme (PEVi), an ambitious program designed to encourage the retreat of illegal farmers from park land and reinstate the park boundaries. Recognition by those on the ground of the 1,150 kilometres of park boundary is essential in being able to identify illegal farmers.

It is difficult to determine the park's precise limits given that most of the boundary markers placed during the 1920s and 1930s have disappeared and that, in any case, there were too few. Most of the boundary needs to be remarked. A further problem is that the legal text stating the boundary limits dates from 1935 (followed by some modifications) and, in many cases, is based on physical characteristics that have either disappeared, been modified, or have local names that are no longer clearly understood. This dual problem is a major challenge for the park's administrators, particularly when it comes to enforcing the integrity of the park's boundaries.

25.2 Legal texts regarding the boundaries of Virunga National Park

25.2.1 Decree of 21 April 1925

Virunga National Park was created on 21 April 1925. It was then called Albert National Park, and the royal decree of this date protects some 20,000 hectares centred on the extinct volcanoes of Mikeno, Karisimbi and Visoke that formed part of the Belgian Congo territory. The park's main vocation (Article 3) was stated as the protection of the mountain gorillas, discovered just 23 years earlier. But the park boundaries were not clearly stated in the legal text, leaving the Governor General to establish the limits once the decree had come into effect. The document did, however, state the general western, eastern, northern and southern limits that should not be exceeded.

25.2 *Trees planted in lines make an effective physical marker of the boundaries of a park, but they do need to be regularly maintained.*

25.2.2 Decree of 18 August 1927

This decree established the extension of Albert National Park within the territory of *Ruanda-Urundi*. It does not directly concern the present Virunga National Park because the extension corresponds to the future Volcanoes National Park.

25.2.3 Decree of 9 July 1929

This decree gave the Albert National Park a legal 'personality'. It increased the park's area appreciably (by around 200,000 hectares) with the addition of the active volcanoes, part of the Rwindi Plain and the southern shore of Lake Edward. Strictly speaking, it is the first definition of the boundaries. The national park was divided into four main sectors: Central, Eastern, Western and Northern. The legal text also delimited some 'annexed territories serving as protection to the sectors of Albert National Park.' In reality these constitute entities that, as early as 1929, correspond to the concept of a Buffer Zone. The decree also created a Commission in charge of the administration of the national park. Although Albert National Park had gained its distinct legal personality, it was still financed and controlled by the Minister of the Colonies until independence in 1960.

25.2.4 Decree of 26 November 1934

The decree of 26 November 1934 modified the 1929 decree in order to enable the institution of 'Albert National Park' to be granted the management of other national parks and territories in the Belgian Congo – creating the new *Institut des Parcs Nationaux du Congo Belge* (IPNCB) for the job. The annex of this decree stipulated the boundaries of the national park which increased in area from 370,120 hectares to 386,120 hectares. This annex would be fully repealed by the decree of 12 November 1935.

25.2.5 Decree of 12 November 1935

The decree of 12 November 1935 is the first proper reference point for defining the limits of Albert National Park. It was modified a further three times, but it is the only text to include all the definitions of the boundaries. The decrees that follow add no more than a few corrections to some of the paragraphs. There has never been an officially sanctioned updated and complete text to replace this one. The new definitions that we use today are derived by modification of this 1935 text. Albert National Park continued to expand with the addition of all of the Belgian waters of Lake Edward, the southern part of the Semliki Plain and the Rwenzori and Tshiaberimu massifs.

The definitions appended to this decree cover the Rwandan section of the park which, upon independence, became Volcanoes National Park. The text of the appendices is organised into three sections:
1. The sectors of Nyamuragira, Rwindi-Rutshuru, Lake Edward, the Semliki and the Rwenzori (of the 149 paragraphs, 148 set the boundaries).
2a. The Mikeno sector, the part located within the territory of Rwanda (18 paragraphs).
2b. The Mikeno sector, the part located within the territory of the Belgian Congo (38 paragraphs).

25.2.6 Order of 4 May 1937

This decree inserted new paragraphs in the definitions of the boundaries. However, these modifications were to be repealed by the order of 15 May 1950.

25.2.7 Order of 17 May 1939

This royal order slightly modified the boundary definitions. These modifications were also to be repealed by the order of 15 May 1950.

25.2.8 Order of 15 May 1950

This royal order followed a law of 6 January 1944 and considerably modified the definitions of the park boundaries by inserting or modifying some 50 paragraphs. The royal order does not provide a complete revised text, however, but stipulates all the corrections that should be acted on.

This order constitutes the last substantial modification to the boundaries of Albert National Park, which later became Virunga National Park, without a change of boundaries. The present boundaries of Virunga National Park were set out, from that point on, in both the decree of 12 November 1935 and the order of 15 May 1950.

It is important to note here that since independence in 1960, only a presidential decree can modify the limits of a national park in the Democratic Republic of Congo (DRC). Even if local agreements were established between the representatives of the ICCN and neighbouring communities, they could only pertain to the administration of resources and the buffer zones in particular but not the national park boundaries themselves. It is for this reason that any agreement given by any representative (of the ICCN, the Minister in charge of Protected Areas or of the Province) authorising cultivation in the national park has no legal value.

25.3 Understanding the order of 15 May 1950

The 15 May 1950 order included some small, localised errors that are covered by Languy (2005). These seem to be errors of transcription rather than to difficulties related to interpretation of the boundary and can therefore easily be corrected. For example, the seventh change in the 1950 text replaces paragraph 139 of the 1935 text with a modification of the south-western boundaries of the park (Kilolirwe zone). It stipulates that the limit passes the foot of Mount Kishusha, then heads due east towards the *eastern* promontory of Mount Kamatombe, then due west to join the M'bili Ravine, located south of Mount Kishusha. There is absolutely no doubt, once such limits are placed on a map, that the legislator meant the *western* promontory of Mount Kamatombe.

25.4 Compilation of the full text defining the boundaries of Virunga National Park

The 1935 and 1950 boundary definitions were not drawn together into one document until the *Feuillet Technique n°1* was drawn up as part of an ICCN/WWF/EU program in April 2005 (Languy, 2005). This text consolidates the full boundaries of Virunga National Park in one document for the first time and constitutes the current reference text. It is included in Annex 1 of this book.

The *Feuillet* combines the modifications made in the 1950 order with the annexes of the 12 November 1935 decree. The corrected interpretations of the 1950 order (mentioned in section 25.3 above) were also incorporated. Each paragraph is assigned a reference number from S001 to S200, in order to express the boundaries in 200 segments.

The section of the old Albert National Park located on Rwandan territory was excised in order to establish the boundaries of the present-day Virunga National Park, located entirely in DRC.

25.5 THE CASE OF RUMANGABO STATION

It is widely accepted that Rumangabo Station and the forest immediately to the west of it are part of Virunga National Park. The station and this forest are in fact administered by the ICCN and the tacit agreements held with the tribal chiefs confirm this situation, such as it is represented on all the recent maps of the Park.

This situation, which is not questioned here, needs to be sanctioned by official documents, preferably by a decree that would formally integrate this area into Virunga National Park.

It is thus recommended that this extension of Rumangabo is considered as a part of Virunga National Park, and that the ICCN secures a legal document on the restitution of the land. This principle was taken into consideration during a workshop organised by the WWF in November 2004, which brought together the ICCN and the neighbouring communities who signed a letter of agreement to this effect.

25.6 BRIEF NOTE ON THE FISHERIES OF LAKE EDWARD

In the various documents relating to the creation and modification of Albert National Park, it is clear that the plan was to guarantee the local populations a limited access and control over 'indigenous' fisheries at Lake Edward (see Chapters 1 and 10). The practical administration of these fisheries is regulated by other documents mentioned in Chapter 10. It is simply noted here that the two fisheries at Vitshumbi and Kyavinyonge have a recognised legal status and that the existence of a third fishery, Nyakakoma, whose status remains a subject of debate, is now a fact and it is tolerated. It is, on the other hand, clear that any other fishery set up on Lake Edward within the Congolese territory is strictly illegal.

It should be pointed out that these fisheries are not enclaves but are an integral part of Virunga National Park. Moreover, the administration of their activities is the responsibility of the ICCN.

All efforts to demarcate the spatial extent of the fisheries would have as only goal the restriction of the human habitations. It could not, in any case, be considered a demarcation of the boundaries of the park.

25.7 MOBILE NATURAL BOUNDARIES OF VIRUNGA NATIONAL PARK: COMMENTS AND RECOMMENDATIONS

Most of the paragraphs concerning boundary definitions refer to natural entities such as mountains, ravines, rivers, hill crest lines and so on. Although most of these natural elements (notably hills and mountains) are immobile and have not changed over the course of time, the same cannot be said for other features, such as rivers in particular. In fact, rivers flowing on sandy plains sometimes have changing courses, especially near their delta. The most vivid example of this is the Ishasha River. Since the border of Virunga National Park corresponds with the lower course of the Ishasha River as far as its mouth at Lake Edward, this boundary is 'mobile'. In fact, a study of recent aerial photographs and satellite images from 2004, shows that several of the river's meanders have changed their course over the years and some have 'captured' others. However, the main course of the river has not changed enormously and it is accepted that it does represent the park boundary and that no modification is necessary.

Other cases are more sensitive. The point of contact between the May-Na-Kwenda River and the Kabaraza River is one of the park boundary's points of reference (see segment S017). Present-day images, as well as aerial photographs taken in 1959, show that the Kabaraza River – which is not a permanent river – no longer flows into the May-Na-Kwenda but has changed course and now flows into the Rutshuru. This is obviously a problem when it comes to fixing boundaries. In this case, the only solution is to find where the two rivers did meet in 1935. It is a matter of the physical disappearance of a point of reference. Other similar cases include the meeting point between the Rutshuru and the Rugera Rivers, which still exists, but it seems to have shifted slightly. In this particular case, it is recommended to take the present-day point as the boundary because the entire Rutshuru River sector forms a natural boundary required by the lawmakers and one that corresponds to the best administration method.

As one can see, there is no general rule to follow in these particular cases, except to keep as closely as possible to the original intention of the legislators and, again where possible, to preserve the original natural borders when they have been chosen as limits in the original boundary descriptions.

25.8 BOUNDARIES BASED ON HUMAN INFRASTRUCTURE: COMMENTS AND RECOMMENDATIONS

Various types of man-made infrastructure are also used as points of reference in the definition of boundaries, such as villages, hamlets, motorable tracks and caravan routes. However, many of them have now disappeared or been modified. In most cases, the problem is at once simpler and more complex than for natural frontiers. Simpler because the course of action is clear; the present-day boundaries should follow limits corresponding to the infrastructure that existed in 1935 or 1950 and not that standing in 2006 (contrary to some natural limits). On the other hand, the situation is more difficult because it is trickier to determine the exact location of this infrastructure in 1935 or in 1950.

An obvious and fairly simple case concerns the Goma-Rutshuru road. Given that the road has hardly changed its trajectory and represents a truly sharp boundary, the recommendation is to follow the present-day layout. Caravan routes present a more complicated case because, for the most part, these tracks are no longer visible. Here historical research is needed in order to relocate the routes, together with former intersections

25.3
Populations who invaded the sector near the Karuruma River leaving the national park after signing an agreement with the ICCN in April 2006.

25.4
One of the farmers of the region of Mavivi shows his card, identifying him as one of the people the ICCN is helping move out of the park. An agreement between the ICCN and the tribal chiefs permitted the evacuation of the Mavivi region (6,483 hectares) in June 2004 after authorising the farmers to collect their last harvest.

and hamlets. Various maps from the 1930s and 1940s (and to a lesser extent, the 1959 aerial photos) prove a very useful tool (see the following section).

25.9 THE NEED TO MARK BOUNDARIES AND THE OPPORTUNITIES AVAILABLE

25.9.1 THE NECESSITY OF PHYSICALLY MARKING THE LIMITS

As most of the boundaries of Virunga National Park were established 73 years ago, the way they were defined and the landmarks to which they referred are no longer completely relevant to the present day.

As a consequence, uncertainties have often led to conflicts over the true position of boundaries. Deliberately, and sometimes unknowingly, over the past few years numerous people have violated the boundaries of Virunga National Park.

It is very important that the park boundaries are physically marked, and that at the same time measures are put in place to assist the 'displaced' farmers who have to regroup outside the park. This is a major objective of the ICCN and the WWF, who began this process in 2002 through the PEVi. With the support of the European Union, UNESCO, USAID, WWF-Sweden and WWF-Netherlands, this ambitious program concentrates on the modernisation of the boundaries. The objective is not to modify the legal boundaries but to mark and document these legal boundaries using modern techniques.

25.9.2 THE OPPORTUNITIES OFFERED

Different opportunities are available today that simply did not exist 80 years ago. In addition to the *Feuillet Technique nº 1* – the consolidated document containing all the boundaries – three types of information are at hand to help determine exactly where these boundaries are on the ground.

25.9.2.1 CONSOLIDATED LEGAL TEXT IN A SINGLE DOCUMENT THAT TAKES INTO ACCOUNT ALL THE PRECEDING TEXTS.

Since the development of the *Feuillet technique nº 1*, a complete text is available, and each paragraph is clearly enumerated, which greatly facilitates communication about boundary sections and the use of databases.

25.9.2.2 THE 1948 MAPS

Maps, with a scale of 1/50.000, drawn in 1934 and in 1948, covering Albert National Park and adjacent areas, have been rediscovered in the archives of the Belgian Congo colony, as well as a nearly complete set, in good condition, at the National Botanic Garden of Belgium in Meise. These maps have been digitised by scanning at very high resolution, and each has been geo-referenced. The set was assembled and a mosaic of maps was integrated into a Geographic Information System (GIS). The primary interest of these maps is that they carry the name and position of many localities, hamlets, hills and streams to which reference is made in the original definitions but which have since been forgotten or have moved. These maps have brought to light crucial information for the placing of boundaries with respect to such references.

25.9.2.3 AERIAL PHOTOS FROM 1958 AND 1959

The Belgian National Geographic Institute (NGI) made a series of aerial photographs of nearly the entire national park in 1958 and in 1959. These photos, in black and white, are of professional quality and present an important tiling, which allows stereoscopic studies. Numerous features, such as houses, paths and trails, are very visible and are a great help in the interpretation of the park's boundaries, since there had been very little change between the period of boundary definition (texts of 1935 and 1950) and the period when the photos were taken. Most of the man-made changes occurred later.

25.9.2.4 SATELLITE IMAGES

High-resolution satellite images enable coverage of the entire park and yield a 'photograph' of its condition and of an entire set of natural and human characteristics in a highly rigorous and objective way, including areas that are hard-to-reach in terms of logistics as well as security. The WWF ordered and acquired a complete coverage of Virunga National Park in 2004 and 2005 in SPOT images (see Chapter 23). The resolution chosen is ten metres in colour (four bands) and five metres in black and white (panchromatic). This level of resolution enables the detection of areas that have been invaded and the assessment of pressures from agriculture in the park, information which helps the ICCN and the PEVi project workers to orient the boundary demarcation teams on the basis of urgency. In addition, the images are superimposed on a Digital Elevation Model (DEM, a kind of digital relief map) which allows the easy location of hills, summits, ravines, etc. and the possibility to adjust them to the 1948 maps so as to find the corresponding toponyms.

25.10 PARTICIPATORY DEMARCATION OF THE BOUNDARIES OF VIRUNGA NATIONAL PARK

As already mentioned in section 25.1, in 2002, the WWF and the ICCN started work through their PEVi program on the participatory demarcation of the boundaries of Virunga National Park. The objective of this initiative is to include the neighbouring local communities so that in the long term they recognise and respect the marked limits. Furthermore the illegally settled farmers would also be sought in this inclusive process in order to encourage them to respect the law and aspirations as the rest of the wider community.

25.10.1 PRINCIPLES

This initiative relies on basic principles to which each participant must adhere:
- Virunga National Park is a national (and universal) asset, and only a presidential decree can modify it. The mandate of the mixed commissions is not to redraw the boundaries of the park.
- The boundaries were published and, as they are set out in the decree of 1935 and the order of 1950, they are the only ones to be recognised.
- The communities have the right to know the exact delineation of the boundaries so as to know where farming and other activities are permitted.
- Any physical demarcation of the boundaries must be accepted by all, hence the need to officially

register these limits (that is, by the recording of GPS co-ordinates, the placement or identification of physical markers and by a physical description of the boundaries – altogether, if possible).
- To the greatest possible extent, the ICCN must return support to the local communities respecting the boundaries of Virunga National Park, and help them find access to land and resources outside the park.

Pertaining to this latter point, it is obviously not, in this case, a matter of 'buying' the displacement of the illegal farmers but of giving support to the communities living along the Virunga National Park boundary and ensuring that everything is done to ensure that the people who leave the park find a motivation for remaining outside its limits rather than returning within them. It is thus important, at this time when tourism is at its lowest level, for the communities to receive some benefits or support from the ICCN.

25.10.2 Method

The method used, which has been tried and tested in the field, is to work to consolidate the boundary, sector by sector. The work is done by mixed commissions, including representatives of all interested parties so they all have an input into the process and a complete consensus can be achieved.
- For each sector, the WWF project team is made up of an extension officer and an agronomist who have already worked with the communities (in order to assure a relationship based on mutual confidence).
- In each of the three sectors, Northern, Central and Southern, it is necessary to ensure high-level political support by means of direct lobbying. This should promote awareness within the higher echelons of the administration of the advantages to be gained from the resolution of the problems of illegal occupation of Virunga National Park. A preliminary phase clearly demonstrated that such support is both essential to the resolution of all potential problems at the local level and guarantees that the work is recognised by all parties.
- In each sector, a mixed commission is established in order to undertake the fieldwork necessary for the identification of boundaries and to document the extent of the intrusions.

A typical mixed commission is composed of:
- A representative of the Governor (the highest authority at the provincial level),
- A survey officer from the administration responsible for the registration of land-ownership titles,
- If possible a surveyor to ensure the official registration of the boundaries,
- A representative of the ICCN at the provincial level and a representative of Virunga National Park (preferably the warden in charge of the sector concerned),
- A representative of the chiefdom concerned (traditional authority),
- A representative of the WWF PEVi team,
- Various reference people, who know the territory and local history well and who remember the placement of old boundary markers and other local details.

The commission is always accompanied by representatives (at the highest level possible) of the local authority or of the traditional tribal communities and the closest village.
- After the commission has noted and documented any disagreement (and whether it needs any clarification) about the boundaries, a consensus is reached, and the personnel in the field start work on the delimitation. When an agreement on the interpretation is reached, signs are placed and recorded, and the ICCN representative and the traditional authorities sign the original certified report of the field.
- The project personnel then work with the communities in defining the kind of support that would be most useful to them to ensure they can continue to respect the boundaries in the long-term. In some cases, the communities wish to receive help in developing nurseries to establish green belts from which they would have access to wood and timber and for agro-forestry activities. They might also be interested in direct aid for basic infrastructure, such as schools and dispensaries.

25.10.3 Results

Between 2003 and 2006 the PEVi established several mixed commissions and resolved several land conflicts within the park. Over 100 field missions took place to establish the boundaries in the most troubled areas. Altogether, 256 km of boundaries have been marked and 464 signs placed using the above methods. The most impressive result is the departure of more than 80,000 people from the park during this period, and the abandonment of their illegal fields.

Figure 25.5 demonstrates the evolution of the integrity of the boundaries between 2002 and 2006. It is possible to see very clear progress as a result of this approach since the proportion of boundaries invaded has dropped from 28 per cent to 13 per cent during this period. There still remain, however, two sizeable areas of invasion: The Kilolirwe region, in the south-west of the park, and the western shore of Lake Edward (Figure 25.6).

25.11 Towards a modern definition of the boundaries of Virunga National Park

25.11.1 Objective

The work described above has proved its worth, but it is costly and relatively slow and not an exercise to undertake lightly. The work of interpreting texts and comparing them to old maps and searching for landmarks in the field requires much time if all those concerned wish for an indisputable and high quality result. However, the ICCN and the WWF have decided to devote all resources necessary to this exercise, recognising that it is essential for the restoration of the over-run areas. It is also vital to ensure this sort of invasion does not take place again (or at least that the documentation on the legal status of any disputed land is readily accessible).

Of course, no one would want this sort of work to be repeated every 10 or 15 years, when the institutional memory fades. For this reason, it is important to produce a modernised formulation of the boundaries in order to ease the task of the managers in 20, 30 or 50 years from now, when today's actors are no longer around.

25.5
Evolution of the proportion of boundaries over-run and intact, between 2002 and 2006.
- ● Town
- ● Village
- ▬ Over-run boundary
- ▬ Intact boundary
- ▬ International border

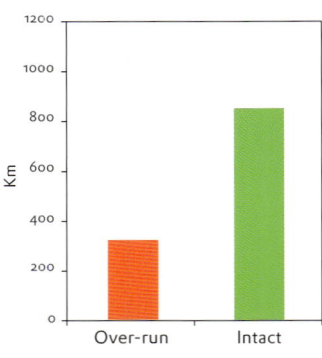

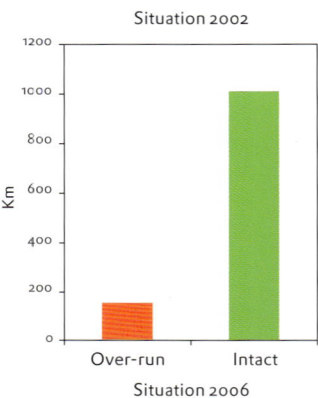

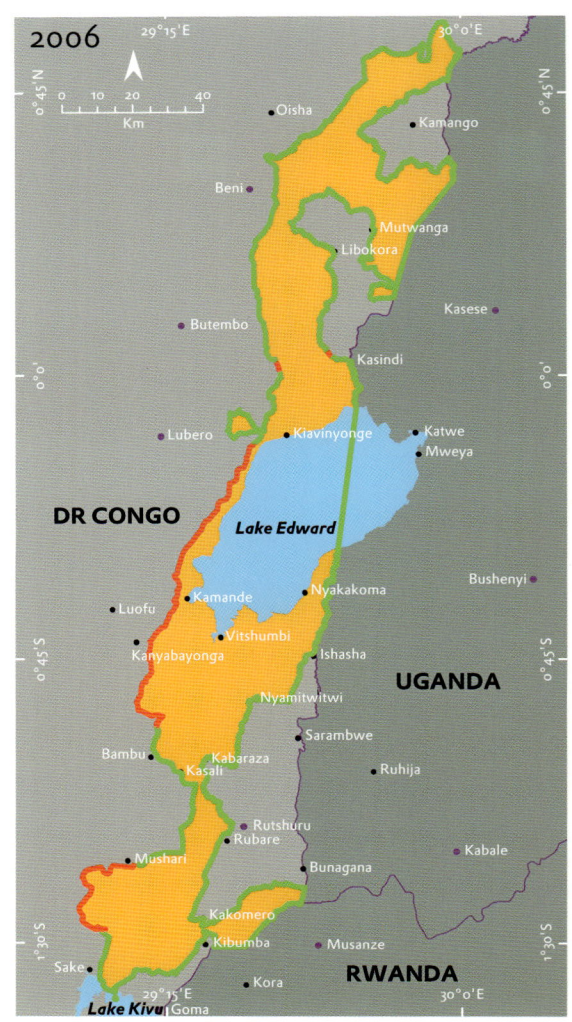

25.11.2 METHOD

Thanks to the different kinds of information detailed above, and following the results of the participatory demarcation work, the ICCN and its partners now have at their disposal a substantial set of data on the exact and legal locations of Virunga National Park's boundaries. This body of information – based on such an old text – is, however, somewhat unwieldy and not user-friendly. The 'translation' of this information into a modern easy-to-use document is a task which the ICCN and the WWF consider important to complete. Taken together, this set of data should constitute the modern formulation of the park boundaries. However it is vital that it should always remain clear that this document is complementary to the legal text and not a replacement of it.

This accessible document must meet a number of criteria:
– The modern formulation must remain entirely true to the original text; it should in no way be a new text.
– The link between the original text and the 'modernised' formulation of the boundaries must be clear and direct.
– The language must be understandable by all the participants.
– Ideally, it should be officially validated and recognised as an annex to the legal text.
– Finally, it should serve as the basis for recording recommendations for possible modifications to the boundaries in the future.

With the help of former rangers and field teams trained in the use of global positioning systems (GPS), each of the boundary segments (such as included in annex 1) is documented according to:
– The text pertaining to the segment as in the legal formulation,
– Its nature: River, hill crest line, meridian, straight line joining two points, path, etc.,
– The geographical co-ordinates of the starting point and the end point, as well as the vector ('shapefile' in GIS terminology),
– Its length (along the contour, and not in a straight line, from the start point to the end point),
– The limits corresponding to this segment, whether it is over-run or not, and if yes, by what type of activity,
– The fact of whether one or several signs were placed and/or whether concrete markers were found; the geographical co-ordinates of these signs and markers, as well as reference to the report having identified the position of these signs,
– If it proves useful, numeric photographs from the

25.7 Levels of invasion of the park

	Invasion in 2000–2005		Situation on 15 November 2006	
Locality	Area (ha)	Persons	Area (ha)	Persons
Kilolirwe	10,200	60,000	10,200	60,000
Tongo	60	0	0	0
Kibirizi	19,000	0	0	0
Kongo	9,000	18,000	0	0
Ishasha	500	15	0	0
Kanyabayonga	2,100	0	1,200	0
Tshiaberimu	3,500	1,800	0	0
Lubylia	4,200	22,000	7	100
Mavivi	19,00	25,000	0	0
Karuruma	2,000	445	0	0
Kyavinyonge	5,000	0	0	0
Kanyatsi	3,000	7,000	0	0
Lume	2,300	4,600	0	0
West coast	11,700	28,000	11,700	28,000
Total	**91,560**	**166,860**	**23,107**	**88,100**

25.6
Invasion areas in the park. Numerous invasion areas were reclaimed between 2003 and 2006 but two important regions remain occupied: Kilolirwe and the west shore of Lake Edward.

- Town
- International border
- Invasion areas at the beginning of the 2000s but reclaimed by the ICCN
- Invasion areas not yet reclaimed by 15 November 2006
- Virunga National Park

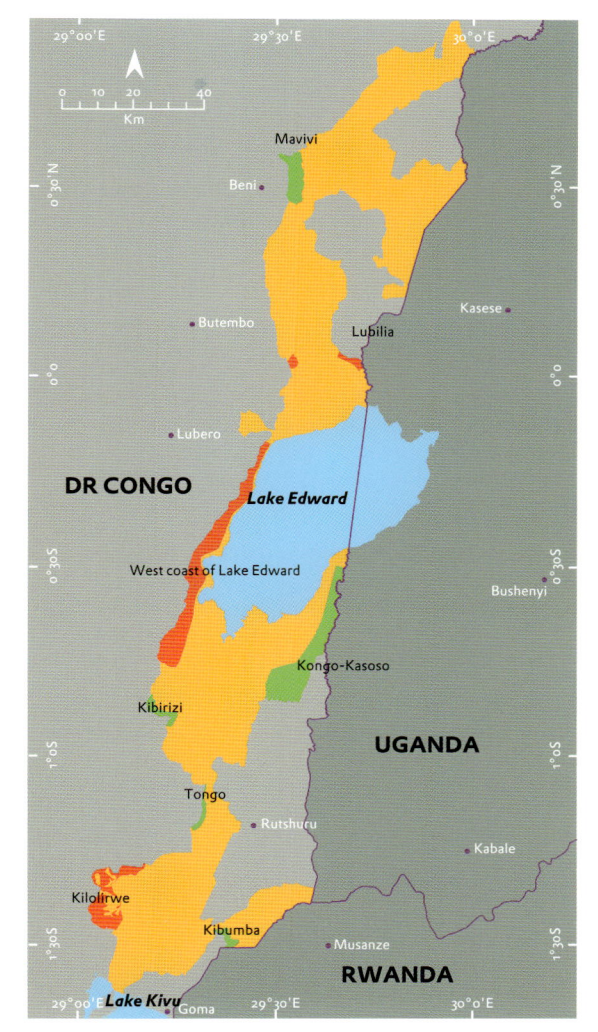

beginning of important intermediate points, easily allowing the restitution of a point in the landscape,
– In addition, all comments on the interpretation of the text or any other useful information on the limits of this segment,
– In parallel, a database taking in all the toponyms of the legal text is organised with the aim of creating a geographical index, or gazetteer, of these toponyms with their geographical co-ordinates. At present, the – still incomplete – data set is kept in an Excel file since it is a simple format relatively easy to use. It is also easily exportable to a Geographical Information System. Although this is quite possible, it is not planned at present to translate it into an Access or other database. If this should be done, it is planned that an Excel version will also be retained for the reasons set out above.

The work required to constitute this kind of database is certainly enormous and as of November 2006, the database was complete for only about 20 boundary segments. Many of the segments still remain out of reach because of insecurity. Others are in secure regions but very remote and in non-priority areas in terms of demarcation. Others, finally, are located in disputed areas for which the participatory demarcation is not yet finished.

The various geographical co-ordinates of boundary points, and notably numerous toponyms (summits of hills in particular), can be obtained on-screen by means of SPOT satellite images at high resolution, with a precision within ten metres. In the database, it is recorded whether these co-ordinates were taken from data on-screen or collected in the field.

The database can be used right now to assist scientific and objective documentation of the boundaries, a major advantage for the ICCN as it gives it the means by which to physically demarcate the limits recognised and respected by the neighbouring communities. In the long-term, the process of validating the database will involve submission to the Ministry in charge of Protected Areas.

Steps leading to the modern formulation of the boundaries of Virunga National Park

The legal texts were compiled and analysed in order to produce a consolidated text taking into account the legal formulations and modifications brought about by various decrees. The paragraphs of this legal text have been numbered in different 'segments' in order to easily refer to each of the sections of the text. Presented here are segments S140 to S144, corresponding to the limits north of Tongo (north-western limits of the Southern Sector).

Compilation of legal text	
Paragraph	**Legal text**
S140	Le ruisseau Kamokanda jusqu'à sa source au pied du mont Tshahi;
S141	Une ligne joignant cette source au sommet des monts Tshahi et Bitingu;
S142	De ce mont une droite joignant la source de la rivière Kalagala, située près du mont Rwanguba;
S143	Cette rivière jusqu'au sentier de Mabenga à Tongo, sentier longeant le pied des monts Kasali;
S144	Ce sentier jusqu'à son point d'intersection avec la rivière Butaku;

The 1948 map of the region is scanned at high resolution, and toponyms taken from the legal texts are located on it (here, underlined in orange). The map shows altitude contours at 25 metre intervals. A few points of reference (A to F) are noted for comparison with the satellite map on the following page. The 1959 photos are assembled in a mosaic that enables a complementary view of the terrain and identification of rivers, hills and other features. The 1948 map and the photos taken in 1959 are integrated in a Geographic Information System that provides approximate geographical co-ordinates of any point and the documents are superimposed for a better reading. Examination of the legal text and of the two historical documents enabled the tracing of the limits corresponding to each one of segments S140 to S144.

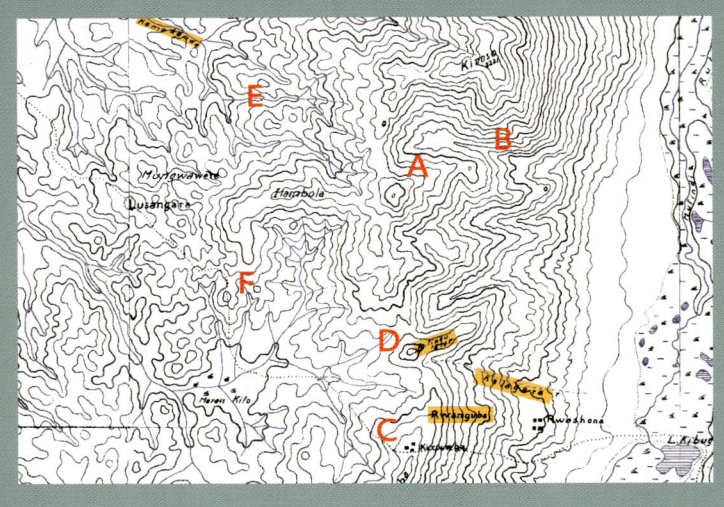

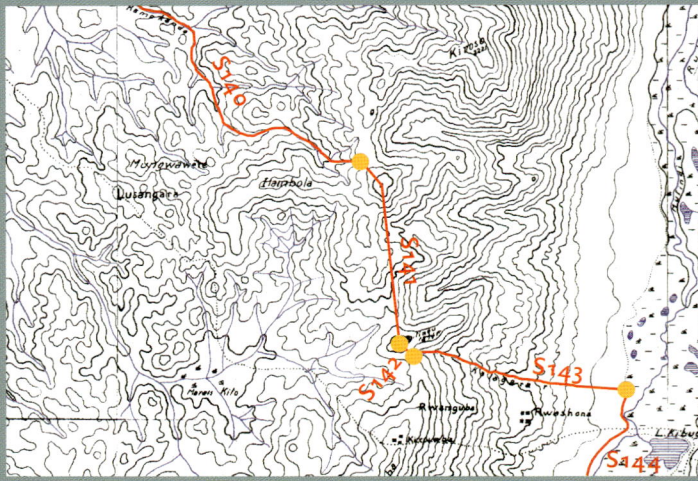

Marc Languy, Peter Banza & Zachary Maritim

High resolution satellite maps (five metres) were specially acquired in 2004 and 2005 for the whole park. The extract of the same region corresponding to S140 to S144 is shown here. It is like an up-to-date aerial photograph of the area but much more precise than a classic photo, which enables the reading of the geographical co-ordinates of any point with very much more accuracy. In addition, these images enable the identification of natural areas (forests, savannas) and areas under cultivation or other use.

Independently, a Digital Elevation Model is obtained (by other satellite data), which gives the altitude of each point, permitting the reconstruction of a relief map and hence the ability to map hills, rivers and ravines. This enables the correction of human errors inherent in the 1948 maps. In general, there is nevertheless a very good correlation with it. Points A to F are indicated here and correspond to the same points on the 1948 map on the preceding page.

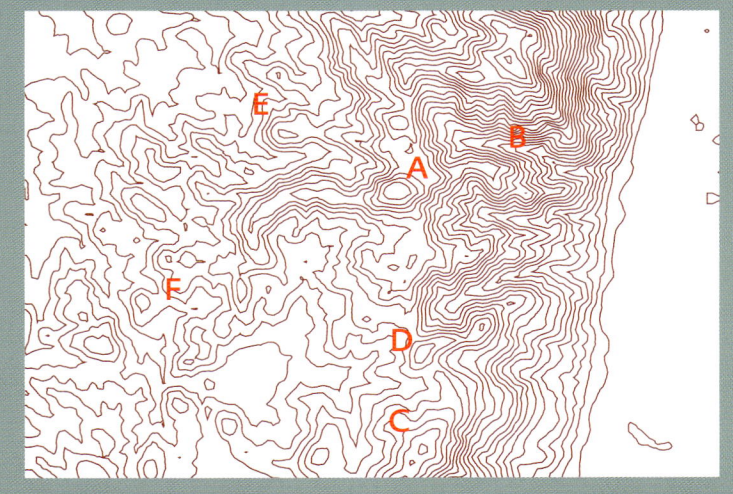

The satellite images of 2004/2005 and the digital model of the terrain are superimposed. This allows a very rigorous reading of the terrain and the establishment of a perfect correspondence with the 1948 map and of photos from 1959.

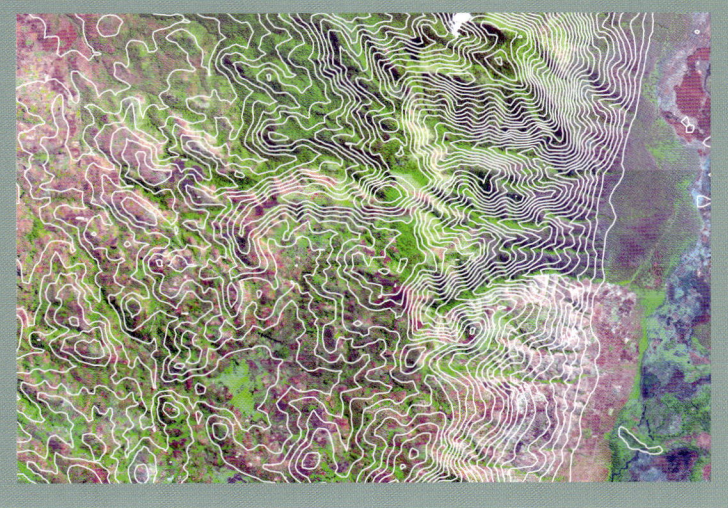

Towards the resolution of boundary conflicts and a modern definition of the limits of Virunga National Park

The documents from 1948, 1959 and 2004/2005 are integrated into one image where segments S140 to S144 are superimposed. It is possible to extract from this document geographical locations of each key point of these limits, which guide the teams in the field and give a preliminary objective interpretation of the limits. These maps are printed for use as a guide for work in the field.

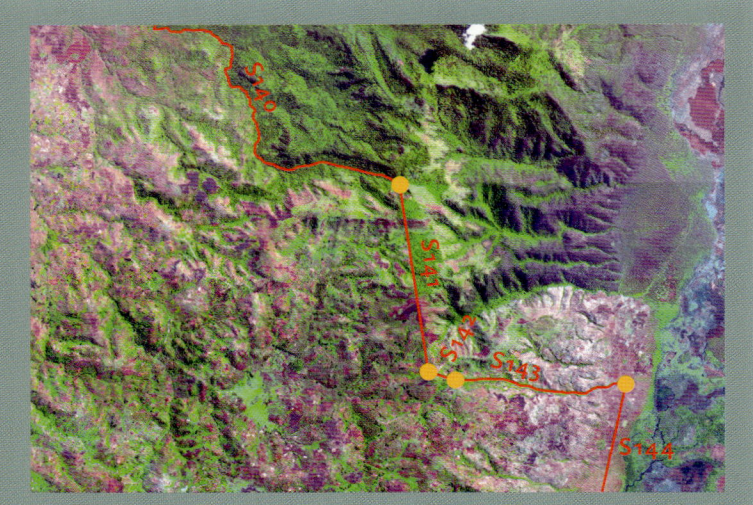

Mixed commissions are then set up, composed of representatives of the province, the ICCN, the WWF and particularly of the tribal chiefs and the villages concerned. These commissions go into the field for the purpose of validating the placement of the boundaries. Validation is done by a common interpretation of the legal text in the event there is confusion (extremely rare), through the identification in the field of hills, rivers, paths or localities mentioned in the text; and by the identification of official markers (most of them placed between the 1930s and 1950s) delimiting the park. In some regions of the park, several days are needed for the validation of only one or two segments, whether for logistical reasons or because the interpretation is not easy, or because of the existence of some old disputes.

This participatory work enables all the parties concerned to come to an agreement on the exact placement of the boundaries. The preliminary work presented on the two preceding pages gives an opportunity to correct numerous false interpretations (voluntary or not) by bringing out historical elements and objectives that guide the interpretation of the text and, above all, that re-establish the true position of the toponyms stated in the legal text. When all the participants agree on the position of the boundaries, a report is signed, the exact co-ordinates are taken by GPS, and the signs are placed in the presence of representatives of each party. Each sign is numbered, and its geographic position is noted. The whole thing is integrated into a database which completes the legal text (following page).

Data derived from analysis of the 1948 maps and 2004/2005 images, from the mixed commissions, and from the reports signed by the parties involved, form a basis that completes the legal text. The various data are integrated in a database (by means of a software program such as Excel). The example shown below is of the region examined in the preceding pages.

The reference number, together with its legal text, is noted for each segment. Various sections are then added to include the results of the fieldwork by the mixed commissions. These may possibly include a comment on the legal text (difficulty of interpretation, notable change in the reference, or other), whether the segment has been marked or not (some segments do not lend themselves to any confusion and do not require any participatory demarcation, or a lack of comment means this exercise has not yet been done). If signs have been placed, their reference numbers are recorded, together with the reference to the reports signed in the field by the various parties; the reference numbers of the digital photos taken in the field (with a direct link that enables the viewing of these photos). The other components of the database are also organised segment by segment and include the following information: the geographic co-ordinates of the points of departure and arrival of each segment and possible intermediary points, the length of each segment and its nature.

At this stage, a new formulation in French is not yet proposed but this could – and should in the long-term – be done integrating, notably, the geographic co-ordinates of some points and/or reformulating the text in a more precise way.

Proposals for boundary modifications can also be supported by this database. No modification could be enacted, however, until after the database has been completed. It is important to control and document the present limits before proposing modifications, whatever they might be. In addition, the consideration of proposed modifications could not take place for at least three to five years since much negotiation would be necessary and any official modification would need a presidential decree. The points where such modifications might be necessary or desirable can nevertheless already be usefully included in the present database.

Final result: Legal text/Characterisation in the field/Modern formulation

Paragraph	S140	S141	S142	S143	S144
Legal texts Compilation	'Le ruisseau Kamokanda jusqu'à sa source au pied du mont Tshahi'	'Une ligne joignant cette source au sommet des monts Tshahi et Bitingu'	'De ce mont une droite joignant la source de la rivière Kalagala, située près du mont Rwanguba'	'Cette rivière jusqu'au sentier de Mabenga à Tongo sentier longeant le pied des monts Kasali'	'Ce sentier jusqu'à son point d'intersection avec la rivière Butaku'
Characteristics in the field Comments on the legal text			Kalagala River is not permanent	The path was later replaced by a hard-surfaced road	The path was later replaced by a hard-surfaced road
Demarcation	no	no	yes	yes	yes
Signs			P037–P038	P038–P040	P040–P049
Report			PVS1/8/05	PVS1/8/06	PVS2/03/06
Photos	S140_B055		S142_A340	S143_B123 S143_B127	S144_A995 S144_B030 S144_A011
Modern formulation Starting point* (Lat/Long)	29°14"15'E 01°00"53'S	29°17"24'E 01°02"40'S	29°17"38'E 01°04"25'S	29°17"49'E 01°04"30'S	29°19"24'E 01°04"31'S
Intermediate points	geographic co-ordinates of selected intermediate signs				
Arrival point* (Lat/Long)	29°17"24'E 01°02"40'S	29°17"38'E 01°04"25'S	29°17"49'E 01°04"30'S	29°19"24'E 01°04"31'S	29°17"11'E 01°11"32'S
Length Nature	8359 m Stream	3244 m Straight line	1003 m Straight line	3098 m River	14.672 m Road

*World Geodesic System 1984 (WGS84)

Emmanuel de Merode & Marc Languy

26 Conclusions
80 years of effort and experience: A vision for the future

Emmanuel de Merode & Marc Languy

Few of the world's protected areas have suffered such sustained and intense pressure as Virunga National Park. The first chapters of this book document the threats the national park has to contend within one of the most densely populated areas of Africa, where demand for land is exceptionally high, compounded by the economic collapse and political turmoil at the heart of the Great Lakes Region. Yet, it is equally true that few sites have benefited from as much commitment and sacrifice on the part of those responsible for its protection. First and foremost are the 120 rangers of the *Institut Congolais pour la Conservation de la Nature* who have died on active service since the outbreak of armed conflict in the region, together with those who preceded them in Congo's turbulent history. The 36 authors of this book are a small sample of those who have witnessed and contributed to the challenge of protecting Virunga through the darkest period of its history, and have risen to the challenge of restoring the park to its former grandeur. *Virunga* was written to commemorate 80 years of conservation effort, but it was also written at a time of renewed effort from a large community of individuals and institutions. Today, the committed investment from donors and the political will of the Congolese state to maintain Virunga for its nationally and globally important resources has never been greater, and for the first time in over a decade, the pattern of natural resource destruction is being reversed. Importantly, the last few years have seen the growth of a highly professional team of conservationists working together in the field towards the shared objective of restoring the integrity of the park's boundaries, strengthening the wildlife authorities' capacity to maintain the rule of law and realising the real benefits that such a park can provide at a local, national and global level. The opportunity for defining a strategy for restoring and securing the future of Virunga National Park has not been a viable option since the outbreak of armed conflict in 1991. Today, that opportunity exists. In this concluding chapter, we draw on the efforts and experiences of all the authors to progress towards a vision and a strategy for Virunga's future.

In articulating a coherent strategy for the park, it is important to frame the discussion by defining specific, agreed and measurable objectives, within a realistic time frame. This is the subject of an ongoing exchange among professionals working in Virunga National Park and will probably culminate in the publication of a management plan in the near future. However, it could be argued that an elaborate management plan is not strictly a requirement for outlining the broad objectives that are being sought in Virunga. The targets are to restore the ecological integrity of the park. Generating economic and social benefits at a local and national level is a precondition for achieving this target. In this chapter, we explore a two-stage strategy. First is the post-war reconstruction of the park's basic functions, and second, developing the conditions for sustaining park management until the integrity of the ecosystem has been fully re-established.

The post-war restoration has been defined by the programmes currently being implemented in the park. These have been discussed and formulated through the park's Coordination Committee, and most of the financial requirements have been secured through the European Development Fund and other donors. It is a three-year plan, covering the re-establishment of the park's boundary, restoration of the basic infrastructure and the reinforcement of ICCN's ability to manage the park. The longer-term strategy has yet to be fully defined, but aims to restore the park's ecological integrity to its state prior to the major resource depletions described in this book. The most significant resource depletion has been that of the large mammals, and we use this to define the recovery period for Virunga National Park. An indication of the time involved is given in Table 26.2, which shows the number of years required for the current populations to grow to the earliest known abundance based on maximum rates of population increase (rates of increase are drawn from the published sources cited in Rowcliffe *et al.*, 2003). This assumes zero mortality from disease or poaching, and does not take into account predation or a logistic reduction in the rate of increase towards the end of the time period. While some populations may partly recover through emigration from nearby Queen Elizabeth National Park, the time estimates can still be taken as a theoretical minimum for the park's recovery.

On the basis of these data, we can safely assume that a strategy for restoring the park must work on a timescale of at least 30 to 40 years. As such, the current project planning paradigm that limits itself to, at best, a three to five year project cycle, is inappropriate if we are seeking to restore the integrity of the ecosystem. Thus, in this chapter we adopt a pragmatic approach of using the existing project framework (three years) for establishing the basis for ecosystem recovery, then explore the conditions for maintaining that recovery over 30 to 40 years.

26.1
The future of Virunga National Park will depend on the efforts of ICCN's rangers, but also on the level of support they receive from government and from the international community.

26.2 Theoretical time required to achieve stable animal populations in Virunga National Park

Species	Earliest recorded population[1]	Current population[2]	Maximum growth rates (annual %)[3]	Minimum no. of years for re-stocking
Buffalo	28,307	3,822	16%	14
Kob	11,218	12,982	25%	0
Elephants	3,425	348	8%	32
Hippopotamus	26,530	887	9%	39
Lowland Gorillas	31	21	7%	6
Mountain Gorillas	274	380	7%	0
Topi	1,732	1,353	18%	2
Warthog	5,939	694	26%	11
Waterbuck	2,223	374	18%	11

1 All estimates are based on Cornet d'Elzius, 1959 except Mountain Gorillas (Webber & Vedder, 1983) and Eastern Lowland gorillas (Aveling, pers comm).
2 All estimates are based on Kujirakwinja *et al.* (2006), except for Mountain Gorillas (Gray *et al.*, 2005) and hippos (de Merode *et al.*, 2005).
3 Based on published maximum intrinsic annual population growth rates (rmax) for specific taxa (Rowcliffe *et al.*, 2003).

26.1 Short-term strategy: post-war reconstruction

Significant damage to the national park's resources only started in the early 1990s. Prior to this, periods of civil unrest, such as the Simba and Mulele rebellions of the mid 1960s together with the subsequent military mutinies, had only an ephemeral impact on the park. Verschuren's account of the post-independence period in Virunga National Park describes a 'rebirth' of the park under a new management paradigm led by an increasingly competent local management. With all of its inevitable weaning difficulties, the basis for sustainable park management was established during this period. Nevertheless, the seeds of longer lasting damage were seen as early as the 1950s, with the building of the first permanent settlements within the National Park. The most damaging period in Virunga's long history began in 1991, with the civil unrest in Katanga and Kivu Provinces and the subsequent suspension of most international aid to Congo. Kalpers and Mushenzi describe the principal events of this period, but also testify to the courage and commitment of those who remained in the park during a period when individual survival became a daily concern. The threats resulting from this period are outlined and examined by d'Huart, Lanjouw and Mushenzi, and can be summed up as the breakdown of the integrity of the park's boundaries, mainly through illegal settlement and agriculture, and the systematic and illegal depletion of the park's resources. Associated with this is the lack of capacity on the part of the wildlife authorities to counter such threats. In this section we explore the first step towards a long-term vision for Virunga: the short-term strategy of post-war reconstruction using a reduction of threats approach (Margoluis and Salafsky, 1998). As such, we examine the three primary threats defined above, and analyse the most viable means of resolving these issues: restoring the integrity of the park's boundaries, re-establishing the state of law within the park, and strengthening ICCN's capacity to manage the park.

26.1.1 Restoring the integrity of the park's boundary

While most biological populations can potentially recover from periods of resource depletion through poaching or logging, the conversion of protected land for consumptive purposes (for example agriculture, urban expansion or resource extraction) followed by the loss of protected area status is generally irreversible. Two key variables determine the loss of protected area status. First is the growing economic and demographic pressure on limited land resources. Second is the capacity of the wildlife agency to enforce conservation laws. Whatever the cause, the irreversible loss of protected land must be considered the most serious of threats to the future of a protected area, and appropriate strategies have to be established well in advance of any *de facto* land conversion within the area's boundaries.

Virunga's experience is not unique. The de-gazetting of significant sections of Akagera and *Parc des Volcans* in Rwanda, and the Lopé Reserve (now National Park) in Gabon, are instructive examples of the irretrievable loss of protected landscapes. In most cases, in Virunga as well as in Rwanda and Gabon, the de-gazetting is preceded by a *de facto*, and by definition illegal, occupation of protected land. In Akagera National Park, land was occupied by internally displaced people in the aftermath of the Rwandan genocide. The human cost of forcing displaced people out of the park was generally considered to be excessive and consequently legislation was passed to de-gazette almost two thirds of the national park. In the Lopé Reserve, a major logging company had already started an illegal logging operation in 'Lot 32' and had built a bridge on the Offoué River, bordering the reserve.

The demographic conditions of North Kivu are described in this book's introduction, and provide a compelling explanation for the pressure on Virunga's boundaries. By late 2006, 15 percent of Virunga's boundaries had been overrun by park invasions. As described by d'Huart *et al.* in Chapter 19 and by Languy *et al.* in Chapter 25, there are broadly three types of land invasion: land invasions caused by internally displaced people following a humanitarian crisis, agricultural expansion by resident populations, and the large-scale sale of land by fraudulent landlords. Drawing on the evidence found by wardens in Virunga, the third case appears to be the most widespread.

26.1.1.1 Illegal commercial land acquisitions

Forced displacement is not the only process through which park invasions occur. In Chapter 18, Muir and de Merode describe the invasion of the southern Mikeno Sector in June 2004. Substantive and conclusive evidence indicates that these invasions were orchestrated by relatively wealthy Rwandan land brokers with Congolese collaborators. The 6,000 cultivators who moved into the park were receiving a daily wage for forest clearance, and the invasion was deliberately timed to coincide with the political turmoil instigated by the two military commanders, Laurent Nkunda and Mutebusi, whose forces had stormed into South Kivu Province only a few days before. The strategy that was adopted to resolve this crisis was successful and merits an analysis of the lessons learned, along with an assessment of the potential for replicating the approach in future crises. The Mikeno Sector land invasion elicited a strong reaction at international level. Pressure was brought to bear at the highest level in a neighbouring country, with immediate consequences on the ground. Equally important was the follow-up on the ground to ensure that the incident was not repeated when the political pressure had subsided. The ability of the conservation community to raise the funds and mobilise the local community to build 20 kilometres of drystone wall within three months was instrumental in providing a long-term solution to the crisis.

Nevertheless, the Mikeno crisis of June 2004 could be considered exceptional. The damage was caused within the range of Virunga's Mountain Gorillas, a situation that commands immediate international attention at the highest levels. Similar land invasions at Kilolirwe, in the south-west of the park, have never been satisfactorily resolved because they failed to generate the same degree of indignation at a local and international level. Similarly, the attempts to resolve land invasion by pastoralists at Kilolirwe have so far failed to yield any significant results on the ground. The transboundary approaches to protected area management documented by Lanjouw et al. provide an insight into the progress that has been made to institutionalise a state-level commitment to securing the integrity of Virunga, together with its contiguous protected areas in Rwanda and Uganda. Certainly, the more structured transboundary approach that has been carved out in recent years provides the most promising way of securing a common commitment to resolving land invasion issues. A second success has been the departure in April 2006 of more than 200 pastoralist families that had settled in the Karuruma area, after protracted negotiations involving the Congolese and Ugandan authorities.

26.1.1.2 Agricultural encroachment by park adjacent populations

Cases of encroachment by people cultivating around the park are concentrated mainly along the western shore of Lake Edward and northward as far as Beni. The two forms of encroachment that Languy et al. describe in Chapter 25 focus on the expansion of existing plots into the park, as a result of ICCN's loss of authority, and on the growth of illegal settlements tied to other economic activities in the park, such as the illegal fishing settlements. The degree to which commercial land-brokers are involved in the extension of agricultural plots into the park is not always clear. The two major factors behind the increase in illegal encroachment have been ICCN's inability to enforce the law and confusion over the national park boundary. The re-establishment of ICCN's authority is essential if all ambiguity is to be removed from the laws governing the protected area and demarcation of the park. The building of a drystone wall around Kibumba is not a deterrent in itself: the wall is only one metre high and can be readily demolished. Its primary role is to provide clarity as to the exact position of the park's boundary. The fact that local community associations were heavily involved in its construction has strengthened the legitimacy of that boundary in the minds of resident populations. Extensive work is under way to demarcate more of the park's boundary with local community representatives.

26.1.1.3 Internally displaced people and refugees

Around Virunga, eight years of civil war and economic collapse have contributed to similar situations in the national park. There have been instances of land occupation by displaced people after intense periods of armed conflict, such as those of Kanyabayonga in December 2004, when Government troops came into contact with mutineers. Fighting lasted for about 12 hours, followed by five days of looting and pillaging of the local populations in northern Masisi and Lubero territories. According to the European Commission's Humanitarian Office in Goma, the humanitarian disaster that ensued led to the displacement of around 40,000 people into the park, towards Rwindi and Vitshumbi (Scotto, pers comm.). This instance is typical of the effects of unrestrained behaviour by armed personnel on the civilian populations in eastern Congo. In contrast to Akagera, there has been a greater range of options available for resolving humanitarian crises that have triggered land invasions, in ways that prevent the permanent loss of protected area land. Drawing on a wide-scale, provincial level approach to planning, and relying heavily on partnerships with organisations working in the humanitarian and development sectors, alternative livelihood strategies have been established with displaced people. These are described by Languy et al. in Chapter 25. At the time of writing, the development of alternative livelihood options for displaced people within the park is contributing to the movement of about 3,000 people per month out of the park. These people are moving to areas where agricultural potential is being developed through improved road access and social infrastructure development (schools, clinics, and so on).

While the features of land invasions in Virunga vary considerably, two significant lessons can be drawn from all the land invasions seen over the past few years. The first concerns the ability to detect and react instantly to the first indications of illegal settlement within the park. The Mikeno crisis was successfully resolved because of the speed with which it was addressed, both at a local and international level. The land invasions at Kilolirwe and Karuruma were detected long after the event, by which time the resources needed to resolve the crisis were substantial, and there was also a degree of inertia at a political level. A priority for the future is the strengthening of ICCN's ability to implement and communicate effective surveillance, and the ability to mobilise support, both in terms of political pressure and through the rapid release of funds, to address crises with immediate effect. The implementation of systematic aerial and ground surveillance, and the establishment of a rapid release fund for crises, is a priority for the park. Both the mobilisation of political pressure and the implementation of long-term solutions through development initiatives are well under way, already, but these will have to be sustained.

The second lesson concerns the ability to consolidate gains made after national park land has been recovered. Again, the outcome achieved in the wake of the Mikeno crisis was conclusive, not just because the land was recovered after the forest clearing teams were ordered to leave, but also because local community associations then built 20 kilometres of drystone wall to demarcate the boundary. Elsewhere in Virunga, considerable effort is going into boundary demarcation through the ICCN/WWF programme. While the demarcation will not result in a physical boundary as pronounced as that of the drystone wall, the participatory nature of the exercise, requiring local authorities and representatives to locate and recognise the legal course of the boundary on the ground, is having much the same effect. This, in conjunction with the construction of concrete bollards, is the beginning of the process of ensuring that the boundary can indeed be re-established along most of the park's periphery.

26.1.2 Re-establishing the rule of law within the park

The rule of law is a national imperative that goes far beyond the need to re-establish Virunga as an effective protected area. The widely cited four million war-induced fatalities incurred by the civilian populations of eastern DRC since 1996 (IRC 2004) were caused not by the proximal effects of war, but by the consequences of the resulting breakdown of law and order and the collapse of basic social services.

Essential infrastructure has to be rebuilt. This does not necessarily imply that all park stations need to be rebuilt to their former state, but certain requirements need to be met: rangers must live in minimally acceptable housing, which is not the case at the present in most stations. Access needs to be improved. Stations need to be able to function properly, with the requisite administrative infrastructure and workshops. Equally important, the management and organisation of park stations needs to be addressed. After years of neglect, there is a need to re-examine the way that stations are managed, how security measures are enforced, and how discipline is maintained.

26.1.2.1 Military presence in the park

A regime of chronic insecurity characterised by systematic pillaging became the norm when the military and the police, the very forces that would normally be expected to uphold the law, were not paid and were poorly supervised, to the extent that they became the perpetrators of the very same illegal acts they were supposed to be suppressing. As a consequence, military personnel are more often associated with pillaging and extortion than with security. Nowhere has this been more apparent than in Congo's national parks, where military personnel often well beyond the reach of their supervisors, have succumbed to a wide range of temptations, including the use of their weapons for poaching. Virunga in particular has seen a gradual increase in the number of permanent military structures within the park since 1996. Virunga has two very sensitive international boundaries on its eastern flank, and for several years it was split in two by the front separating two warring factions, the RCD Goma and the RCD K/ML. Many of the military structures established during this period – such as those at Rwindi, Vitshumbi and Kyavinyonge – still exist. Their impact is in little doubt, given the numerous hippopotamus carcasses that are on view around these military camps. Indeed, the collapse of Lake Edward's hippo populations, described by Languy in Chapter 8, is a direct consequence of the concentration of military personnel within the park boundaries.

For all the difficulties associated with addressing the military question, there have been some notable successes. Until March 2003, Ishango was home to a significant military detachment numbering more than 50 soldiers. For five months the park warden in Virunga north, together with his NGO partners, lobbied the political and military authorities to have the site restored to ICCN authority. After several meetings and site visits, the military were ordered to re-deploy outside the park. Two key factors re-established the park's authority: effective lobbying and the ability to demonstrate ICCN's capacity to fulfil a security function in lieu of the military.

The most significant work still lies ahead. The largest military camps still exist. Some of them, are based at ICCN stations and pose a major obstacle to the re-establishment of ICCN's authority. Others, such as Kanyatse on the Beni-Kasindi road, present the additional challenge of being supported by international donor funding. Kanyatse, most of which is located within the boundaries of Virunga National Park, is one of the biggest military camps in eastern Congo, built with Dutch development assistance funds. To date no significant progress has been made in reversing the development of this military camp, despite letters to the Dutch government from both European Commission and UNESCO representatives.

26.1.2.2 Relationship with the tribunals

Due process can be extremely challenging in the current context. Tribunals have little or no resources. Magistrates are unable to function properly when salaries have not been paid for several years, making them extremely susceptible to graft. Similarly, the *officiers de police judiciaire* have no resources to collect evidence or to make arrests. This has to be provided externally, often by the plaintiff, thereby introducing an immediate bias to the case. Since 2002 there have been several cases brought before the courts by the Institute. These relate primarily to the land invasions around Beni. As described above, the land invasions were often instigated by wealthier landlords laying claim to park land for purposes of leasing the land to peasant farmers.

With respect to Mavivi, the park warden for Virunga instigated court proceedings against two such landlords in March 2005. NGO partners were able to provide legal support to ICCN, and the landlords were convicted *in absentia* (both refused to attend the trial) by the *Tribunal de Grande Instance* of Butembo. The court's conviction was never carried out because of lack of funds for transport to make the arrests.

The lessons drawn from the very real efforts by ICCN and its partners to re-establish due legal process have largely been unsuccessful and will continue to be so unless an independent source of funding can be found to support the tribunals. A precedent exists with the tribunals in Bunia, which received significant support from the European Commission through law-related NGOs that covered their operating costs. Such a programme is a pre-requisite for the re-establishment of the law in Virunga. As the Institute's ability to establish and maintain the law strengthens, so the contribution of Congo's protected areas to the post-war stabilisation and national reconstruction effort become important. This, perhaps more than anything else, will give Congo's national parks a new and decisive role.

26.2 Towards sustainability

In the introduction to this chapter, we define a strategic time-scale of 30 to 40 years for achieving the management objectives of restoring the national park ecosystem. Having established the minimal requirements of re-establishing the integrity of the park, we explore three main themes that need to be addressed for Virunga National Park to survive in the longer term: financial sustainability, human resources for effective management, and the benefits that will have to be generated for Virunga to retain its status as a national and global heritage.

26.2.1 Financial sustainability

A key preoccupation in the long term is to secure essential funding that is not susceptible to the 'boom and bust' regimes of the past. The process of injecting significant donor funding for two to three years followed by a period of financial collapse has been extremely damaging to the park in the past. In Chapters 3 and 4, we see how two major programmes that achieved significant results during their period of implementation, were followed by a longer period of financial and management collapse. It is now increasingly understood, at the level of Government and among donors and other agencies that fund Virunga National Park, that financial planning and budgeting process should support all aspects of financial sustainability. Focusing solely on funding levels will not be enough to overcome the financial constraints that Virunga National Park faces. Financing strategies, and the funds used to achieve them, must cover a range of priorities including the timeliness and administration of funds, the range of beneficiaries to be reached, and overall protected area management effectiveness for biodiversity conservation.

A range of innovative financing mechanisms are being explored, such as finding new donors, including private philanthropists, other government agencies, or tax revenue sharing; sharing the costs and benefits of tourism with local stakeholders; employing new financial tools, such as business planning; and devolv-

26.3
A panorama of Virunga's active volcanoes, featuring Nyiragongo (left) and Nyamulagira (right), as seen from the station at Rumangabo. The spectacular visual qualities of the landscapes in the park, coupled with their immense scientific importance, make Virunga a global heritage site of exceptional value.

ing funding and management responsibilities (to non-governmental partners, local communities, individuals or businesses). In the past few years financial portfolios incorporating a diversity of funding sources and a multiplicity of beneficiaries have been developed, and many of these elements are now discussed and adapted during regular Site Coordination Meetings (CoCoSi). While traditional public and charitable funding is still the most significant source of funds for Virunga National Park, it remains insufficient and inflexible. Building a wider range of funding sources will increase the total amount of funds available and ensure greater flexibility, while spreading risk and enhancing responsiveness to changing park needs and opportunities.

Virunga's managers will need to develop the capacity to take advantage of various mechanisms that capture the willingness-to-pay of those who use, or benefit from, goods and services provided by the protected area. These include mechanisms such as the CoCoSi budgets and financial plans, which have yet to see the light of day in any functional sense. Wildlife staff as well as donors will have to view funding as part of a broader management requirement.

26.2.2 Realising the benefits of Virunga's resources at a local and national level

The Congolese nation has shown an exceptional commitment to the survival of its national parks, especially Virunga. The reasons for this are many: a national pride in its exceptional resources, a commitment to the protection of nature, a belief in the potential role that national parks might one day play in the country's development. While these factors are all important in explaining why Virunga has survived against all odds, it has to be recognised that the park is going to have to generate more tangible benefits if it is to survive in the face of a growing and expectant local population. The historical chapters of this book, together with Lanjouw et al. (Chapter 11), have illustrated just how such benefits as can be derived from Virunga have been subjected to a roller coaster ride. In the 1950s, tourism was largely concentrated on the Rwindi Plains, and while significant employment was created and infrastructure built, African wildlife tourism was still in its infancy and financial returns were relatively limited. There followed 30 years of stagnation, followed by a short, yet highly significant, golden age based on the habituation of gorillas in the Mikeno Sector by Conrad Aveling and the Frankfurt Zoological Society team. This marked the start of the first serious reflection on the role that tourism revenues from Virunga could play at a local and national level. This reflection continues today, while encompassing a range of other potential revenue sources. The WWF's PEVi programme, also established in the late 1980s, set out to mobilise international financial resources for local communities. Since then others, including the International Gorilla Conservation Programme and the Dian Fossey Gorilla Fund, have made their own attempts to extend benefits from the park to the local populations, usually capitalising on the universal charisma of the Mountain Gorillas. While the focus has been largely on the generation and redistribution of resources to local populations, there has been relatively little effort to involve local institutions in management issues. These questions are addressed by Balole et al., who present the rationale for involving local institutions in an argument based not only on principles of social equity, but also on pragmatic conservation management considerations. In some cases, devolving specific aspects of authority to particular non-centralised institutions may provide a more efficient means of managing wildlife.

In collaborative management, such aspects as formal decision-making, responsibility and accountability might rest with ICCN, but the agency would increasingly be required, either by law, policy, or financial constraint, to collaborate with other institutions. In its weak form, 'collaboration' means informing and consulting stakeholders. In its strong form, 'collaboration' means that a multi-stakeholder body develops and consensually approves a number of technical proposals for protected area regulation and management, to be submitted to the decision-making authority. Virunga has established a joint management structure, the CoCoSi, which allows various participants to sit on a management body with decision-making authority, responsibility and accountability. As outlined by Debonnet et al., local institutions have yet to be represented on this body. As local institutions, including local associations, civil administrations, local chiefs or *Bami,* become more or more involved in the decision-making process, so overall management effectiveness will increase. This can develop where the CoCoSi is sufficiently well structured, and where a basic notion of leadership under the authority of ICCN is maintained.

26.2.3 Human Resources

We end this book by examining the staff issues required to manage and protect Virunga National Park. That we can still contemplate the recovery of the national park is largely the result of their efforts during periods of extreme hardship. Now that political and security conditions are slowly returning to normal, the global community of people with a stake in Virunga's future will need to recognise the contribution of those who have continued to manage the park, and the need to improve their effectiveness and efficiency in equal measure. The re-establishment of the park's boundaries and the rule of law within these boundaries is dependant above all on ICCN's capacity. Considerable progress has been made in recent years towards establishing the pre-conditions for re-building ICCN's management capacity, but the implementation of an effective capacity-building programme has yet to be completed. An important step is the just-concluded institutional review of ICCN, which is leading to the post-war restructuring of ICCN that aims to bring the Institute up to speed with modern-day challenges. In Chapter 21, d'Huart and Kalpers pre-empt the review and outline the essential elements of the process, with respect to Virunga National Park.

Important lessons can be drawn from the many recommendations of the 1990-91 institutional review. Of the 12 recommendations, eight involve new or improved strategic directions, such as devolution to provincial headquarters and community conservation, and four involve cost-cutting measures to increase the Institute's efficacy, mainly through the retrenchment of non-essential personnel and administrative structures. In the years that followed, however, the budgets required for IZCN/ICCN to implement most of these recommendations failed to materialise. Characteristic of the period since the last institutional review is a systemic inability on the part of the organisation to relate the scale of operations to a budget. Staff numbers were no longer limited by budget constraints for the simple reason that staff were not paid. As a consequence, personnel in the park increased, instead of decreasing as recommended by the review, provincial headquarters were established without any corresponding reduction of the central authority in the capital, and measures, such as staff regulations, were adopted without any realistic possibility of meeting the financial implications of their implementation. Tragically, the outcome greatly compromised ICCN's ability to address the monumental challenges of the last decade: staff were unpaid, untrained, ill-equipped and profoundly demoralised, at a time when the commitment required from an ICCN ranger or warden was greater than it had been at any time during the park's long history. It is therefore unsurprising that, while rightful praise should be heaped on those who remained loyal during those difficult years, there are still cases of illegal exploitation of the park's resources that have been facilitated, even organised, by wildlife officers. These incidents have remained largely unpunished. The shameful clearance of forests for charcoal production (Chapter 14) is just one example. This culture of impunity needs to end and clear principles of professional integrity need to be restored, without exception. This is the greatest priority, not only to render the Institute more effective, but also to restore the credibility of its officers and rangers with respect to their closest partners: the local populations.

That said, it is remarkable that the evidence shows that the majority of ICCN staff remained committed and loyal to the Institute. There were relatively few desertions, and staff remained surprisingly well organised, if largely incapacitated. This was hardly an adequate prescription for addressing the major threats then facing the integrity of the park. Programmes have since been launched to fill those gaps, including training, financial support, and the provision of equipment and basic infrastructure. The all-important and now long overdue institutional restructuring is expected, once implemented, to help consolidate the gains of the past few years, and to sustain the minimum requirements for securing the park's future.

26.4
ICCN *rangers on patrol on the upper Semliki River. Human resources will be a decisive factor in determining ICCN's ability to re-launch effective management within the park. Enormous professional rigour will be expected from the rangers in restoring the rule of law, while showing sensitivity towards local communities living around the park.*

26.3 FUTURE PRIORITIES

The implementation of the recommendations of the institutional review will be an important test for the commitment of political decision makers to protecting the national and world heritage of Virunga National Park. A number of difficult decisions will have to be taken, not the least as regards the necessary rejuvenation and reduction of staff, giving the park a leaner but more effective, better motivated, and better equipped team of rangers. Such a team would stand a far better chance of success than even a veritable army of poorly equipped, underpaid and demoralised rangers. Similarly, the vast number of patrol posts will have to be reviewed, or at least managed more effectively based on the prescriptions of a management plan that takes full account of the supervisory limitations of the staff, and the chronic need for highly motivated rangers in these remote outposts. Likewise, the function of certain stations, such as Kabaraza, Rutshuru or Lulimbi, will have to be revised based on current priorities, given that they were established 30 to 60 years ago. Demographic conditions affecting this part of the park have dramatically altered the context. Apart from re-defining the roles that some of these stations need to play, it is important to guard against the damage the stations themselves can cause within the park, as has been the case with Lulimbi, which began as a mere patrol post, but which has since become the mainspring for an endless proliferation of schools, churches, shops and other facilities, attracting hordes of civilians whose activities have little or nothing to do with the park.

Next page is a simple table, not exhaustive, of major projects that ICCN and its partners can expect in the coming years.

In a re-unified country with a peaceful environment that provides for economic growth and the rebuilding of infrastructure, it should be possible –with the total commitment of ICCN's rangers and wardens, with continued financial and technical support from NGO partners, and with a well-executed restructuring, coupled with a simple yet realistic management plan – for Virunga National Park to rediscover its former glory, and to confirm its unique position among Africa's protected areas. With the many programmes that seek to establish close ties between protected areas and local populations through tourism, the national park will once again become an engine for the local and national economy, and a source of much needed revenue for ICCN and for local residents.

Priority issues	Main activities
Re-establishing control of the park	– Resettlement of populations illegally settled at Kilolirwe and on the western shores of Lake Edward – Finalisation and consolidation of the boundary demarcation programme – Removal of military camps within the park – Continuation and strengthening of training in law enforcement
Restructuring of park personnel: rejuvenating the ranger force and probably establishing a reduced number of staff in a way that is more manageable.	– Implementation of the recommendations of the Institutional Review; payment of pensions; training programme
Reorganisation of patrol posts	– Redeployment of human resources and their integration into a management plan. Development of sub-stations
Harmonisation of support from partners	– Strengthening of the CoCoSi
Synergy between the various programmes	– Compilation of a management plan
Resolution of the fuelwood crisis	– Launch of an ambitious replanting programme and promotion of alternative energy sources. – Establishment of an anti-deforestation programme.
Re-establish control of the illegal fishing settlements on Lake Edward	– Launch a study on the potential fishing yields; – demolition of illegal fishing settlements, re-negotiation of agreements with COPEVI and other stakeholders
Improving financial sustainability through internal revenue systems	– Re-establish full control of tourism revenues within the park
Improving relations between the park and local populations	– Formalisation of the function within ICCN of community wardens – Re-establishment of the 20% revenue allocation to local communities – Focus on improved integrated conservation and development projects around the park

All the authors of *Virunga* have taken great pride in presenting the many facets of this exceptional national park. They have described the tragedies, as well as triumphs, that define its history, and which can either be rectified or learned from. In documenting the current state of the park in the most realistic and accurate way possible, and in proposing future directions in park management, we hope this book will encourage decision makers, managers, partners, visitors, and readers to take into their hearts the cause of Africa's first and finest national park.

26.5
The forests of the Albertine Rift have enormous value for the whole of humanity, as well as for local communities, which depend – for their supply of clean drinking water – largely on rivers emanating from protected areas, such as Virunga National Park.

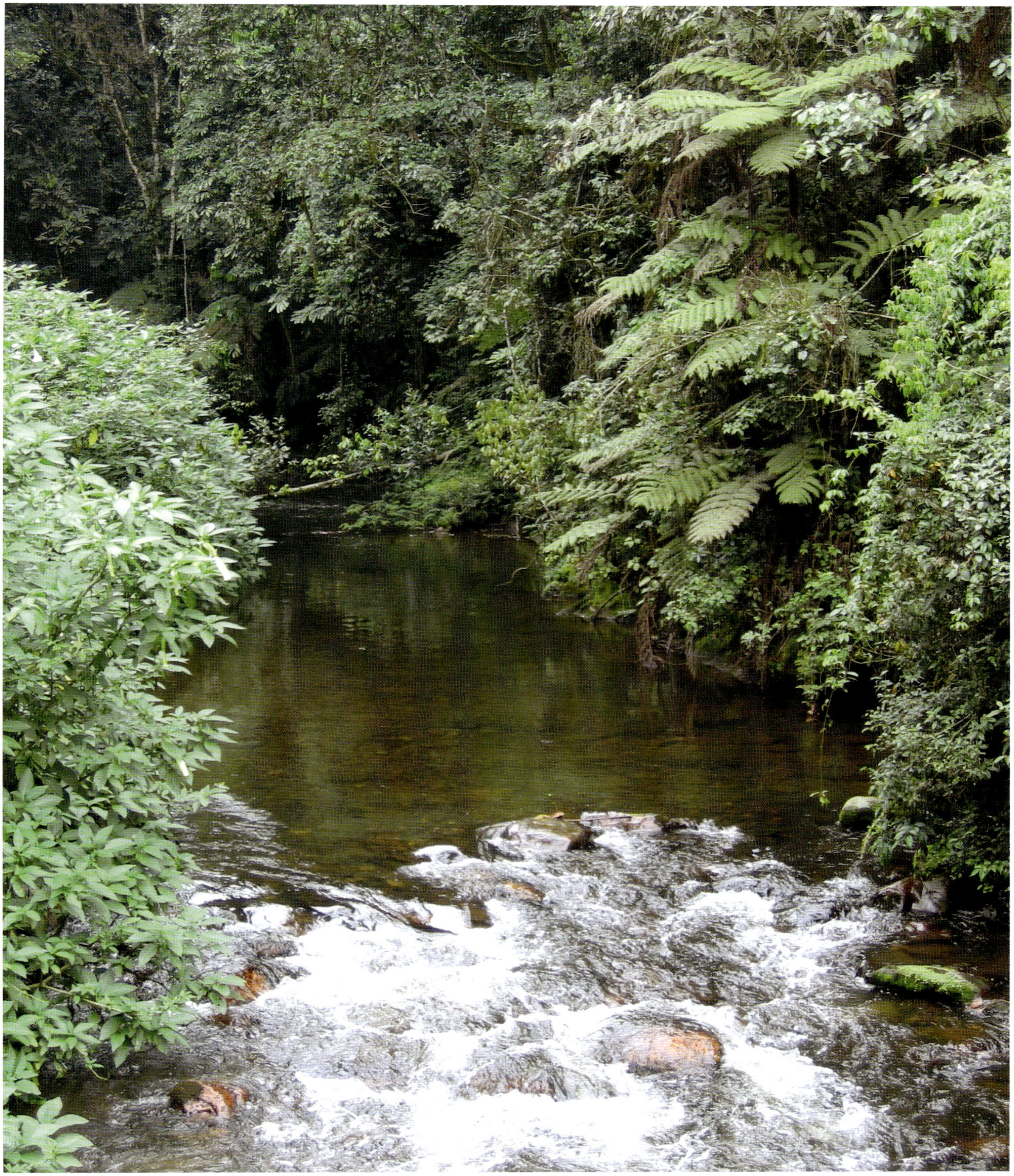

Appendix 1
Legal texts delimiting Virunga National Park

S001 D'un point partant de la rive Nord du lac Kivu et du bord Est de la coulée de lave du volcan Rumoka;

S002 Ensuite le bord oriental de la coulée de lave jusqu'au pied occidental de la colline Nyamutsibu;

S003 De là une droite jusqu'au pied occidental de la colline Nyabusa;

S004 A partir de ce point, une droite joignant la borne 1 située à 300 mètres à l'Ouest du gîte de Rusayo, sur le chemin RusayoSake;

S005 De cette borne, une droite joignant la borne 2 située sur la pente méridionale du mont Mbati;

S006 De cette borne, une droite joignant la borne 3 de Kisagara, près de Rusayo, sur le chemin Rusayo-Kibati;

S007 De ce point, le chemin Rusayo-Kibati jusqu'au carrefour (borne 4) du sentier allant au village Mutaho;

S008 De ce point, une ligne joignant le pied Nord-Ouest de la colline Bubunugu (ou Mutaho) et contournant le pied de cette colline par le Nord, pour rejoindre le sommet le plus septentrional de la colline Bitunguru (arbre isolé);

S009 De ce point, une ligne passant par le pied méridional des monts Katandali et Kanyambuzi, puis par le col qui sépare ce dernier mont du mont Mudjoga pour atteindre la tête du ravin Kavumu, au pied Sud du mont Kavumu;

S010 Ce ravin jusqu'à son embouchure dans la clairière (borne 5) de Kavumu;

S011 Le bord méridional de cette clairière marqué par une piste aboutissant à la route carrossable Ngoma-Rutshuru à 1 km. 400 du village de Kibati (borne 6);

S012 A partir de ce point, le bord occidental de la route carrossable Ngoma-Rutshuru jusqu'au carrefour du chemin Rugari-Kanzenze-Nyamlagira-Mushari;

S013 A partir de ce point, le bord septentrional du chemin Rugari-Kanzenze-Mushari jusqu'à une borne située à environ 4 km. à l'Ouest du carrefour de Rugari;

S014 De cette borne, une ligne joignant la borne située à 2 km à l'Est du pied de la colline Nyasheke-Nord;

S015 Ensuite le pied oriental de la colline Kurushari et une borne située sur le sentier Rutshuru-Tongo, à 4 km. à l'Ouest de son point de jonction avec la route carrossable Rutshuru-Ngoma;

S016 De cette borne, une droite joignant le confluent des rivières Rutshuru et Rugera;

S017 A partir de ce confluent, le bord supérieur oriental (côté rive droite) du ravin de la rivière Rutshuru, jusqu'au parallèle du confluent de la rivière Kabarasa avec la May-Na-Kwenda;

S018 Ce parallèle jusqu'à la May-Na-Kwenda;

S019 La rive Sud de la May-Na-Kwenda jusqu'à l'intersection du sentier Rutshuru-Kalimbo-Kabare;

S020 Ce sentier jusqu'à la rivière Ngesho;

S021 Le bord supérieur méridional du ravin de cette rivière, vers l'amont, jusqu'au marais Nyamborokota;

S022 Le bord de ce marais, par le Sud et l'Est, jusqu'à l'embouchure de la rivière Tshabaganda;

S023 Le bord supérieur méridional du ravin de cette rivière jusqu'au confluent de la rivière Kakoma;

S024 Le bord supérieur méridional du ravin de la Kakoma jusqu'à sa source;

S025 Le parallèle de cette source jusqu'à sa rencontre avec la rivière Kasozo;

S026 Le bord supérieur oriental du ravin de la Kasozo jusqu'à son embouchure dans la rivière Ishasha;

S027 L'Ishasha (frontière de la Colonie) jusqu'à son embouchure dans le lac Edouard;

S028 La frontière de la Colonie, à travers les eaux du lac Edouard, jusqu'à l'embouchure de la rivière Lubilia dans le lac Edouard;

S029 La Lubilia jusqu'à son point d'intersection avec la route carrossable de Beni à Kasindi;

S030 Le bord occidental de cette route jusqu'à l'ancien sentier reliant le village de Mutwanga au mont Libona, sentier coupant les têtes des rivières Butowa et Mahimbi;

S031 Ce sentier vers l'Ouest jusqu'à son intersection avec la route carrossable de Beni à Kasindi;

S032 Le bord méridional de cette route jusqu'à un point situé à un kilomètre à vol d'oiseau de son intersection avec la rivière Semliki;

S033 De ce point, vers le Nord, une ligne parallèle à la rive droite de la Semliki et distante de celle-ci de 1 kilomètre à vol d'oiseau jusqu'à son point de rencontre avec la rivière Musenene ou Lusilube;

S034 La rive gauche de cette rivière jusqu'à son point d'intersection avec le parallèle du confluent des rivières Modidi et Biangolo;

S035 Ce parallèle jusqu'à ce confluent;

S036 La rive gauche de la rivière Biangolo jusqu'à une borne (altitude approximative 1.700 m)

S037 La projection verticale sur le terrain de la polygonale reliant les points de rencontre de la courbe de niveau de cette borne avec les contreforts avancés du massif du Ruwenzori et ce jusqu'à la rivière Talya

S038 La rive gauche de cette rivière jusqu'à son confluent avec la rivière Buliba;

S039 La rivière Talya, vers l'aval, jusqu'à son confluent avec la rivière Bongeya;

S040 Le méridien de ce confluent jusqu'au pied de la colline Bulima;

S041 Le pied méridional des collines Bulima et Ulese, jusqu'à l'extrémité Sud-Ouest de cette dernière;

S042 Le parallèle de cette extrémité jusqu'à son point d'intersection avec la piste caravanière de Mutwanga à Kasindi;

S043 Cette piste jusqu'à une borne située entre le village de Kimene et l'ancien gîte de Mutwanga;

S044 De cette borne une droite de 125 m de longueur, parallèle au village de Kimene;

S045 De l'extrémité de cette droite une perpendiculaire jusqu'à son point d'intersection avec le ravin situé à 50 m au Sud du signal géodésique de Mutwanga;

S046 Ce ravin jusqu'à sa rencontre avec la piste caravanière Mutwanga à Kasindi;

S047 Cette piste vers le Sud, jusqu'à son point d'intersection avec le ruisseau Cokoye;

S048 Ce ruisseau jusqu'au méridien de la source du ruisseau May Ya Moto;

S049 Ce méridien jusqu'à cette source;

S050 Une droite joignant cette source à celle de la rivière Mboa;

S051 Le parallèle de la source de la Mboa jusqu'à sa rencontre avec la rivière Buliba.

S052 La rive droite de cette rivière jusqu'à sa source;

S053 Une droite joignant cette source au sommet du mont Buliki;

S054 Une droite joignant ce sommet à la source du ruisseau Kamesonge;

S055 La rive droite de ce ruisseau jusqu'à son embouchure dans la rivière Lume;

S056 La rive gauche de la rivière Lume jusqu'à une borne (altitude approximative 2.000 m)

S057 Puis vers l'Ouest la projection verticale sur le terrain de la polygonale reliant les points de rencontre de la courbe de niveau de cette borne avec les contreforts avancés du massif du Ruwenzori, et ce jusqu'au point de rencontre avec le méridien de la source la plus septentrionale de la rivière Ulubu.

S058 Ce méridien, vers le Sud, jusqu'à cette source;

S059 Une droite joignant cette source à la source la plus méridionale de cette même rivière;

S060 De cette source, une droite joignant la source de la rivière Tako, affluent de la Lubilia (frontière de la Colonie);

S061 La frontière de la Colonie vers le Nord jusqu'à son point d'intersection avec la rivière Rusege Sud (confluent de la Lamya et de la Rusege Sud);

S062 La rive droite de cette rivière Rusege Sud, vers l'amont, jusqu'à sa source;

S063 Une droite joignant cette source au col situé entre les monts Kirindera et Turwarubere;

S064 Le parallèle passant par ce col jusqu'au point où il rencontre la rivière Ruanoli;

S065 La rive droite de cette rivière, vers l'aval, jusqu'à son point d'intersection avec le sentier qui relie le village Bogemba au mont Teye;.

S066 Ce sentier jusqu'au sommet du mont Teye;

S067 La crête qui se détache, vers l'Ouest, du mont Teye jusqu'à son extrémité;

S068 La droite joignant cette extrémité au point de la rive gauche de la rivière Kombo le plus rapproché de cette extrémité;

S069 La rive gauche de cette rivière jusqu'à son point de rencontre avec une borne (altitude approximative 1.500 m)

S070 Puis vers le Sud-Ouest la projection verticale sur le terrain de la polygonale reliant les points de rencontre de la courbe de niveau de cette borne avec les contreforts avancés du massif du Ruwenzori et ce jusqu'à la rivière Djalele ou Musanonde.

S071 La rive droite de cette rivière, vers l'aval, jusqu'à son point d'intersection avec le prolongement d'une droite joignant le confluent de la Mavea ou Lamya et de la Molingo au point où le sentier Katuka-Pakioma-Botshula coupe la rivière Djobulo;

S072 Cette droite jusqu'au confluent de la Mavea ou Lamya et de la Molingo;

S073 La rive gauche de la rivière Mavea ou Lamya, vers l'aval, jusqu'à un point situé à un kilomètre en aval de son point d'intersection avec la piste caravanière Kapamba-Kinawa;

S074 De ce point, une ligne parallèle à la piste caravanière passant par les villages de Kapamba, Kinawa, Alundja, Kapera, Zoa et distante de cette piste de 1 kilomètre, vers l'Ouest, à vol d'oiseau jusqu'à une borne située à un kilomètre à vol d'oiseau au Nord du village de Djenda (Kasimoto);

S075 Le parallèle passant par cette borne jusqu'au point où il rencontre la rivière Lamya (frontière de la Colonie);

S076 La rive gauche de la Lamya jusqu'à son embouchure dans la Semliki;

S077 La Semliki jusqu'à l'embouchure de la rivière Puemba;

S078 La rive gauche de cette rivière, vers l'amont, jusqu'à son confluent avec la Nyaduguru;

S079 De ce point, une droite jusqu'à la source de la rivière Malibotu;

S080 La rive gauche de cette rivière jusqu'à son confluent avec la rivière Irimba;

S081 La rive gauche de cette rivière jusqu'à son intersection avec le parallèle passant par le confluent des rivières Batonga et Paru (Pulu) (ce dernier étant un cours d'eau temporaire);

S082 Ce parallèle jusqu'à ce confluent;

S083 De ce point une droite jusqu'à la borne sise à la tête du vallon Nyamangose, en bordure de la piste « Tshabi Semliki, chefferie Watalinga »;

S084 De ce point, le lit creusé par les eaux de ruissellement jusqu'au pied du vallon Nyamangose (borne);

S085 Le parallèle passant par la borne sise au pied du vallon Nyamangose jusqu'à son intersection avec la rivière Maginda (borne);

S086 La rive droite de la Maginda vers l'amont jusqu'à son intersection avec le sentier marqué par le point d'altitude 923 et longeant le pied de l'escarpement de Kamariba.

S087 Ce sentier vers l'Ouest jusqu'au carrefour situé près de la rivière Matido;

S088 A partir de ce carrefour, l'ancienne piste caravanière d'Irumu et Boga à Vieux-Beni, piste de 1925–1926 passant successivement par le village actuel de Selemani (Boga) et par les anciens emplacements des villages de Bopo, Kibondo, Alimaci, Adonga, Gamala, Gamalendu, Mutshanga, Baruti, Kartushi, Lupanzula, Molemba, Amici, Kitihire, Kalumendo, jusqu'à son intersection avec la rivière Djuma;

S089 La rive droite de cette rivière, vers l'amont, jusqu'à son point d'intersection avec le méridien qui passe par le point où la rivière Malulu quitte le pied oriental de l'escarpement;

S090 Ce méridien jusqu'à ce point d'intersection de la rivière Malulu et du pied de l'escarpement;

S091 De ce point, vers le Sud, le pied oriental de l'escarpement jusqu'à un point près du mont Luka, à 2 kilomètres au Nord de l'ancien lazaret où le sentier longeant le pied de l'escarpement et venant de Zumbia (gîte) pénètre dans l'escarpement;

S092 Ce sentier, passant par l'ancien lazaret de Beni, par Zumbia (gîte), par Kitero, entre les monts Misebere et Tahamogota, par Kasolenge et Tschamohoma (Kadiadia) jusqu'à son point de rencontre avec la rivière Lusia;

S093 La rive droite de cette rivière vers l'aval jusqu'à son point d'intersection avec le prolongement d'une droite joignant les sommets des monts Katshe et Katundu;

S094 De ce point, une ligne passant par les sommets des monts Katundu, Katshe, Mokondene, Daboma, Kavega, Kasiakake, Itobola et Kebo;

S095 Du sommet du mont Kebo, une droite jusqu'au point où la rivière Kinyamiga est coupée par cette droite joignant les sommets des monts Kebo et Walengiro;

S096 La rive droite de la rivière Kinyamiga vers l'aval jusqu'à son intersection avec la piste caravanière passant au pied du mont Buselio;

S097 Cette piste vers le Sud jusqu'à son point d'intersection avec le prolongement d'une droite joignant les sommets des monts Bikingi et Buselio;

S098 De ce point, une ligne passant par les sommets des monts Buselio, Bikingi, Busoga, Manyoni, Tshanzu et Birimu;

S099 Du sommet du mont Birimu jusqu'au point de la rivière Tambwe le plus rapproché de ce mont;

S100 De ce point, la rive gauche de la rivière Tambwe, vers l'amont, jusqu'à l'embouchure du ruisseau Tshabolere;

S101 La rive gauche de ce ruisseau jusqu'à sa source;

S102 Une droite joignant cette source à celle du ruisseau Kakoko, affluent de droite de la Nyamoisa;

S103 De cette source (au mont Kalumba), la rive droite du ruisseau Kakoko jusqu'au point où il est coupé par le sentier Kinierere à Kasembe (Nkuku);

S104 Ce sentier jusqu'au point où il coupe la crête du mont Kitolu;

S105 Cette crête jusqu'à la tête du ravin Karuasa se détachant de l'éperon Kitolu du mont Mandimba;

S106 De cette tête, une droite joignant le point où la piste caravanière de Gitse à Kinierere rencontre le ruisseau Lutimbi, à l'Ouest du village de Kinierere;

S107 Cette piste Kinierere-Gitse, par la crête Musimba et le mont Kerongo jusqu'au point le plus rapproché de la source du ruisseau Bolekerere;

S108 Une droite joignant ce point à cette source;

S109 La rive gauche de ce ruisseau jusqu'à son confluent avec la rivière Kalibira;

S110 La rive gauche de cette rivière jusqu'à son confluent avec la rivière Talya;

S111 La rive gauche de cette rivière, vers l'amont, jusqu'au point où elle est franchie par le sentier Kitega-Gitse;

S112 Cette piste vers le Sud jusqu'au point où elle rencontre le prolongement d'une droite joignant les sommets des monts Beasa et Katendere;

S113 De ce point, la ligne de crête jalonnée par les monts Katendere, Beasa, Kanei, Niamonindu et Niondo;

S114 Une ligne passant par le sommet du mont Nguli et joignant la source du ruisseau Logese;

S115 La rive droite de ce ruisseau jusqu'à son embouchure dans la rivière Tshondo;

S116 La rive droite de cette rivière, vers l'aval jusqu'à son point d'intersection avec une droite joignant les sommets des monts Musenzeru et Niarusunzu;

S117 Cette droite jusqu'au mont Musenzeru;

S118 De ce mont, une droite jusqu'au mont Kasiiro;

S119 De ce mont, une droite joignant le sommet de la colline Musoti, le village de Nguli restant en dehors du Parc;

S120 De ce sommet, une droite joignant le point où le sentier Kitega-Nguli coupe l'éperon Kateka-Bakole du mont Niondo;

S121 De ce point, une ligne passant par les sommets des monts Kasanga, Busega, Katembo, Loarama, Mapombo, Garara, Bekoha, Kanyiro (un des petits sommets du mont Ikanga), Liassa, Luterero, Bokara, Kabiniri (Kaliniro), Metseka, Kiongoto, Kasanga, Mubiriri et Kalero;

S122 Du sommet de ce mont la crête du mont Kalero jusqu'à son extrémité Sud;

S123 De cette extrémité une ligne joignant les sommets des monts Bukweri, Lunde ou Nyamoninde, Kahungu;

S124 La croupe méridionale, dénommée Bolambo, du mont Kahungu;

S125 Une ligne joignant ce point, dénommé «Bolambo», au confluent de la rivière Butega et de la Talya;

S126 Une ligne joignant ce mont au sommet du mont Mushenge;

S127 De ce sommet une ligne joignant le mont Kalingio, le point Mbulamasi du mont Mushanga, le point Kaboha du mont Kitetsa, le point Kasoso du mont Kilambo;

S128 Ensuite de ce point une droite joignant le mont Ndwale;

S129 Ensuite la crête jalonnée par les monts Ndwale, Kigende, Kaboha (dénommé erronément Kiboha), Kitobo, Shobobia, Lutare, Hahie, Mchembya.

S130 Ensuite une droite joignant ce mont à l'extrémité Nord du mont Lutepa et laissant à l'Est le mont Miegenie;

S131 La chaîne des monts Lutepa-Nyabuki;

S132 Une droite joignant ce mont au sommet des monts Lubwe, Berama, Kasongolere, puis la crête Kasongolelo, Boswekwa, Kierere et Kashwa;

S133 Une ligne joignant ce mont aux monts Kisololwe, Kiniamuyaga, Miholo;

S134 La crête des monts Miholo, Hangira, Kibiru;

S135 Ensuite une ligne suivant la crête jalonnée par les monts Katwa, Matofu, Musima et prolongée jusqu'au ravin Buhula;

S136 A partir de ce point, le bord Sud du ravin Buhula jusqu'à sa rencontre avec la rivière Kibirizi;

S137 Cette rivière jusqu'au bord occidental du ravin de la Ruindi;

S138 La rive gauche de la Ruindi, en amont jusqu'au confluent de la rivière Rwehe;

S139 Le thalweg de cette rivière jusqu'à son confluent avec le ruisseau Kamokanda, en laissant en dehors du Parc la mine de fer du mont Kakorwe, située sur la rive droite de la Rwehe;

S140 Le ruisseau Kamokanda jusqu'à sa source au pied du mont Tshahi;

S141 Une ligne joignant cette source au sommet des monts Tshahi et Bitingu;

S142 De ce mont une droite joignant la source de la rivière Kalagala, située près du mont Rwanguba;

S143 Cette rivière jusqu'au sentier de Mabenga à Tongo, sentier longeant le pied des monts Kasali;

S144 Ce sentier jusqu'à son point d'intersection avec la rivière Butaku;

S145 De ce point une droite au pied occidental du mont Rugomba;

S146 Le méridien de ce point, jusqu'à sa rencontre avec le parallèle passant par l'extrémité Sud de la colline Butambira;

S147 Ce parallèle jusqu'à sa rencontre avec la piste de Tongo à Tshumba;

S148 La piste de Tongo à Tshumba, dans la plaine de lave, jusqu'à son intersection (carrefour de Mariage) avec le sentier Rugari-Kansenze (Nyamlagira)-Mushari;

S149 Ensuite une droite joignant ce carrefour au sommet du mont Mushebele;

S150 Puis la crête Mushebele-Katunda jusqu'au sentier Tshumba-Ngesho-Gandjo;

S151 Ce sentier jusqu'à la limite Est du bloc des concessions de Ngesho;

S152 Ensuite une ligne contournant à l'Est, au Sud et à l'Ouest les concessions du Comité National du Kivu, 26, 71b et 71a jusqu'à l'intersection avec le sentier Tshumba-Katumo;

S153 Ensuite ce sentier vers l'Ouest jusqu'à la limite orientale de la concession Katumo 79e;

S154 De ce point une droite vers le sommet du mont Kishusha;

S155 Ensuite la limite orientale du marais sis au pied du mont Kishusha;

S156 De là un alignement vers une borne située sur le promontoire occidental du mont Kamatombe;

S157 De ce point un alignement vers une borne placée au sommet du ravin M'Bili;

S158 De cette borne un alignement vers la borne se trouvant sur l'ancien sentier Ngandjo-Kingi, excluant des limites du Parc National Albert les collines Shange et Modeya ;

S159 Ensuite cette piste jusqu'à son point le plus proche de la source de la rivière Bulemo;

S160 De ce point une droite jusqu'à cette source;

S161 Ensuite cette rivière en aval jusqu'au bord du marais de la Nyamuragira ;

S162 Le marais jusqu'à l'angle Nord-Ouest de la concession forestière Marchal à Mugando;

S163 Les limites Nord et Est de cette concession jusqu'à son angle Sud-Est;

S164 Ensuite une droite en direction du Sud-Est rejoignant le bord occidental de la coulée de lave de 1938 du Nyamuragira;

S165 Ensuite le bord occidental de cette coulée de lave jusqu'à son intersection avec la parallèle passant par l'endroit où elle se divise en deux bras;

S166 Ensuite ce parallèle jusqu'à l'endroit précité;

S167 Ensuite le bord occidental de la branche orientale de la coulée de 1938 jusqu'au lac Kivu.

S168 Ensuite la rive Nord du lac Kivu vers l'Est, jusqu'au bord Est de la coulée de lave du volcan Rumoka;

S169 De plus, l'île Tshegera sera dans son entièreté comprise dans le Parc National Albert; par contre, les terres habitées et cultivées des collines de Nzuru, Mihonga et Kabazana seront exclues du Parc ; à cet effet, la limite actuelle de ces terres sera abornée définitivement.

Mikeno Sector

S170 A partir de la frontière du Ruanda-Urundi du point où elle est coupée par une droite reliant le sommet de la colline Bugeshi-Mukuru au sommet du mont Arama;

S171	Cette droite jusqu'au sommet du mont Arama;
S172	De ce point, une droite jusqu'au col séparant ce mont du mont Hehu;
S173	Cette droite jusqu'au sommet du mont Hehu;
S174	De ce point une droite jusqu'au ravin Kabagwetu (plaque Parc National Albert);
S175	Ensuite une droite joignant l'abreuvoir de Kikeri, le marais mais non l'abreuvoir restant inclus dans le Parc;
S176	Ensuite un alignement en direction du sommet du mont Mashaye jusqu'à sa rencontre avec le ravin Kanyamagufa (borne);
S177	Ensuite ce ravin jusqu'à sa rencontre avec la route carrossable Ngoma-Rutshuru (borne);
S178	Ensuite le bord oriental de cette route carrossable jusqu'au ravin Masisi (borne);
S179	Ce ravin jusqu'à son origine, au flanc Ouest de la colline Kasenyi;
S180	Ensuite une piste allant au col qui sépare les monts Bushandjogoro et Rwanguba (borne);
S181	De ce point une droite joignant le ravin Mugari au point où il est coupé par la piste caravanière du Rugari au Kibumba;
S182	Ensuite cette piste jusqu'à sa rencontre avec le ravin Kifurura;
S183	Ce ravin jusqu'à l'abreuvoir situé à 1 km. en amont;
S184	De cet abreuvoir une droite joignant le sommet de la colline Nyamariri;
S185	De ce point une droite jusqu'à l'abreuvoir du ravin Kizenga;
S186	Ensuite une droite jusqu'au sommet de la colline Kizenga (rive droite du ravin);
S187	Puis le sentier Kizenga-Katwa jusqu'au point où il rencontre la concession de Kikeri de la Mission de Tongres-Sainte-Marie;
S188	Ensuite les limites méridionale et orientale de cette concession jusqu'à son angle Nord-Est;
S189	De cet angle, une droite joignant le sommet de la colline Kabazogeye;
S190	Ensuite un alignement en direction du sommet du mont Gashole (alignement qui traverse les ravins Kasasa et Margarure) jusqu'à l'endroit où il est recoupé par un autre alignement partant du sommet de la colline la plus orientale du groupe de Bukima et tangent au bord oriental de la mare Kinyamutukura;
S191	Ensuite cet alignement jusqu'au sommet de la colline la plus orientale du groupe de Bukima ;
S192	De ce sommet une ligne joignant les sommets des monts Nyangurube, Nyakiriba et Gugo;
S193	De ce point, une droite jusqu'au pied occidental de la colline Kizunga;
S194	Ensuite une ligne longeant le pied occidental de la crête Tshananke (le sommet de la crête à la courbe de 2.100 m.) jusqu'au point où elle rencontre le ravin Rutabagwe;
S195	Ce ravin jusqu'à l'étang Nyandizima;
S196	Le bord méridional de cet étang jusqu'à la tête du ravin Rukunga;
S197	Ce ravin jusqu'à son point d'intersection avec la droite joignant le sommet le plus septentrional du mont Runyoni au mont Tshanzu;
S198	Cette droite à partir de ce point d'intersection jusqu'au sommet du mont Tshanzu;
S199	Ensuite une droite joignant les sommets des monts Rurindzargwe et Mugongoyindzovu;
S200	Ensuite le parallèle de ce dernier sommet jusqu'à son point de rencontre avec la frontière belgo-anglaise.

APPENDIX 2
LIST OF MAMMALS OF VIRUNGA NATIONAL PARK
Taxonomy and English names follow Wilson & Reeder (2005)
(M. Languy & A. Plumptre)

Order/Family	Species	English Name
TUBULIDENTATA		
Orycteropodidae	*Orycteropus afer*	Aardvark
HYRACOIDEA		
Procaviidae	*Dendrohyrax arboreus*	Southern Tree Hyrax
PROBOSCIDEA		
Elephantidae	*Loxodonta africana*	African Bush Elephant
Elephantidae	*Loxodonta cyclotis*	African Forest Elephant
PRIMATES		
Loridae	*Perodictus potto*	Potto
Galagonidae	*Galago demidoff*	Prince Demidoff's Bushbaby
Galagonidae	*Galago senegalensis*	Senegal Bushbaby
Galagonidae	*Galago thomasi*	Thomas's Bushbaby
Galagonidae	*Otolemur crassicaudatus*	Brown Greater Galago
Cercopithecidae	*Cercopithecus ascanius*	Red-tailed Monkey
Cercopithecidae	*Cercopithecus denti*	Dent's Mona Monkey
Cercopithecidae	*Cercopithecus doggetti*	Silver Monkey
Cercopithecidae	*Cercopithecus hamlyni*	Owl-faced Monkey
Cercopithecidae	*Cercopithecus kandti*	Golden Monkey
Cercopithecidae	*Cercopithecus lhoesti*	L'Hoest's Monkey
Cercopithecidae	*Cercopithecus mitis*	Blue Monkey
Cercopithecidae	*Cercopithecus neglectus*	De Brazza's Monkey
Cercopithecidae	*Chlorocebus tantalus*	Tantalus Monkey
Cercopithecidae	*Lophocebus albigena*	Grey-cheeked Mangabey
Cercopithecidae	*Papio anubis*	Olive Baboon
Cercopithecidae	*Colobus angolensis*	Angola Colobus
Cercopithecidae	*Colobus guereza*	Mantled Guereza
Cercopithecidae	*Piliocolobus foai*	Central African Red Colobus
Cercopithecidae	*Cercocebus agilis*	Agile Mangabey
Hominidae	*Gorilla beringei*	Eastern Gorilla
Hominidae	*Pan troglodytes*	Common Chimpanzee
AFROSORICIDA		
Tenrecidae	*Micropotamogale ruwenzorii*	Rwenzori Otter Shrew
Chrysochloridae	*Chrysochloris stuhlmanni*	Stuhlmann's Golden Mole
MACROSCELIDEA		
Macroscelididae	*Rhynchocyon cirnei*	Checkered Elephant Shrew
LAGOMORPHA		
Leporidae	*Lepus microtis*	African Savanna Hare
ERINACEOMORPHA		
Erinaceidae	*Atelerix albiventris*	Four-toed Hedgehog
SORICIOMORPHA		
Soricidae	*Crocidura hirta*	Lesser Red Musk Shrew
Soricidae	*Crocidura jacksoni*	Jackson's Shrew
Soricidae	*Crocidura lanosa*	Kivu Long-haired Shrew
Soricidae	*Crocidura maurisca*	Northern Swamp Musk Shrew
Soricidae	*Crocidura montis*	Montane White-toothed Shrew
Soricidae	*Crocidura nanilla*	Savanna Dwarf Shrew
Soricidae	*Crocidura nigrofusca*	African Black Shrew
Soricidae	*Crocidura niobe*	Niobe's Shrew
Soricidae	*Crocidura olivieri*	Northern Giant Musk Shrew
Soricidae	*Crocidura tarella*	Tarella Shrew
Soricidae	*Crocidura turba*	Turba Shrew

Family	Species	Common Name
Soricidae	*Paracrocidura maxima*	Greater Large-headed Shrew
Soricidae	*Ruwenzorisorex suncoides*	Rwenzori Shrew
Soricidae	*Scutisorex somereni*	Armored Shrew
Soricidae	*Suncus megalura*	Climbing Shrew
Soricidae	*Sylvisorex granti*	Grant's Forest Shrew
Soricidae	*Sylvisorex lunaris*	Moon Forest Shrew
Soricidae	*Sylvisorex vulcanorum*	Volcano Shrew
Soricidae	*Myosorex babaulti*	Babault's Mouse Shrew
Soricidae	*Myosorex blarina*	Montane Mouse Shrew

CHIROPTERA

Family	Species	Common Name
Pteropodidae	*Eidolon helvum*	African Straw-coloured Fruit Bat
Pteropodidae	*Epomophorus labiatus*	Little Epauletted Fruit Bat
Pteropodidae	*Epomophorus wahlbergi*	Wahlberg's Epauletted Fruit Bat
Pteropodidae	*Epomops franqueti*	Franquet's Epauletted Fruit Bat
Pteropodidae	*Hypsignathus monstrosus*	Hammer-headed Fruit Bat
Pteropodidae	*Lissonycteris angolensis*	Angolan Soft-furred Fruit Bat
Pteropodidae	*Myonycteris torquata*	Little Collared Fruit Bat
Pteropodidae	*Rousettus aegyptiacus*	Egyptian Rousette
Pteropodidae	*Rousettus lanosus*	Long-haired Rousette
Rhinolophidae	*Rhinolophus alcyone*	Halcyon Horseshoe Bat
Rhinolophidae	*Rhinolophus clivosus*	Geoffroy's Horseshoe Bat
Rhinolophidae	*Rhinolophus eloquens*	Eloquent Horseshoe Bat
Rhinolophidae	*Rhinolophus fumigatus*	Rüppell's Horseshoe Bat
Rhinolophidae	*Rhinolophus landeri*	Lander's Horseshoe Bat
Rhinolophidae	*Rhinolophus ruwenzorii*	Rwenzori Horseshoe Bat
Hipposideridae	*Hipposideros caffer*	Sundevall's Leaf-nosed Bat
Hipposideridae	*Hipposideros cyclops*	Cyclops Leaf-nosed Bat
Megadermatidae	*Lavia frons*	Yellow-winged Bat
Emballonuridae	*Taphozous mauritianus*	Mauritian Tomb Bat
Nycteridae	*Nycteris arge*	Bates's Slit-faced Bat
Nycteridae	*Nycteris grandis*	Large Slit-faced Bat
Nycteridae	*Nycteris hispida*	Hairy Slit-faced Bat
Nycteridae	*Nycteris macrotis*	Large-eared Slit-faced Bat
Nycteridae	*Nycteris nana*	Dwarf Slit-faced Bat
Nycteridae	*Nycteris thebaica*	Egyptian Slit-faced Bat
Molossidae	*Chaerephon ansorgei*	Ansorge's Free-tailed Bat
Molossidae	*Chaerephon bemmeleni*	Gland-tailed Free-tailed Bat
Molossidae	*Chaerephon pumilus*	Little Free-tailed Bat
Molossidae	*Mops brachypterus*	Short-winged Free-tailed Bat
Molossidae	*Mops midas*	Mida's Free-tailed Bat
Molossidae	*Mops nanulus*	Dwarf Free-tailed Bat
Molossidae	*Tadarida fulminans*	Malagasy Free-tailed Bat
Verspertilionidae	*Scotophilus nigrita*	Giant House Bat
Verspertilionidae	*Pipistrellus nanulus*	Tiny Pipistrelle
Verspertilionidae	*Pipistrellus rueppellii*	Rüppell's Pipistrelle
Verspertilionidae	*Glauconycteris argentata*	Common Butterfly Bat
Verspertilionidae	*Glauconycteris poensis*	Abo butterfly bat
Verspertilionidae	*Glauconycteris variegata*	Variegated Butterfly Bat
Verspertilionidae	*Hypsugo crassulus*	Broad-headed Pipistrelle
Verspertilionidae	*Mimetillus moloneyi*	Moloney's Mimic Bat
Verspertilionidae	*Neoromicia tenuipinnis*	White-winged Serotine
Verspertilionidae	*Myotis bocagii*	Rufous Myotis
Verspertilionidae	*Myotis tricolor*	Temminck's Myotis
Verspertilionidae	*Myotis welwitschii*	Welwitsch's Myotis
Verspertilionidae	*Miniopterus inflatus*	Greater Long-fingered Bat

PHOLIDOTA

Family	Species	Common Name
Manidae	*Manis tricuspis*	Tree Pangolin
Manidae	*Manis tetradactyla*	Long-tailed Pangolin
Manidae	*Manis gigantea*	Giant Pangolin

CARNIVORA

Family	Species	Common Name
Felidae	*Felis silvestris*	Wild Cat
Felidae	*Leptailurus serval*	Serval
Felidae	*Profelis aurata*	African Golden Cat
Felidae	*Panthera leo*	Lion

Family	Species	Common Name
Felidae	*Panthera pardus*	Leopard
Viverridae	*Civettictis civetta*	African Civet
Viverridae	*Genetta servalina*	Servaline Genet
Viverridae	*Genetta tigrina*	Cape Genet
Viverridae	*Nandinia binotata*	African Palm Civet
Herpestidae	*Atilax paludinosus*	Marsh Mongoose
Herpestidae	*Bdeogale nigripes*	Black-footed Mongoose
Herpestidae	*Crossarchus alexandri*	Alexander's Kusimanse
Herpestidae	*Galerella sanguinea*	Slender mongoose
Herpestidae	*Herpestes ichneumon*	Egyptian Mongoose
Herpestidae	*Ichneumia albicauda*	White-tailed Mongoose
Herpestidae	*Mungos mungo*	Banded Mongoose
Hyaenidae	*Crocuta crocuta*	Spotted Hyena
Canidae	*Canis adustus*	Side-striped Jackal
Canidae	*Lycaon pictus*	African Wild Dog
Mustelidae	*Aonyx capensis*	Cape clawless otter
Mustelidae	*Hydrictis maculicollis*	Spotted-necked Otter
Mustelidae	*Ictonyx striatus*	Zorilla
Mustelidae	*Mellivora capensis*	Honey Badger
Mustelidae	*Poecilogale albinucha*	African Striped Weasel

ARTIODACTYLA

Family	Species	Common Name
Suidae	*Hylochoerus meinertzhageni*	Giant Forest Hog
Suidae	*Phacochoerus africanus*	Common Warthog
Suidae	*Potamochoerus larvatus*	Bush-pig
Hippopotamidae	*Hippopotamus amphibius*	Common Hippopotamus
Tragulidae	*Hyemoschus aquaticus*	Water Chevrotain
Giraffidae	*Okapia johnstoni*	Okapi
Bovidae	*Damaliscus korrigum*	Topi
Bovidae	*Neotragus batesi*	Bate's Dwarf Antelope
Bovidae	*Syncerus caffer*	African Buffalo
Bovidae	*Tragelaphus eurycerus*	Bongo
Bovidae	*Tragelaphus scriptus*	Bushbuck
Bovidae	*Tragelaphus spekii*	Sitatunga
Bovidae	*Cephalophus nigrifrons*	Black-fronted Duiker
Bovidae	*Cephalophus silvicultor*	Yellow-backed Duiker
Bovidae	*Cephalophus dorsalis*	Bay Duiker
Bovidae	*Cephalophus leucogaster*	White-bellied Duiker
Bovidae	*Philantomba monticola*	Blue Duiker
Bovidae	*Cephalophus weynsi*	Weyns's Duiker
Bovidae	*Sylvicapra grimmia*	Bush Duiker
Bovidae	*Kobus ellipsiprymnus*	Waterbuck
Bovidae	*Kobus kob*	Kob
Bovidae	*Redunca redunca*	Common Reedbuck

RODENTIA

Family	Species	Common Name
Sciuridae	*Xerus erythropus*	Striped Ground Squirrel
Sciuridae	*Funisciurus carruthersi*	Carruther's Mountain Squirrel
Sciuridae	*Funisciurus pyrropus*	Fire-footed Rope Squirrel
Sciuridae	*Heliosciurus gambianus*	Gambian Sun squirrel
Sciuridae	*Heliosciurus ruwenzorii*	Rwenzori Sun Squirrel
Sciuridae	*Heliosciurus rufobrachium*	Red-legged Sun Squirrel
Sciuridae	*Paraxerus alexandri*	Alexander's Bush Squirrel
Sciuridae	*Paraxerus boehmi*	Boehm's Bush Squirrel
Sciuridae	*Protoxerus stangeri*	Forest Giant Squirrel
Graphiurinae	*Graphiurus murinus*	Forest African Dormouse
Spalacidae	*Tachyoryctes ruandae*	Rwanda African Mole Rat
Nesomyidae	*Cricetomys emini*	Forest Giant Pouched Rat
Nesomyidae	*Cricetomys kivuensis*	Kivu Giant Pouched Rat
Nesomyidae	*Delanymys brooksi*	Delany's Swamp Mouse
Nesomyidae	*Dendromus insignis*	Montane African Climbing Mouse
Nesomyidae	*Dendromus melanotis*	Grey African Climbing Mouse
Nesomyidae	*Dendromus messorius*	Banana African Climbing Mouse
Nesomyidae	*Dendromus mystacalis*	Chestnut African Climbing Mouse
Nesomyidae	*Dendromus nyasae*	Kivu African Climbing mouse
Muridae	*Deomys ferruginueus*	Congo Forest Rat
Muridae	*Lophuromys aquilus*	Bark-coloured Brush-furred Rat

Family	Species	Common name
Muridae	*Lophuromys luteogaster*	Buff-bellied Brush-furred Rat
Muridae	*Lophuromys medicaudatus*	Western Rift Brush-furred Rat
Muridae	*Lophuromys rahmi*	Rahm's Brush-furred Rat
Muridae	*Lophuromys sikapusi/ansorgei*	Rusty-bellied/Ansorge's Brush-furred Rat
Muridae	*Lophuromys woosnami*	Woosnam's Brush-furred Rat
Muridae	*Gerbilliscus kempi*	Northern Savanna Gerbil
Muridae	*Aethomys kaiseri*	Kaiser's Aethomys
Muridae	*Arvicanthis niloticus*	African Arvicanthis
Muridae	*Colomys goslingi*	African Wading Rat
Muridae	*Dasymys alleni*	Glover Allen's Dasymys
Muridae	*Dasymys incomtus*	Common Dasymys
Muridae	*Dasymys montanus*	Rwenzori Dasymys
Muridae	*Dasymys rwandae*	Rwandan Dasymys
Muridae	*Grammomys dolichurus*	Common Thicket rat
Muridae	*Grammomys dryas*	Albertine Rift Grammomys
Muridae	*Grammomys kuru*	Eastern Rainforest Grammomys
Muridae	*Hybomys lunaris*	Rwenzori Hybomys
Muridae	*Hybomys univittatus*	Peters's Hybomys
Muridae	*Hylomyscus denniae*	Montane Hylomyscus
Muridae	*Hylomyscus stella*	Stella's Hylomyscus
Muridae	*Lemniscomys macculus*	Buffoon Lemniscomys
Muridae	*Lemniscomys striatus*	Typical Lemniscomys
Muridae	*Malacomys longipes*	Common Malacomys
Muridae	*Mastomys erythroleucus*	Reddish-white Mastomys
Muridae	*Mastomys natalensis*	Natal Mastomys
Muridae	*Mus bufo*	Toad Mouse
Muridae	*Mus musculoides*	Subsaharan Pygmy Mouse
Muridae	*Mus triton*	Grey-bellied Pygmy Mouse
Muridae	*Mylomys dybowskyii*	Common Mylomys
Muridae	*Oenomys hypoxanthus*	Common Oenomys
Muridae	*Pelomys fallax*	East African Pelomys
Muridae	*Praomys degraaffi*	De Graaff's Praomys
Muridae	*Praomys jacksoni*	Jackson's Praomys
Muridae	*Praomys verschureni*	Verschuren's Praomys
Muridae	*Rattus rattus*	Roof Rat
Muridae	*Stochomys longicaudatus*	Target Rat
Muridae	*Thamnomys kempi*	Dollman's Thamnomys
Muridae	*Thamnomys major*	Hatt's Thamnomys
Muridae	*Thamnomys venustus*	Thomas's Thamnomys
Muridae	*Zelotomys hildegardeae*	Hildegarde's Zelotomys
Muridae	*Otomys denti*	Montane Groove-toothed Rat
Muridae	*Otomys dartmouthi*	Rwenzori Vlei Rat
Muridae	*Otomys tropicalis*	East African Vlei Rat
Anomaluridae	*Anomalurus beecrofti*	Beecroft's Scaly-tailed Squirrel
Anomaluridae	*Anomalurus derbianus*	Lord Derby's Scaly-tailed Squirrel
Anomaluridae	*Anomalurus pusillus*	Dwarf Scaly-tailed Squirrel
Anomaluridae	*Idiurus macrotis*	Long-eared Scaly-tailed Flying Squirrel
Anomaluridae	*Idiurus zenkeri*	Pygmy Scaly-tailed Flying Squirrel
Hystricidae	*Atherurus africanus*	African Brush-tailed Porcupine
Hystricidae	*Hystrix africaeaustralis*	Cape Porcupine
Thryonomyidae	*Thryonomys gregorianus*	Lesser Cane Rat
Thryonomyidae	*Thryonomys swinderianus*	Greater Cane Rat

Total: 218

Species not yet recorded but that most probably occur within Virunga NP, on the basis of their known range and prefered habitats

Family	Species	Common name
Procaviidae	*Heterohyrax brucei*	Yellow-spotted Rock Hyrax
Soricidae	*Crocidura attila*	Hun Shrew
Soricidae	*Crocidura crenata*	Long-footed Shrew
Soricidae	*Crocidura fuscomurina*	Bicolored Musk Shrew
Soricidae	*Crocidura goliath*	Goliath Shrew
Soricidae	*Crocidura hildegardeae*	Hildegarde's Shrew
Soricidae	*Crocidura luna*	Moonshine Shrew
Soricidae	*Crocidura littoralis*	Naked-tailed Shrew
Soricidae	*Crocidura roosevelti*	Rooselvelt's shrew
Soricidae	*Sylvisorex johnstoni*	Johnston's Forest Shrew
Pteropodidae	*Scotonycteris zenkeri*	Zenker's Fruit Bat

Hipposideridae	*Hipposeridos camerunensis*	Cameroon Leaf-nosed Bat
Hipposeridae	*Hipposeridos gigas*	Giant Leaf-nosed Bat
Hipposeridae	*Hipposeridos ruber*	Noak's Leaf-nosed Bat
Verspertilionidae	*Scotophilus nux*	Nut-coloured House Bat
Verspertilionidae	*Glauconycteris gleni*	Glen's butterfly bat
Vespertilionidae	*Kerivoula lanosa*	Lesser Woolly Bat
Vespertilionidae	*Kerivoula smithii*	Smith's Woolly Bat
Herpestidae	*Herpestes naso*	Long-nosed Mongoose
Muridae	*Hylomyscus aeta*	Beaded Hylomyscus
Muridae	*Hylomyscus parvus*	Lesser Hylomyscus
Muridae	*Mus mahomet*	Mahomet Mouse
Muridae	*Mus sorella*	Thomas's Mouse
Muridae	*Pelomys hopkinsi*	Hopkin's Pelomys
Muridae	*Pelomys minor*	Least Pelomys
Muridae	*Praomys tullbergi*	Jackson's Praomys

Total: 26

Possible additional species (Virunga NP being at the edge of their range)

Leporidae	*Lepus capensis*	Cape Hare
Pteropodidae	*Epomophorus minimus*	Epauletted Fruit Bat
Molossidae	*Chaerephon aloysiisabaudiae*	Duke of Abruzzi's Free-tailed Bat
Molossidae	*Mops thersites*	Railer Free-tailed Bat
Bovidae	*Oreotragus oreotragus*	Klipspringer
Sciuridae	*Funisciurus anerythrus*	Thomas's Rope Squirrel
Muridae	*Lophuromys dudui*	Dudu's Brush-furred Rat
Muridae	*Aethomys hindei*	Northern Bush Rat
Muridae	*Praomys misonnei*	Soft-furred Rat
Hystricidae	*Hystrix cristata*	Crested porcupine

Total: 10

BIBLIOGRAPHY

References cited in the text

Adams, B. & Infield, M., 1998. Community Conservation at Mgahinga Gorilla National Park, Uganda. In: *Community Conservation in Africa. Principles and comparative practice.* AWF. University of Manchester, University of Zimbabwe, University of Cambridge. 36 pp.

Akeley, C., 1923. *In Brightest Africa.* New York, Doubleday. 188–249.

Anderson D., Grove R., (eds) 1999. *Conservation in Africa. People, Policies and Practice.* Cambridge University Press.

Anon., 1964. *Victor Van Straelen : Tel qu'il demeure.* Renson International Marketing. Brussels. 126pp.

Anon. 2002. A. von Beringe: On the Trail of the Man who discovered the Mountain Gorilla. *Gorilla Journal,* June 2002.

Augustine, N.R., 1995. Managing the crisis you tried to prevent. *Harvard Business Review,* 73(6): 147–158.

Aveling, C., 1990. *Comptage aérien total des buffles et éléphants au Parc National des Virunga, octobre 1990.* Unpublished report to Kivu Programme (European Union), sub-programme Virunga. 6 pp.

Aveling, C. & Harcourt, A.H., 1984. A census of the Virunga gorillas. *Oryx,* 18: 8–13.

Balole, E. & Kivunda, J. 2001. *Pression économique sur les ressources du PNVi, secteur Mikeno.* Unpublished report to the International Gorilla Conservation Proramme, Goma, DRC.

Bassolé, A., Brunner J., Tunstall D., 1999. *SIG et appui à la planification et à la gestion de l'environnement en Afrique de l'Ouest.* World Resources Institute.

Bensted-Smith, R. & Cobb S., 1995. Reform of protected area institutions in East Africa. *Parks* 5: 3–19.

Biodiversity Support Programme, 1999. *Study on the Development of Transboundary Natural Resource Management Areas in Southern Africa.* Biodiversity Support Programme, Washington DC.

Bishikwabo, K., 2000. *La situation du Parc National de Kahuzi-Biega au 20 mars 2000.* Unpublished report to ICCN.

Biswas, A.K. & Tortajada-Quiroz, H.C., 1996. Environmental Impacts on the Rwandan Refugees in Zaïre. *Ambio* 25: 403–408.

Biswas, A. K., Tortajada-Quiroz, H. C., Lutete, V., & Lemba, G., 1994. *Environmental impact of the rwandese refugees presence in north and south Kivu (Zaïre).* United Nations Development Programme. 53 pp.

Blaikie, P. & Jeanrenaud, S., 1997. Biodiversity and human welfare. In: Ghimire and Pimbert (eds.). *Social change and conservation.* Earthscan.

Blom, E., 2000. Conclusions and recommendations. In: Blom E., Bergmans, W. *et al.* (eds), Nature in war. *Biodiversity conservation during conflicts* pp. 165–170. Netherlands Commission for International Nature Protection. Mededelingen, 37. Amsterdam.

Blondel, N., 1995. *Un an de présence des camps de réfugiés en périphérie du secteur sud du Parc National des Virunga: bilan des dégâts.* Unpublished report to EDF/PSRR-PNVi. 20 pp.

Blondel, N., 1997. *L'impact des camps de réfugiés sur l'environnement local: étude de cas sur Goma (Nord-Kivu).* Journée d'étude 'l'information satellitaire au service de l'aide humanitaire', 28 mai 1997. Contribution du bureau I-Mage Consult. Unpublished report. 5 pp.

Bourlière, F. & Verschuren, J., 1960. *Introduction à l'écologie des ongulés du Parc National Albert.* Institut des Parcs Nationaux du Congo Belge. Exploration du Parc National Albert. Fasc. 1 and 2. Brussels. 158 pp.

Brabant, S.A.R. le duc de, 1933. Speach delivered at the African Society, London, 16 Novembre 1933: *Les Parcs Nationaux et la protection de la nature,* IPNCB, 1937, p. 17.

Bremer, F., 1996. *Réhabilitation du Parc National des Virunga.* Unpublished report at the ZOPP workshop from 7 to 9 February 1996 in Goma. GTZ/IZCN.

Brockington, D. & Homewood, K., 1996. Wildlife, Pastoralists and Science: debates concerning Mkomazi Game Reserve, Tanzania. In: Leach, M. and Mearns, R. (eds.) *The Lie of the Land: challenging received wisdom on the African environment.* The International African Institute, London.

Cairns, E., 1997. *A safer future: Reducing the human cost of war.* Oxfam Publications, United Kingdom.

Chifundera, K.Z., Nyakabwa, M., Bashonga, M.G., Masumbuka, N.C. & Kyungu, J.C., 2003. *The mount Tshiabirimu in the Albertine Rift: biodiversity, habitat and conservation issues.* Unpublished report to Dian Fossey Gorilla Fund Europe.

Child, G., 1996. Conservation beyond Yellowstone: an economic framework for wildlife conservation. In: *African Wildlife Policy Consultation.* Overseas Development Administration.

Chrétien, J-P., 2000. *L'Afrique des Grands Lacs. Deux mille ans d'histoire,* Paris, Aubier Eds. pp. 31–33 and 47.

Cooper, J. & Cooper, M., 1996. Mountain Gorillas, a 1995 update. *African Primates* 2: 30–31.

Cornet d'Elzius, C., 1964. *Evolution de la végétation dans la plaine au sud du Lac Edouard.* Institut des Parcs Nationaux du Congo et du Rwanda. 23 pp.

Cornet d'Elzius, C., 1996. *Ecologie, structure et évolution des Populations des Grands Mammifères du Secteur Central du Parc National des Virunga (Parc National Albert) Zaïre (Congo Belge).* Fondation pour Favoriser les Recherches Scientifiques en Afrique, Belgium. 231 pp.

Curry-Lindahl, K., 1956. Ecological studies on Mammals, Birds, Reptiles and Amphibians in the Eastern Belgian Congo. *Ann. Musée R. Congo Belge, Tervuren, Sciences Zoologiques,* vol 87. 78 pp.

d'Huart, J.P., 1977. Station de baguage de Lulimbi (Parc National des Virunga, Zaïre): rapport d'activités 1971–1975. *Le Gerfaut* 67: 161–168.

d'Huart, J.P., 1978. *Ecologie de l'hylochère* (Hylochoerus meinertzhageni Thomas) *au Parc National des Virunga.* Exploration Parc National des Virunga, Deuxième Série, Fasc. 25. Fondation pour Favoriser les Recherches Scientifiques en Afrique. Brussels. 156 pp.

d'Huart, J.P., 1987. *Parc National des Virunga (Kivu, Zaïre): état des lieux et recommandations pour un projet d'appui de la Commission des Communautés Européennes.* Mission report to the EEC. 89 pp.

d'Huart, J.P., 2003. *Statut de conservation et proposition de réhabilitation des Parcs Nationaux de la Garamba et des Virunga (RDC).* Rapport d'une mission d'identification et proposition d'intervention pour la préparation d'un projet d'appui GEF/Banque Mondiale. Unpublished report to the World Bank. 59 pp.

Davies, H., & Walters, M., 1998. Do all crises have to become disasters? Risk and risk mitigation. Disaster Prevention and Management, vol 7(5): 396–400.

de Grunne, X., Hauman, L., Burgeon, L. et Michot, P., 1937. *Vers les glaciers de l'Equateur.* Le Ruwenzori, Mission scientifique belge 1932. Eds R. Dupriez, Brussels. 300 pp.

de Witte, G.F., 1937. *Exploration du Parc National Albert.* Mission G.F. de Witte (1933–1935), fasc.1. Institut des Parcs Nationaux du Congo Belge. 39 pp.

de Witte, G.F., 1938. *Mammifères.* Exploration du Parc National Albert. Institut des Parcs Nationaux du Congo Belge, Brussels.

Diamond, J.M., 1975. Assembly of species communities. In: Cody, M.L. and Diamond, J.M. (eds.). *Ecology and evolution of communities.* Belknap, Cambridge, Massachusetts.

Delvingt, W., 1978. *Ecologie de l'hippopotame au Parc National des Virunga.* PhD Thesis, Fac. Sc. Agr. Gembloux, 2 volumes. 333 pp.

DELVINGT, W., 1994. *Etude préparatoire du programme spécial de réhabilitation pour les pays voisins du Rwanda (PSSR), volet environnement. Rapport d'une mission effectuée au Zaïre, 24/11-05/12/94.* Faculté des Sciences Agronomiques de Gembloux, U.E.R. Sylviculture, 28 pp.

DELVINGT, W., 1996. L'intervention de l'Union Européenne dans le Parc National des Virunga. *African Primates* 2: 28–30.

DELVINGT, W., LEJOLY, J. & MANKOTO, M., 1990. *Guide du Parc National des Virunga.* Commission des Communautés Européennes. 192 pp.

DUBLIN, H.T., 1991. Dynamics of the Serengeti-Mara woodlands: an historical perspective. *For. Conserv. Hist.,* 35: 169–178.

DZIEDZIC, W., 2005. *Management of post-conflict situations pertaining to boundaries of Virunga National Park.* WWF report to UNESCO. 16 pp.

FRASER N., 2005. *Qu'est-ce que la justice sociale? Reconnaissance et redistribution.* La Découverte Eds.

FRECHKOP, S., 1938. *Mammifères.* Exploration du Parc National Albert. Mission G.F. de Witte (1933–1935), fascicule 10. Institut des Parcs Nationaux du Congo Belge. Brussels. 103 pp.

FRECHKOP, S., 1941. *Animaux protégés au Congo Belge.* Institut des Parcs Nationaux du Congo Belge. 469 pp.

FRECHKOP, S., 1943. *Mammifères.* Exploration du Parc National Albert. Mission S. Frechkop (1938), fascicule 1. Institut des Parcs Nationaux du Congo Belge. Brussels. 186 pp.

FRECHKOP, S., 1950. Parmi les mammifères de l'est du Congo Belge. *La Terre et la Vie* 1950, 1: 1–15.

GRAY, M., MCNEILAGE, A., FAWCETT, K., ROBBINS, M.M., SSEBIDE, B., MBULA, D. & UWINGELI, P., 2005. *Virunga Volcano Range Mountain Gorilla Census 2003.* Unpublished report to UWA/ORTPN/ICCN.

GREENING, D.W., AND JOHNSON, R.A., 1996. Do managers and strategies really matter? A study in crisis. *Journal of Management Studies,* 33(1): 25–51

GROOM, A.F.G., 1973. Squeezing out the mountain gorillas. *Oryx,* 2: 207–215.

GRUBB, P., BUTYNSKI, T.M., OATES, J.F., BEARDER, S.K., DISOTELL, T.R., GROVES, C.P. & STRUHSAKER, T.T., 2003. Assessment of the diversity of African primates. *International Journal of Primatology,* 24: 1301–1357

GTZ, 2000. *Rapport sur le contrôle de l'état d'avancement du Projet 93.2182.9-003.00, Appui au Renforcement des Capacités Institutionnelles de la DG ICCN.* Institut Congolais pour la Conservation de la Nature, Kinshasa. 26 pp.

HARCOURT, A.H. & CURRY-LINDAHL, K., 1979. Conservation of the Mountain Gorilla and its habitat in Rwanda. *Environmental Conservation,* 6: 143–147.

HARCOURT, A.H. & FOSSEY, D., 1981. The Virunga gorillas: decline of an island population. *African Journal of Ecology,* 19: 83–97.

HARCOURT, A.H., KINEMAN, J., CAMPBELL, G., YAMAGIWA, J., REDMOND, I., AVELING, C. & CONDIOTTI, M. 1983. Conservation and the Virunga gorilla population. *African Journal of Ecology,* 21: 139–142.

HARROY, J.P., 1987. Soixantième anniversaire d'un parc national zaïrois. *Bull. Séanc. Acad. Roy. Sci. Outre-Mer,* 31(4): 507–516.

HENQUIN, B. & BLONDEL, N., 1996. *Etude par télédétection sur l'évolution récente de la couverture boisée du Parc National des Virunga.* Project report PSRR/PNVi, UE-ADG-UNHCR-IZCN. 80 pp.

HENQUIN, B. & BLONDEL, N., 1997. *Etude par télédétection sur l'évolution récente de la couverture boisée du Parc National des Virunga, deuxième partie (période 1995–1996).* Project report PSRR/PNVi, UE-ADG-UNHCR-ICCN. 63 pp.

HILTON-TAYLOR, C. (ed), 2002. *2002 IUCN Red List of Threatened Species.* IUCN, Gland, Switzerland and Cambridge, UK.

HOIER, R., 1950. *A travers plaines et volcans au Parc National Albert.* Institut des Parcs Nationaux du Congo Belge. Brussels. 173 pp.

HOIER, R., 1952. *Mammifères du Parc National Albert.* Collections Lebègue et Nationale n°105. Brussels. 111 pp.

HOMSY, J., 1999. *Ape tourism and human diseases: how close should we get? A critical review of rules and regulations governing park management and tourism for the wild mountain gorilla* Gorilla gorilla beringei. Unpublished report to IGCP. Nairobi.

INCN/IZCN 1973–1977. *Léopard* series, 1 to 5.

INOGWABINI B.-I., OMARI ILAMBU & MBAYMA, A.G., 2005. Protected areas of the Democratic Republic of Congo. *Conservation Biology,* 19: 15–22.

INSTITUT CONGOLAIS POUR LA CONSERVATION DE LA NATURE, 2003. *Stratégie Nationale de la Conservation.* ICCN, October 2003. 22 pp.

INSTITUT CONGOLAIS POUR LA CONSERVATION DE LA NATURE, 1999. *Séminaire sur les Sites du Patrimoine Mondial en danger en République Démocratique du Congo. Naivasha, 12–16 avril 1999. Rapport Final.* Unpublished report to ICCN. 37 pp.

INSTITUT CONGOLAIS POUR LA CONSERVATION DE LA NATURE, 2006. *Liste du personnel de l'ICCN.* ICCN, Kinshasa.

INTERNATIONAL RESCUE COMMITTEE, 2003. *Mortality in the Democratic Republic of Congo: results from a nationwide survey.* International Rescue Committee, New York.

INTERNATIONAL INSTITUTE FOR ENVIRONMENT AND DEVELOPMENT, 1993. *Whose Eden? An overview of community approaches to wildlife management.* International Institute for Environment and Development.

IPNCB, 1935. *Rapport du Conseil Colonial sur le projet de décret organisant l'IPNCB.* Bulletin Officiel, 1935, p. 64.

IPNCB, 1937. *Les Parcs Nationaux et la Protection de la Nature.* Institut des Parcs Nationaux du Congo Belge. 88 pp.

IPNCB, 1942. *Premier rapport quinquennal, (1935–1939).* Institut des Parcs Nationaux du Congo Belge. 75 pp.

ISHWARAN N., 1994. The role of protected areas in promoting sustainable development. *Parks,* 4: 2–7.

JOHNSTON, W.P. & STEPANOVICH, P.L., 2001. Managing a crisis: Planning, Acting and Learning. *American Journal of Health-System Pharmacists;* 58: 1245–1249.

KALPERS, J. & LANJOUW, A., 1997. Potential for the creation of a peace park in the Virunga volcano region. *Parks,* 7: 25–35.

KALPERS, J., 1992. La conservation du gorille de montagne. Le point de la situation en 1993. *Cahiers d'Ethologie,* 12(4): 467–490.

KALPERS, J., 1996. *Suivi systématique de deux sites du patrimoine mondial: PN des Virunga et PN de Kahuzi-Biega. Mission du 22 avril au 9 mai 1996.* Mission report to the World Heritage Committee, UNESCO, Paris. 30 pp.

KALPERS, J., 1998. *Projet de réhabilitation des capacités de gestion de l'ICCN au Parc National des Virunga, secteur sud. Financement UNHCR et WWF-Hollande, deuxième semestre 1997.* Activity report of the International Gorilla Conservation Programme to UNHCR and WWF-Netherlands.

KALPERS, J. & BRAUN, U., 2005. *Programme Environnemental autour des Virunga. Evaluation prospective interne, 7–21 juin 2005.* Final report to WWF. Nairobi. 51 pp.

KALPERS, J. & LANJOUW, A., 1997. Potential for the creation of a peace park in the Virunga volcano region. Parks: *The International Journal for Protected Area Managers,* 7: 25–35.

KALPERS, J. & LANJOUW, A., 1998. *Projet de réhabilitation des capacités de gestion de l'ICCN au Parc National des Virunga, secteur sud. Financement UNHCR, premier semestre 1997.* Activity report of the International Gorilla Conservation Programme. Nairobi.

KALPERS, J., WILLIAMSON, E., ROBBINS, M., MCNEILAGE, A., NZAMURAMBAHO, A., LOLA, N. & MUGIRI, G., 2003. Gorillas in the crossfire: Population dynamics of the Virunga mountain gorillas over the past three decades. *Oryx* 37(3): 326–337.

KAMSTRA, J., 1994. Protected areas: toward a participatory approach. Netherlands Committee for IUCN. NOVIB, Amsterdam. 55 pp.

KI-ZERBO, J., 2004. *A quand l'Afrique? Entretien avec René Holenstein.* De l'Aube Eds. 40pp.

KUJIRAKWINJA, D., PLUMPTRE, A., MOYER, D. & MUSHENZI, N., 2006. *Parc National des Virunga. Recensement aérien des grands mammifères, 2006.* Unpublished report to ICCN, Kinshasa.

KYMLICKA, W., 2003. *Les théories de la justice. Une introduction.* La Découverte Eds.

Languy, M., 1994. *Parc National des Virunga, Zaïre. Stratégie de la conservation à long terme des écosystèmes et ébauche d'un plan directeur du parc national.* WWF/IZCN report to the World Bank. 50 pp.

Languy, M., 1994. *Recensement des hippopotames dans le Parc National des Virunga, 14–17 juillet 1994.* Unpublished report to WWF International, Gland, Switzerland. 15 pp.

Languy, M., 1995. *Problèmes environnementaux liés à la présence des réfugiés rwandais. Identification des interventions réalisées. Coordination entre les organismes et propositions d'interventions complémentaires.* Mission report to UNDP. 29 pp.

Languy, M., 2005. *Compilation et analyse des textes légaux délimitant le Parc National des Virunga. Programme de renforcement des capacités de gestion de l'ICCN et appui à la réhabilitation d'aires protégées en RDC.* Feuillet technique n°1. Institut Congolais pour la Conservation de la Nature, Kinshasa. 20 pp

Lanjouw, A., 2003. Building partnerships in the face of political and armed crisis. *Journal of Sustainable Forestry,* Vol.16, No.3/4.

Lanjouw, A. Kayitare, A., Rainer, H., Rutagarama, E., Sivha, M., Asuma, S. & Kalpers, J., 2001. *Transboundary Natural Resource Management: A Case Study by International Gorilla Conservation Programme in the Virunga-Bwindi Region.* Biodiversity Support Program, Washington. 60 pp.

Lebrun, J., 1947. *La végétation de la plaine alluviale au sud du Lac Edouard.* Exploration du Parc National Albert. Institut des Parcs Nationaux du Congo Belge. 800 pp.

Lebrun, J., 1960. *Etudes sur la flore et la végétation des champs de lave au Nord du lac Kivu (Congo Belge).* Exploration Parc National Albert. Mission J. Lebrun (1937–1938). Institut des Parcs Nationaux du Congo Belge. 352 pp.

Lee, P.C., Thornback, J. & Bennett, E.L., 1988. *Threatened Primates of Africa. The IUCN Red Data Book.* IUCN, Gland, Switzerland.

Leusch, M., 1995. HCR Goma – Unité Environnement. Rapport Intermédiaire du 15 février au 11 août 1995. UNHCR, en collaboration avec les partenaires du Bureau d'information de l'Environnement (Union Européenne; GTZ Goma; PNUD Goma). 105 pp.

Lippens, L. & Wille, H. 1976. *Les Oiseaux du Zaïre.* Lannoo, Tielt (Belgium). 512 pp.

Lyons, A., 1998. A profile of the community based monitoring systems of the Zambian rural development projects. USAID/Zambia, Lusaka.

Mackie, C., 1989. *Recensement des Hippopotames au Parc National des Virunga. Leur impact sur la végétation et sur les sols.* Rapport non publié à l'Institut Zaïrois pour la Conservation de la Nature, Kinshasa. 78 pp.

Maldague M., Mankoto S. & Rakotomavo T., 1997. Notions d'aménagement et de développement intégrés des forêts tropicales, UNESCO, Paris. 94 pp.

Mankoto ma Mbaelele, S., 1978. *Problématique de la mise en valeur globale et intégrée du Parc National des Virunga, Zaïre.* PhD Thesis, Laval University (Canada). 303 pp.

Mankoto ma Mbaelele, S., 1989. *Problèmes d'écologie au Parc National des Virunga. Exploration du PNVi.* Fondation pour Favoriser les Recherches Scientifiques en Afrique, 2ème série, fasc. 28. 63 pp.

Margoluis, R. & Salafsky, N., 1998. *Measures of Success: a systematic approach to designing, managing and monitoring community-oriented conservation projects.* Biodiversity Support Programme: Adaptative Management Series, Washington DC.

Margoluis, R. & Salafsky, N., 2001. *Is our project succeeding. A guide to threat reduction assessment for conservation.* Biodiversity Support Program, Washington DC.

Martin, P. S. & Szuter, C. R., 1999. War zones and game sinks in Lewis and Clark's west. *Conserv. Biol.,* 13: 36–45.

McNeilage, A., Plumptre, A., Brock-Doyle, A., & Vedder, A., 1998. *Bwindi Impenetrable National Park, Uganda. Gorilla and large mammal census, 1997.* WCS Working Paper no 14, Wildlife Conservation Society. 52 pp.

McPherson L., 1991. Etude institutionnelle de l'Institut Zaïrois pour la Conservation de la Nature. Partie 1 : organisation, gestion, ressources humaines. *World Bank.* (Summary). 20 pp.

Mertens, H., 1983. Recensements aériens des principaux ongulés du Parc National des Virunga, Zaïre. *Revue Ecologie Terre et Vie,* 38: 51–64.

Metcalfe, S., 1995. The Zimbabwe Communal Areas Management Programme For Indigenous Resources (CAMPFIRE). In: Western, D. Wright, R.M. and Strum, S.C. *Natural Connections: perspectives in community based wildlife management.* Island Press.

Ministère du Plan de la République Démocratique du Congo, 2002. Document intérimaire de stratégies de réduction de la pauvreté. Kinshasa. 23 pp.

Misonne, X., 1963. *Les rongeurs du Ruwenzori et des régions voisines.* Exploration du Parc National Albert (deuxième série), 14 : 1–164.

Mugangu, S. 2001. *Conservation et utilisation durable de la diversité biologique en temps de troubles armés. Cas du Parc National des Virunga.* Report to IUCN – Central Africa Programme, Yaoundé. 107 pp.

Mugangu, S. & Katembo, V. 2000. *Exploitation conflictuelle et non durable par la gestion séparée de systèmes contigus de production et de conservation de la nature: cas du Mont Tshiaberimu au Parc national des Virunga, République Démocratique du Congo.* Paper presented at the second Pan African Symposium on sustainable use of natural resources in Africa (Ouagadougou, July 2000). IUCN. 14 pp.

Muruthi, P., Soorae, P., Moss, C., Stanley-Price, M., & Lanjouw, A., 2000. Conservation of large mammals in Africa. What lessons and challenges for the future? *In Priorities for the conservation of mammalian diveristy: has the panda had its day? Eds A. Entwistle et N. Dunstone.* Cambridge University Press, Cambridge. 207–219.

Mushenzi, N., 1995. *Rapport Annuel.* Institut Zaïrois pour la Conservation de la Nature. Direction Régionale Nord-Kivu. Goma, RDC.

Mushenzi, N., 1996. *Rapport de mission: état actuel du Parc National des Virunga dans les Secteurs Centre et Sud. Infrastructures, administration et surveillance.* Direction Régionale des Parcs Nationaux, Région du Nord-Kivu. Goma, DRC.

Mushenzi, N., de Merode, E., Smith, F., Hillman-Smith, K., Banza, P., Ndey, A., Bro-Jorgensen, J., Gray, M., Mboma, G., & Watkin, J., 2003. *Aerial Sample Count of Virunga National Park, Democratic Republic of Congo.* Unpublished report to USFWS and ICCN. 17 pp.

Musombwa, H.P., Lukuka, M.S., Kaghuta, M.C., Katselewa, P.W., Kavuya, K.A., Tshoha O.N. & Kalyongo, Z.L., 2005. Evaluation de la consommation des matières ligneuses dans la ville de Goma. Internship report, Université de Goma, DRC.

Nash, R., 1970, The American invention of national parks. *American Quarterly,* 22(3): 22–41.

Nations unies, 2005. *Rapport de synthèse sur l'Evaluation des Ecosystèmes pour le Millénaire.* United Nations, March 2005.

Northon-Griffiths, M., 1978. *Counting Animals.* AWF Handbook n°1, Nairobi. 139 pp.

Oates, J.F., 1999. *Myth and Reality in the Rain Forest. How Conservation Strategies Are Failing in West Africa,* University of California Press, Berkeley.

ORTPN, 2004. *Office Rwandais du Tourisme et des Parcs Nationaux: Plan stratégique 2004–2008. Version provisoire.* ORTPN. Kigali, Rwanda.

Owiunji, I., Nkuutu, D., Kujirakwinja, D., Liengola I., Plumptre, A., Nsanzurwimo, A., Fawcett, K., Gray, M. & McNeilage, A., 2005. *The Biodiversity of the Virunga Volcanoes.* Unpublished report to WCS, DFGF-I, ICCN, ORTPN, UWA and IGCP.

Pearson, C.M., & Mitroff, I.I., 1993. From crisis prone to crisis prepared: A framework for crisis management. *Academy of Management Executive,* 7(1): 48–59.

Pearson, C.M., Clair, J.A. & Kovoor, S., 1997. Managing the unthinkable. *Organizational Dynamics,* 26(2): 51–64.

Perelman C., 1990. De la justice. In *Ethique et droit.* Brussels University Eds.

Plumptre, A., 1991. *Plant-herbivore dynamics in the Birungas.* PhD Thesis, University of Bristol, United Kingdom.

Plumptre A., 2000. *Lessons learned from on-the-ground conservation in*

Rwanda and the Democratic Republic of Congo. Conference on War and Tropical Forests: New Perspectives on Conservation in Areas of Armed Conflict. Yale School of Forestry and Environmental Studies. 30 March–1 April 2000.

PLUMPTRE, A., BEHANGANA, M., DAVENPORT, T., KAHINDO, C., KITYO, R., NDOMBA, E., NKUUTU, D., OWIUNJI, I., SSEGAWA, P., & EILU, G., 2003. *The Biodiversity of the Albertine Rift*. Albertine Rift Technical Reports No. 3. 105 pp.

PLUMPTRE, A., BIZUMUREMYI, J-B., UWIMANA, F. & NDARUHEBEYE, J-D., 1997. The effects of the Rwandan civil war on poaching of ungulates in the Parc National des Volcans. *Oryx* 31: 265–273.

PLUMPTRE, A., COX, D. & MUGUME, S., 2003. *The status of Chimpanzees in Uganda*. Albertine Rift Technical Report Series No. 2. Wildlife Conservation Society, New York, USA.

PLUMPTRE, A. & HARRIS, S., 1995. Estimating the biomass of large mammalian herbivores in a tropical montane forest: a method of faecal counting that avoids assuming a 'steady state' assumption. *Journal of Applied Ecology*, 32: 111–120.

PLUMPTRE, A., KUJIRAKWINJA, D., TREVES, A., OWIUNJI, I., & RAINER, H. (sous presse). Transboundary conservation in the Greater Virunga Landscape: its importance for landscape species. *Biological Conservation*.

PLUMPTRE, A. & WILLIAMSON, E.A., 2001. *Conservation-oriented research in the Virunga region. In Mountain Gorilla: Three Decades of Research at Karisoke* (eds M.M. Robbins, P. Sicotte & K.J. Stewart). Cambridge University Press, Cambridge, UK. 361–389.

PRIGOGINE, A., 1953. Contribution à l'étude de la faune ornithologique de la région à l'ouest du lac Edouard. *Ann. Mus. Congo Tervuren, Zool*, 24. 117 pp.

ROBBINS, M.M., SICOTTE, P. & STEWART, K.J. Eds., 2001. *Mountain Gorillas: Three Decades of Research at Karisoke*. Cambridge University Press, Cambridge.

ROBYNS, W., 1937. *Aspects de végétation des Parcs Nationaux du Congo Belge*. Série 1, Parc National Albert. Institut des Parcs Nationaux du Congo Belge. Vol 1, fasc. 1 and 2. 42 pp.

ROBYNS, W., 1948. *Les territoires biogéographiques du Parc National Albert*. Institut des Parcs Nationaux du Congo Belge. 51 pp.

ROWCLIFFE JM, COWLISHAW G & LONG J. 2003. A model of human hunting impacts in multi-prey communities. *J. Appl. Ecol.* 40: 872–889.

RUTAGARAMA, E., 1999. *Initiatives d'implication des populations dans des micro-projets de gestion du Parc National des Volcans. Rapport préliminaire d'exécution des projets*. Unpublished report to IGCP, Rwanda. 9 pp.

SCHOUTEDEN, H., 1938. *Oiseaux*. Exploration du Parc National Albert. Institut des Parcs Nationaux du Congo Belge, Brussels. 107 pp.

SCLATER, P.L., 1901. (sans titre). *Proc. Zool. Soc. Lond.* 1901(1): 50.

SHAMBAUGH, J., OGLETHORPE, J. & HAM, R., 2001. *The trampled grass: mitigating the impacts of armed conflict on the environment*. Biodiversity Support Program. Washington DC.

SHOLLEY, C., 1991. Conserving gorillas in the midst of guerrillas. In *American Association of Zoological Parks and Aquariums, Annual Conference Proceedings*. 30–37.

SINCLAIR, A.R. & ARCESE, P., 1995. *Serengeti 2: dynamics, management and conservation of an ecosystem*. University of Chicago Press.

SOMMER, V., ADANU, J., FAUCHER, I. & FOWLER, A., 2004. Nigerian Chimpanzees *(Pan troglodytes vellerosus)* at Gashaka: two years of habituation efforts. *Folia Primatol.* 75: 295–316.

STEKLIS, H. D., GERALD, C. N. & MADRY, S., 1997. The mountain gorilla – conserving an endangered primate in conditions of extreme political instability. *Primate Conservation* 17: 145–151.

STUBBART, C.I., 1987. Improving the quality of crisis thinking. *Columbia Journal of World Business*, 22: 89–99.

STULHMANN, F., 1894. *Mit Emin Pascha in Herz von Afrika. Ein Reisebericht mit Beiträge von Dr Emin Pasha, in seinem Auftrage geschildert*, Berlin, Dietrich Reimer, 1894. 205–305.

TELEKI, G., 1989. Population status of wild chimpanzees *(Pan troglodytes)* and threats to survival. In: Understanding Chimpanzees (eds. P.G. Heltne and L.A. Marquardt). Harvard University Press, Cambridge, Massachusetts, USA. 312–353.

TELEKI, G., 1991. *Action plan for the conservation of wild chimpanzees and protection of orphan chimpanzees in the Republic of Burundi*. Unpublished report to the Jane Goodall Institute, Hants, England.

TOMBOLA, J. P. & SANDERS, C., 1994. *Résultats de l'enquête sur l'impact des réfugiés rwandais sur le Parc National des Virunga (secteur sud)*. Unpublished report to UNHCR, Technical Unit.

TURYATUNGA F., 2004. *Tool For Local-Level Rural Development Planning. Combining Use of Participatory Rural Appraisal and Geographic Information Systems in Uganda*. WRI Discussion Brief. World Resources Institute.

UICN, 1994. *Guidelines for protected area management categories*. IUCN and WCMC. Gland and Cambridge.

UNESCO, 1995. Convention concernant la protection du patrimoine mondial, culturel et naturel. Comité du Patrimoine Mondial. 19th session, Berlin, 4–9 December 1995.

UNESCO, 1999. *Biodiversity conservation in regions of armed conflict: protecting World Heritage in the Democratic Republic of the Congo (DRC). Project Review Form*. UNESCO World Heritage Centre, Paris.

UNITED NATIONS, 2001. *Report of the panel of experts on the illegal exploitation of natural resources and other forms of wealth of the Democratic Republic of Congo*. United Nations, New york.

UWA, 2002. *Uganda Wildlife Authority: Strategic plan 2002–2007*. Uganda Wildlife Authority. Kampala, Uganda.

VAKILY, J.M., 1989. *Les pêches dans la partie zaïroise du Lac Idi Amin: analyse de la situation actuelle et potentiel de développement*. Report to DAFECN and the EEC. Kinshasa. 18 pp.

VAN GYSEL, J. & VANOVERSTRAETEN, M., 1982. *Inventaire des potentialités pédo-botaniques pour l'élaboration du plan d'aménagement d'un parc africain*. Comptes rendus du Colloque International sur les productions animales tropicales au bénéfice de l'homme. Institut de Médecine Tropicale d'Anvers, Belgium. 417–422.

VANOVERSTRAETEN, M., 1989. *Apport de la morphopédologie à l'étude de la dynamique des écosystèmes. Application à l'aménagement du Parc National des Virunga*. Phd Thesis. Fac.Sc.Agr. Gembloux, Belgium. 280 pp.

VANOVERSTRAETEN, M., VAN GYSEL, J., MATHIEU, L. & BOCK, L., 1984. Etude intégrée au Parc National des Virunga (PNVi Centre – Zaïre Oriental). Les milieux morphopédologiques, support des écosystèmes. *Bull. Rech. Agron. Gembloux* 19(3/4): 189–225.

VEDDER, A. & AVELING, C., 1986. *Census of the Virunga population of Gorilla gorilla beringei*. Unpublished report, September 1986.

VERHEYEN, R., 1954. *Monographie éthologique de l'Hippopotame*. Institut des Parcs Nationaux du Congo Belge. 91 pp.

VERHAEGHE, M., 1958. *Le volcan Mugogo*. Exploration du Parc National Albert. Mission d'Etudes vulcanologiques, fasc. 3. Institut des Parcs Nationaux du Congo Belge. 29 pp.

VERSCHUREN, J., 1972. *Contribution à l'Ecologie des Primates, Pholidota, Carnivora, Tubulidentata et Hyracoidea (Mammifères)*. Exploration du Parc National des Virunga, Mission F. Bourlière et J. Verschuren, 3: 1–61.

VERSCHUREN, J., 1986. *Observation des habitats et de la faune après soixante ans de conservation*. Fondation pour Favoriser les Recherches Scientifiques en Afrique, Belgique. Fascicule 26. 44 pp.

VERSCHUREN, J., 1993. *Les Habitats et la Grande Faune: Evolution et Situation Récente*. Exploration du Parc National des Virunga (Zaïre), Fascicule 29. 133 pp.

VERSCHUREN, J., 2001. *Ma Vie, Sauver la Nature*. Editions de la Dyle. Gand. 529 pp.

VON GÖTZEN, G., 1899. *Durch Afrika von Ost nach West. Resultaten unde Begebenheiten einer Reise von der Deutsch-Ostafrikanischen Küste bis zur Kongomündung in den Jahren 1893/94,* Berlin, Dietrich Reimer. pp. 201–246.

WALLER, D., 1996. *Rwanda: Which way now? An Oxfam Country Profile*. Oxfam, Oxford.

WATHAUT, A., 1996. *Etat du lieu du PNVi-sud, une année et demi après l'installation des réfugiés rwandais dans et en bordure du parc.* Institut Zaïrois pour la Conservation de la Nature. Parc National des Virunga. Station de Rumangabo, DRC.

WEBER W., 1989. Conservation and Development on the Zaïre-Nile Divide. An analysis of value conflicts and convergence in the management of afro-montane forests. PhD Thesis, University of Wisconsin, Madison. 327 pp.

WEBER, W. & VEDDER, A., 1983. Population Dynamics of the Virunga Gorillas: 1959–1978. *Biological Conservation*, 26: 341–366.

WEBER, W., 1987. *Ruhengeri and its Resources: An Environmental Profile of the Ruhengeri Prefecture, Rwanda.* United States Agency for International Development. 92 pp.

WERIKHE S., MUSHENZI LUSENGE, N. & BIZIMANA, J., 1997. *The impact of war on protected areas in Central Africa. Case study of Virunga Volcanoes Region.* International Conference on Transboundary Protected areas as a vehicle for international co-operation, Cape Town, South Africa, 16–18 September 1997. 8 pp.

WERIKHE S., MUSHENZI LUSENGE, N. & BIZIMANA J., 1998. L'impact de la guerre sur les aires protégées dans la région des Grands Lacs. Le cas de la région des volcans Virunga. *Cahiers d'Ethologie* 18 : 175–186.

WILD, R. G. & MUTEBI, J., 1996. *Conservation through community use of plant resources. Establishing collaborative management at Bwindi Impenetrable and Mgahinga Gorilla National Parks, Uganda.* People and Plants Working Paper n°5, December 1996, Division of Ecological Sciences, UNESCO, 45 pp.

WILLIAM (PRINCE) DE SUÈDE, 1923. *Among Pygmies and Gorillas. With the Swedish zoological expedition to central Africa 1921,* Copenhagen-Berlin-Christiana, Gyldendal. 148–149.

WILSON, D.E. & REEDER, D.M. (EDS), 2005. Mammal species of the world. A taxonomic and geographic reference. Third editions, Vol 1 and 2. The Johns Hopkins University Press, Baltimore, USA. 2141 pp.

GENERAL REFERENCES

ANON., 1934. *Parc National Albert, Congo Belge.* Commission du Parc National Albert. 61 pp.

CHIFUNDERA, K.Z., NYAKABWA, M., BASHONGA, M.G., MASUMBUKA, N.C. & KYUNGU, J.C. (2003). *The mount Tshiabirimu in the Albertine Rift: biodiversity, habitat and conservation issues.* Unpublished report to *Dian Fossey Gorilla Fund Europe*.

D'HUART, J.P., 2002. *Sauvegarder le Parc National des Virunga : comment et par qui?* Unpublished report on the VIP workshop organised by the International Gorilla Conservation Programme and ICCN on nature conservation in the Democratic Republic of Congo in times of war. 22 pp

D'HUART, J.P., 2003. *Renforcement des capacités de gestion et réhabilitation des aires protégées en République Démocratique du Congo.* Rapport d'une mission d'identification et proposition de financement soumis à la Commission européenne. 53 pp.

D'HUART, J.P., 2003. *Statut de conservation et proposition de réhabilitation des Parcs Nationaux de la Garamba et des Virunga (RDC).* Rapport d'une mission d'identification et proposition d'intervention pour la préparation d'un projet d'appui GEF/Banque Mondiale. 59 pp.

DE GRUNNE, X., HAUMAN, L., BURGEON, L. & MICHOT, P., 1937. *Vers les glaciers de l'Equateur.* Le Ruwenzori, Mission scientifique belge 1932. Ed. R. Dupriez, Bruxelles. 300 pp.

DE HEINZELIN DE BRAUCOURT, J., 1935. *Les stades de récession du glacier Stanley occidental (Ruwenzori, Congo Belge).* Exploration du Parc National Albert, série 2(1). Institut des Parcs Nationaux du Congo Belge. 25 pp.

DE HEINZELIN DE BRAUCOURT, J., 1957. *Les fouilles d'Ishango.* Exploration du Parc National Albert, Mission J. de Heinzelin de Braucourt, fasc. 2. Institut des Parcs Nationaux du Congo Belge. 128 pp.

DELVINGT, W. & D'HUART, J.P., 1972. Conservation et recherche scientifique au Parc National des Virunga : la Station de Lulimbi, Zaïre. *Biological Conservation* 4(5) : 397.

DELVINGT, W., 1994. *Etude préparatoire du programme spécial de réhabilitation pour les pays voisins du Rwanda (PSSR), volet environnement. Rapport d'une mission effectuée au Zaïre, 24/11-05/12/94.* Faculté des Sciences Agronomiques de Gembloux, U.E.R. Sylviculture, 28 pp.

DELVINGT, W., LEJOLY, J. & MANKOTO, M., 1990. *Guide du Parc National des Virunga.* Commission des Communautés Européennes. 192 pp.

DE SAEGER, H., 1955. Le Parc National Albert : la Plaine. *Parcs Nationaux* 10(4) : 3–12.

DE WITTE, G.F. (1937). *Exploration du Parc National Albert.* Mission G.F. de Witte, (1933–1935), fasc.1. Institut des Parcs Nationaux du Congo Belge. 39 pp.

DE SAEGER, H., 1958. Le Ruwenzori. *Parcs Nat. Ardenne et Gaume.* 13(4) : 1–12.

HARROY, J.P., 1941. Les parcs nationaux du Congo Belge en 1939 et 1940. *Bull. Agric. Congo Belge* 32(3) : 454–495.

HARROY, J.P., 1987. Soixantième anniversaire d'un parc national zaïrois. *Bull. Séanc. Acad. Roy. Sci. Outre-Mer* 31(4) : 507–516.

HENQUIN, B. & BLONDEL, N., 1996. *Etude par télédétection sur l'évolution récente de la couverture boisée du Parc National des Virunga.* Report from PSRR/PNVi Project, UE-ADG-UNHCR-IZCN. 80 pp.

HENQUIN, B. & BLONDEL, N., 1997. *Etude par télédétection sur l'évolution récente de la couverture boisée du Parc National des Virunga, deuxième partie (période 1995–1996).* Report from PSRR/PNVi Project, UE-ADG-UNHCR-ICCN. 63 pp.

HOIER, R., 1950. *A travers plaines et volcans au Parc National Albert.* Institut des Parcs Nationaux du Congo Belge. Bruxelles. 173pp.

IPNCB, 1935. *Rapport du Conseil Colonial sur le projet de décret organisant l'IPNCB.* Bulletin Officiel, 1935, p. 64.

IPNCB, 1937. *Les Parcs Nationaux et la Protection de la Nature.* Institut des Parcs Nationaux du Congo Belge. 88 pp.

IPNCB, 1942. *Premier rapport quinquennal, (1935–1939).* Institut des Parcs Nationaux du Congo Belge. 75 pp.

KABALA, M., 1976. *Aspects de la conservation de la Nature au Zaïre.* Institut zaïrois pour la Conservation de la Nature, Editions Lokole, Kinshasa. 312 pp.

KALPERS, J., 2005. *Biodiversité et urgence en Afrique subsaharienne : la conservation des aires protégées en situation de conflit armé.* PhD Thesis. Université de Liège (Belgium). Faculté des Sciences. 234 pp.

LANGUY, M., 1994. *Parc National des Virunga, Zaïre. Stratégie de la conservation à long terme des écosystèmes et ébauche d'un plan directeur du parc national.* Report from WWF/IZCN to the World Bank. 50 pp.

LANGUY, M., 2005. *Compilation et analyse des textes légaux délimitant le Parc National des Virunga. Programme de renforcement des capacités de gestion de l'ICCN et appui à la réhabilitation d'aires protégées en RDC.* Feuillet technique n°1. 20 pp

LIPPENS, L., 1937. *Parmi les bêtes de la brousse. Instantanés.* Ed. Raymond Dupriez. Bruxelles. 129 pp.

MANKOTO MA MBAELELE, S., 1978. *Problématique de la mise en valeur globale et intégrée du Parc National des Virunga, Zaïre.* Thèse de Maîtrise, Université Laval (Canada). 303pp.

MANKOTO MA MBAELELE, S., 1989. *Problèmes d'écologie au Parc National des Virunga. Exploration du PNVi.* Fondation pour Favoriser les Recherches Scientifiques en Afrique, 2ème série, fasc. 28. 63 pp.

McPHERSON L., 1991. *Etude institutionnelle de l'Institut Zaïrois pour la Conservation de la Nature. Partie 1 : organisation, gestion, ressources humaines.* Report to the World Bank (summary). 20 pp.

MERTENS, H., 1982. *Etude pour la création d'une exploitation de la faune au Zaïre.* Comptes rendus du Colloque International sur les productions animales tropicales au bénéfice de l'homme. Institut de Médecine Tropicale d'Anvers, Belgium. 274–279.

PLUMPTRE, A., BEHANGANA, M., DAVENPORT, T., KAHINDO, C., KITYO, R., NDOMBA, E., NKUUTU, D., OWIUNJI, I., SSEGAWA, P., & EILU, G., 2003. *The Biodiversity of the Albertine Rift.* Albertine Rift Technical Reports No. 3. 105 pp.

PLUMPTRE, A., KAYITARE, A., RAINER, H., GRAY, M., MUNANURA, I., BARAKABUYE, N., ASUMA, S., SIVHA, M & NAMARA, A., 2004. *The socio-economic status of people living near protected areas in the central Albertine rift.* Albertine Rift Technical reports, 4. 127 pp.

PRIGOGINE, A., 1987. Quelques commentaires à l'occasion du soixantième anniversaire du Parc National des Virunga. Bull. *Séanc. Acad. Roy. Sci. Outre-Mer* 31(4) : 551–554.

VERSCHUREN, J., 1965. Un facteur de mortalité mal connu, l'asphyxie par gaz toxiques naturels au Parc National Albert, Congo. *La Terre et la Vie* 1965 (3): 215–237.

VERSCHUREN, J., 1988. *Problèmes scientifiques et techniques au Parc National des Virunga (Zaïre).* Rapport de mission à l'Administration Générale Belge pour la Coopération au Développement et à l'Institut Zaïrois pour la Conservation de la Nature. 132 pp.

MAMMALS

AVELING, C., 1990. *Comptage aérien total des buffles et éléphants au Parc National des Virunga, octobre 1990.* Rapport non publié au Programme Kivu, sous-programme Virunga. 6 pp.

AVELING, C. & HARCOURT, A.H., 1984. A census of the Virunga gorillas. *Oryx*, 18, 8–13.

BASHONGA, G., 2006. *A propos de la présence des Okapis et autres mammifères sur la rive gauche de la moyenne Semliki, Secteur Nord, Parc National des Virunga.* Mission report to WWF. 19 pp.

BOURLIÈRE, F. & VERSCHUREN, J., 1960. Introduction à l'écologie des ongulés du Parc National Albert. Institut des Parcs Nationaux du Congo Belge. Exploration du Parc National Albert. Fasc. 1 et 2. Bruxelles. 158 pp.

CORNET D'ELZIUS, C., 1996. *Ecologie, structure et évolution des Populations des Grands Mammifères du Secteur Central du Parc National des Virunga (Parc National Albert) Zaïre (Congo Belge).* Fondation pour Favoriser les Recherches Scientifiques en Afrique, Belgique. 231 pp.

CURRY-LINDAHL, K., 1956. Ecological studies on Mammals, Birds, Reptiles and Amphibians in the Eastern Belgian Congo. *Ann. Musée R. Congo Belge, Tervuren, Sciences Zoologiques,* vol 87. 78 pp.

CURRY-LINDAHL, K., 1961. *Contribution à l'étude des Vertébrés Terrestres en Afrique tropicale.* Institut des Parcs Nationaux du Congo et du Ruanda-Urundi, vol. 1. 331 pp.

DE WITTE, G.F., 1938. *Mammifères.* Exploration du Parc National Albert. Institut des Parcs Nationaux du Congo Belge, Bruxelles.

D'HUART, J.P., 1978. *Ecologie de l'hylochère* (Hylochoerus meinertzhageni THOMAS) *au Parc National des Virunga.* Exploration Parc National des Virunga, Deuxième Série, Fasc. 25. Fondation pour Favoriser les Recherches Scientifiques en Afrique. Bruxelles. 156 pp.

DELVINGT, W., 1978. *Ecologie de l'hippopotame au Parc National des Virunga.* PhD Thesis, Fac. Sc. Agr. Gembloux, 2 volumes. 333 pp.

FRECHKOP, S., 1938. *Mammifères.* Exploration du Parc National Albert. Mission G.F. de Witte (1933–1935), fascicule 10. Institut des Parcs Nationaux du Congo Belge. Bruxelles. 103 pp.

FRECHKOP, S., 1941. *Animaux protégés au Congo Belge.* Institut des Parcs Nationaux du Congo Belge. 469 pp.

FRECHKOP, S., 1943. *Mammifères.* Exploration du Parc National Albert. Mission S. Frechkop (1938), fascicule 1. Institut des Parcs Nationaux du Congo Belge. Bruxelles. 186 pp.

FRECHKOP, S., 1950. Parmi les mammifères de l'est du Congo Belge. *La Terre et la Vie* 1950, 1: 1–15.

GRAY, M., MCNEILAGE, A., FAWCETT, K., ROBBINS, M.M., SSEBIDE, B., MBULA, D. & UWINGELI, P., 2005. *Virunga Volcano Range Mountain Gorilla Census 2003.* Unpublished report to UWA/ORTPN/ICCN.

HEDIGER, H., 1951. *Observations sur la Psychologie animale dans les Parcs Nationaux du Congo Belge.* Exploration des Parcs Nationaux du Congo, 1: 1–84.

HILLMAN SMITH, A.K., DE MERODE, E., SMITH, F., AMUBE, N., MUSHENZI, N. & MBOMA, G., 2003. *Parc National des Virunga – Nord: comptages aériens de mars 2003.* Report to ICCN. 38 pp.

HOIER, R., 1952. *Mammifères du Parc National Albert.* Collections Lebègue et Nationale n°105. Bruxelles. 111 pp.

HUBERT, E.J., 1947. *La faune des grands mammifères de la plaine Rwindi-Rutshuru (Lac Edouard). Son évolution depuis sa protection totale.* Institut des Parcs Nationaux du Congo Belge. Bruxelles. 84 pp.

KALPERS, J., WILLIAMSON, E., ROBBINS, M., MCNEILAGE, A., NZAMURAMBAHO, A., LOLA, N. & MUGIRI, G., 2003. Gorillas in the crossfire: Population dynamics of the Virunga mountain gorillas over the past three decades. *Oryx* 37(3): 326–337.

KUJIRAKWINJA, D., PLUMPTRE, A. MOYER, D. & MUSHENZI, N., 2006. *Parc National des Virunga. Recensement aérien des grands mammifères, 2006.* Unpublished report to ICCN, Kinshasa.

LANGUY, M., 1994. *Recensement des hippopotames dans le Parc National des Virunga, 14–17 juillet 1994.* Unpublished report to WWF International, Gland, Switzerland. 15 pp.

MACKIE, C., 1989. *Recensement des Hippopotames au Parc National des Virunga. Leur impact sur la végétation et sur les sols.* Unpublished report to the Institut Zaïrois pour la Conservation de la Nature, Kinshasa. 78 pp.

MERTENS, H., 1983. Recensements aériens des principaux ongulés du Parc National des Virunga, Zaïre. *Revue Ecologie Terre et Vie* 38 : 51–64.

Mertens, H., 1984. Détermination de l'âge chez le topi (*Damaliscus korrigum* Ogilby) au Parc National des Virunga (Zaïre). *Mammalia* 48(3): 425–435.

Mertens, H., 1985. Structures de population et tables de survie des buffles, topis et cobs de Buffon au Parc National des Virunga, Zaïre. *Rev. Ecol. (Terre Vie)* 40: 33–51.

Misonne, X., 1963. *Les rongeurs du Ruwenzori et des régions voisines.* Exploration du Parc National Albert (deuxième série), 14: 1–164.

Mubalama, L., 2000. Population and Distribution of Elephants (*Loxodonta africana africana*) in the Central Sector of the Virunga National Park, Eastern DRC. *Pachyderm* 28: 44–55.

Mushenzi, N., de Merode, E., Smith, F., Hillman-Smith, K., Banza, P., Ndey, A., Bro-Jorgensen, J., Gray, M., Mboma, G., & Watkin, J., 2003. *Aerial Sample Count of Virunga National Park, Democratic Republic of Congo.* Unpublished report to USFWS and ICCN.

Owiunji, I., Nkuutu, D., Kujirakwinja, D., Liengola I., Plumptre, A., Nsanzurwimo, A., Fawcett, K., Gray, M. & McNeilage, A., 2005. *The Biodiversity of the Virunga Volcanoes.* Unpublished report to WCS, DFGF-I, ICCN, ORTPN, UWA et PICG.

Plumptre, A., Behangana, M., Davenport, T., Kahindo, C., Kityo, R., Ndomba, E. Nkuutu, D. Owiunji, I., Ssegawa, P. & Eilu, G., 2003. *The biodiversity of the Albertine rift.* Albertine Rift Technical reports, 3. 104 pp.

Prigogine, A., 1954. Deux nouvelles races d'*Aethosciurus ruwenzorii* du Congo belge. *Ann. Mus. Congo Tervuren, Zool.,* 1: 69–73

Rahm, A., 1960. Les muridés des environs du lac Kivu et régions voisines (Afrique Centrale) et leur écologie. *Rev. Suisse Zool.,* 74(9): 439–519.

Rahm, A., 1972. Note sur la répartition, l'écologie et le régime alimentaire des Sciuridés au Kivu (Zaïre). *Rev. Suisse Zool.,* 85(3–4): 321–339.

Robbins, C. & van der Straeten, E., 1982. A new specimen of *Malacomys verschureni* (Rodentia, Muridae) from Zaïre, Central Africa. *Rev. Zool. Bot. Afr.* Pages inconnues.

Schouteden, H., 1947. De zoogdieren van Belgisch-Congo en van Ruanda-Urundi. *Ann. Museum van Belgisch Congo, Tervuren, Zoologie,* 2(3). 576 pp.

Vedder, A. & Aveling, C., 1986. *Census of the Virunga population of Gorilla gorilla beringei.* Unpublished report, September 1986.

Verheyen, R., 1954. *Monographie éthologique de l'Hippopotame.* Institut des Parcs Nationaux du Congo Belge. 91 pp.

Verschuren, J., 1967. *Introduction à l'Ecologie et à la biologie des Cheiroptères.* Parc National Albert, 2: 1–65.

Verschuren, J., 1972. *Contribution à l'Écologie des Primates, Pholidota, Carnivora, Tubulidentata et Hyracoidea (Mammifères).* Exploration du Parc National des Virunga, Mission F. Bourlière et J. Verschuren, 3: 1–61.

Verschuren, J., 1986. *Relations entre la faune, principalement les vertébrés supérieurs, et les eaux thermales.* Exploration du Parc National des Virunga, série 2, 27. Fondation pour Favoriser les Recherches Scientifiques en Afrique. 21 pp.

Verschuren, J., 1987. L'action des éléphants et des hippopotames sur l'habitat, au Parc National des Virunga, Zaïre. Evolution chronologique de leurs populations. *Bull. Inst. Roy. Sc. Nat. Belg., Biologie* 57: 5–16.

Verschuren, J., 1987. Liste commentée des mammifères des Parcs Nationaux du Zaïre, du Rwanda et du Burundi. *Bull. Inst. Roy. Sci. Nat. Belgique, Biologie,* 57 : 17–39.

Verschuren, J., 1993. *Les Habitats et la Grande Faune: Evolution et Situation Récente.* Exploration du Parc National des Virunga (Zaïre), Fascicule 29. 133 pp.

Verschuren, J., Van der Straeten, E. & Verheyen, W., 1983. *Rongeurs.* Exploration Parc National des Virunga. Fondation pour favoriser les recherches scientifiques en Afrique, 4: 1–121.

Weber, A.W. & Vedder, A., 1983. Population Dynamics of the Virunga Gorillas: 1959–1978. *Biological Conservation,* 26 : 341–366.

Birds

Bataille, J. Bourguignon, L. Pagezy, H. & Trotignon, J., 1972. Dénombrement des Sauvagines et d'aigles pêcheurs (*Cuncuma vocifer*) sur le lac Edouard (R.D. Congo). *L'oiseau et la R.F.O,* 42. 183–192.

Chapin, J.P., 1932. The birds of the Belgian Congo. *Bull. Amer. Mus. Nat. Hist.,* 65. 756 pp.

Chapin, J.P., 1938. The birds of the Belgian Congo. *Bull. Amer. Mus. Nat. Hist.,* 75. 632 pp.

Chapin, J.P., 1954. The birds of the Belgian Congo. *Bull. Amer. Mus. Nat. Hist.,* 75B. 846 pp.

d'Huart, J.P., 1977. Station de baguage de Lulimbi (Parc National des Virunga, Zaïre) : rapport d'activités 1971–1975. *Le Gerfaut* 67: 161–168.

Demey, R., Herroelen, P. & Pedersen, T., 2000. Additions and annotations to the avifauna of Congo-Kinshasa (ex Zaïre). *Bull. Br. Ornithol.,* 120 : 154–172.

Lippens, L., 1938. Les oiseaux aquatiques du Kivu. *Le Gerfaut,* 28. Fasc. spécial. 103 pp.

Lippens, L. & Wille, H. 1976. *Les Oiseaux du Zaïre.* Lannoo, Tielt. 512 pp.

Mertens, H., 1986. Contribution à l'ornithologie du Parc National des Virunga. *Le Gerfaut* 76: 213–219.

Prigogine, A., 1953. Contribution à l'étude de la faune ornithologique de la région à l'ouest du lac Edouard. *Ann. Mus. Congo Tervuren, Zool.l,* 24. 117 pp.

Schouteden, H., 1938. *Oiseaux.* Exploration du Parc National Albert. Institut des Parcs Nationaux du Congo Belge, Bruxelles. 107 pp.

Schouteden, H., 1954. *Faune du Congo belge et du Ruanda-Urundi. III. Oiseaux non passereaux.* Annales du Musée Royal du Congo belge. Série in 8o, Sciences zoologiques, vol. 29. 434 pp.

Schouteden, H., 1957. *Faune du Congo belge et du Ruanda-Urundi. IV. Oiseaux passereaux (1).* Annales du Musée Royal du Congo belge. Série in 8o, Sciences zoologiques, vol. 57. 314 pp.

Schouteden, H., 1960. *Faune du Congo belge et du Ruanda-Urundi. V. Oiseaux passereaux (2).* Annales du Musée Royal du Congo belge. Série in 8o, Sciences zoologiques, vol. 89. 328 pp.

Schouteden, H., 1968. *La faune ornithologique du Kivu. I. Non passereaux.* Musée Royal d'Afrique Centrale. Documents zoologiques, n°12. 168 pp.

Schouteden, H., 1969. *La faune ornithologique du Kivu. II. Passereaux.* Musée Royal d'Afrique Centrale. Documents zoologiques, n°15. 188 pp.

Verheyen, R., 1947. *Oiseaux.* Exploration du Parc National Albert. Institut des Parcs Nationaux du Congo Belge, Bruxelles. 87pp.

Verschuren, J., 1966. *Contribution à l'Ornithologie.* Exploration du Parc National Albert, Mission F. Bourlière–J. Verschuren (1957–1961), fasc. 2. 3–24.

Reptiles, Amphibians and Fish

de Witte, G.F., 1941. *Batraciens et reptiles.* Exploration du Parc National Albert, Mission G.F. de Witte (1933–1935), 1. Institut des Parcs Nationaux du Congo Belge, Bruxelles. 39 pp.

de Witte, G.F., 1965. *Les caméléons de l'Afrique Centrale.* Annales du Musée Royal de l'Afrique centrale, sciences zoologiques, n°142. 200 pp.

Laurent, R, 1972. *Amphibiens.* Exploration du Parc National Albert. Institut des Parcs Nationaux du Congo belge, Bruxelles. 125 pp.

Damas, H., 1937. *Recherches hydrobiologiques dans les lacs Kivu, Edouard et Ndagala. Mission H. Damas (1935–1936).* Exploration du Parc National Albert. Institut des Parcs nationaux du Congo belge. 128 pp

Poll, M., 1938. *Poissons.* Exploration du Parc National Albert, Mission G.F. de Witte (1933–1935), fasc. 33. Institut des Parcs Nationaux du Congo Belge, Bruxelles. 81 pp.

Poll, M., 1939. *Poissons.* Exploration du Parc National Albert, Mission H. Damas (1935–1936), fasc. 6. Institut des Parcs Nationaux du Congo Belge, Bruxelles. 73 pp.

SNOEKS, J., 2000. *How well known is the ichtyodiversity of the large East African Lakes.* In Rossiter, A. and H. Kawanabe, eds. Ancient lakes: Biodiversity, ecology and evolution. Advances in Ecological Research, 31. Academic Press. 17–38.

VAKILY, J.M., 1989. *Les pêches dans la partie zaïroise du Lac Idi Amin: analyse de la situation actuelle et potentiel de développement.* Rapport au DAFECN et à la CEE. 18 pp.

VERSCHUREN, J., MANKOTO MA MBAELELE, S. & LUHUNU K. 1989. L'apparition des crocodiles au Lac ex-Edouard, Parc National des Virunga, Zaïre. *Rev. Ecol. (Terre Vie)* 44: 387–397.

FLORA AND HABITATS

CORNET D'ELZIUS, C.,1964. *Evolution de la végétation dans la plaine au sud du Lac Edouard.* Institut des Parcs Nationaux du Congo et du Rwanda. 23 pp.

LANGUY, M., sous presse. *Utilisation des photos périodiques au PNVi dans le cadre du suivi du parc. Programme de renforcement des capacités de gestion de l'ICCN et appui à la réhabilitation d'aires protégées en RDC.* Feuillet technique n°6. WWF International.

LEBRUN, J., 1942. *La végétation du Nyiragongo. Aspects de végétation des Parcs nationaux du Congo belge.* Série I. Parc National Albert, vol 1, fasc. 3–4–5. Institut des Parcs Nationaux du Congo Belge. 121 pp.

LEBRUN, J., 1947. *La végétation de la plaine alluviale au sud du Lac Édouard.* Exploration du Parc National Albert. Institut des Parcs Nationaux du Congo Belge. 800 pp.

LEBRUN, J., 1960. *Etudes sur la flore et la végétation des champs de lave au Nord du lac Kivu (Congo Belge).* Exploration Parc National Albert. Mission J. Lebrun (1937–1938). Institut des Parcs Nationaux du Congo Belge. 352 pp.

MOLLARET, H., 1961. *Biotopes de haute Altitude : Ruwenzori II et Virunga.* Exploration du Parc National Albert. Institut des Parcs Nationaux du Congo et du Ruanda-Urundi, Bruxelles.

PIERLOT, R., 1966. Structure et composition des forêts denses d'Afrique Centrale, spécialement celles du Kivu. *Académie royale des Sciences d'Outre Mer.* 367pp.

PLUMPTRE, A., 1991. *Plant-herbivore dynamics in the Birungas.* PhD Thesis. University of Bristol, United Kingdom.

ROBYNS, W., 1937. *Aspects de végétation des Parcs Nationaux du Congo Belge.* Série 1, Parc National Albert. Institut des Parcs Nationaux du Congo Belge. Vol 1, fasc.1 et 2. 42 pp.

ROBYNS, W., 1947. *Flore des Spermatophytes du Parc National Albert. 2. Sympétales.* Institut des Parcs Nationaux du Congo Belge. 626 pp.

ROBYNS, W., 1948. *Flore des Spermatophytes du Parc National Albert. 1. Gymnospermes et Choripétales.* Institut des Parcs Nationaux du Congo Belge. 745 pp.

ROBYNS, W., 1948. *Les territoires biogéographiques du Parc National Albert.* Institut des Parcs Nationaux du Congo Belge. 51 pp.

VAN GYSEL, J. & VANOVERSTRAETEN, M., 1982. *Inventaire des potentialités pédo-botaniques pour l'élaboration du plan d'aménagement d'un parc africain.* Comptes rendus du Colloque International sur les productions animales tropicales au bénéfice de l'homme. Institut de Médecine Tropicale d'Anvers, Belgium. 417–422.

VANOVERSTRAETEN, M., 1989. *Apport de la morphopédologie à l'étude de la dynamique des écosystèmes. Application à l'aménagement du Parc National des Virunga.* PhD Thesis. Fac.Sc.Agr. Gembloux, Belgium. 280 pp.

VANOVERSTRAETEN, M., VAN GYSEL, J., MATHIEU, L. & BOCK, L., 1984. Etude intégrée au Parc National des Virunga (PNVi Centre–Zaïre Oriental). Les milieux morphopédologiques, support des écosystèmes. *Bull. Rech. Agron. Gembloux* 19(3/4): 189–225.

VERSCHUREN, J., 1986. *Observation des habitats et de la faune après soixante ans de conservation.* Fondation pour Favoriser les Recherches Scientifiques en Afrique, Belgique. Fascicule 26. 44 pp.

CRISES

BISWAS A.K. & TORTAJADA-QUIROZ H.C., 1996. Environmental Impacts on the Rwandan Refugees in Zaïre. *Ambio* 25: 403–408.

BISWAS, A. K., TORTAJADA-QUIROZ, H. C., LUTETE, V., & LEMBA, G., 1994. *Environmental impact of the rwandese refugee presence in north and south Kivu (Zaïre).* United Nations Development Programme. 53 pp.

BLOM E., 2000. Conclusions and recommendations. In: Blom E, Bergmans W, Dankelman I et al. (eds), Nature in war. *Biodiversity conservation during conflicts* pp. 165–170. Netherlands Commission for International Nature Protection. Mededelingen, 37. Amsterdam.

BLONDEL, N., 1995. *Un an de présence des camps de réfugiés en périphérie du secteur sud du Parc National des Virunga: bilan des dégâts.* Unpublished report to FED/PSRR-PNVi. 20 pp.

BLONDEL, N., 1997. *L'impact des camps de réfugiés sur l'environnement local: étude de cas sur Goma (Nord-Kivu).* Journée d'étude 'l'information satellitaire au service de l'aide humanitaire', 28 mai 1997. Contribution du bureau I-Mage Consult. Unpublished report. 5 pp.

KALPERS, J. 2005. *Biodiversité et urgence en Afrique subsaharienne : la conservation des aires protégées en situation de conflit armé.* PhD Thesis. Université de Liège (Belgium). Faculté des Sciences. 234 pp.

LANGUY, M., 1995. *Problèmes environnementaux liés à la présence des réfugiés rwandais. Identification des interventions réalisées. Coordination entre les organismes et propositions d'interventions complémentaires.* Mission report to UNDP. 29 pp.

MUGANGU, S. 2001. *Conservation et utilisation durables de la diversité biologique en temps de troubles armés. Cas du Parc National des Virunga.* Report to IUCN–Programme Afrique Centrale, Yaoundé. 107pp

Shambaugh, J., Oglethorpe, J. & Ham, R., 2001. *The trampled grass: mitigating the impacts of armed conflict on the environment.* Biodiversity Support Program, Washington D.C.

STEKLIS, H. D., GERALD, C. N. & MADRY, S., 1997. The mountain gorilla – conserving an endangered primate in conditions of extreme political instability. *Primate Conservation* 17: 145–151.

TOMBOLA, J. P. & SANDERS, C., 1994. *Résultats de l'enquête sur l'impact des réfugiés rwandais sur le Parc National des Virunga (secteur sud).* Unpublished report to UNHCR, Unité technique.

WERIKHE S., MUSHENZI LUSENGE N. & BIZIMANA J., 1998. L'impact de la guerre sur les aires protégées dans la région des Grands Lacs. Le cas de la région des volcans Virunga. *Cahiers d'Ethologie* 18: 175–186.

TRANSBOUNDARY ISSUES

BIODIVERSITY SUPPORT PROGRAMME,1999. Study on the Development of Transboundary Natural Resource Management Areas in Southern Africa, *Biodiversity Support Programme*, Washington D.C.

KALPERS J. & LANJOUW A., 1997. Potential for the creation of a peace park in the Virunga volcano region. *Parks* 7: 25–35.

LANJOUW, A. KAYITARE, A., RAINER, H., RUTAGARAMA, E., SIVHA, M., ASUMA, S. & KALPERS, J., 2001. Transboundary Natural Resource Management: A Case Study by International Gorilla Conservation Programme in the Virunga-Bwindi Region. Biodiversity Support Program. 60 pp.

PLUMPTRE, A., KUJIRAKWINJA, D., TREVES, A., OWIUNJI, I., AND RAINER H. (sous presse). Transboundary conservation in the Greater Virunga Landscape: its importance for landscape species. *Biological Conservation.*

Index

Abatupi, River, 28, 189
Abia, River, 28, 189, 190
Acacia gerardii, 132
Acacia kirkii, 38
Acacia mearnsi, 209
Acacia sp, 124, 128, 131
Acinonyx jubatus, 150
ADMADE, 229
Aerial count, 17, 110, 141, 146, 147
Aerial photographs, 14, 110, 123, 131, 154, 198, 200, 201, 282, 283, 290, 291, 293, 301, 302, 307
Aeschynomene sp., 125
Afro-alpine, 32, 34, 46, 53, 175
Afrocrania, 42, 119
Agamas, 57
Akagera National Park, 312, 313
Akeley, Carl, 65, 66, 76
Albert I, King of Belgium, 65, 66, 75, 76, 88, 105, 172, 259
Albert, Lake, 24, 28, 29, 123, 267
Albertine Rift, 21, 24, 28, 29, 39, 50, 53, 55-58, 60, 114, 175, 190, 207, 267-269, 272, 274
Albizzia, 135
Alimbongo, 124
American Museum of Natural History, 65
Amphibian, 58
Ardea goliath, 55
Armed conflict, 8, 17, 65, 95, 96, 101, 103, 180, 185, 191, 195, 225, 234, 242, 280, 289, 311, 313
Armée de Libération du Rwanda (ALIR), 221
Arundinaria alpina, 45, 98
Arusha International Conference, 87
Aveling, Conrad, 90, 91, 142, 143, 200, 279, 312, 316
Baert, 80
Bagrus, 60
Bagurubumwe, 89, 107
Bahatsa, 145, 191
Bakinahe, 86, 260
Balaeniceps rex, 55
Balanga, Victor Njoli, 91, 174
Balegha, 145, 191
Balombi, 101
Bamate, Chiefdom of, 165
Bamboo, 29, 42, 45, 53, 72, 73, 98, 124, 129, 189, 190, 270, 271, 281, 284, 286, 287
Bamundjoma, 145, 191
Banana, Station of, 22
Banyarwanzururu, 223
Bapere, Chiefdom of, 165
Bashali, Chiefdom of, 293
Baswagha, Chiefdom of, 165
Batangi, Chiefdom of, 165
Baudouin, King of Belgium, 75, 89, 171, 174
Bavukahe, 81
Bee-eaters, 55
Beni, town of, 21, 32, 75, 82, 83, 124, 154, 155, 165, 168, 175, 186, 207, 225, 247, 256, 291, 313, 314
Beni-Kasindi, road, 124, 186, 314
Bernhard, Prince of Netherlands, 87, 174
Bersama, 42, 119
Beudels, Roseline, 92
Biangolo, 150
Bibwe, 293, 294
Bikenge, 101
Bilali, 86

Birere, Mwami Buunda, 167
Birwa, 62, 197
Bitis arietans, 57
Biwela, 79, 86
Bizengimana, 87, 111
Blue-headed Tree Agama, 57
Bobelhabi, 88
Bomama, River, 32
Borassus, 34, 123, 124
Botriochloa, 130, 132
Bourlière, F., 21, 106, 141, 150
Brachiaria eminii, 130
Branca, satellite volcano, 46
Breutelia stuhlmannii, 32
Brussels, 6, 24, 69, 76, 79, 80, 82, 83, 86, 88, 92, 105-107, 260
Buffalo, 50, 52, 91, 142, 171, 172, 185, 189, 190
Buffer Zone, 67, 102, 153, 156, 160, 162, 210, 227, 240, 249, 268, 283, 297, 300
Bufo superciliaris, 58
Bugina, 134, 156
Bukima, 62, 101, 174, 177
Bukuku, Mts, 73, 124, 172
Bukumu, Chiefdom of, 72, 165
Bulongo, 101
Bunia, 315
Bush fire, 106, 287
Bushbuck, 52, 189, 190
Butahu, 32, 124
Butambira, 156
Bwana Kitoko, 75
Bweza, 72
Bwindi, Impenetrable Forest National Park, 57, 177, 186, 228, 234, 267, 268, 269, 270, 272, 273, 276, 280, 284
Bwirina, 76
Bwisha, Chiefdom of, 155, 165, 168
Bwito, Chiefdom of, 72, 73, 155, 165
Cameroon, 28, 29, 89, 92, 121, 215, 261
CAMPFIRE, 229
Capparis tomentosa, 126, 130, 131, 132
Carex, 32, 46
Carex monostachya, 32
Cassia, 34, 124
Cassia siamea, 34
Cenchrus ciliaris, 132
Cephalophus dorsalis, 190
Cephalophus sylvicultor, 190
Cercopithecus ascanius, 190
Cercopithecus kandti, 50, 177, 189
Cercopithecus mona denti, 190
Cessna, 186
Chamaeleo carpenteri, 57
Charcoal, 9, 17, 97, 98, 145, 156, 205, 207, 208, 211, 222, 224, 230, 233, 246, 249, 284, 286, 317
Cheetah, 150
Chimpanzee, 50, 145, 189, 191
Chiromantis rufescens, 58
Chlidonias leucoptera, 56
Chloris gayana, 132, 133
Chrysochloris, 50, 121, 129
Chrysochloris stuhlmani, 50
Clarias, 60
Clarias gariepinus, 60
Clematis wightiana, 45
Climate, 21, 24, 28, 35, 279, 284
Cnemaspis quattuorseriatus, 57
Coalition for the Conservation in Congo, 256, 257
Colobus, 190
Colobus angolensis, 190
Colobus guereza, 190

Common Pool, 255, 257
Community Conservation, 230, 234
Conflict, 7, 13, 17, 68, 95, 168, 174, 185, 210, 221, 222, 224, 229, 232, 237, 245, 248-250, 253-255, 259, 265, 268, 271, 274, 275, 281, 311
Congo, River, 28
Coolidge, Harold, 87, 107
COPEVI, 165, 166, 167, 168, 169, 200, 202, 203, 318
COPILE, 72, 165, 167, 202, 245
Core Group, 254, 256, 257
Cornet d'Elzius, Count, 21, 79, 106, 110, 126, 131, 141-143, 145, 150, 171, 312
Corridor, 29, 34, 62, 100, 124, 129, 155, 157, 160, 190
Corruption, 81, 91, 177, 180, 211
Costermansville, 72, 166
Crategostigma, 132
Crocodile, 57
Crocodilus niloticus, 57, 150
Crocuta crocuta, 190
Cropping, 111
Croton, 38, 127
Croton macrostachys, 38
Curry-Lindahl, Kay, 91, 106, 150
Customary right, 229
Cyathea, 124
Cymbopogon, 34, 130, 132, 133, 185
Cynodon, 37, 185
Cynometra alexandri, 29, 34
Cyperus papyrus, 29
Damaliscus korrigum, 142
Danly, 171, 172
Dasypeltis atra, 57
de Grunne, 105
de Leyn, 86
de Saeger, 79, 105, 106
de Waersegger, 165
de Witte, G.F., 67, 79, 83, 105, 106, 123, 134
Decentralisation, 92, 235, 264, 265
Deforestation, 95, 97-100, 134, 135, 153, 154, 159, 160, 208, 211, 238, 246, 247, 250, 284, 289, 290, 291, 318
Delvingt, Willy, 21, 45, 89, 91, 96, 99, 107, 110, 119, 131, 132, 141, 146, 150, 183, 188
Demarcation, 162, 231, 232, 233, 248, 289, 291, 292, 295, 296, 299, 301, 302, 304, 305, 309, 313, 314, 318
Dendroaspis jamesoni, 57
Digital Elevation Model, 302, 307
Displaced people, 156, 246, 254, 255, 294, 312, 313
Djuma, River, 154, 159, 160, 162
Donor, 7, 233, 255, 268, 314, 315
Duiker, 53, 190, 194
Eagles, 55
East African Rift, 113, 114, 130
Ecotourism, 92, 181
Education, 7, 65, 90, 99, 120, 180, 209, 214, 228-231, 248, 250, 251, 261, 265, 270, 295, 297
Eichornia crassipes, 128
Elections, 7, 83
Elephant, 9, 50, 87, 91, 121, 125, 127, 131, 133-136, 142, 143, 150, 151, 171, 185, 189, 190, 247, 248
Encroachment, 227, 260, 268, 313
Energy, 98, 99, 154, 162, 205, 207, 208, 210, 215, 256, 290, 318
Enterprises, public, 93, 223, 264
Erica arborea, 46
Eruption, 42, 48, 113-116, 120, 121, 128, 129, 175, 224
Eucalyptus, 207, 209, 210
Euclea schimperi, 131
Euphorbia, 38, 126, 127, 131, 132, 172

calycina, 132
candelabrum, 126
dawei, 38, 126, 127, 131, 132
European Development Fund (EDF), 311
European Union, 6, 7, 8, 9, 90, 92, 93, 101, 102, 120, 142, 151, 224, 238, 249, 250, 255, 257, 264, 300, 302
Evi, River, 128, 155
Expropriation, 70
Fabiola, Queen of Belgium, 89, 171
Felis aurata, 190
Feyerick, 88
Ficalhoa, 45
Firewood, 96-99, 131, 153, 156, 205, 208, 222, 230, 232, 233, 246, 268
Fish, 13, 17, 37, 55, 60, 70, 72, 87, 125, 147, 165-167, 171, 197, 203, 230, 246-248, 270, 286
Fish-eagle, 55
Fishing, 9, 14, 17, 37, 67, 69-73, 79, 83, 87, 139, 153, 165-175, 197-203, 213, 214, 227, 230, 232, 245, 246, 248, 283, 284, 286, 295, 313, 318
Fofolo, 93
Fontaine, Jean, 75, 82
Forces Armées Rwandaise (FAR), 96
Forces Démocratiques pour la Libération du Rwanda (FDLR), 221
Forest
  Bamboo, 29, 42, 53, 72, 129
  gallery, 34, 38, 53, 57, 124, 126, 127, 128, 131, 135, 136, 137, 138, 139, 185
  Hagenia, 46, 76
  lowland, 28, 29, 32, 67, 70, 190
  montane, 28, 29, 34, 35, 41, 45, 53, 57, 60, 123, 124, 128, 175, 177, 186, 189, 190, 223
  xerophilous, 28, 124, 125, 126, 127
Fossey, D., 107, 189, 255, 279, 286, 316
Foundation for Advancement of Scientific Research in Africa (FFRSA), 79, 82, 110
Franck, Minister, 66
Frankfurt Zoological Society, 142, 151, 238, 250, 253, 262, 286
Frogs, 58
Gahinga, Volcano, 42, 45, 67, 96, 114
Gahuranyi Tanganyka, 88
Garamba National Park, 22, 73, 76, 79, 80, 91, 92, 106, 139, 142, 195, 215, 216, 234, 254
Garoua, Wildlife School of, 89, 92, 215, 248, 249, 261
Gasigwa, 88
Gatovu, 101
Geckos, 57
General Directorate, 262
Genocide, 96, 237, 269, 289, 312
Geographic Information System (GIS), 215, 282, 289, 291-297, 302, 304, 306
Geology, 14, 24, 65, 68
Geomatics, 289, 295, 296, 297
George, Lake, 29, 238
Gerengo, Koya Gialo Basete, 91
Geurden, 88
Giant Forest Hog, 52, 83, 89, 107, 110, 125, 171, 188, 189
Gilman International Conservation, 150, 189
Glaciers, 24, 28, 32, 105, 111, 114, 124, 173, 190
Global Positioning System, 187, 189, 231, 275, 282, 289-297, 303, 304, 308
Golden Cat, 190
Gorilla beringei beringei, 50, 186, 189
Gorilla beringei graueri, 50, 143, 190
Gorilla census, 142, 188, 272
Gorillas, 9, 45, 50, 65, 87, 90, 96, 98-102, 107, 142, 143, 145, 175, 177, 181, 186-191, 224, 225, 238, 249, 250, 255, 267, 268, 270, 272, 276, 280, 286, 316
Gotongo, 62
Great Apes, 92, 107, 175, 181, 182, 185
Great Lakes, 17, 72, 101, 102, 171, 216, 237, 254, 289, 311
Greater Virunga Landscape, 6, 267, 268, 269, 271-275
Green belt, 210, 231, 303
Gregorian, rift, 24
Grevillea, 209
Grzimek, 87
Guano, 107
Habarukira, 85, 86
Hackars, Colonel, Henri, 69, 70, 76, 106
Hagenia, 29, 46, 53, 76, 124, 129, 189
Hagenia abyssinica, 29, 46, 124, 129
Haliaeetus vocifer, 55
Harroy, Jean-Paul, 73, 79, 91, 92, 106, 160, 197, 198
Hediger, 79, 106
Heine, 79
Helichrysum, 32
Hema, pastoralists, 73
Hemeleers, René, 68
Hemidactylus, 57
Herons, 55
Heteropogon contortus, 131, 132
Hippopotamus, 21, 50, 86, 96, 107, 111, 121, 126, 134, 139, 142, 146, 147, 151, 171, 172, 175, 177, 185, 187, 203, 247, 314
Houben, 79
Hoyo, Mt, 96, 107, 175
Hubert, 79, 80, 106, 150
Humanitarian agencies, 97, 99
Hunde, 75, 247
Hylochoerus meinertzhageni, 189
Hymenochirus, 58
Hyparrhenia familiaris, 130, 131, 132
Hypericum, 46, 189
Hypericum revolutum, 46
Hyperolius xenorhinus, 58
Ilehe, Mt, 41, 128, 135
Imperata cylindrica, 37, 130, 132, 133
Improved wood stove, 99, 209, 231
Independence, 13, 67, 73, 75, 79, 83, 85, 86, 88, 93, 105, 107, 168, 227, 229, 249, 250, 259, 260, 300, 312
Indigofera sp., 133
Infrastructure, 62, 82, 88, 89, 90, 100, 169, 180, 181, 183, 198, 223, 224, 230, 245, 247, 250, 293, 295, 296, 301, 303, 311, 313, 314, 316, 317
Insolation, 24
Institut des Parcs Nationaux du Congo Belge (IPNCB), 62, 67-73, 75, 76, 80, 82, 105, 165, 183, 213, 249, 259, 260, 300
Institutional Review, 92, 214, 216, 249, 261, 264, 274, 316, 317
Interahamwe, 221, 222, 225
International Conference for the Protection of the Animals of Africa, 65
International Gorilla Conservation Programme, 92, 93, 99, 102, 224, 250, 255, 262, 270-275, 280, 284, 286
International Monetary Fund (IMF), 264
Irumu, 70
Ishango, bone of, 124, 182
Ishango, man of, 105, 124, 182
Ishango, station, 8, 21, 22, 34, 58, 62, 65, 76, 83, 96, 105, 124, 134, 150, 172, 175, 181, 182, 215, 224, 250, 314
Ishasha, River, 34, 37, 39, 41, 50, 56, 62, 127, 130-134, 141, 145, 150, 155, 162, 172, 173, 175, 182, 186, 191, 200, 202, 301
Ituri, 28, 29, 57, 60, 70, 100, 156, 160, 190
IUCN Red List, 50, 185, 186, 195
Ivory, 91, 95, 96, 121, 143, 150, 151, 222, 247, 272
Ixobrychus minutus, 55
Jamar, Alain, 88
Joachim, 81
Jomba, 62, 72, 73
Kabara, 66, 76, 82, 107
Kabaraza, 62, 93, 189, 286, 301, 317
Kabaraza, River, 301
Kabasha, 124, 126, 135, 136, 138, 155, 172
Kacheche, 93, 231, 250, 262
Kagera, National Park, 68, 79, 86, 171
Kagnero, 126
Kahindo, camp, 96, 97
Kaisian, Lake, 28
Kajuga, 86, 260
Kakomero, 82, 129, 156, 160, 162
Kalinzu, Reserve, 271
Kalonge, 173
Kamande, Bay of, 124-126, 132, 172, 201, 202
Kamango, 160
Kamatembe, 128
Kambo, 62
Kanyabayonga, 237, 240, 241, 247, 313
Kanyatsi, 314
Karisimbi, Volcano, 22, 24, 42, 46, 48, 65, 66, 87, 98, 114, 182, 194, 299
Karuruma, 8, 62, 101, 223, 246, 313
Kasaka, 76, 88, 168
Kasali, 39, 41, 60, 75, 128, 130, 132, 133, 136, 143, 145, 156, 191
Kasenyi, 238
Kashaka, River, 238
Kashwe, 72
Kasindi, 82, 101, 168, 169, 172, 181, 182, 240, 246
Kasiwa, 81
Kasoso, River, 62, 155
Kasyoha-Kitomi, Reserve, 271
Katako, River, 238
Katale, 96, 97, 245
Katanda, 172
Katuku, 221
Katwa, 145, 191
Kayenga Onzi Ndal, 87
Kazinga Channel, 238
Kengo wa Dondo, 92
Kibale National Park, 271
Kibati, 62, 128, 156, 157, 171
Kibirizi, 62, 130, 133, 156, 172, 240
Kibuga, 41, 121, 156
Kibuga, Lake, 41, 121, 156
Kibumba, 8, 70, 96-98, 101, 157, 223, 242, 243, 245, 313
Kigeri, 72
Kigezi, Reserve, 271
Kihangiro, 62
Kikingi, 145, 191
Kilia, 9
Kilolirwe, 9, 162, 246, 281, 293, 300, 303, 313, 318
Kimbumba, 62
Kimya, 245
Kinshasa, 62, 79, 87-95, 100, 166, 169, 197, 216, 221-223, 234, 240, 249, 250, 254, 256, 261, 262
Kint, 79
Kinyonzo, 62
Kiondo, 24, 173
Kisaka, 125, 137, 139, 169

Index 339

Kisharo, 128, 155
Kiswahili, 34, 48, 80, 85, 183, 205
Kitiriba, 145
Kitsimbani, 145, 174
Kivu Programme (EU), 90, 91, 142, 228
Kizi, fault, 132
Kizi, Lake, 38, 39, 127, 130, 132, 133
Kobus (kob) thomasi, 52, 142
Kruger, National Park, 65, 66, 67, 172, 217
Kwenda, River, 128, 155, 158, 301
Kyambura, Reserve, 271
Kyamdura, River, 238
Kyavinyonge, 37, 101, 124, 165-168, 197, 200, 202, 246, 301, 314
Lac Vert, camp, 96, 97, 98
Land use, 14, 153, 157, 160, 223, 229, 264, 282, 283, 290
Landsat, 154, 157, 160
Lantana, 126
Lava flow, 41, 42, 97, 113-116, 119, 121, 128, 190
Laval, University, 89
Lebrun, 21, 24, 105, 126, 128, 130, 131
Lecrenier, 88
Lemna, 132
Léopold, Prince of Belgium, 67, 259
Léopoldville, 79, 82, 86, 87, 183
Leptosiaphos hackarsi, 57
Leptosiaphos meleagris, 57
Lesse, River, 145, 150, 189, 190, 191
Lessedjina, 87
Letiexhe, André, 88, 89
Lichen, 190
Linet, C., 89
Lion, 50, 76, 121, 150, 171
Lippens, 79, 107
Little Bittern, 55
Lizard, 57
Lobelia, 32, 175, 189
Lobelia lanurensis, 32
Local Defence, 222
Looting, 96, 101, 222, 247, 248, 262, 313
Lopé, Faunal Reserve, 312
Lubilya, 34, 62, 191, 124, 172, 185, 246, 272
Lubutu, 81
Luilango, ponds, 132
Lulengo, Simon, 96, 110
Lulimbi, station, 14, 21, 22, 34, 38, 56, 62, 81, 82, 89, 90, 93, 107, 110, 127, 130-133, 150, 171, 172, 189, 215, 250, 286, 317
Lume, 101, 124
Lunyasenge, 81, 125, 145, 166, 168, 169, 191
Lusilube, River, 32, 150
Mabenga, 21, 34, 38, 62, 70, 82, 128, 136, 145, 156, 158, 191, 240
MacArthur, Foundation, 255
Mackie, Charles, 91, 142, 202
Magunga, 88
Mahangu, 88, 173
Maï-Maï, 221, 222, 225, 245, 247
Makala (charcoal), 145, 205, 207, 208, 211, 246
Makerere University, 238
Makoyobo, River, 150, 189
Malacomys verschureni, 50
Mammals, large, 65, 89, 105, 110, 141, 150, 185-189, 192, 246, 250, 279, 286, 311
Management Plan, 160, 240, 249-251, 265, 286, 311, 317, 318
Maniema, Province, 156, 160
Mankoto, ma Mbaelele, Sami, 14, 85, 89, 91, 92, 106, 107, 110, 183
Margherita, peak, 21, 24, 29

Maroba, 145, 191
Marshes, 32, 128, 171
Masisi, 24, 160, 221, 225, 245-247, 255, 293, 313
Matuka, David Kabala, 92
Mayangose, 101
May-ya-Evi, 132
May-ya-Moto, 70, 128, 130, 134, 145, 150, 172, 191
Mbata, 245
Mbau, 160, 165
Mburanumwe, 86, 249, 260
Medical waste, 98
Meropidae, 55
Mertens, Hadelin, 90, 93, 142, 145
Mfumbiro, Mts, 65
Mgahinga Gorilla National Park, 96, 177, 186, 267, 268, 270
Micha, 79
Mikeno crisis, 291, 295, 313
Mikeno, Volcano, 14, 24, 41, 42, 46, 48, 57, 58, 62, 65-69, 73, 76, 87, 96-101, 142, 162, 174, 175, 177, 182, 185, 192, 194, 195, 223, 238, 250, 279, 280, 283, 286-291, 295, 299, 300, 312, 313, 316
Military camp, 151, 247, 284, 314, 318
Mimosaceae, 34
Miruho, 86
Misonne, Xavier, 22, 24, 83, 105
Mitumba, Mts, 24, 34, 35, 39, 41, 76, 125, 126, 130-133, 136, 139, 155, 201
Mobutu, 62, 87, 89, 91, 96, 111, 168, 249, 260
Mokoto, lakes, 128, 155
Mokwa, 86, 88
Molindi, 41, 42, 70, 121, 128, 156
Molindi, River, 41, 42, 70, 121, 128, 156
Monitor lizard, 57
Mosenda, 81, 125, 137, 139, 150
Moss, 32, 46
Muema, Dr., 87
Muembo, 107
Mugunga, camp, 96, 97, 98, 156, 245, 246
Muhaha, River, 126, 133
Muhavura, Volcano, 42, 45, 46, 67, 96
Muhungezi, 145, 191
Mulalamule, 145, 191
Mulango wa Nyama, 34, 190
Mulele, 312
Mundala, 150
Munyaga, 86, 260
Museya, 76, 88
Mushari, 73
Mushenzi, Norbert, 14, 89, 95, 100, 186, 200, 237, 280, 312
Mutebusi, 312
Mutombo, Mwami Atshongya Kasereka, 167
Mutsora, 21, 22, 29, 32, 62, 68, 69, 80, 82, 86, 92, 107, 124, 128, 175, 182, 260, 286
Mutwanga, 173
Muvo, Mt, 128
Muyirimbo, 35
Mwalika, 96
Mwaro, corridor, 100, 157
Mweka, College, 89, 215, 248
Mwenda, 154, 160, 162, 175
Mwiga, Bay of, 52, 138, 175
Mzee Tembo, 92
Naivasha, meeting, 24, 101, 254
Naja nigricollis, 57
Nande, 70, 88, 125
National Geographic Society, 183, 302
Ndeze, Mwami Daniel, 75, 88, 168
Ndeze, Mwami René, 168
Nectophryne afra, 58

Neoboutonia macrocalyx, 45
Ngesho, River, 127, 128, 130, 132
Ngezi, 72
Nguli, 62, 88
Nile, River, 28, 29, 42, 57, 60, 65, 114, 172
Nkunda, Laurent, 312
Nuyten, 79
Nyaleke, 62, 101, 145, 154, 160, 215, 246
Nyamilima, 155
Nyamulagira, Volcano, 14, 24, 41, 42, 62, 67, 68, 70, 82, 97-100, 114, 119, 120, 143, 173-175, 185, 186, 190, 207, 208
Nyamushengero, 22, 62, 126, 127, 150
Nyirafunzo, 132
Nyiragongo, Volcano, 14, 24, 41, 42, 62, 67, 82, 98, 114, 115, 116, 119, 120, 128, 145, 165, 168, 174, 175, 183, 190, 191, 208, 211, 246
Oisha, 154
Okapi, 53, 147, 150, 189, 190, 194, 216, 254
Olea, 41, 42, 128
Ondo, Lake, 41, 156
Ophyoglossum, 132
Opuntia, 126
Oreochromis niloticus, 60
ORTPN, 92, 264, 267-273, 290
Ory, Marie Huguette, 79, 80
Outbreak, 240, 311
Overgrazing, 132
Pan troglodytes, 50, 145, 186, 190
Panicum maximum, 133
Panicum repens, 132
Panthera leo, 50, 150
Patrol Post, 22, 62, 81, 89, 96, 100, 101, 131, 169, 216, 237, 249-251, 262, 280, 317, 318
Pedology, 90, 110
Pelicans, 126
Pelomys hopkinsi, 50
Pelusios rhodesianus, 56
Pelusios williamsi, 56
Pennisetum, 124, 128, 130, 133
Periodic photographs, 110, 134
Peucedanum, 46, 129
Peucedanum kerstenii, 46
Phacochoerus aethiopicus, 145
Philantomba monticola, 190
Philippia johnstoni, 46
Philippia trimera, 32
Phoenix reclinata, 38, 126
Pilipili, bay of, 35
Pipidae, 58
Pironio, Enrico, 92
Pistia stratiotes, 132
Plate, 113, 114
Pleistocene, 28, 60, 114
Poaching, 50, 52, 73, 81-83, 85, 87, 89, 92, 96-101, 110, 125, 127, 133, 143, 146, 150-153, 172, 180, 181, 195, 215, 221-224, 229, 230, 240, 246, 247, 249, 250, 260, 262, 268, 272, 273, 279, 284, 286, 311, 312, 314
Podocarpus, 45, 97, 124
Post-conflict, 7, 17, 214
Potamochoerus larvatus, 52, 190
Poverty, 95, 180, 181, 215, 223, 227, 228, 233, 235, 245, 246, 268, 274, 294
Prigogine, Alexandre, 92, 105, 107
Primates, 175, 177, 270
Provincial Direction, 62
Pterygota, 41, 127
Pterygota mildbraedii, 41
Ptolomy, 65
Puemba, River, 21, 28, 29, 123, 195

Pygmies, 67, 70, 89, 105, 121, 175, 189
Python, 57
Python sebae, 57
Queen Elizabeth National Park, 35, 53, 76, 111, 150, 151, 157, 171, 175, 182, 192, 197, 267, 268, 271, 311
Rainfall, 22, 28, 32, 42, 45, 128, 130-133, 155, 158, 185, 279, 293
Rana ruwenzorica, 58
Rapanea rhododendroides, 32
Rauwolfia, 128, 131
RCD-Goma, 216, 221, 222, 249, 262
RCD-ML, 221, 249, 262
Rebellion, 167, 168, 180, 197, 223, 249
Rebels, 85, 89, 96, 100, 222, 245, 246, 274
Reforestation, 126, 135-139, 159, 211, 230-233, 246
Refugee, 96-102, 145, 151, 154, 160, 168, 208, 221, 235, 289
Refugee camp, 96-101, 145, 221, 289
Regional Information System, 272, 275
Remote sensing, 14, 17, 273, 275, 280, 289-291, 294-297
Research, 14, 17, 62, 66, 67, 81, 83, 85-90, 105-107, 110, 111, 141, 145, 171, 194, 213, 217, 248-251, 259, 261, 264, 265, 268, 279, 284, 286, 301
Rhacophoridae, 58
Ringing, 89
Robyns, 105, 106, 123, 131
Rodent, 50
Rollais, 88
Rousseau, 79
Ruanoli, River, 29
Rubare, 160, 162, 210
Rugari, 65, 72, 73
Ruhengeri, 73, 96, 238
Rukumi, 82
Rumangabo, station, 21, 22, 24, 62, 68, 73, 79, 82, 86, 87, 93, 100, 128, 155, 157, 207, 260, 284, 286, 301
Rumoka, Volcano, 73, 145, 191
Ruti, 62
Rutshuru, 21, 22, 34, 37-42, 50, 55-58, 62, 65, 68-75, 79, 80, 83, 85-89, 100, 107, 110, 116, 121, 123, 125-135, 138, 141, 145, 150, 151, 153, 155-162, 165, 168, 172, 174, 175, 185, 186, 191, 197, 207, 209, 211, 225, 240, 250, 260, 300, 301, 317
Rutshuru Hunting Zone, 128, 153, 155, 157, 160, 162
Ruyange, 86
Rwanda, 21, 45, 70-73, 79, 86, 87, 96, 98, 157, 174, 177, 183, 186-189, 192, 194, 211, 221-225, 238, 240, 242, 246, 249, 250, 254, 255, 262, 264, 267-270, 273-275, 280, 289-291, 300, 312, 313
Rwanda Patriotic Front (RPF/FPR), 96
Rwenzori Mountains National Park, 29, 175
Rwenzori, Mts, 6, 21, 22, 24, 28, 29, 32-35, 45, 46, 53, 57, 58, 60, 62, 67-70, 76, 80, 82, 87, 96, 105, 111, 123, 124, 129, 130, 143, 165, 171-175, 177, 181, 182, 190, 222, 223, 268, 271
Rwere, 158, 159
Rwindi, 14, 21-24, 28, 34, 37-39, 41, 50, 52, 55-57, 62, 66, 68, 80, 81, 82, 85-93, 110, 123, 126, 127, 130-138, 141, 150, 155, 158, 159, 171-175, 182-186, 202, 237, 240, 246, 247, 260, 279, 284, 286, 300, 313-316
Rwindi, River, 126, 127, 130, 132, 133, 134, 135, 171
Ryckmans, Pierre, 72, 106
Saambili, 86
Sabinyo, Volcano, 42, 45, 65, 67, 96, 114, 194
Sake, 70, 114, 121, 156
Salonga National Park, 83, 92, 216
Sauswa, 86

Savanna, 185, 186, 189, 192, 270
Sayer, Jeffrey, 91
Schaller, George, 73, 87, 107, 174
Schouteden, Henri, 66, 105
Schumacher, Father, 67, 105
Semliki, River, 21, 22, 28, 29, 32, 34, 53, 55, 57, 60, 62, 68, 70-76, 80, 83, 88, 89, 105, 123, 124, 142, 143, 145, 147, 150, 159, 160, 172, 175, 181, 182, 185, 186, 189-191, 194, 195, 223, 246, 249, 283, 300
Semuliki National Park, 157, 194, 267
Senecio, 32, 175, 189
Senecio johnstoni, 32
Senkwekwe, 81
Serengeti, 50, 106, 150
Sesbania sesban, 38
Shoebill, 55
Simba, 312
Sinda, 83, 175, 181
Sleeping sickness, 73, 75, 124, 154, 197
South African Wildlife College, 215, 248
Spawning grounds, 167, 203
Special Presidential Division (DSP), 99
Speke, 65
Sporobolus consimilis, 132
Sporobolus pyramidalis, 131, 132
Sporobolus sanguinei, 132
Sporobolus spicatus, 131, 132
Sporobulus, 37
Sporobulus pyramidalis, 37
Stanley, 123
Stanleyville, 70
Stork, 55
Strategic Document on the Reduction of Poverty, 215
Stuhlmann, 50
Syncerus caffer, 50, 142, 189
Système de Gestion d'Information pour les Aires Protégées (SYGIAP), 215, 286
Talia-Kamandi, 191
Talya, bay of, 125, 202
Tamira, 83
Tanganyika, 24, 60, 106, 114
Tervuren, Museum, 88, 105, 107, 188
Themeda triandra, 37, 130, 131
Timber, 17, 67, 72, 153, 210, 222, 230, 233, 271, 303
Tincani, 92
Tongo, 28, 42, 70, 88, 90, 92, 119, 145, 156, 162, 174, 175, 182, 186, 191, 207, 208, 246, 250, 306
Topis, 127, 129, 132
Tourism, 6, 7, 9, 13, 14, 50, 62, 67, 68, 82, 83, 85, 89, 90, 91, 98-101, 119, 141, 145, 147, 150, 151, 171-177, 180-183, 194, 208, 211, 213, 215, 223, 227, 235, 248-253, 256-259, 261, 264, 268, 272, 274, 283, 295, 303, 315-318
Traditional chief, 88
Tragelaphus eurycerus, 189
Transboundary management, 6
Transboundary Protected Areas, 273, 274
Tree heather, 22, 32, 34, 124, 175
Tree nurseries, 99
Tsanzerwa, 155
Tshegera, 21, 60, 76, 128
Tshiaberimu, Mt, 24, 28, 34, 45, 60, 62, 80, 88, 101, 124, 143, 155, 175, 182, 186, 190, 195, 287, 291, 300
Tumbwe, River, 34, 124
Turea nilotica, 131
Twa, pygmies, 70
Uganda, 21, 29, 34, 35, 39, 52, 53, 73, 86, 87, 93, 96, 101, 107, 111, 123, 124, 131, 132, 135, 142, 157, 166, 171, 174, 175, 177, 182, 183, 186, 190-194,

221-223, 238, 240, 246, 249, 250, 254, 255, 262, 264, 267-275, 280, 284, 313
Uganda Kob, 52, 135, 142
Uganda Wildlife Authorities (UWA), 264, 267-273
UNEP, 102
Ungulates, 50, 52, 53, 90, 106, 107, 141-143, 150, 185, 188, 190
UNHCR, 99, 101, 102
United Nations, 39, 68, 76, 86, 100, 101, 130, 167, 183, 187, 242, 246, 250, 254, 259
United Nations Foundation (UNF), 101, 216, 248, 250, 254, 255, 262
United Nations Office for Humanitarian Affairs (OCHA), 242
Upemba National Park, 75
USAID, 9, 102, 238, 302
Usnea, 32
van der Elst, 79, 86
van Gysel, Jan, 14, 110, 131
van Straelen, Victor, 67, 69, 72, 76, 79, 80, 86, 106, 259, 260
Varanus niloticus, 57
Varanus ornatus, 57
Verhulst, 79
Victoria, lake, 60
Vieux-Beni, 29, 32, 150, 154
Vimbao, 96
Viola abyssinica, 45
Virunga Environmental Programme (WWF), 90, 91, 93, 99, 102, 194, 209, 230, 232, 262, 299, 302, 303, 316
Virunga Programme (EU), 90, 91, 250
Visoke, Volcano, 42, 46, 65, 66, 98, 114, 194, 299
Vitshumbi, 37, 62, 72, 79, 81, 83, 125, 126, 130-133, 165-168, 171, 172, 175, 183, 197, 198, 200, 202, 240, 246, 247, 301, 313, 314
Volcanoes National Park, 87, 96, 157, 177, 268, 281, 300
von Beringe, Robert, 65
Wahunde, 75
Walendu, 70
Wanande, 70, 75
Warthog, 52, 130, 145, 171, 312
Watalinga, 76, 96, 123, 124, 160, 162, 165, 195
Waterbuck, 52, 142, 312
White-winged Tern, 56
Widows, 81
Wigny, 69
Wild Dog, 151, 192
William, Prince of Sweden, 65, 66
World Bank, 92, 93, 249, 261, 264
World Conservation Union (IUCN), 50, 88-93, 174, 185, 186, 195, 228, 273, 274
World Heritage Committee, 254
World Heritage Convention, 253, 254, 257
World Wide Fund for Nature (WWF), 6, 9, 60, 89-93, 99, 101, 102, 110, 123, 142, 145, 150, 151, 189, 194, 201, 202, 205, 209, 210, 224, 230-234, 250, 253, 255, 257, 260, 262, 276, 280, 283, 284, 286, 291, 299-304, 308, 314, 316
Xenopus, 58
Yellowstone National Park, 67, 89, 227
Zairian armed forces (FAZ), 93
Zairianisation, 89, 90

Biographies

Marc Languy
Born in Belgium in 1962, from the age of ten Marc Languy could be found wading through the marshes of his native lowlands looking for birds, butterflies and other animals. This early passion for wildlife has remained constant. Marc graduated in 1985 from the Catholic University of Louvain (UCL), where he specialised in population biology and theoretical ecology. After teaching at the University of Lyon I, he left Europe in 1988 for Africa where he taught in a rural school in Zimbabwe. Back at UCL, he taught at the Biometry Unit, leaving in 1992 to join WWF as a project leader in Virunga National Park, based in Goma. The troubles in Rwanda, and their effects in what was Zaire, reluctantly forced him to leave for family reasons in October 1994. Marc continued to work for WWF in Gabon before completing a Masters at UCL in population ecology, development and environment. He then managed the country office for BirdLife International in Cameroon before starting his current work at the WWF Regional Office for East Africa, which brought him back to the Virunga. WWF's Virunga Project is one of the ten or so projects that he manages as coordinator of the Albertine Rift Programme.
C/o WWF–EARPO, PO Box 62440, 00200, Nairobi, Kenya.
Mlanguy@wwfearpo.org

Dr Emmanuel de Merode
Emmanuel de Merode was born in Carthage, Tunisia in 1970. He spent his childhood in Kenya, where he developed a passion for the savannahs and mountains of eastern Africa. He completed his undergraduate studies at the University of Durham, UK, followed by a doctorate in biological anthropology at the University of London. He arrived in DRC for the first time in 1993, managing a research and training programme in Garamba National Park. The assignment was to last six months but, in the end, he spent seven years in Garamba. In 1999, he worked for the Gabonese Government on a western lowland gorilla conservation programme funded by the European Union. After three years, he returned to DRC to launch the Zoological Society of London's programme in Virunga National Park. In 2003 he was appointed in charge of the European Union's office for the coordination of development programmes for eastern DRC, based in Goma. He established and worked as chief executive of a conservation NGO, WildlifeDirect. In 2008, he accepted an appointment as Chief Warden for Virunga National Park, based at Rumangabo. He shares his passion for wildlife with his wife and two daughters.
ICCN Director, Chef de Site Parc National des Virunga, B.P 660, Goma, DRC
edemerode@gorilla.cd

Ephrem Balole
Ephrem Balole holds a post-graduate degree (DEA) in environmental economics from the University of Libreville, and a diploma in the same field from the University of Stanford (USA). Having worked as programme assistant for the European Union's office for the coordination of development programmes in eastern DRC, he was responsible for the monitoring of three conservation projects working with ICCN in Virunga National Park, together with a number of other projects around the park. He worked also as country programme officer for WildlifeDirect in DRC, based at Mutsora, Virunga-North. He is currently the Assistant of the Chief Warden of Virunga NP.
17, av. Pélican, Quartier des Volcans, Goma, DRC
ebalole@gmail.com

Pyther Banza
Pyther Banza is a forester, having obtained his engineer's qualification from the *Institut Supérieur d'Etudes Agronomiques de Bengamisa* in DRC. Pyther also has a diploma from the Regional Centre for Mapping of Resources for Development in Nairobi. He began his career in the humanitarian sector as head of the agronomy section working on the rehabilitation of zones that were damaged by the passage of refugees between 1998 and 2000. From 2002 to 2006, he was with WWF as officer in charge of the Geographic Information Systems at the Virunga Environmental Programme, based in Goma. Since 2007, he is GIS Officer and Focal Point for the development of a Data Base for UNICEF's Eastern Zone in DRC.
253 Av. du Musée, Quartier Himbi, Goma, DRC
pcibangu@unicef.org

Samuel Boendi Lihamba
Samuel Boendi was born on 24 September 1968 at Yabaondo in the *Province Orientale* of the DRC. He holds a forestry degree from the *Institut Supérieur des Etudes Agronomiques de Bengamisa* (1994) and a Masters in natural resource management from the University of Edinburgh, UK (2003). Actively involved with Virunga National Park since 1995, initially as technical assistant to the project manager within WWF's *Programme Environmental autour du Parc National des Virunga,* and from 2003 to 2006, as deputy coordinator, responsible for community conservation. His research activities among the local communities helped prepare a community forestry programme in *Province Orientale;* he is currently working as a consultant for FFI at Garamba National Park on a Community Conservation Programme.
Samuel_Boendi@yahoo.fr

Dr Carlos de Wasseige
Carlos de Wasseige is a forestry engineer from University of Gembloux and has a doctorate in agronomy from the Catholic University of Louvain (UCL). He has worked at UCL as a specialist in the analysis and interpretation of satellite imagery and digital cartography. Since 1995, he has focused his research interest in Central Africa, especially in the forestry sector. From 2003 to 2008, he has worked on the *Système de Gestion de l'Information pour les Aires Protégées* (SYGIAP) funded by the *Politique Scientifique Fédérale* (Belgium) and which covers the five World Heritage Sites of the DRC, including Virunga National Park. He now works for the FORAF project (Observatory of the Forests of Central Africa), based in Kinshasa.
Carlos.dewasseige@aigx.be

Guy Debonnet
Guy Debonnet was born in Aalst in Belgium in 1965 and has had a passion for wildlife and particular birds since the age of 10. He graduated as an agronomist from the University of Gent (Belgium) in 1987 and in 1989 started working in Burundi in an agricultural project. From 1993 he worked with the German Cooperation, first in Burundi as an advisor to the national park institute and from 1996 till 2001 in DRC in Kahuzi Biega National Park. In 1998, was one of the initiators of the joint programme between UNESCO, ICCN and several conservation NGO, to try and preserve the five World Heritage Sites in the DRC, including Virunga National Park. Since 2002, he is working at the World Heritage Centre of UNESCO, as programme specialist for world natural heritage. As such, he continues to be heavily involved in the efforts to protect the DRC's World Heritage Sites.
C/o UNESCO, 7 Place Fontenoy, 75352 Paris 07 SP, France.
g.debonnet@unesco.org

Dr Baudouin Desclée
Baudouin Desclée is a biologist from the Catholic University of Louvain (UCL). After completing a thesis in forest mapping in the southern Philippines in 2001, he completed a doctoral thesis in the *Laboratoire d'Environnemétrie et de Géomatique* (UCL). He developed an interest in establishing new remote sensing techniques for detecting surface changes in tropical and temperate forests. In collaboration with WWF, he worked on different regions of Virunga National Park to establish areas of deforestation.
17 Rue du Hameau de Barges, B–7500 Ere, Belgium.
Baudouin@desclee.net

Dr Jean Pierre d'Huart
Jean Pierre d'Huart has a degree in zoology and a doctorate from the Catholic University of Louvain (UCL). He began his career as scientific warden and researcher at Lulimbi in Virunga National Park (1971–75). For 24 years (1979-04), he held various positions with WWF, including the regional representative for eastern Africa (1992–1997). As an independent consultant, he remains active in his support to protected area institutions, and in the development of conservation projects in DRC and throughout Africa.
14 rue du Monty, 1320 Beauvechain, Belgium.
dhuartjp@yahoo.com

Jacques Durieux
Jacques Durieux, a vulcanologist specialising in eruption trends and risk management, developed a passionate interest for Nyiragongo and the other volcanoes of Virunga from 1970. After spending four years in Goma, he spent the next 25 years working on most of the world's active volcanoes, including several expeditions to Virunga (1982 to 1985, 1994). Summoned hastily by the United Nations to help with the aftermath of Nyiragonogo's eruptions in 2002, he has been based full time at Goma, managing the United Nations' Volcanoes Risk Reduction programme.
jdurieux@chello.fr

Gregory Duveiller Bogdan
Gregory Duveiller Bogdan is an environmental engineer from the Catholic University of Louvain (UCL) who has specialized in remote sensing. His MSc thesis, which was undertaken in collaboration with WWF, focused on the human pressure within and around Virunga National Park in order to evaluate the options to develop new buffer zones. He subsequently worked in the *Laboratoire d'Environnemétrie et de Géomatique* (UCL) on the estimation of deforestation rates over Central Africa using remote sensing. He is currently working on a doctoral thesis on crop growth monitoring at regional scale over different parts of the globe.
5/301 rue de la Sarriette, B–1348 Louvain-la-Neuve, Belgium.
gduveiller@gmail.com

Wojciech Walter Dziedzic
Wojciech Walter Dziedzic is a tropical agronomist trained at the *Gesamthochschule Kassel*, Witzenhausen, Germany. He joined WWF in November 2004 as programme coordinator for the Environmental Programme around Virunga National Park WWF/PEVI, having previously worked for over 20 years in Tanzania, Rwanda and DRC and is now working for the ECOFAC programme in the Central African Republic. He worked primarily in forestry, agro-forestry and soil conservation in the Great Lakes Region.
dziedzicw@yahoo.fr

Henri Paul Eloma Ikoleki
Henri Paul Eloma Ikoleki qualified with a first degree in biology, and a postgraduate degree in tropical forest management, remote sensing and geographic information systems. He became a researcher at the *Institut Congolais pour la Conservation de la Nature* in 2001. He subsequently managed the UNESCO project for the World Heritage Sites of the Democratic Republic of Congo (2003-05) and currently works as national administrator on the GEF-UNDP programme for the rehabilitation of protected areas in DRC.
81/A Avenue de la Justice, Commune de la Gombe, BP 868 Kinshasa 1, DRC.
HenriE@unops.org

Dr Katie Fawcett
Katie Fawcett is a primatologist specialising in Great Ape behaviour and conservation. She has worked in the Virunga Volcanoes since 2002 as director of the Karisoke Research Center. During her tenure she has led the expansion of Karisoke from its core base of research on the life histories of Mountain Gorillas to broader biodiversity research and conservation education.
Karisoke Research Centre, Dian Fossey Gorilla Fund International, BP 105, Ruhengeri, Rwanda.
Katie_Fawcett@yahoo.co.uk

Maryke Gray
Maryke Gray has worked as a biologist in the Mikeno Sector since 2001, as officer in charge of monitoring for the International Gorilla Conservation Programme (AWF/FFI/WWF). This involves providing technical support to ICCN for the ranger-based monitoring programme which has been running in Virunga since 1996. Currently, she is responsible for knowledge and information management for IGCP, based in Kenya, but working in Rwanda, DRC and Uganda.
PO Box 48177, Nairobi, Kenya.
marykegray@gmail.com

Dr José Kalpers
José Kalpers has been active in the Mikeno Sector of Virunga National Park since 1991, as the first regional coordinator of the International Gorilla Conservation Programme (AWF/FFI/WWF). From 1996, he developed the ranger-based monitoring system that collects patrol and gorilla data across the entire Virunga Massif. He is currently working as programme manager with the African Parks Network in sub-Saharan Africa, a conservation organization based in South Africa.
8 rue E. Jacquemotte, B–4020 Liege 2, Belgium.
jkalpers@gmail.com

Frédéric Kasonia
Frédéric Kasonia was born in Lubero in 1928. He studied at the *Ecole Normale Saint Lwanga de Mulo* and qualified as a teacher in 1949. He worked as a teacher between 1950 and 1960, completing a teaching degree in 1956 at Bukavu. From 1960 to 1974, he developed a career as a district administrator in North Kivu. Retired in 1984, he was subsequently hired as the Director of the CoPeVi. He has remained in the post ever since. He has a deep understanding of the situation the national park is in, especially with respect to the fishing enclaves at Vitshumbi, Kyavinyonge and Nyakakoma.
CoPeVi, BP 1, Vitshumbi, DRC.

Anecto Kayitare

Anecto Kayitare has a degree in agronomy from the *Institut Facultaire des Sciences Agronomiques (IFA) de Yangambi* (DRC). He also has a masters in environment and development policy from the University of Sussex, UK. For several years he has been working with the International Gorilla Conservation Programme (AWF/FFI/WWF), where he is responsible for the transboundary collaboration across the Virunga Massif and Bwindi.
BP 931 Kigali, Rwanda.
anecto@yahoo.com

Deo Kujirakwinja

Deo Kujirakwinja graduated in rural development from the ISDR-Bukavu (the *Institut Superieur de Développement Rural*), and trained in bird inventories. Since 2002 he has been working with the Wildlife Conservation Society in Virunga National Park, implementing field activities within the transboundary programme between ICCN and the Uganda Wildlife Authority, the development of ranger-based monitoring and ranger training.
Avenue des Eglises, Keshero, Goma, DRC.
deokujirak@yahoo.fr

Stéphane Laime

Stéphane Laime is an agronomist trained at Gembloux in Belgium. He has worked for WWF as a specialist in Geographic Information Systems (GIS). His responsibilities included the development of a GIS unit within the Albertine Rift Programme which gave him a wide reaching perspective on the diversity of the landscape and also the scale of the social and political problems to which it is subjected. As a consultant he continues to provide support to Marc Languy's Albertine Rift team.
Rue Jean Dor 11, 5070, Vitrival, Belgium.
s.laime@gmail.com

Dr Annette Lanjouw

Annette Lanjouw has been working for over 20 years on various programmes in the Virunga region. She completed her doctoral thesis at *Rijks Universiteit* of Utrecht in 1987, with a study of the Bonobos in the Lomako forest. She then worked as field director of the Tongo Chimpanzee Conservation Project of the Frankfurt Zoological Society, before working with WCS in the Ituri and elsewhere in Central Africa. She became regional director, and then executive director of the International Gorilla Conservation Programme (AWF/FFI/WWF) from 1993 to 2004, working in Rwanda, the Democratic Republic of Congo, and Uganda. Today, she manages international programmes for the Howard G. Buffett Foundation, based in Cambridge, UK.
C/o HGBF, 159 W. Prairie Avenue, Suite 107, Decatur Illinois, 62523, USA.
Annette.lanjouw@flora-fauna.org

Dr Stéphane Leyens

Stéphane Leyens has a degree in biology (University of Namur–FUNDP) and a PhD in philosophy (Catholic University of Louvain–UCL). His main philosophical concerns are related to ethics, social and political philosophy and philosophy of science, with a special focus on the relation holding between sustainable development and social justice. As post-doctoral fellow at the UCL, he developed a research on the social issues of using GIS technologies in the monitoring of the Virunga National Park. He is now Assistant Professor at the department of 'Sciences, Philosophies, Societies' at the University of Namur, Belgium, where he is in charge of the unit of Sustainable Development. Among other things, he is currently directing a research in India ('Sustainable Development, Social Justice and Gender Issues. A Case Study in Tamil Nadu') in collaboration with the Arul Anandar College, Karumathur, Tamil Nadu.
61, rue de Bruxelles, B–5000 Namur, Belgium.
stephane.leyens@fundp.ac.be

Dr Samy Mankoto Ma Mbaelele

Doctor *Honoris Causa* at the Agricultural University of Gembloux, Samy Mankoto managed ICCN between 1985 and 1995. He worked on promoting international cooperation and strengthening ICCN's image to the international community, focussing especially on the development of tourism in Virunga and Kahusi-Biega national parks. In March 1995, he joined UNESCO, with the Division for Ecology and Earth Sciences, from where he launched the ERAIFT, a pilot programme of the Man and Biosphere Programme, for capacity building in Africa. He is also an IUCN regional advisor for Africa. He was nominated as Chairman of the RAPAC (Regional Network for Central African Protected Areas) in 2003.
11, avenue de la Grande Brosse, F–91390, Morsang-sur-Orge, France
s.mankoto@unesco.org

Zachary Kimutai Maritim

Zachary Maritim graduated in geomatics at the University of Nairobi, Kenya. He began his career in computer cartography at the International Society for Geomatics (Geomaps). For six years (1999 to 2005), he worked in various functions at Geomaps including as programme coordinator for the Geographic Information System of the Kigali Town Council in Rwanda (2003–2005). Today, he is responsible for GIS decision support to WWF's field projects.
PO Box 62440-00200 Nairobi, Kenya
Zmaritim@wwfearpo.org

Mbake Sivha

Mbake Sivha graduated in biological sciences at Kisangani University. She also has a Masters degree in Conservation and tourism from the Durell Institute of Conservation and Ecology (DICE) of the University of Kent, UK. Between 1999 and 2004 she worked as country officer for the International Gorilla Conservation Programme in the Democratic Republic of Congo. Prior to that, she was actively involved in the protection of eastern lowland gorillas in Kahusi Biega National Park and in the protection of Mountain Gorillas in Virunga National Park., collaborating closely with park managers and local communities. Since 2005, she has become actively involved in setting up community conservation programmes around Garamba National Park and developing country policies for community conservation in DRC.
mbakesivha@yahoo.fr

Déo Mbula Hibukabake

Déo Mbula graduated in biology at the University of Kisangani. He has been working for the *Institut Congolais pour la Conservation de la Nature* since 1989, initially on the environmental education programmes with WWF's *Programme d'Education Virunga,* and then as a researcher based in Tshiaberimu (Northern Sector Virunga National Park). He headed the technical and scientific office for Virunga National Park, and subsequently became warden responsible for Lulimbi, and is now a warden in Salonga National Park.
deombula@yahoo.fr

Dr Alastair McNeilage

Alastair McNeilage previously worked at the Karisoke Research Centre in Rwanda, and studied the ecology of the Virunga mountain gorillas for his PhD at Bristol University. He assisted with the planning and implementation of gorilla censuses in Bwindi and Virunga, and undertook several studies assessing the impact of human disturbance on the ecology of these sites between 1989 and 2006. He currently works with the Wildlife Conservation Society and manages the Institute for Tropical Forest Conservation in Uganda, where he collaborates closely with colleagues in Rwanda and DRC.
PO Box 7487, Kampala, Uganda
amcneilage@wcs.org

Leonard K. Mubalama

Leonard Mubalama has a masters degree in biological conservation from the University of Kent at Canterbury (UK). Between 1990 and 1994, he was scientific warden at Lulimbi, for the Central Sector of the Park. He is co-author on a number of scientific papers on elephant management in Virunga, as part of his responsibilities as the country officer for the CITES/MIKE programme. He is currently the coordinator for the SYGIAP/ICCN (Information Management System for Protected Areas).
13, Av des cliniques, Kinshasa-Gombe, DRC
Mikedrce@yahoo.co.uk

Robert Muir

Robert Muir has a Masters in Conservation Biology from the Durell Institute of Conservation and Ecology (DICE, UK). He has worked at Virunga National Park since early 2004 where he manages the conservation programme for the Frankfurt Zoological Society in the Democratic Republic of Congo. He is currently developing new initiatives for the enforcement of conservation laws in Virunga National Park,

and putting together a collaborative framework between ICCN, local government, the military, and the UN peace keeping mission in DRC to address some of the most serious challenges faced by ICCN in Virunga National Park.
Carriage Cottage, Southill Road, Broom, Bedfordshire SG18 9NN, UK
robertmuir@fzs.org

Norbert Mushenzi Lusenge
Norbert Mushenzi has a diploma from the Wildlife Training Centre at Garoua in Cameroon in 1971. He began his professional career at ICCN in the Southern Sector of Virunga National Park as a warden. He also worked in other sites including the Okapi Reserve, Garamba National Park and Kahusi Biega National Park. In Kahusi Biega, he collaborated closely with the Conservateur De Schryver on the gorilla habituation programme. He is currently working in Virunga National Park as Director.
ICCN BP 34, Beni, DRC.
mushelusenge@yahoo.fr

Dr Andrew Plumptre
Andrew Plumptre first visited the park in 1984 and developed a passionate interest for the area. He raised the funds to develope a research programme on the ecology of the large mammals of the Virunga Volcanoes between 1988 and 1991, for which he was awarded a PhD. After working leading a research project in the Budongo Forest of Uganda for 5 years, he was appointed as deputy director for the Africa Programme of the Wildlife Conservation Society (WCS), based in New York. From there, he developed a programme focusing on the Albertine Rift for WCS, and returned to the region in 2000. His activities include biological research, training of senior wildlife staff, transboundary protected area management between Uganda and the DRC and partnership development in the Albertine Rift Region.
c/o WCS, PO Box 7487, Kampala, Uganda.
aplumptre@wcs.org

Véronique Tshimbalanga
Véronique Thimbalanga received her post-graduate degree in development studies at the University of Paris-Sorbonne, in Paris, France. In 1997 she began working in conservation with UNESCO as part of a water management programme focusing on gender issues. She joined WWF in 2000 and worked on the Albertine Rift Programme, as a programme officer covering amongst others, the Environmental Programme of Virunga National Park. She is currently the national coordinator for the Central African Regional Programme on the Environment, using an integrated 'landscape' approach to conservation. She coordinates activities on six landscapes, including Virunga.
c/o WWF/DRC, Av. Sergent Moke, Commune Ngaliema BP 2847, Kinshasa 1, DRC.
vtshimbalanga@wwfcarpo.org

Jean Van Gysel
Jean Van Gysel, agricultural engineer, developed and implemented a quantitative method for vegetation mapping on the southern plains of Virunga National Park between 1978 and 1982. Vegetation mapping was correlated with the useful biomass for large herbivores. Cartography was based on aerial photographs using a systematic grid. A subsequent mission in 1995 made it possible to replicate the results. He currently works as coordinator for bilateral aid projects for Belgian Technical Cooperation in the Democratic Republic of Congo.
27 Pl. du Roi Vainqueur, B1040 Bruxelles, Belgium
janvangysel@hotmail.com

Dr Patricia van Schuylenbergh
Patricia Van Schuylenbergh (1964) completed her PhD in history and a post-graduate degree in development from the University of Louvain in Belgium. She currently works at the *Royal Museum for Central Africa* (Tervuren, Belgium) where she is in charge of scientific research and audiovisual collections project management. She wrote several articles exploring the history of protection and conservation of nature in Central Africa (DRC, Rwanda and Burundi), focusing especially on protected areas and on the management of natural resources during and after the Belgian colonial period, the history of colonial natural sciences and knowledge and representations of animals. She is currently preparing a publication on the history of the National Parks in DRC.
Royal Museum for Central Africa, Leuvensesteenweg, 13, 3080 Tervuren, Belgium.
patricia.van.schuylenbergh@africamuseum.be

Dr. Mireille Vanoverstraeten
Mireille Vanoverstraeten is an agricultural engineer and has a PhD from the Agricultural University of Gembloux (1989). As a research warden at the Lulimbi Station in Virunga between 1978 and 1982, she worked on the assessment of soils on the Rwindi-Rutshuru Plains. She covered many aspects of her discipline throughout her career, all tied to the quality of the environment. As a consultant to UNESCO's Man and Biosphere Programme (MAB), within the Department of Ecological and Earth Sciences, she contributed to the creation of several transboundary biosphere reserves on the African Continent. She is a scientific fellow at the Agricultural University at Gembloux.
Route de Meux 58, B–5031 Grand-Leez. Belgium.
mvanoverstraeten@skynet.be

Dr Jacques Verschuren
Jacques Verschuren, born in 1926, studied zoology, in which he completed his PhD. He committed his entire life to wildlife conservation and to scientific research in the field. A Fellow of the *Institut Royal des Sciences Naturelles de Belgique,* he lived in Congo's national parks for some 15 years between 1948 and 1990, especially in Virunga. He was nominated as Director General of ICCN, and oversaw the creation of 4 new protected areas in Congo. He was awarded WWF's golden medal for his work in Virunga National Park. His work in Congo was interspersed with research and conservation work in other countries, ranging from Paraguy to Indonesia, and include Rwanda, Burundi, Tanzania, Senegal, Mauritania, Liberia, Benin, etc. He has written over 250 scientific and popular articles.
82, Avenue de l'Atlantique, 1150 Brussels, Belgium.

ACRONYMS

ADG	Administrateur Délégué Général (General Director)
ADMADE	Administrative Management Design
AFDL	Alliance des Forces Démocratiques pour la Libération du Congo-Zaïre
AGCD	Administration Générale de la Coopération au Développement (Belgian Aid Agency)
ALIR	Armée pour la Libération du Rwanda
AWF	African Wildlife Foundation
BBC	British Broadcasting Corporation
CADAF	Communauté des Assemblées de Dieu en Afrique
CADECO	Caisse d'Epargne du Congo
CAMPFIRE	Communal Areas Management Programme For Indigenous Resources
CARPE	Central African Regional Program for the Environment
CBCA	Communauté Baptiste au Centre de l'Afrique
CDR	Convention pour la Démocratie et la République
CEBCE	Communauté des Eglises Baptistes du Congo-Est
CEI	Commission Electorale Indépendante (Independent Electoral Commission)
CEPAC	Communauté des Eglises Pentecôtistes en Afrique Centrale
CICR	Comité International de la Croix Rouge
CITES	Convention on International Trade in Endangered Species
CoCoSi	Site Coordination Committee
COPEVi	Coopérative des Pêcheurs de Vitshumbi
COPEVi	Coopérative des Pêcheurs des Virunga
COPILA	Coopérative des Pêcheurs Indigènes du Lac Amin
COPILE	Coopérative des Pêcheurs Indigènes du Lac Edouard
CRSN	Centre de Recherche en Science Naturelle (Lwiro)
DCR	Domaine de Chasse de Rutshuru (Rutshuru Hunting Zone)
DFGF-E	Dian Fossey Gorilla Fund- Europe (now Gorilla Organisation)
DFGF-I	Dian Fossey Gorilla Fund -International
DG	Direction Générale (National Headquarters)
DGM/DSR	Direction Générale des Migrations/Direction Sécurité et Renseignement
DRC	Democratic Republic of Congo
DSP	Division Spéciale Présidentielle
EU	European Union
FAO	Food and Agricultural Organisation
FAR	Forces Armées Rwandaises
FARDC	Forces Armées de la République Démocratique du Congo
FAZ	Forces Armées Zaïroises
FDLR	Forces Démocratiques de Libération du Rwanda
FEPACO	Fraternité évangélique pentecôtiste en Afrique au Congo
FFI	Fauna and Flora International
FFPS	Fauna and Flora Preservation Society (now FFI)
FFRSA	Fondation pour Favoriser les Recherches Scientifiques en Afrique
FPR	Front Patriotique Rwandais
FZS	Frankfurt Zoological Society
GIC	Gillman International Conservation
GIS	Geographic Information System
GO	Gorilla Organisation
GPS	Global Positioning System
GTZ	Gesellschaft für Technische Zusammenarbeit
ICCN	Institut Congolais pour la Conservation de la Nature
ICRC	International Comity of the Red Cross
ICDP	Integrated Conservation and Development Project
IIED	International Institute for Environment and Development
IMF	International Monetary Fund
INCN	Institut National pour la Conservation de la Nature
IPNCB	Institut des Parcs Nationaux du Congo Belge
IRScNB	Institut Royal des Sciences Naturelles de Belgique
ITFC	Institute for Tropical Forest Conservation
IUCN	International Union for the Conservation of Nature
IZCN	Institut Zaïrois pour la Conservation de la Nature
LEM	Law Enforcement Monitoring
LWF	Lutheran World Federation
MECNEF	Ministère de l'Environnement, Conservation de la Nature, Eaux et Forêts

MGVP	Mountain Gorilla Veterinary Project
MIKE	Monitoring the Illegal Killing of Elephants
MIST	Management Information System (UWA)
MONUC	Mission des Nations unies au Congo (United Nations Mission in DRC)
MOU	Memorandum Of Understanding
MSF	Médecins sans Frontières
NK	North Kivu (Province)
NP	National Park
NTFP	Non-Timber Forest Products
NYZS	New York Zoological Society
OCHA	Office for the Coordination of Humanitarian Affairs
ONP	Office National des Pêches
ORTPN	Office Rwandais du Tourisme et des Parcs Nationaux
OVG	Observatoire Vulcanologique de Goma
PAA	Protected Areas Authority
PARCID	Projet d'Appui au Renforcement des Capacités Institutionnelles de la Direction Générale de l'ICCN
PDG	Président Délégué Général
PEVi	Virunga Environmental Programme (WWF)
PNG	Parc National de la Garamba
PNKB	Parc National de Kahuzi-Biega
PNS	Parc National de la Salonga
PNUD	Programme des Nations unies pour le Développement (UNDP)
PNUE	Programme des Nations unies pour l'Environnement (UNEP)
QENP	Queen Elizabeth National Park
RBM	Ranger-based Monitoring
RCD Goma	Rassemblement Congolais pour la Démocratie (Goma)
RCDN	Rassemblement Congolais pour la Démocratie (National)
RCD K/ML	Rassemblement Congolais pour la Démocratie (Kisangani/Mouvement de Libération)
RFO	Réserve de Faune à Okapis
RIS	Regional Information System
RME	Revenus en Monnaie Etrangère
RPVA	Réseau des Planteurs pour la Valorisation de l'Arbre
SNIP	Service National d'Intelligence et de Protection
SNV	Stichting Nederlandse Vrijwilligers
SODERU	Solidarité, Développement Rural (French NGO)
SOMIKIVU	Société Minière du Kivu
SPOT	Satellite Pour l'Observation de la Terre
SYGIAP	Système de gestion de l'Information des Aires Protégées
TBPA	Trans-boundary Protected Area
UCL	Université Catholique de Louvain (Louvain University, Belgium)
UE	European Union
UN	United Nations
UNDP	United Nations Development Programme
UNEP	United Nations Environment Programme
UNF	United Nations Foundation
UNHCR	United Nations High Commissioner for Refugees
USA	United States of America
USAID	United States bilateral aid agency
USD	United States dollars
UWA	Uganda Wildlife Authority
ViNP	Virunga National Park
WCS	Wildlife Conservation Society
WCPA	World Commission for Protected Areas
WWF	World Wide Fund for Nature
ZSL	Zoological Society of London

Acknowledgements

A project on this scale could never have been completed without the support of a very large number of people and institutions. The work is the sum of the collective efforts of 36 authors of eight different nationalities and two editors.

Above all, our thanks go to the *Institut Congolais pour la Conservation de la Nature*, which supported this project from the beginning. We are particularly grateful to Madame Eulalie Bashighe Baliruhya and Pasteur Cosma Wilingula and the entire team at headquarters in Kinshasa. Benoît Kisuki and Dr Muamba Tshibasu have supported the field projects on whose data this book is based. It would not be possible to name each member of the ICCN staff working in Virunga National Park individually, but we should like to extend our thanks to the many rangers who have been instrumental, not only in protecting the park, but also in providing data for this book, often working anonymously and under extremely difficult conditions.

Three institutions have provided funding for the publication of this book: WWF, UNESCO, and the European Union. In particular, we should like to thank Filippo Saracco of the European Commission Delegation in Kinshasa, Enrico Pironio at the European Commission in Brussels, and Allan Carlson of WWF-Sweden, for the support and guidance that enabled us to complete the project. These three institutions must be thanked too for their invaluable support for field conservation in Virunga. Dr Sam Kanyamibwa and Dr Kwame Koranteng have both contributed a great deal to this project, and we should like to thank them for their support, while gratefully acknowledging that of the WWF teams in both Nairobi and Goma: Samuel Boendi, Augustin Ndimu, Peter Banza, and Delphin Nganzi, together with Bisidi Yalolo, now based in Bukavu.

A wealth of technical advice was provided on each of the many different aspects covered in this book. For their help with the Introduction, we should like to thank Robert Ducarme, Harald Hinkel, Michel Louette, Danny Meirte, Mireille Vanoverstraeten and François Vleeschouwer. Special thanks are due to Conrad Aveling, to Willy Delvingt, to the Count Cornet d'Elzius, and to Didier Devers, Pierre Dufourny and Geert Lejeune for their various contributions. Jean-Pierre d'Huart and Jacques Verschuren went well beyond their authorship roles in providing editorial support and fact-checking. Louisa Lockwood and Gordon Boy played key roles in helping to finalise the English text. Jean Terschuren provided considerable support with her translation from the original French. We should also like to thank Samantha Newport and Pierre Peron for their support in proof-reading the manuscript. And we are grateful to François-Xavier de Donnea for his help in promoting both the book and Virunga National Park.

Numerous maps, photographs and diagrams illustrate the book. We are grateful to the many institutions that authorised the reproduction of various documents, both original and adapted: the *Musée Royal d'Afrique Centrale* (Tervuren), the National Botanical Gardens at Meise, and the *Institut Royal des Sciences Naturelles de Belgique*. Johan Lavreau, Patricia Van Schuylenbergh, Jan Rammeloo, Jérôme Degreef and Anne Franklin placed their considerable expertise at our disposal in making key historical documents available. Both UNESCO and the European Space Agency played a part in producing some of the maps that appear in the book. Stéphane Laime and Zachary Maritim spent many days developing the cartographic material for the book, and were unfailingly patient even when repeated changes were imposed on their work. The Wildlife Conservation Society and the Frankfurt Zoological Society made substantial amounts of data, photographs, and other records available to us. For their assistance, we are especially grateful to Andy Plumptre, Déo Kujirakwinja and Robert Muir.

We should like to extend a special note of thanks to Jacqueline d'Huart who read, re-read and read a manuscript made up of submissions varying enormously in style. Her role in bringing a measure of consistency to the narrative was immense.

Paule and Guy Languy, together with Charles Guillaume and Hedwige de Merode, are to be thanked for taking care of the logistical and other aspects of this book's production and distribution. Their constant support and encouragement has been invaluable.

We are grateful to the publisher, *Editions Lannoo*, and in particular to Gauthier Platteau, who accepted this project, and to Antoon De Vylder, who so graciously tolerated the many difficulties and delays we experienced during the compilation of the book.

The two editors would especially like to thank their families for their unfailing support, and for accepting the long periods of absence that the years of commitment to this project have entailed.

This book is respectfully dedicated to the memory of the 120 rangers who have given their lives in the service of the park. Their dedication and courage in defending the priceless national and global heritage that is Virunga, Africa's first national park, should serve as an example to us all.

Photo credits

Pictures and their copyrights belong to the following individuals or institutions:

André Kamdem Toham/WWF: 13.3
Andrew Plumptre/WCS: 0.81, 21.2, 21.3, 22.2, 22.6
Augustin Ndimu/WWF: 0.39, 0.88, 4.5, 14.6, 17.11, 17.12, 17.13, 17.14, 17.15, 19.1, 21.1, 21.4
Bruce Davidson/Naturepl.com: p. 2–3, p. 12, p. 15, p. 16, 0.13, 0.15, 0.17, 0.18, 0.19, 0.33, 0.52, 0.66, 0.69, 0.82, 0.89, 4.1, 4.2, 6.8, 8.5, 11.1, 12.4, 17.1, back cover
Carlos de Wasseige/UCL: 18.3, 24.2
Conrad Aveling/ICCE: 0.54, 0.79, 10.6, 15.1, 20.1
David Simpson/ACF: 26.4
Delphin Nganzi/ACF: 24.5, 25.4
Denis-Huot/BIOS/Wildlife pictures: 8.1, 8.9
Déo Kujirakwinja/WCS: 0.34, 10.2, 10.3, 10.4, 13.1, 15.7, 16.1
Henri-Paul Eloma/ICCN: 20.2
ICCN: 3.4, 3.6, 11.3
Institut Royal des Sciences Naturelles de Belgique: flyleaves, 1.3, 1.4, 1.5, 1.6, 2.2, 2.4, 2.6, 2.8, 3.3, 5.3, 7.1, 11.4, 15.3, section 7.3.3, back cover
Jacques Durieux: 6.2, 6.3, 6.4, 6.5, 6.6, 6.7, 6.9
Jacques Verschuren/IRSCNB: 0.20, 0.31, 0.78, 2.1, 2.3, 2.5, 3.1, 3.2, 5.5, 7.1, 7.3, 9.11b, 11.5
Marc Languy/WWF: p. 4/5, 0.5, 0.8, 0.12, 0.23, 0.24, 0.29, 0.30, 0.36, 0.38, 0.41, 0.42, 0.44, 0.45, 0.46, 0.47, 0.49, 0.51, 0.55, 0.56, 0.57, 0.58, 0.59, 0.71, 0.73, 0.74, 0.75, 0.83, 0.85, 0.86, 3.5, 4.4, 4.5, 5.1, 5.4, 7.4, 7.5, 7.6, 8.4, 9.2, 9.3, 9.5, 9.7, 9.11d, 9.15, 10.5, 11.2, 11.6, 11.9, 12.1, 12.7, 12.14, 13.8, 14.1, 14.2, 14.5, 14.8, 15.4, 15.5, 17.2, 17.4, 17.5, 17.6, 17.7, 17.16, 18.2, 19.2, 22.1, 23.1, 23.8, 24.1, 24.6, 25.1, 25.2, 26.3, 26.5
Maryke Gray/AWF-FFI-WWF: 5.2, 23.9
Musée royal de l'Afrique centrale, Tervuren: 1.1, 1.2, 1.7, 1.8, 2.7, 2.9, 9.1, 9.11a, 9.14a
Robert Muir/SZF: 4.6, 15.2, 16.2, 16.3, 16.4, 16.5, 18.1, 18.4, 19.6, 24.7, 25.3
Samy Mankoto/ICCN: 3.7, 3.8, 3.9
Stéphane Laime: 0.67, 0.68, 0.70, 0.72, 0.80, 8.7, 12.16, 12.8, 20.3
Steve Turner/OSF/Wildlife Pictures: 12.3
Svein Erik Harklåu/WWF: 0.26, 0.27, 0.6, 0.61, 0.62, 7.2
WWF: p. 308, 0.76, 17.3, 19.5
WWF-Canon/Martin Harvey www.wildimagesonline.com: front cover, p. 1, p.10, 9.1, 9.6, 11.7, 12.12, 21.5, 26.1
WWF-Canon/Sandra M. Obiago: 10.1
Yolente Delaunoy/WWF: 0.21, 18.5

Maps that are not listed above belong to WWF. Graphs and figures belong to the authors of the respective chapters in which they appear.

Metadata of maps

Maps for this book were produced under the following systems:

Geographic coordinates and geodesic system: WGS 84
Projection parameters: WGS 84, Zone 35S UTM
False Easting: 500,000.000
False Northing: 10,000,000.000
Central meridian: 27.00000
Scale factor: 0.9996000
Latitude of origin: 0.00000

Origin of satellite images and of vectors (given in parentheses) for the following maps:

0.2	GeoTiff DEM, IUCN database (IUCN, WWF)
0.3	BEGO Geotiff DEM (WWF)
0.14	Africa DEM, Africover (IUCN, WWF)
0.16	SPOT 5mC, images JK:123-348, 123-349 of 11 Feb. 2005. True color composite 4,1,3 (WWF)
0.22	Mosaic ASTER 15mC of 21 Feb. 2005. True color composite 2,3,1 (WWF)
0.25	Mosaic ASTER 15mC of 21 Feb. 2005. True color composite 2,3,1 (WWF, BEGO)
0.30	SPOT 5mC Images JK:123-350 of 11 February 2005. True color composite 4,1,3 (WWF)
0.32	SPOT 5mC Images JK:123-351 of 27 June 2004. True color composite 4,1,3 (WWF)
0.35	Adapted from Vakily (1989) (WWF)
0.37	SPOT 5mC, images JK:123-351 and 123-352 of 27 June 2004. True color composite 4,1,3 (WWF)
0.40	SPOT 5mC, images JK:123-352 of 27 June 2004. True color composite 4,1,3 (WWF)
0.43	Mosaic SPOT 5mC, images JK:123-352 of 27 June 2004, JK 123-353 of 3 July 2004, JK 122-352 and 122-353 of 21 Feb. 2005. True color composite 4,1,3 (WWF)
0.53	SPOT 5mC, images JK:123-353 of 3 July 2007, 124-353 of 22 Feb. 2005. True color composite 4,1,3 (WWF)
0.87	Based on interpretation of SPOT 5mC of 2004 and 2005 (WWF)
9.4	SPOT 5mC, images JK:123-348, 123-349 of 11 Feb. 2005. True color composite 4,1,3 (WWF, UCL)
9.9	SPOT 5mC, images: JK: 123-351/352/353 (WWF, UCL)
9.10	SPOT 5mC, images: JK: 123-351/352/353 SRTM UCL 90m (WWF, UCL)
9.12	SPOT 5mC, images JK:123-348, 123-349 of 11 February 2005. True color composite 4,1,3 (WWF, UCL)
9.13	SPOT 5mC, images JK:123-348, 123-349 of 11 February 2005. True color composite 4,1,3 (WWF, UCL)
9.14	Aerial pictures of 1959; SPOT image of 2004 JK:123-353 with a false color composite 4,2,1 and image Landsat7 JK:173-060 in false color 5,3,3 (WWF)

13.2 to 13.6	Aerial pictures of 1959 (MRAC); aerial pictures of 1987 and 2005 (WWF, UCL)
13.9	Images SPOT JK: 123-351 (WWF)
18.2	SPOT 5mC: JK 123-353. False color composite 4,2,1 (UCL)
23.2 to 23.6	Original data from ICCN, vectors from WWF.
24.2	Landsat ETM (15m) of 31 January 2003, SPOT (5m) of 07 June and 03 July 2004 (UCL)
24.3	Mosaic SPOT (5mC) images JK: 123-350/349 of 11 February 2005 with true color composite 4,1,3 (WWF)
24.4	Mosaic SPOT (5mC) images JK: 123-353/352 of 11 February 2005, 122-353/352 of 21 February 2005 with true color composite 4,1,3 (WWF)

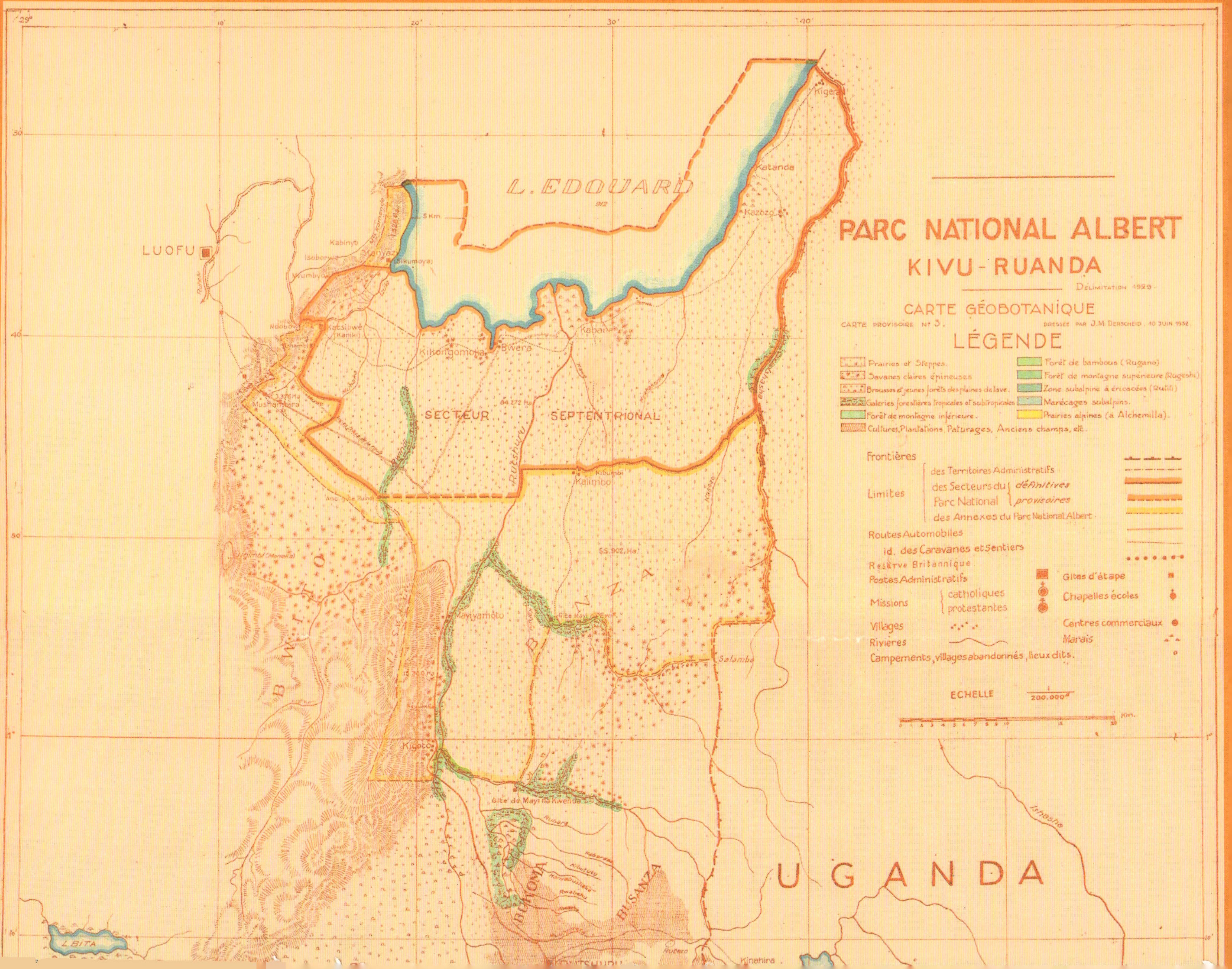

PARC NATIONAL ALBERT
KIVU-RUANDA
Délimitation 1929.
CARTE GÉOBOTANIQUE
CARTE PROVISOIRE Nº 3.
DRESSÉE PAR J.M DERSCHEID. 10 JUIN 1932
LÉGENDE
Prairies et Steppes.
Savanes claires épineuses
Brousses et jeunes forêts des plaines de lave.
Galeries forestières tropicales et subtropicales
Forêt de montagne inférieure.
Cultures, Plantations, Paturages, Anciens champs, etc.
Forêt de bambous (Rugano)
Forêt de montagne supérieure (Rugeshi)
Zone subalpine à éricacées (Rutili)
Marécages subalpins.
Prairies alpines (à Alchemilla).
Frontières
Limites
des Territoires Administratifs
des Secteurs du Parc National définitives
provisoires
des Annexes du Parc National Albert.
Routes Automobiles
id. des Caravanes et Sentiers
Réserve Britannique
Postes Administratifs
Missions
catholiques
protestantes
Villages
Rivières
Campements, villages abandonnés, lieux dits.
Gîtes d'étape
Chapelles écoles
Centres commerciaux
Marais
ECHELLE
1/200.000e
Km.
L. EDOUARD
912
LUOFU
SECTEUR
SEPTENTRIONAL
UGANDA
L. BITA
Kiger
Katanda
Kazizo
Kabare
Bwera
Kikongomoka
Sikumoya
Kabinyi
Isoborwe
Kasibwe
Kamiki
Nduboni
Mushashoro
Kibumbi
Kalimbo
55.902 Ha
Salamba
Nyiyamoto
Kigoto
Gîte de Mayi na Kwenda
Rutshuru
BUHOMA
BUSANZA
Kinehira
Ruteko
Ishasha
5 Km.